Molecular Biology
Biochemistry and Biophysics
24

Chemical Relaxation in Molecular Biology

Edited by

I. Pecht and R. Rigler

With Contributions by
M. Ehrenberg · E. Grell · D. N. Hague
G. Ilgenfritz · T. M. Jovin · D. Lancet · D. Magde
I. Oberbäumer · I. Pecht · F. M. Pohl · D. Pörschke
R. Rigler · P. Schuster · G. Striker · D. Thusius
K. Tortschanoff · W. Wintermeyer · P. Wolschann

With 141 Figures

Springer-Verlag Berlin · Heidelberg · New York 1977

Dr. ISRAEL PECHT
The Weizmann Institute of Science
Department of Chemical Immunology
Rehovot/Israel

Dr. RUDOLF RIGLER
Department of Medical Biophysics
Karolinska Institutet
S-10401 Stockholm 60

ISBN 3-540-08173-9 Springer-Verlag Berlin Heidelberg New York
ISBN 0-387-08173-9 Springer-Verlag New York Heidelberg Berlin

Library of Congress Cataloging in Publication Data. Main entry under title: Chemical relaxation in molecular biology. (Molecular biology, biochemistry, and biophysics; 24) Includes bibliographical references. 1. Chemical reaction, Rate of—Addresses, essays, lectures. 2. Molecular biology—Addresses, essays, lectures. I. Pecht, I., 1937— . II. Rigler, Rudolf. III. Series. QD502.C47. 574.8'8. 77-7167

Offsetprinting and bookbinding: Brühlsche Universitätsdruckerei, Lahn-Gießen.
2131/3130-543210

Dedicated to Manfred Eigen

Contents

Contributors

EHRENBERG, M.	Department of Medical Biophysics, Karolinska Istitutet, 104 01 Stockholm, Sweden
GRELL, E.	Max-Planck-Institut für biophysikalische Chemie, 3400 Göttingen-Nikolausberg, FRG
HAGUE, D.N.	University of Kent at Canterbury Kent, University Chemical Laboratory, Canterbury CT2 7NH, England
ILGENFRITZ, G.	Institut für physikalische Chemie, 5000 Köln 41, FRG
JOVIN, T.M.	Max-Planck-Institut für biophysikalische Chemie, 3400 Göttingen-Nikolausberg, FRG
LANCET, D.	Department of Chemical Immunology, The Weizmann Institute of Science, Rehovot, Israel
MAGDE, D.	University of California, San Diego, La Jolla, CA 92037, USA
OBERBÄUMER, I.	Max-Planck-Institut für Biophysikalische Chemie, 3400 Göttingen-Nikolausberg, FRG
PECHT, I.	Department of Chemical Immunology, The Weizmann Institute of Science, Rehovot, Israel
POHL, F.M.	Fachbereich Biologie der Universität Konstanz, 7750 Konstanz, FRG
PÖRSCHKE, D.	Max-Planck-Institut für Biophysikalische Chemie, 3400 Göttingen-Nikolausberg, FRG
RIGLER, R.	Department of Medical Biophysics, Karolinska Istitutet, 104 01 Stockholm, Sweden
SCHUSTER, P.	Institut für Theoretische Chemie und Strahlenchemie der Universität, 1090 Wien, Austria

STRIKER, G.	Max-Planck-Institut für Biophysikalische Chemie, 3400 Göttingen-Nikolausberg, FRG
THUSIUS, D.	Laboratoire D'Enzymologie Physico-chimique et Moléculaire, Bâtiment 433, Université de Paris-Sud, Centre D'Orsay, 91405 Orsay, France
TORTSCHANOFF, K.	Institut für Theoretische Chemie und Strahlenchemie der Universität, 1090 Wien, Austria
WINTERMEYER, W.	Institut für Physiologische Chemie, Physikalische Biochemie und Zellbiologie der Universität, 8000 München, FRG
WOLSCHANN, P.	Institut für Theoretische Chemie und Strahlenchemie der Universität, 1090 Wien, Austria

Theory and Simulation of Chemical Relaxation Spectra

G. ILGENFRITZ

I. Introduction

Chemical relaxation methods proved to be important tools for the study of kinetics in solution and are widely used for the investigation of inorganic, organic, and biochemical reaction mechanisms, and the characterization of their elementary steps. The principle of these methods is that the reaction is initiated by a small perturbation of an external thermodynamic parameter. The response of the system after the external perturbation, i.e., the time course of the concentration changes of the reacting species, is measured by suitable physical parameters. It is the essential advantage of the relaxation methods (temperature-jump, pressure-jump, electric-field-jump, ultrasound and dielectric dispersion methods) that the reactions can be followed into the micro- and nanosecond time range, an essential requirement for the study of elementary reaction steps. Methods and theory of chemical relaxation have been extensively discussed in literature. It is the aim of this chapter to summarize the theoretical approach and to provide a computer program for the simulation of relaxation kinetic progress curves.

The kinetics of the reaction system near the thermodynamic equilibrium state can be described by linear differential equations, leading to characteristic time constants (relaxation times) for the reequilibration process. Since it is neither possible in general, nor very reasonable even in simple reaction schemes, to obtain exact analytical expressions for relaxation times and relaxation amplitudes, approximate relationships, derived under various simplifying assumptions, are mostly used for analyzing relaxation experiments. It is assumed, for instance, that certain reaction steps are always in equilibrium while the reaction proceeds, or that steady-state conditions can be introduced for certain reacting species. While such approximations are useful and necessary for the qualitative discussion of a postulated reaction mechanism, a rigorous comparison of theory and experiment can only be made if exact solutions of the relaxation equations are available. Simulation of the relaxation spectrum for a postulated reaction mechanism will be useful for a discussion of the expected kinetic behavior and the degeneration in the relaxation spectrum, which arises if some relaxation times appear with negligible amplitude or if some relaxation times lie very close together in time. It may also be used for testing the adequacy of various simplifying assumptions made in the kinetic analysis. A computer program has therefore been developed, which allows the user to calculate numerically, in a quick and simple way, the relaxation spectrum, i.e. relaxation times and amplitudes, of a given chemical reaction mechanism. Although for the mathematical procedure, the solution of an eigenvalue problem, appropriate routines can be found in program libraries, it may be convenient to have a program at hand which directly meets the requirements of relaxation kinetics.

A. The Relaxation Kinetic Progress Curve

The chemical reaction system, containing S reacting species and R elementary steps can be formulated:

$$\sum_{i=1}^{S} \nu_{i\alpha} X_i = 0 \qquad \alpha = 1,\ldots,R \qquad (1)$$

Equation (1) applies for each reaction step, α, whereby X_i denotes the reacting species, i, and $\nu_{i\alpha}$ the stoichiometric coefficient of species i in the αth reaction. Stoichiometric coefficients which appear on the left and right side of the chemical reaction equation are taken as negative and positive respectively.

For the quantitative description of the relaxation process when a perturbation is applied to the chemical system, one may consider either the temporal changes of the concentration of the reacting species (EIGEN and DE MAEYER, 1963, 1973) or the temporal changes of the advancements for each individual reaction step (CASTELLAN, 1963; SCHIMMEL, 1971). The corresponding kinetic equations are explicitly given in Section C. The time course of the concentrations or the reaction parameters depends on the time course of the applied perturbation. At each time t there is an "equilibrium value" of the concentrations corresponding to the value of the perturbing parameter. It describes the situation when equilibrium is established at each time interval after the applied perturbation, i.e. if the chemical reaction were infinitely fast. The actual concentration values lag behind the equilibrium values due to the finite rate of the chemical reactions. If all concentration changes of the reacting species are given with respect to the (time-independent) reference state before the perturbation is applied, the symbols $\bar{x}_i(t)$ may denote the equilibrium concentrations and $x_i(t)$ the actual concentration values (see Fig. 1).

The equilibrium concentration changes can be written in the form

$$\bar{x}_i(t) = x_i^o\, f(t) \qquad (2)$$

defining a "strength of perturbation", x_i^o, and a "forcing function", f(t). If the perturbation is initiated at $t = \theta$, the unperturbed equilibrium state is characterized by $f(t) = 0$, i.e.

$$x_i(t) = \bar{x}_i(t) = 0 \qquad t < \theta \qquad (3)$$

The deviations of the actual concentrations at a time t from the reference state, $x_i(t)$, which can be obtained as solutions of the

Fig. 1. Definition of concentration changes x. Upper curve: forcing function; lower curve: relaxation function. T, P, E Perturbation parameters (temperature, pressure, electric field strength), $\bar{c}$ equilibrium concentration corresponding to the perturbation at time t, c actual concentration at time t

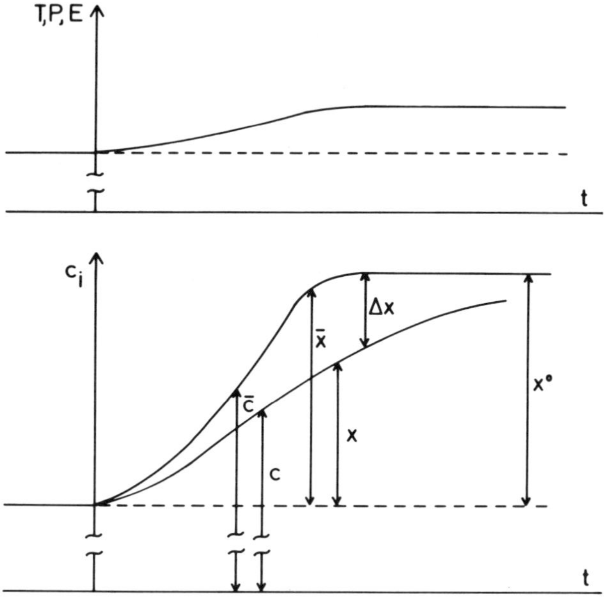

relaxation equations (see Sect. C), are given by:

$$x_i(t) = \sum_{k=1}^{n} A_{ik}\psi(t,\tau_k) \qquad \begin{matrix} i = 1,\ldots S \\ t \geqslant \theta \end{matrix} \tag{4}$$

The relaxation times τ_k and the relaxation amplitudes A_{ik}, i.e. the change of the concentration species i with respect to the relaxation time τ_k, depend on the kinetic and thermodynamic properties of the chemical reaction system. The quantity n denotes the number of relaxation times, which is equal to the number of independent concentration variables or the number of thermodynamically independent reactions of a particular mechanism. The "relaxation function", $\psi(t,\tau)$ depends on the forcing function, f(t):

$$\psi(t,\tau) = \int_{\theta}^{t} f(\delta)\frac{1}{\tau}\exp\left[(\delta-t)/\tau\right]d\delta \tag{5}$$

Its explicit form is necessary for obtaining relaxation times and amplitudes from the experiment. For a stepwise perturbation at a time $\theta = 0$ one obtains $f(t) = 1$; $\psi(t,\tau) = 1 - \exp(-t/\tau)$ and for a stationary periodic perturbation, i.e. $\theta \to -\infty$ $f(t) = \sin\omega t$; $\psi(t,\tau) = \sin\omega t/(1+\omega^2\tau^2) - \cos\omega t \cdot \omega\tau/(1+\omega^2\tau^2)$. Thus, for a stepwise perturbation the relaxation function is characterized by an exponential time course. For a periodic perturbation with frequency ω, the relaxation function is also periodic with the same frequency, the phase however being shifted and the amplitude depending on the frequency applied. Since the relaxation function for $\tau \to 0$ is equal to the forcing function, Eqs. (2) and (4) lead to

$$x_i^o = \sum_{k=1}^{n} A_{ik} \qquad i = 1,\ldots,S \tag{6}$$

The amplitude factors A_{ik} represent the concentration changes of the species i in the kth "normal reaction". They can furthermore be correlated with the advancement of each individual reaction step. If $P_{\alpha k}$ denotes to which extent each individual reaction step α contributes to the kth normal reaction, the following equation applies[1]

$$\sum_{\alpha=1}^{R} \nu_{i\alpha} P_{\alpha k} = A_{ik} \qquad \begin{array}{l} i = 1,\ldots,S \\ k = 1,\ldots,n \end{array} \tag{7}$$

The values A_{ik} or $P_{\alpha k}$ can be used to calculate any physical parameters of a normal reaction, e.g. the quantity actually observed in the experiment, from the properties of the individual reacting species or the individual reaction steps. The change of optical extinction δE_k, for instance is given by

$$\delta E_k = \sum_{i=1}^{S} \varepsilon_i A_{ik} = \sum_{\alpha=1}^{R} \Delta\varepsilon_\alpha P_{\alpha k} \tag{8}$$

where ε_i denote the molar extinction coefficient of species X_i and $\Delta\varepsilon_\alpha$ the molar extinction change in the αth reaction step.

The normalized quantities, $p_{\alpha k}$, defined as

$$p_{\alpha k} = \frac{P_{\alpha k}}{\sum_{\beta=1}^{R} |P_{\beta k}|} \qquad \begin{array}{l} \alpha = 1,\ldots,R \\ k = 1,\ldots,n \end{array} \tag{9}$$

are very useful for the discussion of the relaxation spectrum, since they allow to correlate quantitatively the reaction process for a particular relaxation time with individual reaction steps in the reaction mechanism.

B. Optical Detection Signals

In general, relaxation methods which make use of stepwise perturbation for the initiation of the reaction, differ in the mode of following the concentration changes from those relaxation methods with periodic perturbation. Whereas in jump-methods a physical property is measured as a function of time which is specifically related to the individual reacting species, e.g. optical parameters or conductivity, the periodic methods use the extensive thermodynamic parameter which is conjugate to the perturbing parameter for following the relaxation process as a function of frequency. Since a phase shift between conjugate thermodynamic parameters results in dissipation of energy, a suitable way of monitoring the reaction is obtained by measurement of excess sound-absorption in ultrasound methods or the excess dielectric loss in dielectric methods.

---

[1]For reaction systems containing thermodynamically dependent steps there exist additional relations which describe the constraints between the advancements for the dependent steps (see Sect. C).

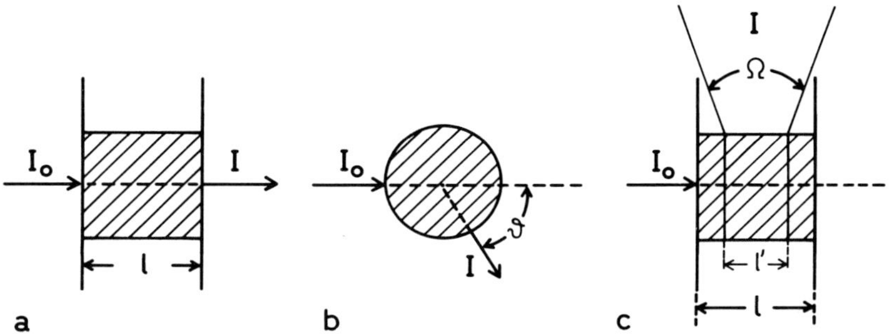

Fig. 2a-c. Scheme of optical path for the measurement of (a) optical absorption, (b) light scattering, (c) fluorescence

Optical properties of the solution, such as optical absorption, light scattering, or fluorescence, are the most important physical parameters in biochemical reaction systems since they are specifically related to the individual reacting species. The most sensitive one of these parameters is fluorescence emission; it allows the use of low concentrations and is furthermore highly specific, since quantum yield and polarization depend critically on the molecular environment of the fluorescent chromophores. The relationship between the change of the optical signal and the changes of the individual reacting species is summarized below for measurements of optical absorption, light scattering, and fluorescence. The optical path is shown schematically in Figure 2, where I_o and I denote respectively the light intensities before and after the light has passed the measuring cell and interacted with the reacting species.

Optical Absorption. If ε_i denote the molar decadic extinction coefficients of species i, the light intensity of the monitoring light is given by: (LAMBERT-BEER'S law)

$$I = I_o \exp(-2.3A); \quad A = \sum_{i=1}^{S} \varepsilon_i c_i l \tag{10}$$

For small concentration changes one obtains for the change in light intensity

$$\delta I = -2.3 I \delta A, \quad \delta A = \sum_{i=1}^{S} \varepsilon_i \delta c_i l \tag{11}$$

Thus, for optical absorption measurements it is the relative change in light intensity which is directly proportional to the concentration changes. The evaluation of relaxation amplitudes requires the knowledge of the extinction coefficients.

Light Scattering. The amount of light scattered from a solution of macromolecules depends on the weight average molecular weight and the total weight concentration.

If θ denotes the angle under which scattering is observed, M_i the molecular weight of the reacting species i, and B the second virial coefficient, the scattered light intensity is given by:

$$I(\theta) = \frac{I_o K(\theta)\,\Sigma M_i^2 c_i}{1 + 2B\Sigma M_i^2 c_i} \tag{12}$$

$K(\theta)$ may be regarded as an instrument constant, which can be obtained by suitable calibration. It depends on the wavelength and distance of observation and incorporate macroscopic properties of the solution (refractive index and its increment with weight concentration) and microscopic properties of the scattering particles (radius of gyration). For small concentration changes, i. e. neglecting variations in $K(\theta)$ and B, the change in scattered light intensity results in

$$\delta I(\theta) = \frac{I_o K(\theta)\,\Sigma M_i^2 \delta c_i}{(1 + 2B\Sigma M_i^2 c_i)^2} \tag{13}$$

In case of light scattering it is the absolute change in light intensity which is directly proportional to the concentration changes. In the special case of a polymerizing system the molecular weight of the polymer M_i is i times the molecular weight of the monomer. The scattered light intensity is thus proportional to

$$\delta I(\theta) \propto \Sigma i^2 \delta c_i$$

and an evaluation of relaxation amplitudes requires no additional specific parameters.

Fluorescence. If ε_i denote the molar decadic extinction coefficients of species i at the excitation wavelength, and φ_i the integrated quantum yield of the emission spectrum, which is observed at a space angle Ω, the fluorescence intensity is given by: (RIGLER et al., 1974)

$$I = I_o \frac{\Omega}{4\pi} \frac{F}{E} \exp(-2.3El/2) \cdot 2 \sinh (2.3El'/2)$$
$$E = \Sigma \varepsilon_i c_i ; \quad F = \Sigma \varepsilon_i \varphi_i c_i \tag{14}$$

For small concentration changes one obtains for the change in light intensity:

$$\delta I = I \cdot \left[\frac{\delta F}{F} - \delta E \left(\frac{2.3l}{2} - \frac{(2.3l')^2}{12} E + \ldots \right) \right]$$
$$\delta E = \Sigma \varepsilon_i \delta c_i ; \quad \delta F = \Sigma \varepsilon_i \varphi_i \delta c_i \tag{15}$$

where the first term represents the changes in fluorescence yield, and the second term reflects the inner filter effect at the excitation wavelength.

The total fluorescence yield arises from contribution of light polarized parallel and perpendicular to the direction of the excitation light vector

$$I = I_{\parallel} + 2I_{\perp}$$

Further information on the flexibility of the fluorescent chromophor and the rotational movement of its carrier is obtained from measurements of the fluorescence anisotropy, r, which is defined as

$$r = \frac{I_{\parallel} - I_{\perp}}{I_{\parallel} + 2I_{\perp}} = \frac{\Sigma \varepsilon_i c_i \varphi_i r_i}{\Sigma \varepsilon_i c_i \varphi_i} \qquad (16)$$

where r_i denotes the anisotropy of the ith species.

C. Theoretical Description of Relaxation Kinetics

In the following summary the relaxation kinetics is discussed using both concentrations and advancements for the description of the kinetic equations. The summary is restricted to a consideration of the phenomenological kinetic equations. The thermodynamic formulation of chemical reactions near the equilibrium state will not be discussed.

Each elementary step of the reaction system Eq. (1)

$$\sum_{i=1}^{S} \nu_{i\alpha} X_i = 0 \qquad \alpha = 1,\ldots,R \qquad (17)$$

is characterized by the two rate constants k_α and $k_{-\alpha}$. The kinetic and thermodynamic properties of each reaction step at equilibrium are described by the exchange rate, r_α, and the equilibrium constant, K_α

$$r_\alpha = k_\alpha \prod_q \bar{c}_q^{\,|\nu_{q\alpha}|} = k_{-\alpha} \prod_r \bar{c}_r^{\,|\nu_{r\alpha}|} \qquad (18)$$

$$K_\alpha = \frac{k_\alpha}{k_{-\alpha}} = \prod_i \bar{c}_i^{\,\nu_{i\alpha}} \qquad \alpha = 1,\ldots,R \qquad (19)$$

The product over q is over reactants ($\nu_{q\alpha} < 0$) and the product over r is over reaction products ($\nu_{r\alpha} > 0$) of the αth reaction step; $\bar{c}_i$ denotes the equilibrium concentrations of the reacting species. Activity coefficients are not considered here.

The definition of the concentration variables and their changes during the reaction process are shown in Figure 1; $c_i(t)$ and $\bar{c}_i(t)$ denote respectively the actual concentrations of species i at time t and the equilibrium concentrations which correspond to the perturbation applied. $x_i(t)$, $\bar{x}_i(t)$ denote the difference of the corresponding concentration values from the time-independent reference state. The parameters for the advancements of each individual reaction step α will be denoted by $\xi_\alpha(t)$ and $\bar{\xi}_\alpha(t)$ respectively. Concentration and reaction variables are related by the definition equation:[2]

[2]The definition of advancements, ξ', is actually based on mole numbers. The variable ξ introduced in Eq. (20) ff. is thus related to the advancements by $\xi = \xi'/V$, where V is the volume.

$$x_i = \sum_{\alpha=1}^{R} \nu_{i\alpha}\xi_\alpha \qquad i = 1,...,S \tag{20}$$

The kinetics of the reaction system near equilibrium, i.e. from an initial equilibrium state to a perturbed equilibrium state nearby, is described by the relaxation equations which can be obtained by linearizing the phenomenological rate equations.

The rate equations for the reaction parameters ξ_α read ($\nu_{q\alpha} < 0$; $\nu_{r\alpha} > 0$)

$$\frac{d\xi_\alpha}{dt} = k_\alpha \prod_q c_q^{|\nu_{q\alpha}|} - k_{-\alpha} \prod_r c_r^{|\nu_{r\alpha}|} \qquad \alpha = 1,...,R \tag{21}$$

and the rate equations for the concentration variables can be obtained by introducing Eq. (20) into Eq. (21). The relaxation equations for the temporal change of the concentrations of the reacting species have the general form

$$\frac{d}{dt} x_i(t) + \sum_{k=1}^{S} a_{ik}x_k(t) = \sum_{k=1}^{S} a_{ik}x_k^o f(t) \qquad i = 1,...,S \tag{22}$$

Eq. (22) and the "kinetic matrix" a_{ik} can be obtained by linearizing the phenomenological rate equations for each reacting species, where the actual concentrations are substituted by (see Fig. 1)

$$c_i = \bar{c}_i + (x_i - \bar{x}_i) \tag{23}$$

Consideration of only linear terms in x_i and introduction of the equilibrium conditions Eq. (18) results in:

$$a_{ik} = - \sum_{\alpha=1}^{R} r_\alpha \frac{\nu_{i\alpha}\nu_{k\alpha}}{c_k} \qquad i,k = 1,...,S \tag{24}$$

The strength of perturbation, x_i^o, is related to the change of the equilibrium constants of each individual reaction step. The relations can be written in the general form

$$\sum_{k=1}^{S} c_{\alpha k}x_k^o = \delta \ln K_\alpha \qquad \alpha = 1,...,R \tag{25}$$

Eq. (25) and the "thermodynamic matrix" $c_{\alpha k}$, can be obtained by linearizing the mass-action Eq. (19), i.e.

$$K_\alpha + \delta K_\alpha = \prod_i (\bar{c}_i + x_i^o)^{\nu_{i\alpha}} \tag{26}$$

resulting in

$$c_{\alpha k} = \frac{\nu_{k\alpha}}{c_k} \qquad \begin{matrix} \alpha = 1,...,R \\ k = 1,...,S \end{matrix} \tag{27}$$

The corresponding relaxation equations for the temporal change of the advancements of the individual reaction steps have the form:

$$\frac{d}{dt}\xi_\alpha(t) + \sum_{\beta=1}^{R} b_{\alpha\beta}\xi_\beta(t) = \sum_{\beta=1}^{R} b_{\alpha\beta}\xi_\beta^o f(t) \qquad \alpha = 1,...R \tag{28}$$

This equation arises from the linearization of the phenomenological rate equations for the reaction parameters, Eq. (21), whereby Eqs. (20) and (23) are introduced. The kinetic matrix $b_{\alpha\beta}$ is given by

$$b_{\alpha\beta} = -r_\alpha \sum_{k=1}^{S} \frac{\nu_{k\alpha}\nu_{k\beta}}{c_k} \qquad \alpha,\beta = 1,...,R \tag{29}$$

The strength of perturbation, ξ_β^o, is related to the change of the equilibrium constants by linear equations of the form

$$\sum_{\beta=1}^{R} g_{\alpha\beta}\xi_\beta^o = \delta \ln K_\alpha \qquad \alpha = 1,...,R \tag{30}$$

which are obtained by linearizing Eq. (26) and introducing Eq. (20). The thermodynamic matrix $g_{\alpha\beta}$ reads:

$$g_{\alpha\beta} = \sum_{k=1}^{S} \frac{\nu_{k\alpha}\nu_{k\beta}}{c_k} \qquad \alpha,\beta = 1,...,R \tag{31}$$

It is the special advantage of this treatment that only one matrix has to be introduced since $b_{\alpha\beta} = r_\alpha\, g_{\alpha\beta}$ and that the matrix $g_{\alpha\beta}$ is symmetrical. The relative change of the equilibrium constants, due to the perturbation, $\delta \ln K_\alpha$, which enters in Eqs. (25) and (30), is related to the change of the external thermodynamic parameters through the VAN'T HOFF equations.

Eqs. (22) are not independent, i.e. matrix a_{ik} is singular, due to the conditions of mass conservation for the reacting species, and Eqs. (28) are not independent, i.e. matrix $b_{\alpha\beta}$ is singular, if there are parallel reaction paths. The number, n, of independent concentration variables x_i is equal to the number of independent reaction variables ξ_α and is given by $n = S - M = R - T$ where M denotes the total number of mass conservation conditions for the reacting species and T the number of thermodynamically dependent reactions, i.e. the number of "closed loops" in the reaction system. Only n independent Eqs. of (22) or (28) need to be considered; it will be assumed in the following description that the first n equations represent such an independent set.

The conditions of mass conservation show up in a relationship between the stoichiometric coefficients such that the stoichiometric coefficients for species X_{n+1} to X_S can be expressed as a linear combination of the stoichiometric coefficients for the species X_1 to X_n:

$$\nu_{mk} = \sum_{l=1}^{n} \gamma_{ml}\nu_{lk} \qquad \begin{matrix} m = n+1,...,S \\ k = 1,...,n \end{matrix} \tag{32}$$

A corresponding relation holds for the matrix coefficients a_{ik}, i.e.

$$a_{mk} = \sum_{l=1}^{n} \gamma_{ml} a_{lk} \qquad \begin{array}{l} m = n+1,\ldots,S \\ k = 1,\ldots,n \end{array} \qquad (33)$$

resulting in the mass conservation relations

$$x_m(t) = \sum_{l=1}^{n} \gamma_{ml} x_l(t) \qquad m = n+1,\ldots,S \qquad (34)$$

The constraints of the reaction parameters for thermodynamically dependent reactions show up in a relationship between the stoichiometric coefficients such that the coefficient for each reaction step $\beta = n+1$ to R can be expressed by a linear combination of the stoichiometric coefficients of reaction steps $\beta = 1$ to n.

$$\nu_{\alpha\beta} = \sum_{\gamma=1}^{n} \nu_{\alpha\gamma} \rho_{\gamma\beta} \qquad \begin{array}{l} \alpha = 1,\ldots,n \\ \beta = n+1,\ldots,R \end{array} \qquad (35)$$

A corresponding relation holds for the matrix coefficients $b_{\alpha\beta}$, i.e.

$$b_{\alpha\beta} = \sum_{\gamma=1}^{n} b_{\alpha\gamma} \rho_{\gamma\beta} \qquad \begin{array}{l} \alpha = 1,\ldots,n \\ \beta = n+1,\ldots,R \end{array} \qquad (36)$$

The relations between the dependent and independent reaction parameters are given by:

$$\frac{\xi_\beta(t)}{r_\beta} = \sum_{\gamma=1}^{n} \rho_{\gamma\beta} \frac{\xi_\gamma(t)}{r_\gamma} \qquad \beta = n+1,\ldots,R \qquad (37)$$

as can be verified by inspection of the relaxation Eq. (28).

The procedure of solving Eq. (22) leads to an eigenvalue problem, which is treated extensively in mathematical textbooks: There exists a transformation matrix m_{ik} such that the kinetic matrix a_{ik} is diagonalized by a similarity transformation. The procedure is as follows: by introducing a new set of variables, y_k, ("normal variables") defined by

$$x_i = \sum_{k=1}^{S} m_{ik} y_k \qquad i = 1,\ldots,S \qquad (38)$$

into Eq. (22), the system of coupled differential equations for the variables x_i can be transformed to a system of uncoupled differential equations for the variables y_i, which then reads

$$\frac{d}{dt} y_i(t) + \lambda_i y_i(t) = \lambda_i y_i^o f(t) \qquad i = 1,\ldots,n \qquad (39)$$

Footnote 3 continued
matrix (v_{ik} denotes the eigenvector with respect to the eigenvalue λ_k). The eigenvectors of the original matrix a_{ik} are then given by $m_{ik} = c_i^{1/2} v_{ik} S_k$, where S_k is an (arbitrary) normalization constant. As a consequence the inverse of m_{ik} can be expressed by the transposed.
In matrix notation: S * S matrices $A \equiv a_{ik}$, $M \equiv m_{ik}$, $V \equiv v_{ik}$, $E \equiv \delta_{ik}$, $\Lambda \equiv \lambda_k \delta_{ik}$, $S \equiv S_k \delta_{ik}$, $C \equiv c_i \delta_{ik}$ (Λ, S and C are diagonal matrices). Since $M = C^{1/2} VS$ and $VV' = E$, one obtains: $M^{-1} = S^{-1} V' C^{-1/2}$. Combination of this relation with $M' = SV'C^{1/2}$ leads to $M^{-1} = S^{-2} M' C^{-1}$. The square of matrix S is determined by the condition $M^{-1}M = E$ resulting in $S^2 = M'C^{-1}M$; Similarity transformation: $\Lambda = M^{-1}AM = V'C^{-1/2}AC^{1/2}V$.

and which can readily be integrated, yielding (see Sect. 1)

$$y_i(t) = y_i^o \psi(t,\lambda_i) \qquad i = 1,\ldots,n \qquad (40)$$

The values λ_i are the eigenvalues of the matrix a_{ik}, and are obtained by the determinantal relationship

$$\mathrm{Det}(a_{ik} - \lambda_j \delta_{ik}) = 0 \qquad (41)$$

Eq. (41) is equivalent to a polynominal of the order S, exhibiting S roots, whereby only n roots have a finite (negative) value. The remaining (S-n) roots are identical zero. The transformation matrix m_{ik} is given by the eigenvectors of the matrix a_{ik}. They are determined by the system of the homogeneous linear equations:

$$\sum_{k=1}^{S} (a_{ik} - \lambda_j \delta_{ik}) m_{kj} = 0 \qquad i,j = 1,\ldots,S \qquad (42)$$

Each eigenvector m_{kj} for an eigenvalue $\lambda_j \neq 0$ can freely be normalized. For $\lambda_j = 0$ there are (S-n) components which can freely be chosen for each eigenvector. Only n differential equations (Eq. (39)) need to be considered, since the remaining (S-n) equations reflect the time-independent mass conservation conditions as expressed in Eq. (34) which are characterized by eigenvalues equal to zero (infinite relaxation time), i.e.

$$\frac{d}{dt} y_i(t) = 0; \quad y_i(t) = 0 \qquad i = n+1,\ldots,S \qquad (43)$$

The solution of Eq. (38) can then be written in the form

$$x_i(t) = \sum_{k=1}^{n} A_{ik} \psi(t,\tau_k) \qquad i = 1,\ldots,S \qquad (44)$$

The reciprocal relaxation times are the eigenvalues of the kinetic matrix, i.e.

$$\frac{1}{\tau_k} = \lambda_k \qquad k = 1,\ldots,n \qquad (45)$$

and the amplitude factors are the eigenvectors, normalized such that Eq. (6) is fulfilled, i.e.

$$A_{ik} = m_{ik} y_k^o; \quad \sum_{k=1}^{S} m_{ik} y_k^o = x_i^o \qquad (46)$$

Denoting the elements of the inverse of matrix m_{ik} by m_{ik}^* the amplitude factors read explicitly:[3]

---

[3]The amplitude factors can be written in a form, which does not require the explicit determination of the inverse of m_{ik}, since

$$m_{ik}^* = \frac{m_{ki}/c_k}{\sum_l (m_{li})^2 c_l}$$

This relation arises from the consideration of the S*S matrix $a_{ik}(c_k/c_i)^{1/2}$, which is similar to the original matrix a_{ik} and thus exhibits the same eigenvalues. Since it is symmetrical, there exist S orthogonal eigenvectors, v_{ik}, i.e. $v'_{ik} = v_{ik}^*$, where $v'_{ik} = v_{ki}$ denotes the transposed and v_{ik}^* the inverse

$$A_{ik} = \sum_{l=1}^{S} m_{ik} m^{*}_{kl} x^{o}_{l} \qquad \begin{array}{l} i = 1,\ldots,S \\ k = 1,\ldots,n \end{array} \qquad (47)$$

The appearance of zero roots can be avoided by considering the set of n independent differential equations and substituting for the values of x_i with $i = n+1$ to S using the mass conservation relations Eq. (34). The system of n independent differential equations has the form

$$\frac{d}{dt} x_i(t) + \sum_{k=1}^{n} \hat{a}_{ik} x_k(t) = \sum_{k=1}^{n} \hat{a}_{ik} x^{o}_{k} f(t)$$

$$\hat{a}_{ik} = a_{ik} + \sum_{l=n+1}^{S} a_{il} \gamma_{lk} \qquad i,k = 1,\ldots,n$$

The total changes of the independent concentration variables are determined, by introducing Eq. (34) into Eq. (25), from the system of linear equations

$$\sum_{k=1}^{n} \hat{c}_{\alpha k} x^{o}_{k} = \delta \ln K_{\alpha}$$

$$\hat{c}_{\alpha k} = c_{\alpha k} - \sum_{l=n+1}^{S} c_{\alpha l} \gamma_{lk} \qquad \alpha,k = 1,\ldots,n$$

The matrix $\hat{a}_{ik}$ has the same eigenvalues λ_j as the matrix a_{ik} (except for the zero eigenvalues) and the elements of the jth eigenvector $\hat{m}_{ij}$ of the matrix $\hat{a}_{ik}$ are the same as the first n elements of the eigenvector m_{ij} $(i,j = 1,\ldots,n)$ of the matrix a_{ik}, i.e.

$$m_{ij} = \hat{m}_{ij} \qquad \begin{array}{l} i = 1,\ldots,n \\ j = 1,\ldots,n \end{array}$$

The elements of the eigenvectors m_{ij} for the dependent concentration variables are related to those of the independent concentration variables by the mass conservation relations Eq. (34), i.e.

$$m_{ij} = \sum_{l=1}^{n} \gamma_{il} \hat{m}_{lj} \qquad \begin{array}{l} i = n+1,\ldots,S \\ j = 1,\ldots,n \end{array}$$

The same relation holds for the amplitude factors A_{ij}.

Footnote 4 continued
of the original matrix $b_{\alpha\beta} = r_{\alpha}\, g_{\alpha\beta}$ are then given by $q_{\alpha\beta} = (r_{\alpha})^{1/2}\, u_{\alpha\beta}\, T_{\beta}$, where T_{β} is an (arbitrary) normalization constant. As a consequence, the inverse of $q_{\alpha\beta}$ can be expressed by the transposed.

In matrix notation: R * R matrices $B \equiv b_{\alpha\beta}$, $G \equiv g_{\alpha\beta}$, $Q \equiv q_{\alpha\beta}$, $U \equiv u_{\alpha\beta}$, $E = \delta_{\alpha\beta}$, $\Lambda = \lambda_{\alpha}\delta_{\alpha\beta}$, $T = T_{\alpha}\delta_{\alpha\beta}$, $R = r_{\alpha}\delta_{\alpha\beta}$ (Λ, T, R are diagonal matrices). The matrices B and G are related by B=RG. Since $Q = R^{1/2}UT$ and $UU' = E$ one obtains $Q^{-1} = T^{-1}U'\,R^{-1/2}$; Combination of this relation with $Q' = TU'R^{1/2}$ leads to $Q^{-1} = T^{-2}Q'R^{-1}$. The square of the matrix T is determined by the condition $Q^{-1}Q = E$ resulting in $T^2 = Q'R^{-1}Q$. Similarity transformation: $\Lambda = Q^{-1}BQ = U'R^{-1/2}BR^{1/2}$ using kinetic matrix B, or using the thermodynamic matrix G, $\Lambda = Q^{-1}RGQ = U'R^{1/2}GR^{1/2}$.

The turnover in the chemical system may be interpreted as arising from n uncoupled reactions ("normal reactions") as is seen from Eq. (39) where the kinetics is described by n uncoupled differential equations, each of which is characteristic for a single reaction step and described by a single relaxation time.

The amplitude factors A_{ik} then represent the concentration changes of the reacting species X_i in the kth normal reaction, which is associated with the relaxation time τ_k, and each normal reaction may be explicitly formulated — analogous to Eq. (17) — by the reaction equations (HAYMAN, 1969)

$$\sum_{i=1}^{S} N_{ik}X_i = 0 \qquad k = 1,\ldots,n \qquad (48)$$

whereby the "generalized" (non integer) stoichiometric coefficients are given by

$$N_{ik} \propto A_{ik} \propto m_{ik}$$

with an arbitrary proportionality constant (normalization) for each normal reaction k.

The advancement of each individual reaction step α, associated with the relaxation time τ_k, which will be denoted by $P_{\alpha k}$, can be obtained — in an analogous way as indicated for the concentration variables — by solving Eq. (28):

$$\xi_\alpha(t) = \sum_{k=1}^{n} P_{\alpha k}\psi(t,\tau_k) \qquad \alpha = 1,\ldots,R \qquad (49)$$

Reciprocal relaxation times and amplitude factors are determined by the eigenvalues and eigenvectors of the kinetic matrix $b_{\alpha\beta}$. The zero-roots of the determinantal relationship reflect here the time-independent relations between the advancements of thermodynamically dependent reactions as expressed by Eq. (37). If $q_{\alpha k}$ (α, k = 1 to R) denotes the kth eigenvector of the matrix $b_{\alpha\beta}$, the explicit expression for the advancements is given by[4] (analogous to Eq. (47)):

$$P_{\alpha k} = \sum_{\beta=1}^{R} q_{\alpha k}q^*_{k\beta}\xi^0_\beta \qquad \begin{matrix}\alpha = 1,\ldots R\\ k = 1,\ldots n\end{matrix} \qquad (50)$$

---

[4]The amplitude factors can again be written without the explicit introduction of the inverse of $q_{\alpha\beta}$, since

$$q^*_{\alpha\beta} = \frac{q_{\beta\alpha}/r_\beta}{\sum_\gamma (q_{\gamma\alpha})^2/r_\gamma}\,.$$

This relation arises from the consideration of the R * R matrix $b_{\alpha\beta}(r_\beta/r_\alpha)^{1/2} = (r_\alpha)^{1/2}\, g_{\alpha\beta}(r_\beta)^{1/2}$ which is similar to the original matrix $b_{\alpha\beta}$. Since it is symmetrical, there exist R orthogonal eigenvectors $u_{\alpha\beta}$ ($u_{\alpha\beta}$ denotes the eigenvector with respect to the eigenvalue λ_β), i.e. $u'_{\alpha\beta} = u^*_{\alpha\beta}$. The eigenvectors

The presence of zero eigenvalues can again be avoided by reducing the system of R dependent differential equations (Eq. (28)) to a system of n independent equations by substituting for the values ξ_β, $\beta = n+1$ to S according to Eq. (37)

It can be shown that Eq. (50), together with Eq. (30), can be written in the following form (SCHIMMEL, 1971)[5]

$$P_{\alpha k} = \Gamma_k \sum_{\beta=1}^{R} q_{\alpha k} q_{\beta k} \delta \ln K_\beta \tag{51}$$

where the Γ-value which is a measure of the chemical turnover in the particular normal reaction, is given by

$$\frac{1}{\Gamma_k} = \sum_{l=1}^{R} \sum_{m=1}^{R} q_{lk} q_{mk} g_{lm} \tag{52}$$

Alternatively, $P_{\alpha k}$ may be obtained from the solutions of the concentration variables by using Eq. (20) and (37) resulting in

$$\sum_{\alpha=1}^{R} \nu_{i\alpha} P_{\alpha k} = A_{ik} \qquad i,i = 1,...,n \tag{53}$$

and

$$\frac{1}{r_\beta} P_{\beta k} - \sum_{\alpha=1}^{n} \frac{\rho_{\alpha\beta}}{r_\alpha} P_{\alpha k} = 0 \qquad \begin{matrix}\beta = n+1,...,R \\ k = 1,...,n\end{matrix} \tag{54}$$

The same correlation exists between the eigenvectors of the matrices a_{ik} and $b_{\alpha\beta}$, i.e. between m_{ik} and $q_{\alpha k}$ respectively, Eq. (53) itself is sufficient for determining the reaction parameters from the concentration variables if the reaction system contains no thermodynamically dependent reaction steps, i.e. if $R = n$. According to the signs chosen for the stoichiometric coefficients, the turnover in reaction α is from the left to the right side of the corresponding reaction equation (Eq. (17)), if $P_{\alpha k} > 0$ and is from the right to the left side if $P_{\alpha k} < 0$.

---

[5]This is because the R * R transformation matrix $q_{\alpha\beta}$ which diagonalizes the kinetic matrix $b_{\alpha\beta}$ by a similarity transformation, diagonalizes simultaneously the thermodynamic matrix $g_{\alpha\beta}$ by a congruence transformation: Eq. (52) reads in matrix notation $\Gamma^{-1} = Q'GQ$. Introducing the relations given in footnote 4, i.e. $Q'GQ = TU'R^{1/2}GR^{1/2}UT = T\Lambda T$, it is seen that Γ is diagonal, since T and Λ are diagonal matrices. If vectors ξ^o and $\delta\ln K$ are defined as $\xi^o \equiv \xi^o_\alpha$, $\delta\ln K \equiv \delta\ln K_\alpha$ the sum which appears in Eq. (50) can be written as $Q^{-1}\xi^o = \Gamma Q' G \xi^o$. Using Eq. (30) in the form $G\xi^o = \delta\ln K$ results in $Q^{-1}\xi^o = \Gamma Q' \delta\ln K$.

The relative contribution of each reaction step α to the kth normal reaction, denoted by $p_{\alpha k}$, may be defined by normalizing the quantities $P_{\alpha k}$ or $q_{\alpha k}$ such, that $\sum_{\alpha=1}^{R} |p_{\alpha k}| = 1$ resulting in

$$p_{\alpha k} = \frac{P_{\alpha k}}{\sum_{\beta=1}^{R} |P_{\beta k}|} = \frac{q_{\alpha k}}{\sum_{\beta=1}^{R} |q_{\beta k}|} \qquad \begin{array}{l} \alpha = 1,\ldots,R \\ k = 1,\ldots,n \end{array} \qquad (55)$$

The eigenvectors $q_{\alpha k}$ must be used in Eq. (55) if the reaction system degenerates such, that there is no turnover for a particular normal reaction, k, i.e. $A_{ik} = 0$ for all i = 1 to n and correspondingly $P_{\alpha k} = 0$ for all α = 1 to R. The eigenvectors $q_{\alpha k}$, as well as the normalized quantities $p_{\alpha k}$ are uniquely determined by the reaction system, independent of the strength of perturbation.

The relative contribution, $p_{\alpha k}$, of an individual reaction step α to the kth normal reaction, as given in Eq. (55), defines the generalized stoichiometric coefficients of the normal reaction, N_{ik} (see Eq. (48)), as

$$N_{ik} = \sum_{\alpha=1}^{R} \nu_{i\alpha} p_{\alpha k} \qquad \begin{array}{l} i = 1,\ldots,S \\ k = 1,\ldots,n \end{array} \qquad (56)$$

Any physical parameter for a normal reaction can be calculated from the corresponding parameters of the individual reaction steps of the mechanism Eq. (17). For instance, if ΔH_α denotes the reaction enthalpies, and $\Delta\varepsilon_\alpha$ the molar extinction change for each reaction step α, the reaction enthalpy and the optical extinction change for the kth normal reaction, ΔH_k and $\Delta\varepsilon_k$ respectively are given by

$$\Delta H_k = \sum_{\alpha=1}^{R} p_{\alpha k} \Delta H_\alpha ; \quad \Delta\varepsilon_k = \sum_{\alpha=1}^{R} p_{\alpha k} \Delta\varepsilon_\alpha$$

The physical quantity actually observed in the experiment can be determined either from the properties of the individual reacting species or from the properties of the individual reaction steps. If the optical absorption, E, is chosen as an example, the change in the kth normal reaction is determined by:

$$\delta E_k = \sum_{i=1}^{S} \varepsilon_i A_{ik} = \sum_{\alpha=1}^{R} \Delta\varepsilon_\alpha P_{\alpha k} \qquad k = 1,\ldots,n \qquad (57)$$

where ε_i denote the molar extinction coefficients of species i and $\Delta\varepsilon_\alpha$ the molar extinction change in the αth reaction step, i.e.

$\Delta\varepsilon_\alpha = \sum_{i=1}^{S} \nu_{i\alpha}\varepsilon_i$. Introducing Eq. (51), the relaxation amplitudes can be written as

$$\delta E_k = \sum_{\alpha=1}^{R}\sum_{\beta=1}^{R} \Phi_{\alpha\beta k}\Delta\varepsilon_\alpha \delta \ln K_\beta$$
$$\Phi_{\alpha\beta k} = \frac{p_{\alpha k} p_{\beta k}}{\sum_l \sum_m p_{lk} p_{mk} g_{lm}} \qquad (58)$$

The sum of all reactions, which appears in Eq. (58) can be reduced – if R > n – to a sum over only the independent reactions by substituting for the dependent quantities:

$$\Delta\varepsilon_\beta = \sum_{\gamma=1}^{n} \Delta\varepsilon_\gamma \rho_{\gamma\beta}; \quad \delta \ln K_\beta = \sum_{\gamma=1}^{n} \delta \ln K_\gamma \rho_{\gamma\beta} \qquad \beta = n+1,\ldots,R \qquad (59)$$

In the reduced form Eq. (58) can be used to determine with a regression procedure the quantities ($\Delta\varepsilon_\alpha \delta \ln K_\beta + \Delta\varepsilon_\beta \delta \ln K_\alpha$) from the observed relaxation amplitudes (THUSIUS, 1972). It is this combination of molar extinction changes and reaction enthalpies for a set of thermodynamically independent reactions which is a unique quantity characteristic for the individual relaxation amplitudes. The total change of the observed parameter is given by the sum of all normal reactions

$$\delta E^o = \sum_{k=1}^{n} \delta E_k = \sum_{i=1}^{S} \varepsilon_i x_i^o = \sum_{\alpha=1}^{R} \Delta\varepsilon_\alpha \xi_\alpha^o \qquad (60)$$

Eq. (60) can again be written as a sum over the products of extinction changes and reaction enthalpies by applying Eq. (58). It can be expressed as a sum of only thermodynamically independent reactions by applying directly Eq. (30).

D. Single Reaction Steps

Relaxation times and amplitudes for single reaction steps can immediately be calculated from the equations given in Section C. When using the matrix coefficients which correspond to the reaction parameters (see Eqs. 29, 31), the relaxation time reads: (index α = 1, is omitted, i.e. $r = r_1$, $g = g_{11}$)

$$1/\tau = rg \qquad (61)$$

and the relaxation amplitudes:

$$\frac{x_1^o}{\nu_1} = \frac{x_2^o}{\nu_2} = \ldots\ldots = \frac{x_S^o}{\nu_S} = \xi^o = \frac{\delta \ln K}{g} \qquad (62)$$

The quantity Γ, introduced in Eq. (51) as a measure of the chemical turnover, is given by

$$\Gamma = 1/g \qquad (63)$$

Table 1

Reaction $\underset{k_{-1}}{\overset{k_1}{\rightleftharpoons}}$	Relaxation times $1/\tau$	Relaxation amplitudes Γ^{-1}
$A \rightleftharpoons B$	$k_1 + k_{-1}$	$1/c_A + 1/c_B$
$A + B \rightleftharpoons C$	$k_1(c_A + c_B) + k_{-1}$	$1/c_A + 1/c_B + 1c_C$
$2A \rightleftharpoons A_2$	$k_1 \cdot 4c_A + k_{-1}$ $= \sqrt{(k_1)^2 + 8k_1k_{-1}c_o}$; $c_o = c_A + 2c_{A_2}$	$4/c_A + 1/c_{A_2}$
$A + B \rightleftharpoons C + D$	$k_1(c_A + c_B) + k_{-1}(c_C + c_D)$	$1/c_A + 1/c_B + 1/c_C + 1/c_D$
$A + B + C \rightleftharpoons D$	$k_1(c_Ac_B + c_Ac_C + c_Bc_C) + k_{-1}$	$1/c_A + 1/c_B + 1/c_C + 1/c_D$
$A + 2B \rightleftharpoons C$	$k_1c_B(c_B + 4c_A) + k_{-1}$	$1/c_A + 4/c_B + 1/c_C$
$A + C \rightleftharpoons B + C$ (C catalyst)	$(k_1 + k_{-1})c_C$	$1/c_A + 1/c_B$
$A + B \rightleftharpoons C$ (B buffered)	$k_1c_B + k_{-1}$	$1/c_A + 1/c_C$

List of relaxation times and amplitudes for single reaction steps of different reaction order (all concentrations are equilibrium values).

In a straightforward way of determining the relaxation time for a single reaction step one may start from the kinetic equation of one of the reacting species. The rate equation may then be linearized using the simple rules to form differentials, e.g. the linearization of products (c_ic_k) in the kinetic equations yields $\delta(c_ic_k) = \bar{c}_i\delta c_k + \bar{c}_k\delta c_i$ $(\delta c_i \equiv x_i)$. Introduction of the mass conservation relations yields the relaxation equation in the form $\frac{d}{dt}\delta c_i = -1/\tau\,\delta c_i$ for any species i.

If the concentration of one of the species, e.g. species k, does not change at all during the progress of the reaction, i.e. if $\delta c_k = 0$, the concentration terms associated with δc_k drop out in the above equation. Such a situation exists, for instance, if species k is (infinitely) buffered. In the same way the relaxation amplitudes may be obtained from the linearized mass action law.

Table 1 lists the expressions for relaxation times and amplitudes for reactions of different order. It is seen that the concentration dependence of relaxation times can be used as a diagnostic tool for determining the order of the observed chemical reaction. Table 1 contains also steps of higher order which will certainly not reflect elementary steps in solution. However, it may be possible to reduce more complex systems to a single reaction step by introducing steady-state assumptions for intermediates (see Sect. D). The overall reaction may then be, for instance of third order etc., characterized by apparent (pseudo-third order, etc.) rate constants. A corresponding reduction to a single step has been made for reactions in which a particular species acts as a catalyst or is (infinitely) buffered; in both cases the actual reaction systems are necessarily more complex. The measurement of relaxation amplitudes can be used to determine the equilibrium constant of a binding process; this method is of special advantage in the case of very large stability constants when the static titration curve, i.e. the relative amount of complex formed as a function of total ligand concentration, fails to give information about the equilibrium constant, but only reflects the total concentration of the species which is being titrated (see e.g. EIGEN and WINKLER, to be published). Alternatively, if the equilibrium constant is known independently, the Γ-curves may be used, in addition to the kinetic measurements, in order to test whether the system investigated is indeed adequately described by a single reaction step (see e.g. ILGENFRITZ and SCHUSTER, 1974).

How the equilibrium constant K enters the relations for static binding curves and relaxation amplitudes will be compared for the simple binding reaction $A + X \rightleftharpoons AX$. If the relative amount of complex formed, Y, is determined as a function of total ligand concentration, c_X^o, at constant concentration of A, c_A^o, the following relation holds for the concentration of ligand added, at half-saturation Y = 0.5

$$c_X^o|_{Y=0.5} = \frac{c_A^o}{2} + \frac{1}{K} \tag{64}$$

From the expression in Table 1 it can easily be derived that for $c_X^o = c_A^o \equiv c_o$ the Γ-factor is given by

$$\Gamma(c_X^o = c_A^o) = \frac{1}{2K}\left[\frac{1+2Kc_o}{(1+4Kc_o)^{1/2}} - 1\right] \tag{65}$$

It is seen that for large stability constants, i.e. $Kc_A^o \gg 1$, the static titration experiment does not allow to determine K, since it appears only as an additive (small) term in Eq. (64). On the other hand, the value of Γ is directly proportional to the equilibrium constant:

$$\Gamma(c_X^o = c_A^o) = \frac{1}{2}(c_o/K)^{1/2}; \quad Kc_A^o \gg 1 \tag{66}$$

Moreover, in determining $\Gamma(c_X^o)$ as a function of total ligand concentration at constant total concentration of A, a sharp maximum appears for $c_X^o = c_A^o$, as long as $Kc_A^o \gg 1$. The condition of equal concentrations, which must be fulfilled for applying Eq. (65), can thus easily be found experimentally, as long as the experimental arrangement is sensitive enough to allow the detection of the relaxation signals. Relaxation amplitudes allow therefore to determine the equilibrium constant in the very cases where static titration is not applicable. However, since the relaxation amplitudes are given by $\Gamma\delta\ln K$ an independent knowledge of the reaction enthalpy is required.

E. Approximations for Complex Reaction Systems

The complexity of the correlation of relaxation times and amplitudes with the thermodynamic and kinetic parameters of the reaction system is reduced considerably if certain conditions regarding the rate of individual reaction steps or the concentration of particular species are fulfilled. Three such conditions, which often apply also to biophysical reaction systems will be discussed.

Assumption 1: The equilibrium concentrations of the intermediate species in a consecutive reaction path are very small compared to the concentrations of the reactants and the products. In this case, the consecutive series of reaction steps is effectively reduced to one single reaction step, exhibiting apparent rate constants. As a consequence, the relaxation spectrum shows a degeneracy such, that all but one relaxation times appear with negligible amplitude and thus do not contribute to the relaxation spectrum. The formal description of the (relaxation) kinetics is based on the "steady-state" assumption for those intermediates, which are present in very small concentrations.

Assumption 2: The concentration of a ligand or substrate, binding to an acceptor, e.g. a protein, is large compared to the concentration of the acceptor. In this case, the ligand concentration does not change effectively during the progress of the reaction; the ligand behaves as if being buffered ("quasi-buffered"). As a consequence, in parallel reactions involving the same ligand there is no more coupling via the common ligand.

Assumption 3: Certain reaction steps are very fast compared to the overall reaction process. If this situation is present, there is an effective decoupling between the group of fast reactions and the group of slowly equilibrating reactions. The species involved in the fast reactions equilibrate before there is any significant change of the concentrations according to the slow reaction processes.

In general, the conditions as quoted above will not strictly be fulfilled for an actual reaction system. But even in the cases where the assumptions are not really justified, i.e. where the concentrations of intermediates are not sufficiently small, or in cases where certain reaction steps do not equilibrate fast enough, one may still obtain quite a good approximation for the kinetics of the actual system, which may help to interpret the relaxation spectrum and to understand its degeneracies and the concentration dependence of the relaxation time(s). Whether or not it is reasonable to introduce the steady-state assumptions in the first analysis of the relaxation spectrum, depends to some extent on independent information about the reaction system. In cooperative systems, for instance, the intermediates will be present only in small concentrations and the observed slow relaxation process may be quite well approximated by a single reaction – even in the cases where the presence of faster relaxation times indicates that intermediate species are present in detectable concentration. Examples for this situation are given in the helix-coil-transition of oligonucleotides (POERSCHKE, 1974), or in the oxygen binding to hemoglobin (ILGENFRITZ and SCHUSTER, 1974). The classical example of the application of the steady-state condition in biophysics is enzyme catalysis at low levels of enzyme concentration (MICHAELIS-MENTEN kinetics) and thus at conditions where the enzyme substrate complex(es) is only populated to a very small extent.

The assumption of fast equilibrating reaction steps can already be made as an approximation if relaxation times are found which are separated in time by at least one order of magnitude, i.e. in praxi, if the relaxation curve can reliably be analyzed in terms of more than one time constant.

The Steady-State Assumption. The steady-state assumption for the intermediates $X_1, X_2, \ldots$ in a consecutive reaction path leading from reactants, R_i, to products, P_i

$$\Sigma R_i \underset{}{\overset{(1)}{\rightleftharpoons}} X_1 \overset{(2)}{\rightleftharpoons} X_2 \overset{(3)}{\rightleftharpoons} X_3 \rightleftharpoons \quad \ldots \Sigma P_i \tag{67}$$

is introduced into the kinetic equations by putting (the dot denotes differentiation with respect to time)

$$\begin{aligned} \dot{c}_1 &= k_1 \Pi c_{R_i} - (k_{-1}+k_2) c_1^{(st)} + k_{-2} c_2^{(st)} &&= 0 \\ \dot{c}_2 &= \qquad k_2 c_1^{(st)} - (k_{-2}+k_3) c_2^{(st)} + k_{-3} c_3^{(st)} &&= 0 \\ &\ldots\ldots \end{aligned} \tag{68}$$

Eq. (68) is not to be understood that $c_1, c_2, \ldots$ have time-independent constant values, but that there is a time-independent relation between the concentrations of these species and the concentration of the reactants or products. This relation is obtained by solving Eq. (68) for the steady-state concentrations $c_1{}^{(st)}$, $c_2{}^{(st)}, \ldots$ If advancements are used for the kinetic description, the steady-state assumption results in

$$\begin{aligned} \dot{c}_1 &= \dot{\xi}_1 - \dot{\xi}_2 &&= 0 \\ \dot{c}_2 &= \quad \dot{\xi}_2 - \dot{\xi}_3 &&= 0 \\ &\ldots\ldots \end{aligned} \tag{69}$$

It is seen that the steady-state condition for the intermediates corresponds to the situation, where the reaction rates $d\xi_\alpha/dt$ are equal for all reaction steps in the consecutive path. As a consequence the system can be described by one single advancement and the relaxation spectrum degenerates such, that only one relaxation time has a finite amplitude. It corresponds to the overall reaction

$$\Sigma R_i \rightleftharpoons \Sigma P_i \tag{70}$$

which is characterized by apparent rate constants, $\vec{k}$, $\overleftarrow{k}$. It is obvious that no more detailed information about the intermediate steps is available. This information can only be obtained if, with other – non-equilibrium – methods, the concentration of the

Table 2

	Overall reaction	Intermediates	$\vec{k}$	$\overleftarrow{k}$	
	R_i: reactants P_i: products				
a)	$\Sigma R_i \rightleftharpoons \Sigma P_i$	$\underset{(1)}{\rightleftharpoons} X_1 \underset{(2)}{\rightleftharpoons}$	k_1k_2/N	$k_{-1}k_{-2}/N$	$N=k_{-1}+k_2$
b)	$\Sigma R_i \rightleftharpoons \Sigma P_i$	$\underset{(1)}{\rightleftharpoons} X_1 \rightleftharpoons X_2 \underset{(3)}{\rightleftharpoons}$	$k_1k_2k_3/N$	$k_{-1}k_{-2}k_{-3}/N$	$N=k_{-1}k_{-2}+k_{-1}k_3+k_2k_3$
c)	$\Sigma R_i+Y \rightleftharpoons \Sigma P_i$	$\underset{(1)}{\rightleftharpoons} X_1 \overset{+Y}{\underset{(2)}{\rightleftharpoons}}$	k_1k_2/N	$k_{-1}k_{-2}/N$	$N=k_{-1}+k_2\bar{c}_Y$
d)	$\Sigma R_i+2Y \rightleftharpoons \Sigma P_i$	$\underset{(1)}{\rightleftharpoons} X_1 \overset{+Y}{\underset{(2)}{\rightleftharpoons}} X_2 \overset{+Y}{\underset{(3)}{\rightleftharpoons}}$	$k_1k_2k_3/N$	$k_{-1}k_{-2}k_{-3}/N$	$N=k_{-1}k_{-2}+k_{-1}k_3\bar{c}_Y+k_2k_3\bar{c}_Y^2$
e)	$\Sigma R_i+Y \rightleftharpoons \Sigma P_i+Z$	$\overset{(1)}{\underset{+Z}{\rightleftharpoons}} X_1 \overset{+Y}{\underset{(2)}{\rightleftharpoons}}$	k_1k_2/N	$k_{-1}k_{-2}/N$	$N=k_{-1}\bar{c}_Z+k_2\bar{c}_Y$

Apparent rate constants $\vec{k}$, $\overleftarrow{k}$ for the overall reaction where the intermediates X_1, X_2 are assumed to be in a steady state. The types of reaction apply e.g. to (a, b) enzyme catalysis, (c, d) (cooperative) ligand binding to protein, (e) exchange reactions.

intermediates is populated significantly. The apparent rate constants for reaction Eq. (70) are obtained by comparing the kinetic equation for the reactants or products of scheme Eq. (70) with the equation of scheme (Eq. (67)) whereby the steady-state values $c_1^{(st)}$, $c_2^{(st)}$,... obtained from the solution of Eq. (68), are introduced. Table 2 lists the apparent rate constants $\vec{k}$, $\overleftarrow{k}$ for several simple reaction paths. The relaxation times are obtained by introducing these values into the expression for the relaxation time for one single reaction step. For a more detailed discussion of the steady-state relaxation time for enzymatic mechanisms see e.g. HAMMES and ALBERTY (1960).

Fast Reaction Steps. The assumption that certain reaction steps equilibrate rapidly, i.e. that the exchange rates r_α are large compared to the exchange rates of other steps, implies that these steps are uncoupled from the slower equilibrating steps. Thus, the relaxation kinetic behavior of the group of fast reactions is unaffected by the slow reactions, and relaxation times and amplitudes for this class of reaction can be obtained as described above. During the turnover in the slow normal reactions an equilibrium distribution can then be assumed for the species involved in the fast reactions. This condition is introduced in the kinetic Eq. (21) by putting

$$\dot{\xi}_\alpha = k_\alpha \prod_q c_q^{|\nu_{q\alpha}|} - k_{-\alpha} \prod_r c_r^{|\nu_{r\alpha}|} = 0 \qquad (71)$$

and is to be understood that the condition

$$\prod_i c_i^{\nu_{i\alpha}} = K_\alpha \qquad (72)$$

is fulfilled for all reactions which are assumed to be fast (c_i are not equilibrium concentrations here). In order to obtain the (slow) relaxation times for this case, methods have been devised, which are based on the kinetic matrix for the advancements (see Eq. (29)), i.e. the determinantal relationship $\mathrm{Det}(b_{\alpha\beta} - \lambda_i \delta_{\alpha\beta}) = 0$ (CASTELLAN, 1963; HAMMES and SCHIMMEL, 1966). If there is only one slowly equilibrating step, one may start from the kinetic equations for the concentration variables and proceed in a straightforward way as follows: if these equations are written in the form (see Eq. (20))

$$\begin{aligned} \dot{c}_1 &= \nu_{11}\dot{\xi}_1 + \nu_{12}\dot{\xi}_2 + \ldots \\ \dot{c}_2 &= \nu_{21}\dot{\xi}_1 + \nu_{22}\dot{\xi}_2 + \ldots \\ \dot{c}_3 &= \ldots\ldots \end{aligned} \qquad (73)$$

one arrives at the differential equation for the slow reaction process by forming a linear combination

$$y = \sum_i s_i c_i \qquad (74)$$

such that the advancements of the fast equilibrating steps drop out in Eq. (73). It is, thus, a sum of the concentrations of the rapidly interconverting species which is considered. Introduction of the explicit rate expression for $\dot{\xi}_\alpha$ (Eq. (21)) and linearization lead to an equation of the form ($\delta c_i \equiv x_i$)

$$\delta\dot{y} = \sum_i w_i \cdot \delta c_i \qquad (75)$$

which is brought into the form of the relaxation equation, i.e. $\dot{y} = -1/\tau\,\delta y$, by multiplying and dividing the right side by δy:

$$\delta\dot{y} = \left[\sum w_i \left(\frac{\delta c_i}{\delta y}\right)\right] \delta y \qquad (76)$$

The ratios which appear in Eq. (76) do not contain any kinetic parameters. They are uniquely determined by the linearized "equilibrium" conditions, Eq. (72), i.e.

$$\sum_k \frac{\nu_{k\alpha}}{c_k} \cdot \delta c_k = 0 \qquad (77)$$

and by the conditions of mass conservation (the quantity δy represents the normal variable for the slow normal reaction).

This easy procedure may be illustrated by an example: two fast binding reactions coupled via slow isomerization. R, T, may denote two structure forms of a protein; for an actual application of this procedure see, for instance, the treatment of the allosteric model as applied to the binding of nicotinamide-adenine-dinucleotide to glyceraldehyde-3-phosphate dehydrogenase (EIGEN, 1968; KIRSCHNER et al., 1971).

Example:

$$T + X \underset{(1)\ \text{fast}}{\rightleftharpoons} TX \underset{(2)\ \text{slow}}{\rightleftharpoons} RX \underset{(3)\ \text{fast}}{\rightleftharpoons} R+X$$

The Eqs. (73) to (76) read explicitly for this example:

Eq. (73):

$$\begin{aligned}
\dot{c}_X &= -\dot{\xi}_1 + \dot{\xi}_3\\
\dot{c}_T &= -\dot{\xi}_1\\
\dot{c}_{TX} &= +\dot{\xi}_1 - \dot{\xi}_2\\
\dot{c}_{RX} &= +\dot{\xi}_2 - \dot{\xi}_3\\
\dot{c}_R &= +\dot{\xi}_3
\end{aligned}$$

Eq. (74):

$$y = c_T + c_{TX} = -(c_R + c_{RX})$$

Eq. (75)

$$\delta\dot{y} = -\dot{\xi}_2 = -k_2\,\delta c_{TX} + k_{-2}\,\delta c_{RX}$$

Eq. (76): $$\delta\dot{y} = -\left[k_2 \frac{\delta c_{TX}}{\delta c_T + \delta c_{TX}} + k_{-2} \frac{\delta c_{RX}}{\delta c_R + \delta c_{RX}}\right] \cdot \delta y$$

Eq. (77), applied to the reaction steps (1) and (3), and the mass conservation relations for total protein and total ligand lead to the system of linear equations:

$$\begin{array}{llllr} K_1\bar{c}_T\delta c_X & + K_1\bar{c}_X\delta c_T & - \delta c_{TX} & & = 0 \\ K_3\bar{c}_R\delta c_X & & & + K_3\bar{c}_X\delta c_R - \delta c_{RX} & = 0 \\ & \delta c_T & + \delta c_{TX} & \delta c_R + \delta c_{RX} & = 0 \\ \delta c_X & & \delta c_{TX} & \delta c_{RX} & = 0 \end{array}$$

with the solution

$$\frac{\delta c_T}{\delta c_{TX}} = \frac{\gamma_T}{K_1\bar{c}_X} \qquad \gamma_T = \frac{1 + K_1\bar{c}_T + K_3(\bar{c}_R+\bar{c}_X)}{1 + K_3\bar{c}_T + K_3(\bar{c}_R+\bar{c}_X)}$$

$$\frac{\delta c_R}{\delta c_{RX}} = \frac{\gamma_R}{K_3\bar{c}_X}\,; \qquad \gamma_R = \frac{1 + K_3\bar{c}_R + K_1(\bar{c}_T+\bar{c}_X)}{1 + K_1\bar{c}_R + K_1(\bar{c}_T+\bar{c}_X)}$$

and the relaxation time is given by

$$\frac{1}{\tau} = k_2\frac{1}{1+\gamma_T/K_1\bar{c}_X} + k_{-2}\frac{1}{1+\gamma_R/K_3\bar{c}_X}$$

The expression for the (slow) relaxation time becomes very simple if the concentration of ligand is very large, i.e. $\bar{c}_X \gg \bar{c}_T$, $\bar{c}_R$. In this case the γ-factors become unity. The expression for the relaxation time under this additional condition can be derived directly, if quasi-buffering for the ligand is assumed, i.e. if the ratio of concentration changes according to Eq. (72) is determined under the condition $\delta c_X = 0$.

The fast relaxation times are to be determined from the two reaction steps $T+X \rightleftharpoons TX$, $R+X \rightleftharpoons RX$ which are coupled via the common ligand X.

F. Average Relaxation Times

For complex reaction systems, it will be practically impossible to determine the complete spectrum of relaxation times and amplitudes from the observed relaxation progress curve. Even if the number of relaxation times is considered to be a fixed parameter it will not be possible to analyze the relaxation curve in terms of a unique set of relaxation times. Some normal reactions may appear with negligible amplitude and other normal reactions may exhibit time constants which lie very close together, but even in the case where there are not true degeneracies, it will often be found, due to the limited signal-to-noise-ratio, that an intrinsically complex relaxation spectrum looks rather "simple" and can be described within the experimental errors by only a few, e.g. one or two, time constants. However, the straightforward analysis in terms of a few time constants only will in general lead to values which have no physical meaning; these time constants correspond to undefined averages of the true relaxation times. True degeneracies may be evident, if independent information about the kinetic properties of certain reaction steps and/or about the concentration of intermediate species is available. They may also become apparent if the relaxation spectrum is simulated by assuming "reasonable" values of rate and equilibrium parameters. From considerations of this kind it may then be justified to analyze the observed relaxation curve in terms of a few time constants only.

An alternative to the analysis of the relaxation curve in terms of n relaxation times and n amplitudes consists in the evaluation of well-defined averages of relaxation times.

For the discussion of average relaxation times we consider only relaxation curves which arise from a stepwise perturbation (jump-methods) and are thus characterized by a sum of exponential terms. If the optical absorption is taken as an example, the relaxation curve is given by

$$\delta E(t) = \sum_{k=1}^{n} \sum_{i=1}^{S} \varepsilon_i A_{ik}\left[1-\exp(-t/\tau_k)\right] = \sum_{k=1}^{n} \sum_{\alpha=1}^{R} \Delta\varepsilon_\alpha P_{\alpha k}\left[1-\exp(-t/\tau_k)\right] \tag{78}$$

A set of 2n moments, μ_m ($m = 0,1,\ldots,2n-1$), which correspond to average relaxation times, can be obtained in a unique way from the relaxation curve by an integration procedure:

$$\mu_{m-1} = \int_0^\infty t^{m-1}\left[\delta E(t) - \delta E^o\right] dt \qquad m = 1,\ldots,2n \tag{79}$$

Introducing Eq. (78), it is seen, that μ_o corresponds to the average relaxation time, μ_1 to the average of the square of the relaxation times, etc.:

$$\frac{1}{\delta E^o}\,\frac{\mu_{m-1}}{(m-1)!} = \langle\tau^m\rangle = \frac{\sum\limits^{n}\sum\limits^{S}\varepsilon_i A_{ik}\tau_k^m}{\sum\limits_{k}\sum\limits_{i}\varepsilon_i A_{ik}} = \frac{\sum\limits^{n}\sum\limits^{R}\Delta\varepsilon_\alpha P_{\alpha k}\tau_k^m}{\sum\limits_{k}\sum\limits_{\alpha}\Delta\varepsilon_\alpha P_{\alpha k}} \tag{80}$$

However, these moments do not lend themselves to an easy correlation with the kinetic parameters of the individual reaction steps. A method for obtaining relaxation times and amplitudes from the moments, according to Eq. (80), has been devised by DYSON and ISENBURG (1971). Averages of reciprocal relaxation times can be obtained in a similar way (SCHWARZ, 1968):

$$\left\langle\left(\frac{1}{\tau}\right)^m\right\rangle = (-1)^m \frac{1}{\delta E^o}\,\frac{d_m}{dt^m}\delta E(t)\Big|_{t=o} \qquad m = 1,2\ldots \tag{81}$$

The determination of the average reciprocal relaxation time $\langle\frac{1}{\tau}\rangle \equiv \frac{1}{\tau^*}$ from the initial slope of the relaxation curve is shown in Figure 3. This quantity allows a simple correlation with the thermodynamic and kinetic parameters of the reaction system without the need to solve the relaxation equations explicitly. It may therefore prove to be useful if the complexity of the system does not allow a more detailed analysis of the relaxation curve.

Introducing Eqs. (21, 23) into Eq. (81) with m = 1, the average reciprocal relaxation time is given by

$$\frac{1}{\tau^*} = \frac{1}{\delta E^o}\sum_{\alpha=1}^{R}\Delta\varepsilon_\alpha \frac{d\xi_\alpha}{dt}\Big|_{t=o} = \frac{1}{\delta E^o}\sum_{\alpha=1}^{R}\Delta\varepsilon_\alpha r_\alpha \delta\ln K_\alpha \tag{82}$$

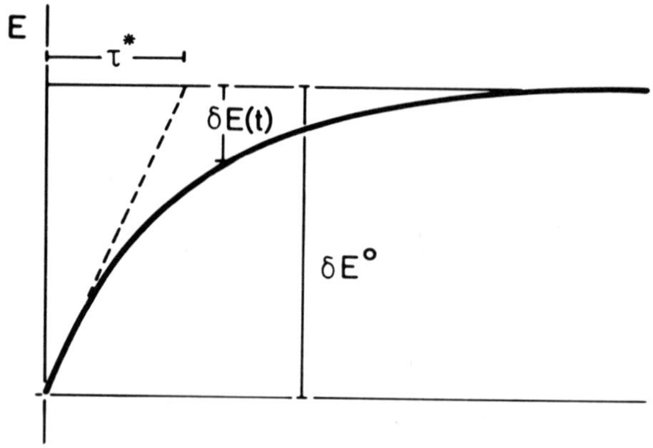

Fig. 3. Determination of the average reciprocal relaxation time, τ , from the initial slope of the relaxation curve

It should be noted, that – in contrast to the individual relaxation times – the average relaxation time τ^* contains the specific signal changes of the observed quantity. Different values for τ^* will therefore be obtained at different wavelengths, unless the ratio of the quantities $\Delta\varepsilon_\alpha$ at different wavelengths is equal for all reaction steps.

G. Computer Program for Simulation of Relaxation Spectra (FORTRAN IV)

The program is used to calculate for a given reaction mechanism the relaxation times, the relaxation amplitudes, and the relative extent to which each individual reaction step contributes to a normal reaction, as defined in Eqs. (45, 46 and 55). A MAIN-program has to be written by the user for each particular reaction mechanism, according to the instructions given below. The subroutines – START, AB, ABC, ABCD, CALC, EIGEN, REDUKT, VECTOR, EQUAT – apply generally. The reaction mechanism is characterized by specifying, for each elementary reaction step, the reacting species, the equilibrium concentrations of the reacting species, the rate constants, and the relative change of the equilibrium constant. The corresponding parameter values are arguments of the subroutines START and AB, ABC, ABCD, which are called from the MAIN-program. The calculation of the relaxation spectrum is initiated by a call to subroutine CALC. The results appear as arguments of the subroutine CALC. The applicability of the program given below is restricted to mechanisms containing as elementary reaction steps only monomolecular and bimolecular reactions of the form:

$$A \rightleftharpoons B,\ A+B \rightleftharpoons AB,\ (2A \rightleftharpoons A_2),\ A+B \rightleftharpoons C+D$$

It is assumed that all kinetic and thermodynamic parameters, which characterize the mechanism, are known. All rate and equilibrium constants, relative changes of equilibrium constants, and all equilibrium concentrations have to be given as input or to be calculated before. Only the values of independent parameters should be given as data input. Dependent parameters should generally be calculated with double precision accuracy. This is important to avoid inconsistencies within the various quantities. Arrays and matrices are dimensioned such, that maximally 13 individual reaction steps or reacting species – whichever is the larger number – can be introduced. The SUBROUTINES of the program have been tested extensively for various mechanisms on the UNIVAC 1108. (Gesellschaft für wissenschaftliche Datenverarbeitung, Göttingen; Specifications for a double precision word: range 10^{-308} to 10^{+308}, 17 decimal digits). Extension of dimensioning is easily possible. However, the accuracy of calculation should then be tested.

The mathematical procedure is as follows: (cf. Sect. C) In subsequent calls of the subroutines AB, ABC, or ABCD from the MAIN-program the S×S matrices a_{ik}, c_{ik} are constructed. These matrices are available through COMMON blocks in subroutine CALC, where they are reduced to the n×n matrices $\hat{a}_{ik}$ and $\hat{c}_{ik}$. The calculation of eigenvalues is initiated by a call to subroutine EIGEN; the array of n eigenvalues appears as an argument of subroutine EIGEN. The calculation of eigenvectors is initiated by a call to subroutine VECTOR; the matrix of eigenvectors appears as an argument of subroutine VECTOR. (The largest component of each eigenvector is thereby normalized to 1).

The concentration changes A_{ik} $(i = 1,...,S)$ and the relative advancements $p_{\alpha k}$ $(\alpha = 1,...,R)$ are calculated in subroutine CALC, whereby the solution of linear equations is achieved by calls to subroutine EQUAT. Eigenvalues are determined in subroutines EIGEN and REDUKT according to the iteration procedure of V.MISES (see ZURMÜHL, 1958): in subroutine EIGEN the largest eigenvalue of a matrix of dimension N is determined; in subroutine REDUKT the matrix is then reduced to a matrix of dimension N-1, which has the same remaining eigenvalues as the original matrix. The solution of a system of linear equations is performed in subroutine EQUAT according to the elimination procedure of GAUSS. In order to handle homogeneous systems a lower limit for matrix coefficients has been introduced. (Due to the limited accuracy of the numerical computations it is not possible to decide whether a matrix element is identical zero or has a finite value).

The subroutines EIGEN, REDUKT, VECTOR, and EQUAT thus perform standard mathematical computations. Any program which performs corresponding calculations may be used instead.

Instructions for the MAIN-Program. Two series of numbering have to be chosen (S: total number of reacting species, R: total number of individual reaction steps, n: number of independent variables, i.e. number of relaxation times):

1. Numbering of Reacting Species

The reacting species are numbered subsequently from i = 1 to S and a corresponding numbering is applied for the concentration changes, x_i, and the equilibrium concentrations, c_i, of these species. The numbering must be chosen such, that the first n concentration variables are independent with respect to mass conservation. Whether a set of n concentration variables is an independent set can easily be decided by inspection of the relationship of mass conservation. The condition requires that the concentration variables x_{n+1} to x_S can be expressed by the concentration variables x_1 to x_n.

2. Numbering of Individual Reaction Steps

The individual reaction steps are numbered subsequently from $\alpha = 1$ to R. The numbering must be chosen such, that the first n reaction steps are thermodynamically independent. Whether a set of n reactions is thermodynamically independent, can easily be decided by inspection of the reaction mechanism. The condition requires that the set of reactions does not contain parallel reaction paths between reacting species ("loops"). The reaction mechanism is characterized by specifying the arguments of the subroutine START and, for each individual reaction step, the arguments of the subroutines AB, ABC, or ABCD.

START. The number of relaxation times, n, the number of reacting species, S, the number of individual reaction steps, R., and the equilibrium concentrations of the reacting species, c_i, $i = 1,\ldots,S$, are specified in the statement:

```
CALL     START     (N,    NS,    NR,    V)
```

where integers N, NS, NR, and double precision array V correspond to n, S, R, c_i, $i = 1,\ldots,S$ respectively.

AB, ABC, ABCD. The kinetic and thermodynamic properties of each individual reaction step of the mechanism may be characterized in a symbolic form:

1. For a reaction step, monomolecular in both directions, i.e. a reaction of the type

$$X_i \underset{s_{-\alpha}}{\overset{s_\alpha}{\rightleftharpoons}} X_k$$

for which the mass action law has the form

$$\frac{c_k}{c_i} = K_\alpha \qquad (\delta \ln K_\alpha = \Delta H_\alpha / RT^2)$$

the notation is (α, i, s_α, k, $s_{-\alpha}$, $\delta \ln K_\alpha$) where s_α, $s_{-\alpha}$, denote the rate constants and $\delta \ln K_\alpha$ the relative change of the equilibrium constant due to the perturbation. The number α specifies the number of the reaction step and the numbers i, k, specify the numbers of the reacting species.

2. For a reaction step, bimolecular in one direction and monomolecular in the other, i.e. a reaction of the type

$$X_i + X_k \underset{s_{-\alpha}}{\overset{s_\alpha}{\rightleftharpoons}} X_l$$

for which the mass action has the form

$$\frac{c_l}{c_i \cdot c_k} = K_\alpha \qquad (\delta \ln K_\alpha = \Delta H_\alpha / RT^2)$$

the notation is (α, i, k, s_α, l, $s_{-\alpha}$, $\delta \ln K_\alpha$) where s_α, $s_{-\alpha}$, denote the rate constants of recombination and dissociation respectively. The number α specifies the number of the reaction step and the numbers i, k, l, denote the numbers of the reacting species. The above notation holds also for a dimerization reaction, i.e. for $i = k$: (α, i, i, s_α, l, $s_{-\alpha}$, $\delta \ln K_\alpha$).

3. For a reaction step bimolecular in both directions, i.e. a reaction of the type

$$X_i + X_k \underset{s_{-\alpha}}{\overset{s_\alpha}{\rightleftharpoons}} X_l + X_m$$

for which the mass action law has the form

$$\frac{c_l \cdot c_m}{c_i \cdot c_k} = K_\alpha \qquad (\delta \ln K_\alpha = \Delta H_\alpha / RT^2)$$

the notation is (α, i, k, s_α, l, m, $s_{-\alpha}$, $\delta \ln K_\alpha$).

The numbers of the reaction steps, the numbers of the reacting species and the corresponding rate constants and relative changes of equilibrium constants are specified – according to the symbolic notation given above – in the following statements:

For a monomolecular/monomolecular reaction

```
CALL AB (NR, I, S1, K, S2, DLNK).
```

For a bimolecular/monomolecular reaction

```
CALL ABC (NR, I, K, S1, L, S2, DLNK).
```

For a bimolecular/bimolecular reaction

```
CALL ABCD (NR, I, K, S1, L, M, S2, DLNK).
```

Integers NR, I, K, L, M, and double precision variables S1, S2, DLNK, correspond respectively to the variables α, i, k, l, m, and s_α, $s_{-\alpha}$, $\delta \ln K_\alpha$) as explained above. There has to be a statement of this form for each individual reaction step in the particular mechanism, leading to R statements totally. (If the reaction system contains parallel reaction paths the values of $\delta \ln K_\alpha$ for the dependent reaction steps, i.e. for $\alpha = n+1$ to R, need not be specified).

The results of the calculation appear as arguments of the subroutine CALC:

$x_i^o \quad i = 1,\ldots,S$

the total concentration change of the species X_i

$1/\tau_k \quad k = 1,\ldots,n$

the reciprocal relaxation times, numbered according to increasing values (τ_1 corresponds to the slowest, τ_n to the fastest relaxation process)

$A_{ik} \quad i = 1,\ldots,S, \quad k = 1,\ldots,n$

the concentration change of species x_i with respect to relaxation time τ_k

$p_{\alpha k} \quad \alpha = 1,\ldots,R, \quad k = 1,\ldots,n$

the relative contribution of reaction step α to the normal reaction k (normalization $\sum_\alpha |p_{\alpha k}| = 1$)

Signs are such, that $p_{\alpha k} < 0$ ($p_{\alpha k} > 0$) denotes the reaction process from left to right (right to left) in the corresponding α^{th} reaction equation.

The subroutine is called from the MAIN-program by the statement

```
CALL    CALC    (XO, E, A, P)
```

where double precision arrays XO, E, and double precision matrices A, P correspond to x_i^o, $1/\tau_k$, and A_{ik}, $p_{\alpha k}$ respectively.

Example for the MAIN-PROGRAM . The procedure of writing the MAIN-program is shown in an example which is of interest in biophysics, namely the "ADAIR-model" for ligand binding to a tetrameric protein. The complete MAIN-program, together with test calculations, is listed below.

The ADAIR-model describes (cooperative) ligand binding in terms of – phenomenologically – four consecutive bimolecular ligand binding steps. If B_o denote the tetrameric protein species with i ligands bound and F the ligand, the ADAIR-model is given by the reaction scheme

$$B_o+4F \rightleftharpoons B_1+3F \rightleftharpoons B_2+2F \rightleftharpoons B_3+F \rightleftharpoons B_4$$

(For the numbering of reacting species and reaction steps see list of the program). The mechanism contains 6 reacting species, for which 2 relationships of mass conservation – for total protein and total ligand concentrations – must be fulfilled:

$$x_{B_o} + x_{B_1} + x_{B_2} + x_{B_3} + x_{B_4} = 0$$

$$x_F + x_{B_1} + 2x_{B_2} + 3x_{B_3} + 4x_{B_4} = 0$$

The four individual reaction steps are thermodynamically independent. The relaxation kinetics of the ADAIR-scheme is characterized by 4 relaxation times. The numbering of the reacting species and the reaction steps may be chosen arbitrarily. The set of variables x_F, x_{B_o}, x_{B_1}, x_{B_2} numbered from 1 to n in the program is an independent set since x_{B_3} and x_{B_4} can be expressed as a linear combination of these variables. The definition of the various parameters may be seen from the comment cards of the computer list. A test calculation is added which represents the degenerate case of equal intrinsic properties for the individual steps. In this case, the relaxation times are given by simple and analytical expressions

$$1/\tau_1 = k_D + k_R(\Phi + F)$$

$$1/\tau_i = i(k_D + k_R F) \qquad i = 2,3,4$$

whereby $\Phi = 4c_{B_0} + \ldots c_{B_3}$ denotes the concentration of free binding sites. Only the relaxation process associated with τ_1 has a finite amplitude.

References

CASTELLAN, G.W.: Calculation of the spectrum of chemical relaxation times for a general reaction mechanism. Ber. Bunsenges. Phys. Chem. 67, 898-908 (1963)

DYSON, R., ISENBERG, I.: Analysis of exponential curves by a method of moments. Biochemistry 10, 3233-3241 (1971)

EIGEN, M.: Kinetics of reaction control and information transfer in enzymes and nucleic acids. In: NOBEL-Symposium V on Fast Reactions and Primary Processes in Chemical Kinetics, pp. 333-369. Stockholm: Almquist and Wiksell 1968

EIGEN, M., DE MAEYER, L.: Relaxation methods. In: Techniques of Organic Chemistry. WEISSBERGER, A. (ed.), Vol. VIII, Part 2, pp. 895-1054. New York: Wiley 1963

EIGEN, M., DE MAEYER, L.: Theoretical basis of relaxation spectrometry. In: Techniques of Chemistry. WEISSBERGER, A., HAMMES, G.G. (eds.), Vol. VI, Part 2, pp. 63-146. New York: Wiley 1973

EIGEN, M., WINKLER, R.: To be published

HAMMES, G.G., ALBERTY, R.: The relaxation spectra of simple enzymatic mechanisms. J. Am. Chem. Soc. 82, 1564-1569 (1960)

HAMMES, G.G., SCHIMMEL, P.R.: Chemical relaxation spectra: Calculation of relaxation times for complex mechanisms. J. Chem. Phys. 70, 2319-2324 (1966)

HAYMAN, H.J.G.: Orthonormal chemical reactions and chemical relaxation. Trans. Farad. Soc. 65, 2918-2928 (1969)

ILGENFRITZ, G.: SCHUSTER, T.M.: Kinetics of oxygen binding to human hemoglobin. J. Biol. Chem. 249, 2959-2972 (1974)

KIRSCHNER, K., GALLEGO, E., SCHUSTER, I., GOODALL, D.: Cooperative binding of nicotinamide-adenine dinucleotide to yeast glyceraldehyde-3-phosphate dehydrogenase. J. Mol. Biol. 58, 29-50 (1971)

POERSCHKE, D.: Model calculations on the kinetics of oligonucleotide double helix coil transition. Biophys. Chem. 2, 83-96 (1974)

RIGLER, R., RABL, C.R., JOVIN, T.M.: Temperature-jump apparatus for fluorescence measurements. Rev. Sci. Instrum. 45, 580-588 (1974)

SCHIMMEL, P.R.: On the calculation of chemical relaxation amplitudes. J. Chem. Phys. 54, 4136-37 (1971)

SCHWARZ, G.: Kinetic analysis by chemical relaxation methods. Rev. Mod. Phys. 40, 206-218 (1968)
THUSIUS, D.: Relaxation amplitudes for systems of two coupled equilibria. J. Am. Chem. Soc. 94, 356-363 (1972)
ZURMÜHL, R.: Matrizen. Berlin: Springer 1958

```
C       PROGRAM RELAXATION KINETICS. MAIN PROGRAM
C
C       CALCULATION OF RELAXATION TIMES AND RELAXATION
C       AMPLITUDES FOR AN ADAIR MODEL
C
C       MECHANISM   B0+4F = B1+3F = B2+2F = B3+F = B4
C
C       N=4,  NS=6,  NR=4
C       NUMBERING OF REACTING SPECIES
C       1. F, 2. B0, 3. B1, 4. B2, 5. B3, 6. B4
C       NUMBERING OF REACTION STEPS
C       1. B0+F=B1 , 2. B1+F=B2 , 3. B2+F=B3 , 4. B3+F=B4
C       REACTION STEPS                  NOTATION
C     B0+F/4.*RK(1)/=B1/    DK(1)/   (1, 2,1,4.*RK(1),3,     DK(1), D(1))
C     B1+F/3.*RK(2)/=B2/2.*DK(2)/   (2, 3,1,3.*RK(2),4,2.*DK(2), D(2))
C     B2+F/2.*RK(3)/=B3/3.*DK(3)/   (3, 4,1,2.*RK(3),5,3.*DK(3), D(3))
C     B3+F/    RK(4)/=B4/4.*DK(4)/   (4, 5,1,     RK(4),6,4.*DK(4), D(4))
C
C       SYMBOLS
C     NS                NUMBER OF REACTING SPECIES
C     NR                NUMBER OF INDIVIDUAL REACTIONS STEPS
C     N                 NUMBER OF RELAXATION TIMES
C     GK(I) I=1,4       INTRINSIC EQUILIBRIUM CONSTANTS
C     D(I) I=1,4        RELATIVE CHANGE OF EQUILIBRIUM CONSTANTS
C     DK(I) I=1,4       INTRINSIC DISSOCIATION RATE CONSTANTS
C     RK(I) I=1,4       INTRINSIC REKOMBINATION RATE CONSTANTS
C     C0                TOTAL PROTEIN CONCENTRATION (MONOMER UNITS)
C     F,V(1)            FREE LIGAND CONCENTRATION
C     B0,...B4          CONCENTRATION OF PROTEIN TETRAMER WITH
C       V(I) I=2,6        0,1...4 LIGANDS BOUND
C     X0(I) I=1,NS      TOTAL CHANGE OF CONCENTRATION VARIABLE I
C     E(K) K=1,N        RECIPROCAL RELAXATION TIMES
C     A(I,K)            CHANGE OF CONCENTRATION VARIABLE I WITH
C       I=1,NS K=1,N      RESPECT TO RELAXATION TIME K
C     P(I,K)            RELATIVE CONTRIBUTION OF REACTION STEP I
C       I=1,NR K=1,N      TO NORMAL-REACTION K
C                       OBSERVED PARAMETER
C     Y                 SATURATION OF PROTEIN WITH LIGAND
C     DY0               TOTAL CHANGE OF SATURATION
C     DY(K) K=1,N       CHANGE OF SATURATION WITH RESPECT TO
C                         RELAXATION TIME K
C     RDY(K) K=1,N      RELATIVE AMPLITUDES FOR THE CHANGE OF
C                         SATURATION WITH RESPECT TO RELAX. TIME K
C
      IMPLICIT DOUBLE PRECISION (A-H,O-Z)
      DIMENSION X0(13),E(13),A(13,13),P(13,13),V(13)
      DIMENSION DK(4),RK(4),GK(4),D(4),DY(4),RDY(4)
      INTEGER NAME(7)
      EQUIVALENCE (V(1),F), (V(2),B0), (V(3),B1)
      EQUIVALENCE (V(4),B2), (V(5),B3), (V(6),B4)
C
   10 FORMAT(5D10.2)
   20 FORMAT(1H1/ 12X,11HADAIR MODEL,10X,4HSTEP,
     *              7X,2HDK,8X,2HRK,8X,2HGK,6X,1HD )
   30 FORMAT(35X, I2, 1P3E10.3, 0PF7.3 )
   40 FORMAT(1H0/ 10X, 1PD10.2,                17X, 2X, 1P4D12.3)
   50 FORMAT(      10X, 2PF10.3, 3X,A2, 2PF12.3, 2X, 2P4F12.4)
   60 FORMAT(                                   37X, 2X, 2P4F12.3)
   70 FORMAT(      10X, 1PD10.2, 3X,A2, 1PD12.2, 2X, 1P4D12.3)
   80 FORMAT(                     25X,       I12, 2X,    4F12.3)
```

```
C
      DATA NAME/2H F,2HB0,2HB1,2HB2,2HB3,2HB4,2H Y/
      DATA N,NS,NR/4,6,4/
C
C       INPUT -CALCULATION- OF EQUILIBRIUM AND RATE CONSTANTS
C
   90 CONTINUE
      READ(5, 10 ) (DK(I),I=1,4)
      READ(5, 10 ) (RK(I),I=1,4)
      DO 100 I=1,4
  100 GK(I)= RK(I)/DK(I)
      READ(5, 10 ) (D(I),I=1,4)
C
      WRITE(6, 20 )
      DO 110 I=1,4
  110 WRITE(6, 30 ) I,DK(I),RK(I),GK(I),D(I)
C
C       INPUT -CALCULATION- OF EQUILIBRIUM CONCENTRATIONS
C
      READ(5, 10 ) F,C0
      DIV= 1.+4.*F*GK(1)+6.*F*GK(1)*F*GK(2)+4.*F*GK(1)*
     1     F*GK(2)*F*GK(3)+F*GK(1)*F*GK(2)*F*GK(3)*F*GK(4)
      B0= C0/(4.*DIV)
      B1= 4.    *B0*F*GK(1)
      B2= 3./2.*B1*F*GK(2)
      B3= 2./3.*B2*F*GK(3)
      B4= 1./4.*B3*F*GK(4)
      Y= (B1+2.*B2+3.*B3+4.*B4)/C0
C
C       SPECIFICATION OF THE MECHANISM
C
      CALL START(N,NS,NR,V)
      CALL ABC(1, 2,1,4.*RK(1),3,    DK(1), D(1))
      CALL ABC(2, 3,1,3.*RK(2),4,2.*DK(2), D(2))
      CALL ABC(3, 4,1,2.*RK(3),5,3.*DK(3), D(3))
      CALL ABC(4, 5,1,    RK(4),6,4.*DK(4), D(4))
C
C       CALCULATION -OUTPUT- OF RESULTS
C
      CALL CALC(X0,E,A,P)
      DY0= -X0(1)/C0
      SUM= 0.
      DO 120 K=1,4
      DY(K)= -A(1,K)/C0
  120 SUM= SUM+DY(K)
      DO 130 K=1,4
  130 RDY(K)= DY(K)/SUM
C
      WRITE(6, 40 )    C0,                    ( E(I),I=1,4)
      WRITE(6, 50 )     Y, NAME(7),   DY0, ( DY(I),I=1,4)
      WRITE(6, 60 )                         (RDY(I),I=1,4)
      WRITE(6, 70 )
      DO 140 I=1,6
  140 WRITE(6, 70 ) V(I), NAME(I), X0(I), ( A(I,K),K=1,4)
      WRITE(6, 80 )
      DO 150 I=1,4
  150 WRITE(6, 80 )                       I, ( P(I,K),K=1,4)
C
      GO TO 90
      END
```

ADAIR MODEL	STEP	DK	RK	GK	D
	1	4.000+02	1.000+07	2.500+04	-.150
	2	4.000+02	1.000+07	2.500+04	-.150
	3	4.000+02	1.000+07	2.500+04	-.150
	4	4.000+02	1.000+07	2.500+04	-.150

1.00-004			1.300+003	1.600+003	2.400+003	3.200+003
50.000	Y	-2.308	-2.3077	.0000	-.0000	-.0000
			100.000	-.000	.000	.000

4.00-005	F	2.31-006	2.308-006	-2.227-035	1.045-036	1.706-037
1.56-006	B0	2.88-007	2.885-007	6.951-021	-2.046-021	4.808-022
6.25-006	B1	5.77-007	5.769-007	1.284-035	4.092-021	-1.923-021
9.37-006	B2	-5.08-024	1.101-020	-1.390-020	-4.939-036	2.885-021
6.25-006	B3	-5.77-007	-5.769-007	-3.014-035	-4.092-021	-1.923-021
1.56-006	B4	-2.88-007	-2.885-007	6.951-021	2.046-021	4.808-022

1	-.125	-.250	.250	-.125
2	-.375	-.250	-.250	.375
3	-.375	.250	-.250	-.375
4	-.125	.250	.250	.125

```
C         PROGRAM RELAXATION KINETICS. SUBROUTINE 1
C
          SUBROUTINE START(N1,N2,N3,CONC)
          IMPLICIT DOUBLE PRECISION (A-H,O-Z)
          DIMENSION A(13,13),C(13,13),T(13,13)
          DIMENSION CONC(13),V(13),CB(13),PH(13)
          COMMON /MX/A,C,CB   /CO/V   /ZH/N,NM,NC,NS   /STM/T,PH
C
          N= N1
          NC= N2
          NS= N3
          NM= NC-N
          DO 10 I=1,NC
          V(I)= CONC(I)
          DO 10 K=1,NC
          A(I,K)= 0.
       10 C(I,K)= 0.
          DO 20 I=1,NS
          DO 20 K=1,NS
       20 T(I,K)= 0.
          RETURN
          END
```

```
C         PROGRAM RELAXATION KINETICS. SUBROUTINE 2
C
          SUBROUTINE AB(NR,M1,RK1,M2,RK2,DLNK)
          IMPLICIT DOUBLE PRECISION (A-H,O-Z)
          DIMENSION A(13,13),C(13,13),T(13,13)
          DIMENSION V(13),CB(13),PH(13)
          COMMON /MX/A,C,CB   /CO/V   /ZH/N,NM,NC,NS   /STM/T,PH
C
          A(M1,M1)= A(M1,M1)-RK1
          A(M1,M2)= A(M1,M2)+RK2
          A(M2,M1)= A(M2,M1)+RK1
          A(M2,M2)= A(M2,M2)-RK2
          T(M1,NR)= T(M1,NR) -1.
          T(M2,NR)= T(M2,NR) +1.
          PH(NR)= RK1*V(M1)
          IF(NR.GT.N) RETURN
          C(NR,M1)= C(NR,M1) -1./V(M1)
          C(NR,M2)= C(NR,M2) +1./V(M2)
          CB(NR)= DLNK
          RETURN
          END
```

```
C        PROGRAM RELAXATION KINETICS. SUBROUTINE 3
C
         SUBROUTINE ABC(NR,M1,M2,RK,M3,DK,DLNK)
         IMPLICIT DOUBLE PRECISION (A-H,O-Z)
         DIMENSION A(13,13),C(13,13),T(13,13)
         DIMENSION V(13),CB(13),PH(13)
         COMMON /MX/A,C,CB  /CO/V  /ZH/N,NM,NC,NS  /STM/T,PH
C
         A(M1,M1)= A(M1,M1)-RK*V(M2)
         A(M1,M2)= A(M1,M2)-RK*V(M1)
         A(M1,M3)= A(M1,M3)+DK
         A(M2,M1)= A(M2,M1)-RK*V(M2)
         A(M2,M2)= A(M2,M2)-RK*V(M1)
         A(M2,M3)= A(M2,M3)+DK
         A(M3,M1)= A(M3,M1)+RK*V(M2)
         A(M3,M2)= A(M3,M2)+RK*V(M1)
         A(M3,M3)= A(M3,M3)-DK
         T(M1,NR)= T(M1,NR) -1.
         T(M2,NR)= T(M2,NR) -1.
         T(M3,NR)= T(M3,NR) +1.
         PH(NR)= RK*V(M1)*V(M2)
         IF(NR.GT.N) RETURN
         C(NR,M1)= C(NR,M1) -1./V(M1)
         C(NR,M2)= C(NR,M2) -1./V(M2)
         C(NR,M3)= C(NR,M3) +1./V(M3)
         CB(NR)= DLNK
         RETURN
         END
```

```
C        PROGRAM RELAXATION KINETICS. SUBROUTINE 4
C
         SUBROUTINE ABCD(NR,M1,M2,RK1,M3,M4,RK2,DLNK)
         IMPLICIT DOUBLE PRECISION (A-H,O-Z)
         DIMENSION A(13,13),C(13,13),T(13,13)
         DIMENSION V(13),CB(13),PH(13)
         COMMON /MX/A,C,CB  /CO/V  /ZH/N,NM,NC,NS  /STM/T,PH
         A(M1,M1)=A(M1,M1) -RK1*V(M2)
         A(M1,M2)=A(M1,M2) -RK1*V(M1)
         A(M1,M3)=A(M1,M3) +RK2*V(M4)
         A(M1,M4)=A(M1,M4) +RK2*V(M3)
         A(M2,M1)=A(M2,M1) -RK1*V(M2)
         A(M2,M2)=A(M2,M2) -RK1*V(M1)
         A(M2,M3)=A(M2,M3) +RK2*V(M4)
         A(M2,M4)=A(M2,M4) +RK2*V(M3)
         A(M3,M1)=A(M3,M1) +RK1*V(M2)
         A(M3,M2)=A(M3,M2) +RK1*V(M1)
         A(M3,M3)=A(M3,M3) -RK2*V(M4)
         A(M3,M4)=A(M3,M4) -RK2*V(M3)
         A(M4,M1)=A(M4,M1) +RK1*V(M2)
         A(M4,M2)=A(M4,M2) +RK1*V(M1)
         A(M4,M3)=A(M4,M3) -RK2*V(M4)
         A(M4,M4)=A(M4,M4) -RK2*V(M3)
         T(M1,NR)= T(M1,NR) -1.
         T(M2,NR)= T(M2,NR) -1.
         T(M3,NR)= T(M3,NR) +1.
         T(M4,NR)= T(M4,NR) +1.
         PH(NR)= RK1*V(M1)*V(M2)
         IF(NR.GT.N) RETURN
         C(NR,M1)= C(NR,M1) -1./V(M1)
         C(NR,M2)= C(NR,M2) -1./V(M2)
         C(NR,M3)= C(NR,M3) +1./V(M3)
         C(NR,M4)= C(NR,M4) +1./V(M4)
         CB(NR)= DLNK
         RETURN
         END
```

```
C        PROGRAM RELAXATION KINETICS. SUBROUTINE 5
C
         SUBROUTINE CALC(X0,EIG,AM,P)
         IMPLICIT DOUBLE PRECISION (A-H,O-Z)
         DIMENSION X0(13),EIG(13),AM(13,13),P(13,13)
         DIMENSION A(13,13),C(13,13),T(13,13),X(13,13),COE(6,13)
         DIMENSION CB(13),PH(13),S(13),R(13)
         COMMON /MX/A,C,CB  /ZH/N,NM,NC,NS  /STM/T,PH
C
         DO 20 L=1,NM
         DO 10 I=1,N
         S(I)= A(N+L,I)
         DO 10 K=1,N
      10 X(I,K)= A(K,I)
         CALL EQUAT(N,X,S,R)
         DO 20 K=1,N
      20 COE(L,K)= R(K)
C
         DO 30 I=1,N
         DO 30 K=1,N
         DO 30 L=1,NM
         A(I,K)= A(I,K)+COE(L,K)*A(I,N+L)
      30 C(I,K)= C(I,K)+COE(L,K)*C(I,N+L)
         CALL EIGEN(N,A,R)
         CALL VECTOR(N,A,R,X)
         DO 40 K=1,N
      40 EIG(K)= -R(K)
C
         CALL EQUAT(N,C,CB,S)
         CALL EQUAT(N,X,S,R)
         DO 50 I=1,N
      50 X0(I)= S(I)
         DO 60 I=1,N
         DO 60 K=1,N
      60 AM(I,K)= R(K)*X(I,K)
         DO 70 L=1,NM
         X0(N+L)= 0.
         DO 70 K=1,N
      70 X0(N+L)= X0(N+L)+COE(L,K)*X0(K)
         DO 80 L=1,NM
         DO 80 ME=1,N
         AM(N+L,ME)= 0.
         DO 80 K=1,N
      80 AM(N+L,ME)= AM(N+L,ME)+COE(L,K)*AM(K,ME)
C
         IF(N.EQ.NS) GO TO 130
         DO 90 I=1,NM
         DO 90 K=1,NS
      90 T(N+I,K)= 0.
         N1= N+1
         DO 120 M=N1,NS
         DO 100 I=1,N
     100 S(I)= T(I,M)
         CALL EQUAT(N,T,S,R)
         DO 110 K=1,N
     110 T(M,K)= R(K)/PH(K)
     120 T(M,M)= -1./PH(M)
     130 CONTINUE
         DO 170 K=1,N
         DO 140 I=1,NS
     140 S(I)= 0.
         DO 150 I=1,N
     150 S(I)= AM(I,K)
         CALL EQUAT(NS,T,S,R)
         SUM= 0.
         DO 160 I=1,NS
```

```
  160 SUM= SUM+DABS(R(I))
      DO 170 I=1,NS
  170 P(I,K)= R(I)/SUM
      RETURN
      END
```

```
C       PROGRAM RELAXATION KINETICS. SUBROUTINE 5
C
C       CALCULATES THE EIGENVALUES EIG(K) OF THE
C       MATRIX A(I,K) OF DIMENSION N ACCORDING
C       TO THE ITERATION PROCEDURE OF MISES.
C       USES SUBROUTINE REDUKT
C       NOTE = A VALUE 'ACCY' IS INTRODUCED AS AN ACCURACY
C               LIMIT FOR THE EIGENVALUES
C       NOTE = A VALUE 'ITMAX' IS INTRODUCED AS AN UPPER
C               LIMIT FOR THE NUMBER OF ITERATIONS
C
      SUBROUTINE EIGEN (N,A,EIG)
      IMPLICIT DOUBLE PRECISION (A-H,O-Z)
      DIMENSION AR(13,13),Z0(13),Z1(13),Z11(13)
      DIMENSION A(13,13),EIG(13)
C
      ACCY= 1.D-16
      ITMAX= 20000
      NR= N
      DO 10 I=1,NR
      DO 10 K=1,NR
   10 AR(I,K)= A(I,K)
C
   20 CONTINUE
      IF (NR.EQ.1) GO TO 100
      DO 30 K=1,NR
   30 Z0(K)= 1.
      EIG1= 0.
      IT= 1
   40 DO 50 I=1,NR
      Z1(I)= 0.
      DO 50 K=1,NR
   50 Z1(I)= Z1(I)+AR(I,K)*Z0(K)
      NB=1
      BIG= DABS(Z1(1))
      DO 60 I=2,NR
      IF(DABS(Z1(I)).GT.BIG ) NB=I
   60 BIG= DMAX1(BIG,DABS(Z1(I)))
      EIG2= Z1(NB)/Z0(NB)
      DO 70 I=1,NR
   70 Z11(I)= Z1(I)/Z1(NB)
      IF (DABS((EIG2-EIG1)/EIG2).LE.ACCY)   GO TO 90
      DO 80 I=1,NR
   80 Z0(I)= Z11(I)
      EIG1= EIG2
      IT= IT+1
      IF (IT.LE. ITMAX) GO TO 40
   90 EIG(NR)= EIG2
      CALL REDUKT(NR,AR,NB,Z11)
      GO TO 20
C
  100 EIG(1)= AR(1,1)
      RETURN
      END
```

```
C       PROGRAM RELAXATION KINETICS. SUBROUTINE 7
C
C       REDUCES A MATRIX A(I,K) OF DIMENSION N OF WHICH ONE
C       EIGENVECTOR X(K) IS KNOWN (WHEREBY X(J)=1) TO A MATRIX
C       A(I,K) OF DIMENSION N-1, WHICH HAS THE SAME REMAINING EIGEN-
C       VALUES AS THE ORIGINAL MATRIX
C
      SUBROUTINE REDUKT (N,A,J,X)
      IMPLICIT DOUBLE PRECISION (A-H,O-Z)
      DIMENSION A(13,13),X(13),AL(13)
C
      IF (J.EQ.N) GO TO 70
      IF (J.EQ.1) GO TO 90
      L= J-1
      M= J+1
      DO 10 K= 1,L
   10 AL(K)= A(J,K)
      DO 20 K= M,N
      X(K-1)= X(K)
   20 AL(K-1)= A(J,K)
C
      DO 30 I= 1,L
      DO 30 K= M,N
   30 A(I,K-1)= A(I,K)
      DO 50 I= M,N
      DO 40 K= 1,L
   40 A(I-1,K)= A(I,K)
      DO 50 K= M,N
   50 A(I-1,K-1)= A(I,K)
      N= N-1
      DO 60 I=1,N
      DO 60 K=1,N
   60 A(I,K)= A(I,K)-X(I)*AL(K)
      RETURN
C
   70 N= N-1
      DO 80 I= 1,N
      DO 80 K= 1,N
      AL(K)= A(J,K)
   80 A(I,K)= A(I,K)-X(I)*AL(K)
      RETURN
C
   90 DO 100 K=2,N
      X(K-1)= X(K)
  100 AL(K-1)= A(J,K)
      DO 110 I=2,N
      DO 110 K=2,N
  110 A(I-1,K-1)= A(I,K)
      N=N-1
      DO 120 I=1,N
      DO 120 K=1,N
  120 A(I,K)= A(I,K)-X(I)*AL(K)
      RETURN
      END
```

```
C       PROGRAM RELAXATION KINETICS. SUBROUTINE 8
C
C       CALCULATES A MATRIX OF EIGENVECTORS X(I,K) FOR
C       THE MATRIX AS(I,K) OF DIMENSION N, WHICH HAS THE
C       EIGENVALUES EIG(K). THE LARGEST COMPONENT OF AN
C       EIGENVECTOR IS NORMALIZED TO 1
C       USES SUBROUTINE EQUAT
C
      SUBROUTINE VECTOR(N,AS,EIG,X)
      IMPLICIT DOUBLE PRECISION (A-H,O-Z)
      DIMENSION A(13,13),B(13),XI(13)
      DIMENSION AS(13,13),EIG(13),X(13,13)
C
      IF(N.EQ.1) GO TO 60
      NV= N
C
   10 DO 20 I=1,N
      B(I)= 0.
      DO 20 K=1,N
   20 A(I,K)= AS(I,K)
      DO 30 I=1,N
   30 A(I,I)= A(I,I)-EIG(NV)
      CALL EQUAT(N,A,B,XI)
      BIG= 0.
      DO 40 I=1,N
   40 BIG= DMAX1(DABS(XI(I)),BIG)
      DO 50 I=1,N
   50 X(I,NV)= XI(I)/BIG
      IF(NV.EQ.1) RETURN
      NV=NV-1
      GO TO 10
C
   60 X(1,1)= 1.
      RETURN
      END
```

```
C        PROGRAM RELAXATION KINETICS. SUBROUTINE 9
C
C        CALCULATES THE SOLUTION XI(K) OF A SYSTEM OF N
C        LINEAR EQUATIONS ACCORDING TO THE ELIMINATION
C        PROCEDURE OF GAUSS
C        NOTE  A VALUE 'ZERO' IS INTRODUCED AS A LOWER
C        LIMIT OF ALL MATRIX COEFFICIENTS
C
      SUBROUTINE EQUAT(N,AE,BE,XI)
      IMPLICIT DOUBLE PRECISION (A-H,O-Z)
      DIMENSION A(13,13),B(13),S(13)
      DIMENSION AE(13,13),BE(13),XI(13)
C
      ZERO= 1.D-36
      DO 10 I=1,N
      B(I)= BE(I)
      DO 10 K=1,N
   10 A(I,K)= AE(I,K)
      IF(N.EQ.1) GO TO 90
      J= 1
C
   20 MAX= J
      J1= J+1
      DO 30 M=J1,N
   30 IF( DABS(A(M,J)).GT.DABS(A(MAX,J)) )  MAX=M
      DO 40 K=J,N
   40 S(K)= A(MAX,K)
      SB= B(MAX)
      DO 50 K=J,N
      A(MAX,K)= A(J,K)
   50 A(J,K)= S(K)
      B(MAX)= B(J)
      B(J)= SB
      IF (DABS(A(J,J)).LE.ZERO )  GO TO 80
      DO 70 I=J1,N
      S(I)= A(I,J)/A(J,J)
      DO 60 K=J,N
   60 A(I,K)= A(I,K)-S(I)*A(J,K)
   70 B(I)= B(I)-S(I)*B(J)
   80 J=J+1
      IF(J.LT.N) GO TO 20
C
   90 CONTINUE
      DO 100 I=1,N
      IF (DABS(A(I,I)).LE.ZERO) A(I,I)=ZERO
  100 CONTINUE
      J= N
  110 DO 120 I=1,J
  120 IF( DABS(B(I)).LT.ZERO) B(I)=ZERO
      XI(J)= B(J)/A(J,J)
      IF(J.EQ.1) RETURN
      J1=J-1
      DO 130 I=1,J1
  130 B(I)= B(I)-A(I,J)*XI(J)
      J=J-1
      GO TO 110
      END
```

Concentration Correlation Analysis and Chemic

D. MAGDE

I. Introduction

The molecular world is a world of discrete entities which are in ceaseless motion, even when a system is at equilibrium. This motion, although it is correctly described as "random" and "chaotic", nevertheless may be studied systematically by Concentration Correlation Analysis (CCA).

Consider a small volume of solution containing several sorts of molecules, which may react chemically and/or be free to more into or out of the volume. Molecular chaos ensures that the number of molecules of a particular sort within the volume (the occupation number N_j) will fluctuate about some mean value – rarely straying far from the mean, but also rarely assuming exactly the mean value. Although each fluctuation is random, the "average fluctuation" must obey well-defined laws. CCA is a class of new experimental procedures which measure the magnitude and time behavior of the fluctuations and relate these to equilibrium and kinetic parameters – chemical rate constants, molecular transport coefficients, and so on – the study of which is the goal of the molecular scientist. In brief, the fluctuations relax according to the same laws which govern macroscopic perturbations.

The biochemist, in particular, is often concerned with molecular processes which occur in response to small changes. The many elegant kinetic studies relying on macroscopic perturbations, which are discussed throughout the rest of this volume, may, in many cases, be considered an attempt to approximate the conditions inherently present in a CCA experiment. In the vast majority of cases, one may raise no objections whatsoever to such approximation, but in principle it is often the behavior of the unperturbed system which is of more fundamental interest. Hence, it is proper to compare the two approaches. Moreover, there are instances in which a suitable macroscopic experiment is difficult or impossible to construct, while microscopic fluctuations are always present. Again, this is particularly true of biological systems. Of course, the existence of the fluctuations does not guarantee that they will be easy to measure. The theoretical possibility of CCA was long ago recognized. It is recent progress in developing experimental probes able to monitor such small fluctuations which justifies calling CCA to the attention of a wider audience of practicing kineticists at this time.

This is the readership I address here. I begin at the beginning, proceed far enough to permit a quantitative estimate of the feasibility of proposed measurements, paying particular attention to signal vs. noise considerations, outline a variety of experimental approaches, and identify references helpful in more detailed analyses.

In emphasizing these concerns, I regretfully abandon two other objectives. Even though the references cited would permit reconstructing the history of the development of CCA, I have not done justice in making clear the debt of CCA to very extensive prior work both in the theory of fluctuations and in the evolution of experimental probes sensitive enough to study such microscopic behavior. And again, despite a conscious effort to emphasize the breadth of possible experimental probes with which one may implement CCA, I have totally ignored all the magnetic resonance techniques whose line broadening studies are intimately related to CCA. Magnetic resonance kinetic studies enjoy a large literature of their own and it is probably wise to allow CCA to mature further before attempting a comprehensive comparison.

II. Properties of Thermodynamic Fluctuations

A. Magnitude of Occupation Number Fluctuations

I begin by deriving the magnitude of occupation number fluctuations from general thermodynamic principles. The final result for fixed volumes of ideal, dilute, solutions bounded by permeable walls, Eq. (15), will appear familiar and one may indeed skip to that point. However, I include this derivation because it emphasizes the assumptions on which CCA is based, and because a thermodynamic formulation is more easily extendable to non-dilute, non-ideal solutions, when CCA is ready for such extension.

A thermodynamic system may be specified by a set of extensive parameters: the internal energy U, the volume V, and the number of moles of each component N_A, N_B,... In preparing a particular thermodynamic system, one may constrain some of the parameters X_i to constant values X_i^c, for example V^c. By assumption these are *exactly* determined and do *not* fluctuate at all. The remaining variables X_j are said to be unconstrained, free, or "in contact with a reservoir" with which they may exchange energy, volume, or matter as the case may be. It is the task of thermodynamics to calculate equilibrium values X_j^o for any particular set of X_i^c. This goal is achieved by defining a function (the entropy) of the extensive parameters $S(X_0, X_1, X_2, \ldots)$ and calculating the maximum of S over possible equilibrium states subject to the constraints. The maximum occurs at $X_j = X_j^o$ for all j. The justification for this surprisingly simple procedure is, ultimately, agreement with experiment. But it is possible also to ground it in statistical mechanics, which may be justified by additional or different experiments.

If a measurement of X_j is made very carefully, we find that although X_j lies "very close to" X_j^o, in fact there will be a small discrepancy $\delta X_j = X_j - X_j^o$. If we then repeat the experiment, we will find a different discrepancy. Alternatively, if we watch our prepared system over a period of time, making successive measurements to determine $\delta X_j(t) = X_j(t) - X_j^o$, we find that $\delta X_j(t)$ fluctuates in time. To summarize, the constrained X_i^c do *not* fluctuate; the X_j in contact with the reservoir do fluctuate about their equilibrium values X_j^o.

These fluctuations are typically very small, but they are not always remote from everyday experience with macroscopic systems. They account for the blue color of the sky. And in the laboratory they set an irreducible minimum noise level in electronic devices which, in fact, is quite commonly reached as the limiting factor in modern instrumentation. In special circumstances, near thermodynamic critical points, the fluctuations become enormous, leading to such dramatic effects as critical opalescence.

It is the task of a generalization of thermodynamics to provide us a general means of predicting both the magnitude and the time behavior of these fluctuations $\delta X_j(t)$. Naturally thermodynamics will only describe the behavior of an "average" or "typical" fluctuation. CCA experiments do, in fact, observe $\delta X_j(t)$ for a long time in order to measure this "average" behavior. Let us confine our attention first to the magnitude of $\delta X_j(t)$. The time average of $\delta X_j(t)$, had better vanish, since on the average our system is, by assumption, at equilibrium. The mean square fluctuation does not vanish:

$$([\delta X_j(t)]^2)_{AV} = \lim_{T\to\infty} \frac{1}{T} \int_o^T [\delta X_j(t)]^2 dt. \tag{1}$$

Nor, in general, do higher moments vanish. It is convenient to identify the experimentally accessible time average with the theoretically more convenient ensemble average, which for one free variable reads

$$([\delta X_j(t)]^2)_{AV} = \langle [\delta X_j]^2 \rangle = \int dX_j [X_j - X_j^o]^2 W(X_j). \tag{2}$$

The probability density $W(X_j)$ is the probability of observing a particular X_j. More generally, an expression in obvious analogy to Eq. (2) relates the cross correlation of any order $\langle [\delta X_j]^n [\delta X_k]^m \ldots \rangle$ to a joint probability density $W(X_j, X_k \ldots)$.

A thermodynamic approach to fluctuation theory involves postulating a particular $W(X_j, X_k \ldots)$, calculating the consequence thereof, and demonstrating consistency of these consequences with other thermodynamic postulates and with observed experimental behavior. (The choice of a particular W could, of course, be motivated by considerations of a particular microscopic theory of matter, that is, by statistical insights). Such a program was carried out systematically more than twenty years ago (GREENE and CALLEN, 1951), generalizing earlier work. I cite the one result we require.

The $\langle \delta X_j \delta X_k \rangle$, which give the mean square amplitude of fluctuations (for $j = k$) and the correlation of the fluctuation of one variable with that of another ($j \neq k$), are

$$\langle \delta X_j \delta X_k \rangle = -k \left(\frac{\partial X_j^{\circ}}{\partial F_k^{\circ}} \right)_{F_0 \ldots F_{k-1},\ F_{k+1} \ldots F_s,\ X_i^{c}.} \tag{3}$$

The coefficient k is Boltzmann's constant.

This Eq. (3) relates the fluctuations to a single partial derivative evaluated at equilibrium, calculated while holding constant all constrained X_i^c and all intensive parameters except the one in the derivative. Recall that the intensive parameters in the entropy representation $F_k = \partial S^{\circ}/\partial X_k^{\circ}$ are $1/T$, P/T, $-\mu_k/T$, ... where T, P, and μ_k are the temperature, pressure, and electrochemical potential of the k'th component. Note also that Eq. (3) is symmetric in j and k.

Let us consider the particular system which has been employed in CCA experiments. This constrains V, but leaves free U and the mole numbers N_j of one or more components j. We could, but will not, consider fluctuations in energy or the cross correlation of δU with δN_j. We will consider here only fluctuations in the occupation numbers of the molecular constituents:

$$\begin{aligned} \langle \delta N_j \delta N_k \rangle &= -k \left(\frac{\partial N_j^{\circ}}{\partial(-\mu_k/T)} \right)_{1/T,\ V,\ \mu_{\ell} \neq k} \\ &= kT \left(\frac{\partial \mu_k}{\partial N_j^{\circ}} \right)^{-1}_{T,\ V,\ \mu_{\ell} \neq k.} \end{aligned} \tag{4}$$

A difficulty with Eq. (4) is that we would prefer to consider partial derivatives at constant P and N_{ℓ}° rather than V and μ_{ℓ}. The necessary transformation is effected by means of the Jacobian:

$$\begin{aligned} \left.\frac{\partial \mu_k}{\partial N_j^{\circ}}\right|_{T,V,\mu_{\ell} \neq k} &= \left.\frac{\partial(\mu_k, V, \mu_1 \ldots \mu_{k-1}, \mu_{k+1} \ldots \mu_s)}{\partial(N_j^{\circ}, V, \mu_1 \ldots \mu_{k-1}, \mu_{k+1} \ldots \mu_s)}\right|_T = \\ &= \frac{\left.\dfrac{\partial(\mu_k, V, \mu_1 \ldots \mu_{k-1}, \mu_{k+1} \ldots \mu_s)}{\partial(N_j^{\circ}, P, N_1^{\circ} \ldots N_{j-1}^{\circ}, N_{j+1}^{\circ} \ldots N_s^{\circ})}\right|_T}{\left.\dfrac{\partial(V, \mu_1 \ldots \mu_{k-1}, \mu_{k+1} \ldots \mu_s)}{\partial(P, N_1^{\circ} \ldots N_{j-1}^{\circ}, N_{j+1}^{\circ} \ldots N_s^{\circ})}\right|_{T,N_j^{\circ}.}} \end{aligned} \tag{5}$$

Thermodynamics tells us nothing more, in general, about the dependence of μ_k on N_j°. However, the dependence can always be measured if necessary. If we restrict our attention to a perfect gas or an ideal solution, or otherwise assume a particular form of μ_k, we may proceed. The ideal solution is a very plausible approximation for the experiment described below. Then, by definition, for all components,

$$\mu_k = \mu_k^* \ (T,P) + RT \ln \ (N_k^o/N) \tag{6}$$

where $N = \Sigma N_k^o$ and R is the gas constant.

Now the partial derivative implied in Eq. (5) are:

$$\left.\frac{\partial\mu_k}{\partial N_j^o}\right|_{T,P,N_\ell \neq j} = -\frac{RT}{N} + \frac{RT}{N_k}\delta_{jk} \qquad \left.\frac{\partial V}{\partial P}\right|_{T,N_\ell} = -V\kappa$$

$$\left.\frac{\partial\mu_k}{\partial P}\right|_{T,N_\ell} = \left.\frac{\partial V}{\partial N_k}\right|_{T,P,N_\ell = k} = v_k^* \qquad \left.\frac{\partial V}{\partial N_j^o}\right|_{T,P,N} = v_j^* \tag{7}$$

Here κ is the isothermal compressibility and v_j^* is the molar specific volume. A Maxwell relation was invoked to evaluate $(\partial\mu_k/\partial P)_{T,N_\ell}$.

As an illustration, consider a solution of two components N_C and N_S obeying Eq. (6) and find the mean square amplitude of fluctuations in N_C:

$$\langle [\delta N_C]^2 \rangle = kT \left(\frac{\partial\mu_C}{\partial N_C^o}\right)^{-1}_{T,V,\mu_S}. \tag{8}$$

When we expand the determinants implied in Eq. (5) and use Eq. (7), we find

$$\frac{1}{RT}\left.\frac{\partial\mu_C}{\partial N_C^o}\right|_{T,V,\mu_S} = \frac{1}{N_C^o} - \frac{1}{N} + \frac{v_C^*(v_S^* + V/N_S) - RTV\kappa/N}{v_S^* v_S^* N + RTV\kappa N_C^o/N_S^o}. \tag{9}$$

This expression makes no assumptions beyond those of Eq. (6). In particular, it applies at any concentration for which ideal behavior holds. Furthermore, extension to non-ideal solutions by way of appropriate activity coefficients is obvious.

Below I will discuss an experiment involving the pure diffusion of a dilute solution of solute C in solvent S. The first term in Eq. (9) dominates and we have, substituting Eq. (9) into Eq. (8),

$$\langle [\delta N_C]^2 \rangle = kN_C^o/R = N_C^o/N_o \tag{10}$$

where N_o is Avogadro's number and both N_C^o and δN_C are in moles.

It is of some interest to consider the remaining terms of Eq. (9). Even for a perfect gas, the terms involving κ are only comparable to the terms to which they are added; for a liquid they are much smaller. For a liquid, then, the leading terms of Eq. (9) become

$$\frac{1}{RT}\left.\frac{\partial \mu_C}{\partial N_C{}^o}\right|_{T,V,\mu_S} = \frac{1}{N_C{}^o} - \frac{1}{N}\left[1 - \frac{v_C{}^*}{v_S{}^*}\left(1 + \frac{V}{N_S v_S{}^*}\right)\right] + \ldots \tag{11}$$

As should be expected, the "corrections" are significant if $N_C{}^o \simeq N_S{}^o$. Furthermore, the "correction" depends really on $N_C{}^o v_C{}^*$ and $N_S{}^o v_S{}^*$. This should have been anticipated for incompressible fluids. In fact, we could have developed the whole theory in those terms. However, we often would not know the appropriate $v_C{}^*$ and for dilute solutions the approach usually taken, as in Eq. (10), is adequate. In any case, Eq. (9) is exact.

In developing Eq. (9), at no time did we introduce any asymmetry between the two species S and C. Hence, we can write $\langle [\delta N_S]^2 \rangle$ immediately. The cross correlation, however, requires a new calculation:

$$\langle \delta N_C \delta N_S \rangle = kT\left(\frac{\partial \mu_C}{\partial N_S{}^o}\right)_{T,V,\mu_S} = kT\left(\frac{\partial \mu_S}{\partial N_C{}^o}\right)^{-1}_{T,V,\mu_C} \tag{12}$$

Using the same definitions Eq. (7), expanding the determinants as before, and also invoking again the assumption of an incompressible fluid, we have

$$\frac{1}{RT}\left.\frac{\partial \mu_C}{\partial N_S{}^o}\right|_{T,V,\mu_S} = -\frac{1}{N}\left[2 + \frac{v_S{}^*}{v_C{}^*}\frac{N_S}{N_C} + \frac{v_C{}^*}{v_S{}^*}\frac{N_C}{N_S}\right] + \ldots \tag{13}$$

The implications of Eq. (13) are simple, as may be seen from a consideration of the special case $v_S{}^* = v_C{}^*$. Then $\langle [\delta N_C]^2 \rangle = \langle [\delta N_S]^2 \rangle = -\langle \delta N_C \delta N_S \rangle$. That is, a fluctuation of δN_C implies that there must be a fluctuation of equal magnitude but opposite sign $-\delta N_S$ in the other component. This followed from the assumption the assumption of incompressibility. Given a fixed volume with permeable walls, excess of one component may and will flow in, but that implies that the other has flowed out.

The next situation we consider involves a four component system including solvent S and three solutes reacting according to:

$$A + B \underset{k_b}{\overset{k_f}{\rightleftharpoons}} C. \tag{14}$$

For future use, define $C_j{}^o = N_j{}^o/V$ and $K = k_f/k_b = C_C{}^o/C_A{}^o C_B{}^o$. For this case, the Jacobian Eq. (5) leads to a 5x5 determinant. This becomes complicated. But for dilute solutions a leading term Eq. (10) holds approximately for each of the dilute components. Furthermore, the cross correlations among the solutes are still of order 1/N and, therefore, negligible. It is true that incompressibility ensures that fluctuations in one component do affect the other components by way of excluded volume. However, for a dilute solution these are "taken up" by the solvent. Hence, for dilute solutions of solutes j, k... in an open but fixed volume

$$\langle \delta N_j \delta N_j \rangle = N_j^{\,o}/N_o$$
$$\langle \delta N_j \delta N_k \rangle = 0 \quad j \neq k. \tag{15}$$

B. Dissipation of Number Fluctuations

Let us next turn our attention to the behavior in time of the fluctuations $\delta X_j(t)$. Again we can only consider the "average" time behavior, and the appropriate characterizations are the auto-correlation ($j = k$) and cross correlation ($j \neq k$) functions

$$\Phi_{jk}(\tau) = \lim_{T\to\infty} \frac{1}{T} \int_0^T \delta X_j(t)\,\delta X_k(t+\tau)\,dt. \tag{16}$$

I assume a stationary process; that is, I assume that $\Phi(\tau)$ is independent of what time I begin my measurement. Note that the whole measurement process Eq. (16) only defines $\Phi_{jk}(\tau)$ for one particular value of the "lag time" τ. The correlation function of interest is a complete set of $\Phi_{jk}(\tau)$. Microreversability requires $\Phi_{jk}(\tau) = \Phi_{jk}(-\tau)$. Furthermore $\Phi_{jk}(\tau) = \Phi_{kj}(\tau)$.

As before, it is convenient to invoke the equality of the experimentally measured $\Phi_{jk}(\tau)$ with the theoretically more convenient ensemble average

$$\begin{aligned}\Phi_{jk}(\tau) &= \langle \delta X_j(0)\,\delta X_k(\tau)\rangle \\ &= \int dX_o' \ldots \int dX_s' W(X_o' \ldots X_s')\,\delta X_j' \langle \tau, X_o' \ldots X_s' \mid \delta X_k \rangle.\end{aligned} \tag{17}$$

Here W is the same joint probability density function for the free variables we encountered above in Eq. (2) and $\langle \tau, X_o' \ldots X_s' \mid \delta X_k \rangle$ is the expectation value of δX_k at any given time under the condition that each of the free variable assumed specified values X_j' at a time τ earlier. That the integration over the ensemble of all such possibilities will calculate $\Phi_{jk}(\tau)$ is plausible. A detailed discussion has been given both for the case of one free variable (CALLEN and GREENE, 1952) and for the case of mulitple free variables (GREENE and CALLEN, 1952). The important point is that each conditional probability $\langle \tau, X_o' \ldots X_s' \mid \delta X_k \rangle$ corresponds exactly to the question posed in a perturbation relaxation experiment: What value will be observed for the extensive parameter δX_k now given, first, that the equilibrium value now is $X_k = X_k^{\,o}$ and, second, that at time τ ago, the extensive parameters assumed particular values $X_o' \ldots X_s'$?. The principle invoked is that the system does not care whether it started from $X_o' \ldots X_s'$ because an experimenter contrived to force it to do so or because of a chance fluctuation. This is the "fluctuation dissipation theorem".

The development of perturbation kinetic analysis normally considers cases where the kinetic behavior is assumed independent of the size of the perturbation. We are greatly aided in evaluating Eq. (17) if we too assume that the conditional probability is independent of the magnitudes of $\delta X_o' \ldots \delta X_s'$, at least over the range for which W has significant weight. Since in CCA the fluctuations are always very small, at least away from critical points, this is a safer assumption than it is for finite perturbation experiments. When the scheme Eq. (17) is carried out, the result is

$$\Phi_{jk}(\tau) = \langle \delta X_j(0)\delta X_k(0)\rangle \; f(\tau) \tag{18}$$

where $\langle \delta X_j(0)\delta X_k(0)\rangle$ is given by Eq. (3) and $f(\tau)$ is the solution of the appropriate dynamic problem. In all cases $f(0) = 1$ and for dissipative processes $f(\infty) = 0$.

Our task has been reduced, therefore, by the application of the fluctuation dissipation theorem, to the problem of calculating the kinetics by which a very small macroscopic increment of concentration would dissipate. Since this calculation of the dissipation is the heart of the problem, it is unfortunate that it is not possible to provide a general treatment which is sufficiently specific to be useful. I will simply outline the considerations involved (ELSON and MAGDE, 1974).

In certain cases the fluctuation may relax solely by chemical reaction. This could be true if the mole numbers N_j in a solution were monitored inside a sample cell with impermeable walls. It would also apply if the reacting molecules were immobilized within a solid or bound tightly to fixed sites on a surface. In these cases, of course, Eq. (15) does not apply. Furthermore, these situations would inhibit many chemical reactions; but others, an isomerization, for example, might still occur. When a chemical reaction provides the dissipative relaxation, the kinetics is obtained systematically by the matrix formulation of chemical rate theory. The relaxation of each component is governed by

$$\frac{\partial \delta N_j(\tau)}{\partial \tau} = \sum_k T_{jk}\delta N_k(\tau). \tag{19}$$

Since the analysis of this equation is fully discussed by EHRENBERG in another Chapter of this Volume we need not elaborate here. For the simple bimolecular reaction (Eq. (14)) the result is that the relaxation occurs via a single exponential decay with rate constant

$$R = k_f\left[C_A{}^{o}+C_B{}^{o}\right] + k_b. \tag{20}$$

The even simpler instances of isomerization and dimerization are special cases of this reaction. For more complex reactions, the observed decay becomes a sum of exponential terms of varying amplitudes and decay rates which, although expressible in terms of the k_f and k_b assigned to elementary steps of the reaction, do not bear in general any simple relation thereto. The decay rates are simply the eigenvalues of the governing matrix. This will be familiar to the perturbation kineticist.

A formal expression for the correlation functions in reactive solution subject, essentially, to Eq. (19) was published by another group (CHEN, 1973; CHEN and HILL, 1973), independent of, and simultaneous with, our own work. The approach and even the notation of the two treatments is largely parallel and the treatments are in part complementary. CHEN and HILL present a more exhaustive discussion of the effects of different constraints on the chemical kinetic portion of the problem, including interaction with a reservoir, but do not treat the complication, crucial in practice, that material transport to and from the reservoir during the observation must be described by appropriate hydrodynamic equations, as well as by the much simpler chemical kinetic expressions.

In general, and in fact for all the CCA experiments I will discuss below, at least some of the walls bounding V are permeable and it is possible for a fluctuation $\delta N_j(r,t)$ of species j, at position r, at time t to dissipate by means of the actual transport of molecules away from or toward r. Such transport will always be possible by means of diffusion

$$\frac{\partial N_j(r,\tau)}{\partial \tau} = D_j \nabla^2 \delta N_j(r,\tau), \tag{21}$$

although the diffusion coefficient (with units $cm^2 s^{-1}$) may vary over a wide range.

It is also possible that transport may occur by a collective flow through defined volume V. This could be thermally driven (convective), gravity driven (by a pressure head, when two reservoirs are connected by a channel in which V is located), or mechanically driven (by a pump deliberately used to produce a flow, by the translation of a membrane, and so on). Many times such flows are present only as undesired experimental artifacts, but deliberately induced flows can be extremely useful to the investigator.

The implications of dissipation by actual transport of molecules are extensive. Now a fluctuation at one position r is correlated with fluctuations at other positions r', so that we must refine the definitions of the correlation functions. I define a symbol $C_j(r,t)d^3r$ to be the number of molecules of species j found at time t in a small volume element d^3r located at position r. Then $\int_V C_j(r,t)d^3r = N_j(t)$. The $C_j(r,t)$ amount to local, fluctuating concentrations. The concentration correlation functions become

$$\Phi_{jk}(r,r',t) = \langle \delta C_j(r,0)\,\delta C_k(r',\tau)\rangle . \tag{22}$$

This is quite general, assuming only stationary behavior. The δC_j required to evaluate Eq. (22) must be obtained as solutions of Eq. (21) or other applicable transport equation, subject to boundary conditions appropriate to a particular experiment. Note that the characteristic times of relaxation will not be strictly molecular properties but will depend also on the size and shape of the volume under observation. If V has a characteristic dimension ℓ, then a characteristic time of order ℓ^2/D is expected for diffusion or ℓ/v for flow, where v is the fluid velocity.

In general, of course, for V bounded by rigid but permeable walls, dissipation governed by both chemical and diffusive mechanisms will occur. (There may also be present flow, which I will not consider further.) Then δC_j obeys the general differential equation

$$\frac{\partial \delta C_j(r,\tau)}{\partial \tau} = D_j \nabla^2 \delta C_j(r,\tau) + \sum_k T_{jk} \delta C_k(r,\tau). \qquad (23)$$

This may be written conveniently in matrix form with only the diagonal terms involving the diffusion. Then Eq. (23) may be solved by standard techniques. Both complications mentioned above will occur: first, the solution must be consistent with specific boundary conditions. And second, the solution will result in a sum of relaxation processes, each of which is a "normal mode" of relaxation which, in general, not only mixes multiple chemical relaxations but mixes diffusive relaxation with these. The main point I wish to emphasize is that, in general, diffusive and chemical relaxation are inextricably coupled; they do not separate. Even the single reaction Eq. (14), when combined with diffusion, leads to a relaxation which has three components.

This concludes a qualitative survey of the general features of the time dependence of correlation functions describing the dissipation of occupation number fluctuations in open systems. Logically we should next consider possible experimental approaches to the measurement of $N_j(t)$, which will define particular boundary conditions, and then return to various particular solutions of Eq. (23). I will, in fact, turn next to particular experiments. But I shall not return to a further systematic discussion of Eq. (23). On the one hand, boundary value problems constitute a large field of applied mathematics which is far too extensive to repeat here; on the other, several particular and detailed solutions of Eq. (23) for CCA have appeared together in the literature (ELSON and MAGDE, 1974) and it seems most efficient to refer the reader who is interested in the details of how the τ dependence is actually calculated to that comprehensive treatment. I will list partial results below in connection with particular experiments.

III. Measurement of Number Fluctuations

Occupation number fluctuations might be detected and characterized by a variety of experimental methods. All these methods involve two distinguishable steps: the translation of the time dependent occupation number information $N_j(t)$ into a related time dependent electric current $i(t)$ and the subsequent computation of a current autocorrelation function $G(\tau)$. For the particular translation scheme which I will discuss in most detail, fluorescence, the relation is direct proportionality, possibly involving a summation over different constituent species:

$$i(t) = \int_V \sum_j g_j(r) C_j(r,t) d^3r. \qquad (24)$$

The constants of proportionality $g_j(r)$, unfortunately, may in real experiments vary with r, even inside V. In general, an expression more complicated than Eq. (24) relates i(t) to the $C_j(r,t)$. However, small $\delta i(t)$ often have a simple linear dependence on $\delta C_j(r,t)$. The subsequent processing of i(t) may proceed by any one of several equivalent methods, which differ in detail, but all of which generate an experimental correlation function of $\delta i(t)$, defined

$$G(\tau) = \langle \delta i(0)\delta i(\tau)\rangle = \lim_{T\to\infty} \frac{1}{T}\int_0^T \delta i(t)\delta i(t+\tau)dt. \quad (25)$$

The interpretation of $G(\tau)$ then proceeds via Eq. (24) or an analogous relation back to the various $\Phi_{j\ell}(r,r',\tau)$.

Equation (25) characterizes the fluctuating $\delta i(t)$ in terms of the persistence in time of fluctuations. It is equally possible to think in terms of the frequencies $\omega = 2\pi\nu$ present in the fluctuating i(t). The equivalence of the two representations is formally stated in the Wiener-Khinchin theorem:

$$G(\tau) = (2\pi)^{-1/2} \int_{-\infty}^{\infty} d\omega\ G(\omega)e^{i\omega\tau} \quad (26a)$$

where

$$G(\omega) = (2\pi)^{-1/2} \int_{-\infty}^{\infty} d\tau\ G(\tau)e^{-i\omega\tau}. \quad (26b)$$

Until recently it was much easier to measure $G(\omega)$, which is the power passed by an electronic filter centered at ω. (Recall that electrical power is proportional to $[i(t)]^2$). Today either $G(\tau)$ or $G(\omega)$ may be measured. Given either, it is a simple computation to transform to the other. The choice of experimental approach depends on no fundamental principle but on details of how efficiently the signal processing in a particular machine operates. As computer processing becomes more intimately spliced into the electronic measurement, it becomes an increasingly subtle question how displayed output actually relates to some fundamental measurement. As to which function is more convenient to use for subsequent analyses, the distinction is largely a matter of prejudice based on past practice. In certain cases, however, one choice is clearly more convenient for human visualization. For example, when $G(\omega)$ consists of "lines" at discrete frequencies bearing no particular relation to each other, then $G(\tau)$ becomes complicated to picture.

It is helpful to remember that when $G(\tau)$ is exponential, $G(\omega)$ has a Lorentzian (Cauchy) form; when $G(\tau)$ is Gaussian (normal), $G(\omega)$ is also Gaussian; when $G(\tau)$ is a spike (delta function) at $\tau = 0$, $G(\omega)$ is a constant for all ω; when $G(\tau)$ is a sine curve, $G(\omega)$ has a spike at a single ω.

I will return below to the complex question of what considerations are involved in actual practice in designing a machine to evaluate Eq. (25). But first let us examine one partiuclar translation method which can be used to convert occupation number fluctuations into a fluctuating electric current. I select the fluorescence approach to discuss in detail because in that case i(t) is most directly, linearly, and obviously proportional to $N_j(t)$.

A. Fluorescence Correlation Analysis (FCA)

One of the most obvious methods for extracting information about fluctuating numbers of molecules in a sample volume V and converting it into a fluctuating electric current would employ irradiation of V by a "light" source and detection of reemitted photons with a photomultiplier. The incident radiation could be electromagnetic radiation of any frequency or conceivably even a beam of particles, but I will consider only radiation near optical frequencies. The reemitted light may occur at essentially the same frequency as the incident light, shifted in energy only as required to conserve momentum, or it may appear shifted by a discrete energy change corresponding to quanta of energy given up to or taken up from the molecules in V. The former case I shall call quasi-elastic light scattering (QELS). I will defer to a later section a few remarks on possible advantages of QELS as a probe for CCA. The principle disadvantages is, of course, limited specificity.

In contrast, inelastic interactions, in which there occur discrete shifts in frequency and wavelength, are specifically related to molecular excitations and consequently provide superior means of distinguishing fluctuations in one particular constituent of a complex mixture. (A truly elegant experiment would analyze emitted radiation into several bands, each of which is separately but simultaneously detected leading to a set of time dependent electric currents. The autocorrelation of each separate current and the cross correlation of one with another would yield simultaneous data on many fluctuating concentrations.)

Regretably, all the many varieties of Raman scattering, which is both highly specific and yet very generally applicable, exhibit very small interaction cross sections for nonresonant conditions. We are led, therefore, to consider as most feasible excitation at a resonant optical frequency. In solutions, for conditions of biological interest, this means ordinary fluorescence. Phosphorescence is possible in principle but rarely applicable in practice and, then, limited to slow fluctuations.

Fluorescence emission occurs over a time of order 10^{-8} s following excitation. During the 10^{-8} s, the excited molecule undergoes many collisions with solvent which disrupt all memory of the phase of the exciting light. Hence, in the fluorescent process each molecule acts as an independent emission source and it is correct to sum the *intensities* of emission, as expressed in Eq. (24).

Detection of molecular concentrations — or occupation numbers — by fluorescence methods has five advantages which render it useful for CCA:

1. A beam of light may be focused and used itself to define the volume observed. If a laser provides the beam, the flux of photons may be extremely high in a very small area, which maximizes our ability to "observe" every molecule.

2. Selectivity can be quite high.

3. Photons of optical energy are sufficiently energetic that they can be individually counted. Since, in the absence of blank fluorescence, there is no signal except that due to N_j, we expect sensitivity to small number of molecules to be high. Fluorescent methods are, along with radiometric techniques which involve even more energetic photons and particles, among the most sensitive of analytic methods.

4. The Stokes shift to longer wavelengths of the fluorescence is, in practice, sufficient to permit discrimination against scattered radiation at the exciting frequency. (However, care must be exercised not to introduce spurious fluorescence generated in optical filters.) Hence, blank fluorescence can be fairly low.

5. The 10^{-8}s time delay between absorption and emission is sufficiently short as to impose no serious limit to the range of correlation times which may be studied.

The disadvantages of fluorescence — or more accurately, its limitations — are principally two:

1. The high energy of visible photons raises the risk of photochemical degradation of the sample being observed. Short of permanent degradation, there is the risk of generating metastable transient species.

2. The shift to lower energies of the fluorescence emission usually corresponds to at least some Raman shifts for molecular vibrations of the solvent and of the sample cell. This can set a lower limit to the blank fluorescence and, hence, to how dilute the solutions used may be. This limit lies in the range 10^{-9} mol ℓ^{-1}, but varies by orders of magnitude.

I will not elaborate here on these disadvantages. I previously discussed their practical implications at some length with emphasis on the reasons why even a very low quantum yield for photodegradation (in the range 10^{-5}) can be troublesome (MAGDE et al., 1974).

One other feature of fluorescence appears to be ambiguous: Fluorescence methods are quite widely applicable. Yet as constraints on selectivity and sensitivity are tightened and limits on permissable photoreactivity reduced, the candidate systems are rapidly restricted. Particular fluorescent probes may be discovered or synthesized and particular conditions optimized for a variety of biochemical studies — but how far this can be extended is not yet clear. Whether such "special" solutions to particular requirements are evidence for or against asserted general applicability for FCA becomes moot.

1. Design of the FCA Experiment

Let us consider how an FCA experiment could be implemented. A schematic of the apparatus appears in Figure 1. We assume that a solution is contained in a more or less ordinary fluorescence cell whose dimensions we calculate below. We illuminate a small

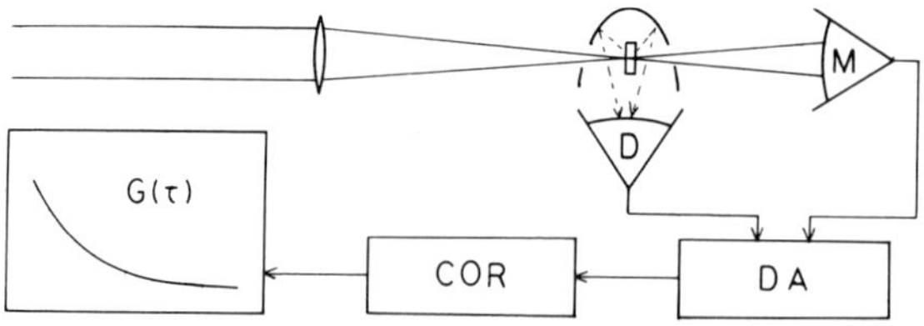

Fig. 1. Schematic of apparatus for an FCA experiment. A laser beam is focused into a small sample volume, which is surrounded by a mirror which collects the fluorescence and sends it to the detector D. The incident laser power is monitored by M. The two signals are subtracted in a difference amplifier DA, whose output is analyzed in a correlation computer COR. The correlation function may be displayed or sent directly to a general purpose computer for curve-fitting. Optical and electronic filters, not shown here, will be needed

volume within this cell by passing a laser beam through it along an axis $\hat{z}$. The solution may be assumed to be optically thin, so that the excitation beam is unattenuated along $\hat{z}$. The beam has some intensity distribution in the transverse $(\hat{x},\hat{y})$ directions. The most convenient intensity distribution to employ is radially symmetric of the form

$$I(r) = I_o \exp[-2r^2/w^2]. \quad (27)$$

This is called a Gaussian beam. The intensity reaches a maximum I_o on the axis and falls off radially such that at a radius $r = w$, $I/I_o = \exp|-2|$. This is the output obtained from a laser which is operating in its lowest order, or so-called Tem_{oo}, mode. When a Gaussian beam is focused by a lens, its distribution remains Gaussian, with new parameters I_o' and w'. This convenient property does not hold for any other distribution $I(r)$. The total power in the Gaussian beam Eq. (27) is

$$P = \int_o^\infty 2\pi r\, I(r)dr = I_o \pi w^2/2 \ . \quad (28)$$

If we define the distance between the walls of the fluorescence cell along $\hat{z}$ to be L, the observation volume V is quasi-cylindrical with length L and area of order πw^2. We must be careful, however, not to take the "area" too seriously; we have to retain the actual $I(r)$, given by Eq. (27), and perform the integration over V only at the appropriate time.

Some small fraction of the laser light will be absorbed; a fraction of that will be emitted as fluorescence; a fraction of that will be collected, filtered, and permitted to fall upon a sensitive photocathode to generate, finally, a cathode current $i(t)$ given by a particular case of Eq. (24):

$$i(t) = 2.3g\sum_j \varepsilon_j Q_j \int I(r)C_j(r,t)d^3r \quad (29)$$

Here I allow for the possibility that several species j may absorb and emit. I still assume optically thin samples. The coefficients ε_j and Q_j are respectively the decadic extinction coefficient at the excitation wavelength and the quantum yield of the j component. The local concentration is $C_j(r,t)$. The parameter g accounts for geometrical losses in collecting the fluorescent emission, losses involved in spectral filtering of the fluorescence, and losses due to the limited quantum yield of the photocathode.

The mean cathode current for the excitation intensity distribution described by Eq. (27) and Eq. (28) is

$$\begin{aligned} i(t) &= 2.3g \sum_j \varepsilon_j Q_j \langle C_j(t) \rangle \int_V I(r) d^3r \\ &= 2.3gPL\sum_j \varepsilon_j Q_j C_j^{o} \end{aligned} \tag{30}$$

where C_j^{o} are ordinary equilibrium concentrations. The photocurrent is independent of beam area, as we would expect.

The fluctuations of the current Eq. (29) about the mean are very simply

$$\delta i(t) = i(t) - \langle i(t) \rangle = g\sum_j \varepsilon_j Q_j \int I(r) \delta C_j(r,t) d^3r \tag{31}$$

The mean fluctuation, of course, vanishes: $\langle \delta i(t) \rangle = 0$. The autocorrelation of the current fluctuations constitutes the primary experimental observation. It is

$$G(\tau) = \delta i(0) \delta i(\tau) = \sum_{jk}\sum G_{jk}(\tau) \tag{32a}$$

where

$$G_{jk}(\tau) = 2.3^2 g^2 \varepsilon_j \varepsilon_k Q_j Q_k \int I(r) I(r') \Phi_{jk}(r,r',\tau) d^3r d^3r'. \tag{32b}$$

Observe that although there is only one detected current and one observed autocorrelation function $G(\tau)$, the latter is, in general, related to all possible auto- and crosscorrelations Φ_{jk} of all solutes present. However, it will often be experimentally desirable to arrange that either ε_k or Q_k or both be very small for all $k \neq j$. Then only one term Eq. (32b) contributes to the sum Eq. (32a). I will make that assumption from now on. The full set Eq. (32b) were retained in the original theoretical treatment of FCA and some implications particularly of the cross-correlation terms noted (ELSON and MAGDE, 1974).

To proceed further we need an explicit expression for $\Phi_{jj}(r,r',\tau)$, which I discussed above but did not develop explicitly for all τ. However, we may calculate G(0), which is simply the mean square fluctuation of the current, in terms of microscopic occupation number fluctuations available from thermodynamic considerations, without knowledge of time dependent hydrodynamic or chemical kinetic behavior. We rely on the fact

$$\Phi_{jj}(r,r,0) = \langle [\delta C_j(0)]^2 \rangle \, \delta(r-r') = C_j^{o} \delta(r-r')/N_o \tag{33}$$

This is in accord with Eq. (22) and adds the further assumption, expressed in the vector delta function, that a fluctuation is local, that is, whatever happens within the small volume d^3r in which we define $C_j(t)$ is uncorrelated with what happens outside. If $I(r)$ varies rapidly over molecular dimensions, Eq. (33) will require reconsideration. That $\langle[\delta C_j(0)]^2\rangle = C_j^o/N_o$ follows Eq. (15) since the "concentrations" $C_j(t)$ are identified with the occupation numbers of a small volume element. Substituting Eq. (33) into Eq. (32) yields

$$G(0) = 2.3^2 g^2 L(P^2/\pi w^2)\varepsilon_j{}^2 Q_j{}^2 C_j{}^o/N_o. \tag{34}$$

The relative mean square fluctuation is a key experimental parameter. Using Eq. (30) and Eq. (34),

$$\frac{G(0)}{\langle[i(t)]^2\rangle} = \frac{1}{\pi w^2 L N_o C_j{}^o}\,. \tag{35}$$

Recalling that $\pi w^2 L$ is an effective volume, N_o is Avogadro's number, and $C_j{}^o$ is the concentration in moles per volume, we note that Eq. (35) is simply the reciprocal of the number of molecules $N_j N_o$ of type j in the volume. This is expected. However, it was necessary to proceed carefully in order to ascertain what is the appropriate effective volume.

An effective volume of $\pi w^2 L = 10^{-8}\,cm^3$ is readily attainable. It could be achieved by focusing a laser beam with a lens of focal length a few centimeters through a fluorescence cell of $L \simeq 10^{-2}$ cm. Such a volume requires $C_j{}^o < 10^{-7}$ mol ℓ^{-1} if we desire the relative mean square fluctuations to exceed 10^{-6}, or equivalently the relative rms (root mean square) current fluctuations to exceed 10^{-3}. Fluctuating currents considerably smaller than 10^{-3} of a d.c. level may be detected with ease. However, we must not just detect $G(\tau)$ but measure it with high precision in order to permit curve fitting to a sum of nonexponential terms. We will see in a moment that statistical uncertainty (and this surprises some) is independent of the size of the fluctuations. However, as the rms fluctuations drop, there exists ever greater risk of encountering undesired, correlated current fluctuations due to systematic errors somewhere in the measurement. Clearly, it will always be desireable to maximize the ratio Eq. (35). Unfortunately, we are not free to vary arbitrarily the parameters of Eq. (35). Dimensions of the volume determine characteristic times for diffusion, while molecular concentrations affect chemical reaction rates. Hence any experimental changes made in the ratio Eq. (35) will also affect the time behavior of $G(\tau)$. This is not necessarily harmful, but it must be considered. Moreover, good experimental practice requires that kinetic measurements be carried out over as wide a range as possible of experimental parameters. As the parameters are varied from initial values selected to maximize the relative mean square fluctuations, the fluctuations must inevitably decrease.

Throughout the foregoing discussion, I have assumed implicitly that the translation Eq. (29) proceeds with no loss of information, that is that $i(t)$ is truly a faithful and precise measure of $N_j(t)$. This is not necessarily so. The cathode current $i(t)$ is, of course, a flow of discrete electrons. A more precise formulation of Eq. (29) would relate a probability distribution for observing a number of electrons within an observation period Δt to the fluctuating concentrations. Our second principal concern of experimental design (after optimizing Eq. (35)) involves assuring that there are enough electrons within Δt to adequately define i(t). The appropriate Δt to consider depends on the times $\Delta\tau$ which we want to resolve in $G(\tau)$. The number of electrons is $N_e = \langle i(t)\rangle \Delta\tau$. (Equation (30) gives $\langle i(t)\rangle$ in electrons per second, if P is expressed as photons per second.) Electrons emitted from a photocathode obey Poisson statistics; hence the uncertainty in $N_e^{1/2}$. Our goal is to use these N_e electrons to measure the number of fluorescent molecules N_jN_o with sufficient precision to discern fluctuations of magnitude $(N_jN_o)^{1/2}$. Hence, we need to have N_e at least as large as N_jN_o, if the uncertainty of the measurement is to be no worse than the fluctuation we seek to measure. This is not absolutely necessary; we can proceed even if $N_e<N_jN_o$: correlation analysis is inherently an averaging process which extracts characteristic behavior from a noisy record, and it will eventually pick out the concentration fluctuations, even if they are individually very poorly measured. This is the "shot-noise" limit. On the other hand, the time required for analysis eventually becomes prohibitive. Hence, we would like $N_e > N_jN_o$. One might suppose it would help if $N_e \gg N_jN_o$. However, a point of diminishing return sets in. Since we have to examine many fluctuations δN_j anyway in order to characterize $G(\tau)$, it does very little good to measure each particular fluctuation to very high precision. This matter has been treated rigorously (KOPPEL, 1974). The principal conclusions are two:

1. When $N_e \ll N_jN_o$, the precision S/N of measurement of $G(\tau)$ increases as

$$S/N \sim N_e(T/\Delta\tau)^{1/2}. \tag{36a}$$

2. When $N_e \gg N_jN_o$, the precision improves as

$$S/N \sim (T/\tau_c)^{1/2}. \tag{36b}$$

In Eq. (36), T is the total record length processed and τ_c is the characteristic relaxation time of $G(\tau)$.

Our experimental plan, then should seek to increase N_e until it is equal to (or a few times larger than) N_jN_o, unless N_e is limited to lower values by the onset of photodegradation. (Thermal problems are usually not important in FCA, because the solutions are so dilute.) We would like, therefore:

$$\frac{N_e}{N_jN_o} = 1 = \frac{\langle i(t)\rangle \Delta\tau}{N_jN_o} = \frac{2.3gPL\varepsilon_jQ_jC_j{}^{o}\Delta\tau}{\pi w^2 LC_j{}^{o}N_o}\ . \tag{37}$$

This determines the necessary laser power

$$P = \pi w^2 N_o/2.3g\varepsilon_jQ_j\Delta\tau. \tag{38}$$

Plausible values are $\pi w^2 = 10^{-6}$ cm^2 and $g = 2.5 \cdot 10^{-2}$. In very favorable cases, ε_j and Q_j could be 10^5cm^{-1}mol$^{-1}\ell$ and 1 respec-

These values require $P = 10^{11}(\Delta\tau)^{-1}$ photons per second. [illegible] of comparison, a 1mW HeNe laser emits just about $3 \cdot 10^{15}$ [illegible] per second. Hence, we could have a temporal resolution [illegible]μs even with such a modest laser, assuming noise free amplification and processing of the cathode current. However, more realistic values of ε_j and Q_j would each be a factor of 10 smaller. And signal processing is never perfect and may sometimes involve substantial loss of information, which is exactly equivalent to failing to detect some photons. This must be compensated by increasing P if photodegradation permits. Note further that g was evaluated assuming 10% photoemissive quantum yield (obtainable today over most of the spectrum) and 25% probability that each fluorescent photon reaches the cathode. Such high collection efficiency requires collection by ellipsoidal or parabolic mirrors or integrating spheres (certainly not lenses) and permits only the most rudimentary spectral processing. Introduction of even a very fast dispersive monochromator would cause g to plunge to 10^{-3} or less.

Equation (38) may be derived or interpreted from another point of view. P is the photon flux over the area πw^2 required so that on the average each molecule of j absorbs and emits enough photons so that one is detected every $\Delta\tau$. This implies that our correlation method is truly observing the kinetics of individual molecules. On reflection, this must be so; for there is absolutely no correlation in the behavior of these independent molecules with each other in our assumed ideal solution.

In this sense CCA constitutes the ultimate in molecular kinetic analysis. Perturbation approaches appear in this light essentially as schemes to force a macroscopic number of molecules to behave in concert, thereby facilitating the study of their behavior.

This viewpoint emphasizes why g, ε, and Q are extremely important parameters, deficiencies in which may not always be counteracted by increasing P. The product $(gQ)^{-1}$ is the number of photons which must be absorbed by each molecule in order to detect on the average only one photon from that molecule per $\Delta\tau$. As $\Delta\tau$ becomes shorter and gQ is reduced, the molecule must cycle between its excited state and its ground state at a high rate. Even a small probability for photodegradation (or the existence of a metastable excited state) will complicate matters appreciably. On this basis a small ε is not quite so troublesome. To some extent we can compensate for a low probability of interaction by increasing the photon flux. However, this neglects the fact that there is inevitably some level of blank fluorescence, Raman scattering or otherwise, which increases linearly with P.

If it appears that I am unduly pessimistic about photodegradation, observe that even a 1mW laser, focused to an area 10^{-6} cm^2, implies a continuous intensity of $1KWcm^{-2}$.

2. Correlation Computers

I have just argued that given constraints on the permissible P, it is essential to arrange efficient collection of fluorescence by maximizing g. I added in passing that it is equally important to ensure efficient and noise-free processing of the signal after initial acquisition. This becomes a very subtle problem in practice. It is impossible to build an ideal correlator capable of working over the entire range of experimental conditions that may be encountered. In particular, the range of times of interest extends from nanoseconds to hours. No single processing method can serve all purposes; comprise is necessary. Given prior knowledge of some properties of the signal, compromises are possible which perform almost ideally despite great simplification of devices. However, a procedure which is almost ideal for certain signals may become very inefficient for others. Commercial instruments, in particular, must lay heavy emphasis on versatility, even at the expense of efficiency under particular conditions. A quantitative, yet practical, review of this entire problem does not exist — and might well be permature, given the rapidly evolving state of the electronic art. There do exist excellent articles on important aspects of the problem, particularly from the point of view of the demands of QELS. These, together with careful study of commercial literature and persistent attention to details, can mean to the experimenter differences of orders of magnitude in processing time or equipment costs. Here I will outline in extremely elementary fashion what is involved and make appropriate reference to helpful original literature.

Before we come to calculating $G(\tau)$, we face some preliminary decisions.

First, note that in Eq. (25) $G(\tau)$ is written in terms of $\delta i(t)$. In most experiments, it is $i(t)$ which is available. We have an immediate choice: we can either perform some analog processing on $i(t)$ to generate a $\delta i(t)$ which fluctuates about zero or we can compute the autocorrelation function of the full $i(t)$ to obtain the more complete expression

$$G'(\tau) = \langle i(0)i(\tau)\rangle = \langle i(t)\rangle^2 + G(\tau). \qquad (39)$$

The separation of the two terms then depends, usually, on the fact that $G(\tau) \to 0$ as $\to \to \infty$.

The use of Eq. (39) instead of analog processing followed by autocorrelation just of $\delta i(t)$ would seem more elegant and, in principle, more efficient. Occasionally, it is possible to have $\delta i(t) \sim \langle i(t)\rangle$. In this case, it is simpler and much preferable to work with the correlation of the full $i(t)$.

If, however, $\delta i(t)/\langle i(t)\rangle << 1$ (It can be 10^{-4} or less.), then efficient calculation of Eq. (39) can be obtained only if $i(t)$ can be measured at each t with very high precision and if many significant figures are retained in all products and sums. Since this is either impossible, or, at least, expensive, one normally must resort to one of two methods or preprocessing $i(t)$ to eliminate the average or d.c. component $\langle i(t)\rangle$. The two methods are: first, electronic high pass filtering, which blocks the d.c. component, and, second, substracting from $i(t)$ a current equal to $\langle i(t)\rangle$ in a differential amplifier. The term "differential" is a misnomer; the instrument actually is a "difference" amplifier.

This second method is preferable if the current which is subtracted can be obtained from a monitor of some excitation intensity to which i(t) itself is proportional. This will always be the case for optical measurements. Then one automatically corrects fluctuations in the source intensity, even when they are within the frequency range of the kinetic data. If the signal and the monitor are not exactly balanced, or perhaps drift apart as the calculation proceeds, then $G(\tau)$ will be distorted. If the imbalance is slow drift (as is likely), the solution is to combine the difference measurement with high pass filtering.

Our second decision is whether to compute $G(\tau)$ by Eq. (25) or to measure the power spectrum Eq. (26b). Occasionally one hears, or even reads, of a dramatic discovery in which the correlation function approach represents a great breakthrough in processing efficiency. This is nonsense, based on contrasting a more or less refined correlation computer to a swept frequency spectrum analyzer which measures only one frequency at a time. It is entirely possible to design an instrument capable of measuring all frequencies simultaneously. Indeed, I believe that the present generation of multiplex spectrum analyzers operate somewhat faster than most current commercial correlators. Such an instrument was used in one of the studies discussed below (FEHER and WEISSMAN, 1973). These instruments, however, show the problems of analog devices: limited linearity, uncertain baseline flatness, and so on. The current trend is definitely toward computing first $G(\tau)$ and then, if desired, calculating $G(\omega)$ by Fourier inversion.

Our third decision involves whether to use "real time" analysis or "remote" processing. In the latter case, the signal record is recorded and preserved, in toto, usually on magnetic tape and usually, in the past, in analog form. If low frequencies (below $\sim$ 20Hz) are of interest, FM recording is indicated. For very high frequencies (above some value in the range 2kHz to 50kHz) up to about 1MHz, AM recording may be the sole choice. Today a digital record could be tape recorded. Once this record exists, processing may proceed at "full efficiency", evaluating Eq. (25) or Eq. (39) exactly as expressed. This may require a large computer and it may take a long time, but it is the most efficient possible use of all the data. This approach is indicated when the cost of data collection is high (radioastronomy, oceanography), or the phenomena are inherently transient and difficult to repeat (ecological studies, economics), or the subject of the study is degraded by the measurement and limited in supply (biological molecules). The existence of the permanent record allows a variety of statistical checks. Remote processing was implemented in one of the experiments to be considered below (ZINGSHEIM and NEHER, 1974).

In many cases, however, it is desireable to be able to observe the computed $G(\tau)$ as modifications are made in experimental procedure. This requires real time computation of $G(\tau)$. Although under exceptional circumstances involving rapid fluctuations and a short total record it would be possible to use batch processing and still have results in a few seconds, we normally mean by real-time processing that calculations of $G(\tau)$ proceeds while data are collected and that only the improving approximation to

$G(\tau)$ is retained. Sample points themselves are discarded after some processing, not retained for reexamination at a later time. The object, then, is to decide what information to keep and how long to keep it. Roughly, we need remember $\delta i(t)$ only for times comparable to the τ_c of interest. But most procedures in fact do not measure accurately all $\delta i(t)$ and keep all of them for τ_c. Compromises are necessary for reasons of cost, if not because of technical constraints.

Let me describe four imaginary machines, four computational algorithms, which will make clear some possible compromises:

1. We could measure $\delta i(t_1)$, wait to measure $i(t_1 + \tau')$, multiply the pair, and add the result to a memory labeled $G(\tau')$. We could then start over with $\delta i(t_2)$ and do the same thing. If we repeat this N times we will generate an estimate of $G(\tau')$. We can then start again and do the same for $G(\tau'')$. And so on, for all τ of interest. Typically this is at least 100. For 100 evenly spaced τ, spaced $\Delta\tau$ apart and N samples of each, we will need a total record $T = (99)(50)N\Delta\tau$. I know of no instances where this procedure has been used.

2. We could measure $\delta i(t_1)$ and then $\delta i(t_1 + \Delta\tau)$, $\delta i(t_1 + 2\Delta\tau)$, etc. We could form the product of $\delta i(t_1)$ with itself and with 99 subsequent samples and store the results in 100 memories, appropriately labeled. Then we could start again for $\delta i(t_2)$. The total record required is only $T = 100N\Delta\tau$ for the same precision (N samples at each τ') we obtained in the first scheme. This is a commonly implemented scheme and under certain rather unlikely conditions (no high frequency noise) it is a reasonable strategy.

3. We can measure $\delta i(t_1)$ and 99 subsequent samples as above, form products and store the results. But then, rather than beginning again at $t_2 = t_1 + 100\Delta\tau$, we could use $t_2 = t_1 + \Delta\tau$. The same record length used above, now generates 100N samples at each τ'. This sounds better, and often is, but these samples are no longer independent samples. If there is not high frequency noise present and if we are interested in $\tau_c \sim 100\Delta\tau$, then we still get only N independent measurements. However, this is not the more commen case. In the shot noise limit, this third algorithm is the best possible. (The distinction here is the same one involved in Eq. (36a,b). But algorithm 3 involves 100 products and sums every $\Delta\tau$. For $\Delta\tau = 1\mu s$, we could keep a very large computer busy. Interestingly, a machine using analog multiplication and capacitive integration shares some of the features of algorithm 3 – although the information about $\delta i(t)$ is not fully available for all products; it must be "physically" divided up among the 100 channels introducing extra noise. This, together with the inherent problems of analog processing and storage, limit the analog processor most just where algorithm 3 itself is most useful – namely, in the extreme shot noise limit.

4. We could keep the procedure of algorithm 3 but reduce the computation needed by modifying the "measurement of δi" to generate only a very approximate measurement – perhaps only 1 significant figure (in binary, this would be levels 1, 2, 4, 8, 16...)

or even just "positive or negative". This not only speeds analog to digital conversion but, more importantly, it makes the multiplications easier. One just shifts the "binary point". This is done, one way or another, in most commercial instruments. Unfortunately, these procedures distort $G(\tau)$. Not much, in some cases, but possibly enough to be of concern when fitting multiple relaxation processes. One method involves no distortion, but involves yet additional loss of efficiency. Of course, if the functional forms of the signal and noise are known, the distortion may be computed and the signal corrected.

Now the challenge confronting the kineticist is clear. For long $\Delta\tau$, we would like to use 3. For short $\Delta\tau$, where that becomes impossible, we would like to choose either 2 or 4 depending on the type of noise in our signal. We might even be driven to using both compromises, 2 and 4, together.

This is only the beginning of the problem of making judicious trade-offs. Consider procedure 2. One reason for its inefficiency is that if $\delta i(t_1)$ should happen to be zero – and that is the most likely value – then the next 100 $\Delta\tau$ are wasted performing multiplications which just add zero to all the memory locations. We can do better if we do not choose t_1 ahead of time, but wait until $\delta i(t)$ assumes, by chance, a large value. Call that $t = t_1$ and proceed. This sounds like – and is – a fluctuation analysis experiment in which our machine waits for a large – but still noisy – fluctuation, which it records and stores in memory. Then it waits for much such "useable" fluctuations until it accumulates enough to characterize an "average" fluctuation. This trick may be combined with the ideas of procedure 4.

In all the above discussion, I have not mentioned the possibility that if $i(t)$ is derived from a photomultiplier output, it is often possible to process in the photon counting mode. It is not easy to process just $\delta i(t)$ in this case; one must work with $i(t)$. But the entire discussion then applies if one converts "measure $i(t_1)$" into "count the photon arrivals between $i(t_1)$ and $i(t_1 + \Delta\tau)$." The instrumentation and, hence, the compromises needed may, of course, be very different.

A number of detailed treatments have appeared on aspects of the problem of the uncertainty in $G(\tau)$ derived from photocount data. Their results may, in part, be translated into $i(t)$ terminology. I cite four: (DEGIORGIO and LASTOVKA, 1971; KELLY, 1971; HUGHES et al., 1973; KOPPEL and SCHAEFER, 1973).

3. Experimental Results

I will illustrate FCA by summarizing three experimental measurements: (1) the determination of the translational diffusion coefficient D for a fairly small molecule, (2) the evaluation of the kinetic parameters of a bimolecular reaction of the sort Eq. (14), and (3) the characterization of the lateral diffusion of fluorescent tagged lipid molecules in a membrane. The first two were reported together with a brief but essentially complete

Fig. 2. The correlation function, with experimental data points and continuous theoretical curve (discussed in the text), for the fluorescent species rhodamine 6G undergoing free diffusion. The experimental parameters were: $C_D^o = 5.2 \cdot 10^{-10}$mol l^{-1}, L = 0.15 mm, w = 8.5 μm. (Adapted from MAGDE et al., 1974)

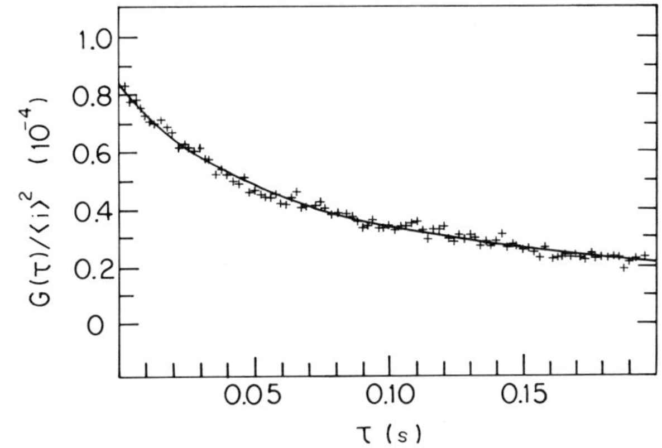

theoretical interpretation some time ago (MAGDE et al., 1972) and fully described subsequently (ELSON and MAGDE, 1974; MAGDE et al., 1974). The latest experiment (HERBERT et al., to be published) has not yet been reported as I write.

The photocurrent autocorrelation function $G(\tau)$ for diffusion of a single fluorescent spcecies D is calculated, as outlined above, by solving Eq. (21) for cylindrical boundary conditions and substituting that solution into Eq. (22) and then Eq. (32). The result is

$$G(\tau) = (g\varepsilon_D Q_D)^2 P^2 L C_D^o (\pi w^2)^{-1} \left[1 + \tau/\tau_D\right]^{-1}. \tag{40}$$

This expression Eq. (40) involves G(0), which we discussed before for FCA, multiplied by a single, τ-dependent factor. The equation is exact and its simple form is the direct result of using the Gaussian profile for the excitation beam. A beam with uniform intensity over a circular profile, for example, leads to a $G(\tau)$ which is a sum of Bessel functions (ELSON and MAGDE, 1974). The characteristic time for diffusion in Eq. (40) is $\tau_D = w^2/4D$. The relative mean square fluctuation is, as given by Eq. (35):

$$G(0)/ \langle i(t) \rangle^2 = (\pi w^2 L C_D^o)^{-1}.$$

Experiments were performed to measure D_D for the laser dye rhodamine 6G (R6G) in water and to demonstrate the parametric dependence of Eq. (40) and Eq. (35). Excitation was by an argon ion laser at 514.4nm.

Results for one particular set of experimental parameters, listed in the caption, are shown in Figure 2. The curve drawn obeys Eq. (40). Two variables, τ_D and G(0), were varied to optimize fit. G(0) is not, in principle, a free variable but may be calculated from Eq. (35) for the particular experimental conditions. However, the parameters of Eq. (35) were only known to about 10% each, while the measured precision of G(0) is much better. The measured and calculated G(0) agreed well within the uncertainty of the latter. From the fit, $\tau_D = 65 \pm 5$ms. From this and the parameter w^2, $D_D = 2.8 \pm 0.7 \cdot 10^{-6} cm^2 s^{-1}$. Again the major uncertainty involved not the measured $G(\tau)$ but w^2.

A series of similar measurements of $G(\tau)$ under different experimental conditions demonstrated:

1. That τ_D varied with w^2 but with no other parameter.

2. That $G(0)/\langle i(t)\rangle^2$ varied as expected with w^2, with L, and with $C_D{}^o$.

3. That changing P over the range 0.25mW to 4mW had no effect.

Also measured was D_D for a second molecule of similar size, the drug ethidium bromide (EB). This molecule had low $\varepsilon_D = 3.8 \cdot 10^3$ and very low $Q_D \sim 7 \cdot 10^{-3}$ and was somewhat photosensitive, and so demonstrated marginal conditions for a FCA-diffusion measurement. D_D was the same as for R6G — which would be expected.

The more dramatic of the original FCA experiments was the measurement of the kinetic constants for a bimolecular chemical reaction — a long-standing and somewhat elusive goal of fluctuation studies. Selected for investigation was the reaction of binding sites (species A) on a DNA strand with the almost nonfluorescent drug ethidium bromide (EB) (species B) to form a fluorescent ligand complex (species C). The reaction presumably involves intercalation of the dye between adjacent base pairs. It was known previously, and verified for the particular materials used, that there is about one site per 2.5 base pairs. Calf thymus DNA was used, broken into short pieces so that there were about 100 available binding sites per strand. Almost all experiments were carried out with very low dye concentrations, so that only about one EB was bound per DNA strand. The kinetics were analyzed according to the simplest possible model Eq. (14).

$G(\tau)$ was calculated by solving Eq. (23) for cylindrical boundary conditions and for the assumed reaction kinetics and substituting the solution into Eq. (22) and then Eq. (32). Although the solution was obtained exactly, the data were discussed in terms of a simplified expression involving assumptions which would apply to most measurements of ligand binding to a macromolecule. The simplified $G(\tau)$ was a series expansion. For our purposes here, I will truncate it even earlier, retaining only the leading term:

$$G(\tau) = \frac{(g\varepsilon_C Q_C P)^2 L C_C{}^o}{\pi w^2}\left[A_o(\tau) + A_+(\tau) + A_-(\tau)\right], \tag{41}$$

where

$$A_o(\tau) = KC_B{}^o\,(1 + \tau/\tau_A)^{-1}$$

$$A_+(\tau) = \frac{KC_A{}^o}{1 + KC_A{}^o} \cdot \frac{1}{1 + \tau/\tau_+}$$

$$A_-(\tau) = \frac{1}{1 + KC_A{}^o} \cdot \frac{\exp\,(-R\tau)}{1 + \tau/\tau_-}\,.$$

The pure diffusion times are $\tau_{A,B} = w^2/4D_{A,B}$ and $\tau_A = \tau_C$. The modified diffusion times are

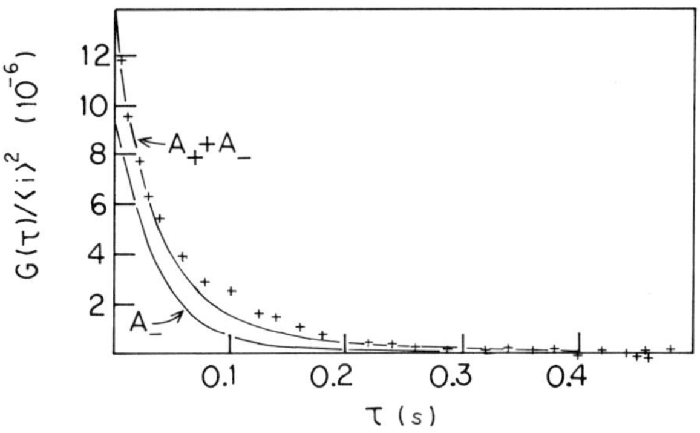

Fig. 3. The correlation function, with experimental data points and continuous theoretical curves, for the chemical reaction of a ligand, ethidium bromide, with DNA to form a fluorescent complex. The conditions were: $C_A{}^o = 2.0 \cdot 10^{-7}$, $C_B{}^o = 2.7 \cdot 10^{-8}$, $C_C{}^o = 3.0 \cdot 10^{-9}$mol ℓ^{-1}; L = 0.15mm; w = 5.7μm. The full correlation function is represented as the sum of two terms, discussed fully in the text. The A_- term in this instance is nearly a simple exponential for the chemical rate. The discrepancy near 0.1 s, due to a cross correlation term, was discussed in the original literature. (Adapted from MAGDE et al., 1974)

$$\tau_+ = \tau_B(1 + KC_A{}^o) \text{ and } \tau_- = \tau_B(1 + KC_A{}^o)/KC_A{}^o. \qquad (42)$$

The relative mean square total fluctuation is still given by Eq. (35): $G(0)/\langle i(t)\rangle^2 = (\pi w^2 L C_C{}^o)^{-1}$. Now, however, this fluctuation magnitude is "divided up among" three relaxation processes, whose amplitudes relative to each other vary as reactant concentrations change.

The three terms of Eq. (41) have physical interpretations. $A_o(\tau)$ involves the diffusion time of the DNA strands into and out of the beam. $A_+(\tau)$ involves the effective time for the diffusion of the small EB molecules into and out of the beam, allowing for the fact that they are slowed by the fraction of the time they spend complexed to the DNA. $A_-(\tau)$ is dominated by the chemical relaxation rate. The $A_o(\tau)$ term conveys no chemical information. It was not given detailed treatment in interpreting the data.

Experimental data points and a curve of best fit are shown in Figure 3 for the $A_+(\tau) + A_-(\tau)$ portion of $G(\tau)$ for a particular set of conditions where chemical relaxation dominated. There were enough free parameters so that a single $G(\tau)$ could not determine all to useful precision. However, by requiring consistency for several measured $G(\tau)$ as $C_A{}^o$ was varied, it was possible to arrive at a unique solution. The results obtained at 22°C were $k_f = 1.8 \pm 0.8 \cdot 10^7$ $s^{-1}mol^{-1}\ell$, $k_b = 20 \pm 7$ s^{-1}, $\tau_B = 0.036 \pm 0.01$ cm^2s^{-1}. The value K can be obtained from k_f/k_b, but was better determined from the variation of τ_+ as given in Eq. (42). This yielded $5.4 \pm 1 \cdot 10^5 mol^{-1}\ell$. This K was consistent with a value obtained from equilibrium fluorescence titration. A conventional temperature - jump study was undertaken for comparison and reasonable agreement noted.

ire kinetic analysis required changing only C_A°. Also var- separate studies, were L, w^2, and C_C° in order to verify e relative mean square fluctuation magnitude varied as expected with all of them, that the diffusion times varied only with w^2, and that R was unaffected (since, at all times, $C_A^{\circ} >> C_B^{\circ}$). Finally there was no effect as P was altered, for P near 1mW.

After proving correct parametric dependence, it was surprising that $G(0)/\langle i(t)\rangle^2$ was consistently too small by a factor nine, well outside the combined uncertainties of the determination of all the experimental parameters. It was suggested that this might be due to an inadequacy in the model Eq. (14) leading to incompleteness in Eq. (41).

The above described measurements show in detail what is involved in CCA experiments. They were designed to test aspects both of the theory and of the instrumentation and not primarily to ascertain molecular parameters. It is likely that the advantages of CCA will prove most significant not for studies in homogeneous solution but for investigations of molecular processes in situ in unusual environments.

HERBERT's investigation, using fluorescent labeled lipid analogs, of lateral diffusion in membranes, is a most exciting step in that direction. A focused laser beam was used to illuminate a micron-sized disk-shaped region on the membrane. The characteristic time for motion into or out of that region is obtained from $G(\tau)$ in close analogy to the above described R6G diffusion experiment. Of particular interest were mixed lecithin-cholesterol membranes. Measurements of the diffusion constant as a function of temperature permitted investigation of phase transitions and phase separations. Aside from the importance at this early stage of our understanding of transport in membranes (for a review, see SINGER and NICHOLSON, 1972) of testing models with multiple, independent techniques, FCA has a unique capability of studying transport in localized areas of a membrane. With some experimental refinement, one might hope to probe local regions of a living cell.

B. Resistance Correlation Analysis (RCA)

A second prime candidate to be considered as a means of monitoring the behavior of concentration fluctuations is (ionic) specific conductivity σ or, equivalently, specific resistivity $\rho = \sigma^{-1}$. Electrical resistance measurements can be extremely sensitive to the existence of minute concentrations as is demonstrated by recent experience in the semiconductor industry with both bulk and surface conductivity and, equally well, by recent physiological studies.

The conductivity of a medium may be expressed as the product of a number of mobile units (electrons, ions, etc.) n times as charge per carrier e times a mobility μ (which is just the mean velocity per unit applied field):

$$\sigma = n\ e\ \mu = \rho^{-1}$$

For polyionic macromolecules e can be many electron charges – and this might be exploitable to yield large fluctuations. For ionic solutions, the total conductivity is the sum of two terms Eq. (43), one each for positive and negative carriers.

OHM's law relates the current i between electrodes to a potential difference V by way of a resistance, V = iR. The resistance R is related to the specific resistivity ρ and the geometry of the medium between the electrodes. For a channel of constant area A and length ℓ, $R = \rho\ell/A$.

An RCA experiment is performed by constraining either V or i to a constant value and measuring the fluctuations $\delta R(t)$ by way of their effect on the other parameter. For the constant current case, $i = I_{dc}$ is fixed and

$$G(\tau) = \langle \delta V(0)\delta V(\tau)\rangle = I_{dc}^2 \langle \delta R(0)\delta R(\tau)\rangle . \tag{44}$$

Then $G(\tau)$ is related to microscopic properties by way of Eq. (43). We will concerned with fluctuations primarily of n, but e and μ might also fluctuate. Either or both, for example, could be sensitive to conformational change of a macromolecule.

Advantages of RCA appear to be the following:

1. Electrical measurements are very highly developed and very widely applicable. V and I themselves can be measured to high precision over many orders of magnitude. By changing the geometry of the measurement, the relation between R and ρ is widely variable. Hence a very wide range of conductivities may be investigated.

2. There is already evidence that RCA is applicable over a wide range of fluctuation magnitudes. One of the experiments I will describe below involved very small fluctuations $(G(0)/\langle V(t)\rangle^2$ 10^{-14} or $\delta V/V \sim 10^{-7})$, much smaller than FCA so far has achieved; while the other achieved very large fluctuations near unity.

3. RCA is extremely specific for processes involving charged species.

4. RCA avoids the problems of photodegradation that plagues FCA.

The principal disadvantages of RCA are:

1. RCA has its own problems with Joule heating. The current itself assumes a role analogous to the incident laser beam power of FCA. In both cases it is advantageous to increase the power dissipated in the observation volume until some destructive process sets in.

2. RCA is limited to the study of processes involving charged species.

3. It may be very difficult to measure rapid fluctuations, es-

easonably large fluctuations are desired to overcome ematic noise. For then n should be small. But small ge R, which in the presence of inevitable capacitance arge RC time constants.

difficult to measure slow fluctuations if one en- noise.

An excellent and demanding test of RCA would involve measuring the chemical dissociation of a weak electrolyte. FEHER (1970) and FEHER and WEISSMAN (1973) have accomplished this. They examined the dissociation of beryllium sulfate, analyzing the kinetics according to the reaction scheme

$$Be^{+} + SO_4^{-2} \underset{k_{21}}{\overset{k_{12}}{\rightleftharpoons}} Be^{+2}(H_2O)SO_4^{-2} \underset{k_{32}}{\overset{k_{23}}{\rightleftharpoons}} BeSO_4 . \tag{45}$$

The equilibrium constants were taken to be $K_o = k_{12}/k_{21} = 3.1\ \ell\ mol^{-1}$, $K_i = k_{23}/k_{32} = 0.7$. The measurements were carried out at a total salt concentration of $0.03\ mol\ \ell^{-1}$.

FEHER and WEISSMAN used a small capillary connecting two reservoirs to define a volume equal to $6.6 \cdot 10^{-6} cm^3$ (corresponding to a total number of ions in the range 10^{14} or relative rms current fluctuations of 10^{-7}!)

A "four terminal" measurement was made: a constant current source (in the range 10^{-4} amps) was connected to a pair of electrodes, one in each reservoir. Then the potential difference between a second pair of electrodes, one in each reservoir, was monitored. The resistance — and particularly fluctuations in the resistance — should be small throughout the reservoirs. But the connecting capillary had a substantial resistance R (in the range 10^6 ohms), which fluctuated as the occupation numbers of ionic carriers $n_1(t)$ each of Be^{+2} and SO_4^{-2} varied in time.

The relation between macroscopic resistance fluctuations $\delta R(t)$ and occupation number fluctuations $\delta n_1(t)$ was treated phenomenologically:

$$\langle \delta R^2 \rangle = \left(\frac{\partial R}{\partial n_1} \right)^2 \langle \delta n_1 \rangle^2 \tag{46}$$

In the spirit of Eq. (43), one would expect $(\partial R/\partial n_1)^2 = (R/n_1)^2$. A test of this under the particular conditions of the experiment showed the need for a small multiplicative correction factor equal to 0.8. Future RCA experiments, therefore, must be cautious on this point. A second correction factor is needed to account for the spreading resistance at the ends of the capillary; this too was small, but was carefully included. Apart from these corrections, which I will ignore here, the mean square fluctuation in voltage expected for an imposed current I_{dc} was simply

$$G(0) = \langle \delta V^2 \rangle = I_{dc}^2 \langle \delta R^2 \rangle = \langle V_{dc}^2 \rangle \langle \delta n_1^2 \rangle / \langle n_1 \rangle^2 \tag{47}$$

The magnitude and time behavior of the fluctuations $\delta n_1(t)$ must then be expressed in terms of the proposed kinetic model and possible hydrodynamic processes. A flow through the channel between reservoirs was one troublesome spurious process, but was carefully eliminated. That leaves diffusion and reaction. According to Eq. (23) a capillary open at the ends together with the sequential reaction kinetics of Eq. (45) would show dissipation involving four relaxation modes. The full mean square fluctuation $\langle \delta n_1{}^2 \rangle = \langle n_1 \rangle$ would be distributed among the modes. FEHER and WEISSMAN did not attempt to solve the full coupled kinetics problem. They utilized a rather large volume ($6.6 \cdot 10^{-6}\,cm^3$) which, moreover, was long ($5.8 \cdot 10^{-2}\,cm$) and narrow. Since diffusion occured only out the ends, characteristic times were long (30 s). Electronic filtering eliminated these slow fluctuations before analysis. They then calculated magnitudes and relaxation rates as though the volume were closed. In this approximation, one should observe two modes of relaxation dominated by the two chemical reaction rates, which are required by the sequential reaction Eq. (45). Of these two modes, only the slower was observed. It was analyzed as a simple exponential. (In fact, the processing was carried out in the frequency domain, so the power spectrum, a single Lorentzian, was recorded and analyzed.) Similarly the mean square amplitude as calculated was predicted to account for about 10% of the total fluctuation.

The experimental result justified neglect of detailed calculation of the full, exact $G(\tau)$. The power spectrum of the one relaxation detected did fit a Lorentziam quite nicely and the magnitude was as predicted. Since concentration was not varied, detailed kinetic parameters were not independently derivable. But the relaxation rate R did agree with a previously determined value. Perhaps most convincing was the succesful extraction of the activation energy (again, in agreement with previous data) from an Arrhenius plot of data recorded over a range of temperature corresponding to a three-fold change in relaxation rate from $\sim 600\,s^{-1}$ to $\sim 200\,s^{-1}$.

Somewhat disappointing was the inability to resolve the fast relaxation mode of Eq. (45). This matters not in the least as far as demonstrating the principle of RCA, but is worrisome in that consecutive reactions like Eq. (45) must be anticipated for ionic dissociative processes and both relaxations are needed to provide an independent check on the validity of the kinetic scheme assumed. The inability to resolve the rapid fluctuation may be due to the high internal resistance of the capillary (10^6 ohms) in conjunction with inevitable capacitive loading. The resistance could be reduced by further increasing the concentration of ions — but the concentration was already rather high. We may be confronting here a fundamental limit of RCA. But recall that the "missing magnitude" in the FCA experiment was also attributed, tentatively, to a similar failure to detect all of the kinetics.

The choice of capillary dimensions and ionic concentrations in an RCA experiment cannot be determined solely to favor observation of the chemical fluctuations. FEHER and WEISSMAN give considerable attention to the problem of Johnson noise which assumes a role very similar to that of shot noise in the FCA experiment.

gh the authors do not point this out, I think it is illu- ng to note that it can be attributed to inherent fluctuations in $\mu(t)$ in Eq. (43). Johnson noise has the correlation function

$$G_j(\tau) = \langle \delta V(0)\delta V(\tau)\rangle = 4kTR \exp(-\tau/\tau_c) \tag{48}$$

Here k is Boltzmann's constant and T is the absolute temperature. The inherent correlation time τ_c of the Johnson noise is so short ($<10^{-13}$ s) that the actual spectrum is determined by the measurement circuity. Johnson noise, like shot noise, is uncorrelated. The measure of the feasibility of an RCA experiment becomes the ratio γ of the two voltage fluctuation Eq. (47) and Eq. (48).

$$\gamma = \frac{I_{dc}^2}{4kT}\frac{\langle \delta R^2\rangle}{R} \approx \frac{I_{dc}^2}{4kT}\frac{\langle \delta n_1{}^2\rangle}{n_1} \approx \frac{I_{dc}^2}{4kT} \tag{49}$$

So we have the simple result that we must increase the impressed current I_{dc} as far as possible. This has an easy interpretation in terms of Eq. (43): Since μ is essentially the carrier velocity, increasing the velocity diminishes the relative fluctuation in μ. Interesting, γ is independent of R and n_1 in Eq. (49). This is because R is proportional to $n_1{}^{-1}$. FEHER and WEISSMAN write Eq. (49) using the full form of Eq. (46). Then there is some variation of γ with n_1, but it would seem to be secondary, not fundamental.

Furthermore, my γ in Eq. (49) refers to the ratio of the total fluctuation of n_1 in an open volume. Only a fraction of that magnitude is involved in the chemical mode of relaxation (cf. my Eq. (41) or the treatment of FEHER and WEISSMAN from the point of view of a closed volume). That fraction depends very much on n_1. The point of Eq. (49), then, is that γ remains constant as we vary n_1 to maximize the fraction of the fluctuation which involves the chemical term.

FEHER and WEISSMAN lay considerable stress on the technical, but important, fact that to maximize I_{dc} it is necessary to maximize the transfer of heat to the walls of the capillary. A long narrow channel is indicated.

It is impressive that FEHER and WEISSMAN were able to design their experiment to achieve $\gamma>1$. This is not necessary, of course, as I discussed earlier, leading to Eq. (36). The FCA experiments were carried out with the equivalent ratio (signal to shot noise) considerably less than one. On the other hand, the instrumentation used by FEHER and WEISSMAN was not optimal for working in the limit $\gamma<<1$. They used a spectrum analyzer. In that case, uncorrelated noise appears as a constant at all frequencies; whereas in a correlator, it appears as a spike at $\tau = 0$. Although in principle there is no difference, in practice the spectrum analyzer, if it is basically an analog device, may be more susceptible to problems of nonlinearities.

FEHER and WEISSMAN discuss a variety of possible spurious noice sources. One which they did not discuss originally involves excess

1/f noise, which is clearly a matter for concern. More recently they did direct attention to this problem (WEISSMAN and FEHER, 1975).

To summarize this aspect of RCA, a capability to measure at least some parameters of the dissociation of an electrolyte has been convincingly demonstrated. However, a strong motivation for using RCA for such studies on a homogeneous bulk sample was not intended and has not been made apparent. But it is likely – as seemed probable aslo for FCA – that fruitful exploitation of the principle will follow extension to heteregeneous systems – films, membranes, layers, and so on. Indeed, just such a study has very recently been reported (ZINGSHEIM and NEHER, 1974).

This study was again an RCA experiment, although ZINGSHEIM and NEHER employed a constant potential V_{dc} = 100mV and detected the fluctuating i(t) as the resistance R(t) between the electrodes changed. In this case again the resistance fluctuated in response to a chemical reaction, specifically assumed to be a dimerization

$$A + A \rightleftharpoons A_2. \tag{50}$$

Unlike the example of RCA discussed above, in this case the $\delta R(t)$ has nothing to do with the number of ions (typically NaCl at 1 mol ℓ^{-1}) in solution. They serve only as a probe – and one offering tremendous amplification of a chemical process. A black lipid membrane was positioned over a small ($\sim$1mm) opening connecting two reservoirs in which were located the measuring electrodes. The resistance offered by the membrane was high. Gramicidin A was added to one reservoir. Some of the gramicidin A molecules entered the membrane. This was observed to reduce R. The interpretation (well-established before this study) is that in the membrane the gramicidin A molecules are in equilibrium between monomers and dimers in accord with Eq. (50) and that the dimers (and only the dimers) facilitate conduction through the membrane – perhaps by forming a pore to serve as an ionic channel. Detection of fluctuations in the ionic current by measurement of its autocorrelation function, then, permits a determination of both the number of channels present and the persistence time of each channel, which should be consistent with the hypothesized mechanism Eq. (50).

Particularly pleasing in the experiment of ZINGSHEIM and NEHER was their detection of the formation and dissociation of a single molecular dimerization! One channel gave an ion current of order 10^{-12} amp corresponding to a total R $\sim 10^{11}$ ohm. In this case, of course, δR was also $\sim 10^{11}$ ohm. The lifetime of single channels was measured to be of order 1s.

Extension of measurements to conditions under which many channels were present demonstrated that the conductivity no longer behaved as a sum of independent channels.

The authors discussed and excluded the lateral diffusion of gramicidin A in the membranes as a detectable process. They did observe convective phenomena. They concluded that their results were consistent with the dimerization mechanism but not sufficient lude other possibilities. They, correctly in my view, argued nary that the two-dimensional geometry of membranes and interfaces makes them particularly well-suited to study by CCA. This is not a revolutionary idea. The magnitude of G(O) in such systems, as opposed to the full G(τ), has long been studied. But the full G(τ) should be much more informative.

Despite my obvious enthusiasm for RCA studies of membrane conductivity, I have here confined my discussion to reviewing a particular, successful example of such a study and have not formulated a general analysis of fluctuations in such systems. There are many considerations involved in such a formulation. And it will be more useful to the reader if I refer him directly to a paper by CHEN and HILL (1973), who have independently formulated a treatment of membrane conductivity very similar in spirit and even in notation to that employed for other systems earlier in this chapter.

C. Absorbance Correlation Analysis (ACA)

UV-visible absorption measurements are perhaps the most common and convenient probe for chemical kinetic studies, particularly for characteristic times in the range we are interested in here, roughly microseconds to seconds. Absorption methods are reasonably sensitive, particularly when we are concerned with low concentrations (as oposed to small numbers); for then it is often possible to compensate with long pathlengths. But absorption methods are not generally regarded as possessing the exceptional sensitivity to very small numbers of molecules which CCA requires. In comparison with FCA, the essential difference is that with FCA one detects a few photons against a background which, in the absence of blank fluorescence, should be zero. With ACA, one seeks to measure small changes in a very intense beam. One cannot simply try to use fewer and fewer molecules to obtain larger relative fluctuations. The situation, however, is by no means hopeless.

I will demonstrate that systematic noise seems likely to be a greater problem than with FCA, but that this may be partly compensated by reduced problems with shot noise.

No ACA experiment has succeeded, or even been attempted, so far as I am aware. No detailed calculation has been presented for any proposed experiment. Such a calculation, however, follows very closely the treatment of FCA, as outlined above in Eqs. (24-35), and we can quickly write down the relevant expressions in order to check the feasibility of such a technique. I assume just one absorbing solute j. For the reason just mentioned, we anticipate that the optimum configuration may not involve optically thin samples, so we define an incident total laser power (in photons per second) P_i and a transmitted power P_t:

$$P_t = P_i \exp(-2.3\ \varepsilon_j C_j L) \tag{51}$$

Our photodectector will intercept all P_t, so

$$i(t) = gP_t(t). \tag{52}$$

Here g involves only the quantum yield of the detector. If we can use a silicon photodiode, g may exceed 0.5! (This 20-fold increasing compared with FCA implies correspondingly reduced problems with photodegradation or, for comparable degradation, an improvement of $(20)^{1/2}$ in our S/N.) As before, the primary experimental observation is the autocorrelation of the fluctuations of the cathode current $G(\tau) = \langle \delta i(0)\ \delta i(\tau)\rangle$. To test the feasibility of ACA, we seek conditions which maximize the relative mean square fluctuation, which is

$$\frac{G(0)}{\langle i(t)\rangle^2} = (2.3\ \varepsilon_j C_j{}^{o} L)\ \frac{2.3\ \varepsilon_j}{\pi w^2 N_o}. \tag{53}$$

This says that the optical density $\varepsilon_j L C_j{}^{o}$ should be large, that the probability of interaction between photons and molecules ε_j should be large, and that the beam area should be as small as possible. Recall that N_o is Avogadro's number. This is plausible: if we wish to measure few molecules, we should use a long path length and a small cross sectional area.

But will the relative fluctuations really increase for increasing $C_j{}^{o}L$? Yes, the relative current fluctuations will (as opposed to the relative number fluctuations) but each fluctuation becomes poorly characterized as the number of photons reaching the detector falls. The limit on the permissable optical density is set by the same requirement that we encountered before – that to keep uncertainty of the measurement comparable to the magnitude of the concentration fluctuation being measured, we would like the number of electrons emitted by the cathode $N_e = \langle i(t)\rangle\ \Delta\tau$ to be equal to or somewhat greater than the number of molecules under observations:

$$gP_i \Delta\tau\ \exp\left[-2.3\varepsilon_j C_j{}^{o} L\right] > \pi w^2 L C_j{}^{o} N_o. \tag{54}$$

Although this appears formidable, it actually is not very restrictive. For milliwatt lasers and millisecond times, quite ordinary optical densities (1 to 2) are appropriate. Equation (53) shows that there is little harm in erring by using slightly too low a $C_j{}^{o}L$. Besides, up to a point, that will help reduce shot noise. The main use of Eq. (54) will be to check that the inequality is not so badly violated as to make the uncertainty of measurement overwhelming. This will be a problem primarily for short $\Delta\tau$. The main point of Eq. (53) is that we must not use extremely transparent samples. There is a broad middle ground.

With ε_j fixed by the system of interest and $C_j{}^{o}L$ determined by Eq. (54), the only variable really available in Eq. (53) for optimization is πw^2, and even this may be determined by constraints on characteristic diffusion times. Let us, then estimate the relative mean square fluctuations for possible but favorable parameters: $\varepsilon_j = 10^5 \mathrm{cm^{-1} mol^{-1}} \ell$, $\pi w^2 = 10^{-6} \mathrm{cm}^2$, $\varepsilon_j C_j{}^{o} L = 2$. Then $G(0)/\langle i(t)\rangle^2 \sim 0.16 \cdot 10^{-8}$. Equivalently, the rms current fluctuations are $0.4 \cdot 10^{-4}$. These are considerably larger fluctuations than were studied in the electrolyte dissociation RCA experiment.

act, they compare with some conditions of the FCA experiment.

One more point must be raised. Diffraction limits L, for a given πw^2 and wavelength of light, to $L < (2\pi/\lambda)w^2$. For $\pi w^2 = 10^{-6}cm^2$ and visible light, $L < 3 \cdot 10^{-2}$cm. This means, for the above parameters, $C_j^{\circ} = 6.7 \cdot 10^{-4}$mol ℓ^{-1}. This is certainly reasonable for moderately small molecules. The principle limit, as I see it, is that we have been rather optimistic in setting $\varepsilon_j = 10^5$. Even so, diligence will be required to reduce spurious modulation of the transmitted power to the requisite low levels. Dissipation as heat of the absorbed light may be a problem.

D. Quasi-Elastic Light Scattering (QELS) and Turbidity Correlation Analysis (TCA)

Quasi-elastic light scattering depends on variations in the refractive index of a medium. When these variations are time dependent fluctuations, QELS may be employed to make kinetic measurements. This has become a highly developed and fruitful area of physical biochemistry and has been reviewed regularly and recently (PECORA, 1972; BERNE, 1973; FLEURY and BOON, 1973).

Conventional QELS experiments are deliberately designed not to be sensitive to occupation number fluctuations. Recent work has demonstrated that a change in experimental design permits QELS to be used to study occupation number fluctuations, which I will term turbidity correlation analysis (TCA), and that this extends the range of molecular – and cellular – properties which may be studied by QELS.

The advantages of QELS and TCA are:

1. Widespread applicability for molecules and particles sufficiently large to scatter light (roughly, of dimensions comparable to a wavelength, but extendable with patience to dimensions about two orders of magnitude below that) without regard to chemical composition.

2. Greatly reduced problems with photodegradation and thermal dissipation of absorbed energy or Joule heating.

The disadvantage is just the converse of 1 above: most objects in the beam appear pretty much the same; it is hard to distinguish among multiple constituents or even to distinguish the objects of interest from dust or other accidental contamination. From this point of view, biological fluids are not the most desireable solutions to study by QELS.

To understand QELS, consider a small object, with at least one dimension comparable to the wavelength of light, placed in a laser beam. The object might be a slit or a wire, a pinhole or a tiny disk, a viral particle, a spermatazoon, a DNA strand, or a globular protein. The object need not be opaque – any difference in refractive index will do. In any of these cases, we would observe a diffraction pattern of scattered light.

If we now place a large number of these objects in a regular array in the beam, we will see a regular pattern of interference fringes — bands or spots, depending on the symmetry of the array, of light and dark — made up by the superposition of interfering diffracted — or scattered — light from all the objects. If the objects are placed in the beam at random, rather than in an ordered array, there is still an interference pattern of spots of light and dark, only now the pattern is random — speckled — rather than regular. Finally, if the objects move, the speckled pattern, of course, changes. A photodetector may be positioned in the pattern; ideally, it should have an area just about large enough to encompass an "average" speckle spot of light. Then as the objects move, shadows, as it were, move across the detector. A correlation analysis of the photocurrent reveals a characteristic persistence time for the passing of light and dark regions over the detector. This characteristic time is directly related to the rate of motion of the objects in the beam. If there are different types of objects, there will be multiple times. The motion may be diffusive or otherwise. Over what distance must the objects move to change the interference pattern? Half the wavelength of the light is the answer. The fact that occasionally an object enters or leaves the beam is irrelevant to the QELS experiment I have just described.

But if we reduce the number of objects in the beam far enough, eventually the changes in the total scattered light due to changes in the number of particles in the beam become detectable as an additional modulation added to the moving interference shadows. The photocurrent autocorrelation function then shows both relaxation processes. This relatively complex case was treated by SCHAEFER and BERNE (1972) and by SCHAEFER and PUSEY (1972).

It is possible, however, to change the experimental design to suppress the interference fluctuations and emphasize the number fluctuations. By making the photodector large enough to encompass many speckle spots (coherence areas), the modulation of the signal current as the spots move is reduced. The modulation of the total scattered light by number fluctuations is unaffected. In fact, this can be carried to the point of enveloping the illuminated volume with a reflector to collect all the scattered light. This, of course, leads to problems with "blank scattering" from cell walls. The usual collection optics for QELS involves focusing the illuminated region onto a plane in which is located a slit. The observation volume V is determined in cross section by the laser beam and in length not by the cell but by the occulation of the image by the slit. With care, this gives excellent rejection of stray light. For QELS, where a single coherence area is studied, that "focusing" simply involves a pinhole camera. For number fluctuation studies, one would use a large aperture lens.

There remains one more possible experimental design. Rather than detect the total scattered light, one might detect modulation of the transmitted beam as scatteres move into or out of the beam. The considerations parallel those of ACA discussed above.

of these QELS experiments modified to emphasize measure- what is essentially just the time modulation of the total ed light, the name turbidity correlation analysis (TCA) is descriptive.

The two procedures, QELS and TCA, can provide complementary length scales. The former could be used to probe, for example, bacterial motility on a short scale (set by the wavelength of light), and the latter to investigate longer range behavior (determined by beam dimensions). SCHAEFER (1973) has successfully used TCA to show that bacteria move as random walkers and to characterize their mean step length by adjusting the dimensions of his scattering volume until coincidence with step length was achieved. There are a large number of investigations by QELS of the motion of everything from swimming "bugs" to diffusing proteins to ions moving under the influence of an electric field (electrophoresis). (For reviews see: PECORA, 1972; BERNE, 1973; FLEURY and BOON, 1973).

Granted that for analyzing particle and molecular motion QELS has had undisputed success and TCA is beginning to show promise, does either technique have application to the study of chemical reaction rates? This question has received both theoretical (BLOOMFIELD and BENBASAT, 1971, and references therein) and experimental (YEH and KEELER, 1969a,b; YEH, 1970) attention.

There are two possible approaches: one might distinguish chemical reactants – and therefore be able to monitor reaction kinetics – by way of changes in the refractive increment of the molecules. This corresponds to the changes in other indicators exploited by other CCA methods: ionic charge for RCA, absorptivity for ACA, and either absorptivity or fluorescence yield or both for FCA. However, refractive index is rather insensitive to the chemical state of a molecule. A second possibility involves detecting chemical reaction indirectly by its influence on molecular mobility. This possibility, of course, applies also to all other forms of CCA. But it is possible that assembly of units into an oligomer and other similar reactions, which would show up reasonably well as changes in diffusion constants, are particularly interesting over the range of molecular sizes for which QELS is at its best.

E. Orientation Correlation Analysis (OCA)

Consider a molecule in solution. At a particular time it has a certain orientation. If we continue to watch, its orientation will wander due to rotational Brownian motion. From a time record just of the orientation, we can evaluate the molecule's rotational diffusion constant or constants. These in turn give us information about the molecular size and shape. Since rotational diffusion constants vary much more rapidly with molecular size and shape than do translational diffusion constants, they are much more characteristic indicators of specific molecular properties. This can be useful in itself as a structural probe. In addition, this sensitivity means that changes in rotational diffusion constants can serve as reasonably sensitive indicators of chemical reaction.

If the molecule under observation isomerizes or dissociates, or whatever it does, that fact will be much more noticeable in its rotational diffusion than in its translational diffusion.

Conversely, if the molecular diffusion constants are known, a study of the time dependence of the random reorientation of the observed molecule yields information about the viscosity of the fluid medium. This may be of limited interest for homogeneous fluids, but it offers a very exciting prospect in probing the local environment of membrane-bound proteins or molecules otherwise located in unusual environments. In such circumstances the molecular reorientation dynamics might not even follow Brownian diffusion laws. A study of the orientation dynamics would reveal this. Finally, we should note that it is sometimes possible to resolve the orientation of a part of a molecule relative to the remainder.

It is impractical to measure the orientation in time of a single molecule. When we have an ensemble in solution, there is on average no preferred direction. Again thermodynamic fluctuations can be exploited. If we measure carefully at a particular time, we will find a small excess of molecules aligned in some direction. This fluctuation must dissipate subsequently according to the rotational dynamics of the molecule under study. (I assume now that no chemical reactions occur and that translational diffusion out of the beam is much slower than rotational motion). All we need is an experimental probe which can translate the number of molecules aligned along a particular orientation into a time dependent electric current. Two groups have recently demonstrated in detail how fluorescence intensity measurements can be used to do just this (EHRENBERG and RIGLER, 1974; ARAGON and PECORA, 1975). Absorbance fluctuations of a polarized probe beam could also be used, in principle. Probably the easiest technique, for relatively large objects, is QELS. Both QELS and TCA can be carried out with polarized light to monitor the orientation of large molecules or other objects which have anisotropic polarizability tensors – in effect, any anisotropic species – without the need for natural or labeled fluorescence. This has already been demonstrated for diffusive reorientation of tobacco mosaic virus and DNA (CUMMINS et al., 1969; SCHMITZ and SCHURR, 1973; SCHURR and SCHMITZ, 1973) and for reorientation of tobacco mosaic virus driven by fluid turbulence (JOHNSON and WEBB, 1972). In the former case, the properties of the particles were the focus of attention; in the latter it was the properties of the medium which were sought, studied by their influence on the particle rotation.

These studies all involve orientation correlation analysis (OCA). By this time, I trust that I have made my point that analogous principles appear repeatedly. It will suffice to consider briefly just one case, that of fluorescence OCA (FOCA).

There are several possible experimental arrangements. Consider first a design in which a polarized light beam illuminates a very small volume containing the molecules of interest. The probability that any particular molecule absorbs a photon depends on the angle between the polarization vector of the light and the transition dipole moment of the molecule (which has some fixed direction in the molecule). Therefore, if a fluctuation occurs such that extra molecules are "favorably" aligned, more photons are absorbed. If the molecules are fluorescent, this extra absorption will lead to extra fluorescence over the next 10^{-8} s. The experiment depends, really, just on the anisotropy of absorption during the fluctuation; the fluorescence serves only to measure the extent of that absorption. The polarization and the time dependence of the fluorescence need not be measured. This experiment is capable of measuring rotional diffusion on any time scale, unlike two classical optical procedures (time dependence of the anisotropy of emission following pulsed, polarized excitation; and magnitude of the emission polarization ratio for steady, polarized excitation), both of which are only effective when the orientation correlation time is comparable to the fluorescence life time. In fact, FOCA will be easier to implement for slow rotation because of the fact discussed in the section on FCA that efficient evaluation of the correlation function requires that we detect about one, or preferably more, photons per molecule per characteristic time. An estimate of the excitation flux necessary to do this is still given by Eq. (38).

The experiment works equally well, in principle, for unpolarized excitation and polarization-sensitive detection of the emission. However, in practice this would appear to be much more difficult to implement with high efficiency due to the problem of constructing a polarization sensitive detector which collects emission in all directions. Finally, it is also possible to use both polarized excitation and polarized detection.

Besides the complications of these various experimental configurations; there is some complexity in relating the rotation of an arbitrary rotor to an experimentally measured current autocorrelation function. General expressions are not particularly transparent. The two studies referred to just above describe various particular instances and are, to some extent, complementary as regards the situations they choose to emphasize and, correspondingly, the simplifying assumptions they must introduce. These have appeared very recently as I write, and I refer the reader interested in the detailed expressions for the correlation functions to the original reports.

IV. Summary and Conclusions

In this chapter, I have offered an introduction to concentration correlation analysis. In this new experimental technique, the thermodynamic fluctuations in the numbers of various chemical species within a small observation volume are analyzed. The magnitude of the fluctuations is a direct measure of the number of independent molecules (or other objects) engaged in a particular process, while the persistence time of the fluctuations is determined by the dynamics of the process — reaction rate constants, translational or rotational diffusion constants, velocity of fluid flow, and so on.

I emphasized the assumptions made and the principles of the derivations involved in recent calculations of $G(\tau)$ — the autocorrelation function, which is the sole experimental measure of the fluctuations. I identified the references where various $G(\tau)$ are calculated. Since, in general, exact expressions are somewhat complicated and different approximate expressions are required for specific experimental conditions, I did not catalog here all the presently known forms of $G(\tau)$. Two specific cases were illustrated in relation to actual experimental data obtained by fluorescent monitoring of the fluctuations.

I discussed the problem of detecting small fluctuations and emphasized the generality of the principle by describing methods based on fluorescence, resistance, absorbance, and light scattering. More briefly, I outlined the potential of studies utilizing the polarization properties of any of the three optical methods. Of these, all but the absorbance approach have already been successfully implemented in one form or another. For the first three methods, I offered quantitative guides to calculating the statistical uncertainty (signal to noise ratio) of proposed experiments. I stressed also the importance of efficient processing of $G(\tau)$ from the primary experimental record so as not to degrade the statistics. Finally, I compared and contrasted the first four methods, listing both the inherent limitations of each and the likely systematic noise. The various polarization methods are unique in the information they yield about rotational motion, so they are not directly comparable. Their relative advantages parallel those of the basic methods from which they are derived. I took pains to distinguish number fluctuations detected by light scattering from the now classical quasi-elastic light scattering. The latter measures the persistence of an interference pattern and, therefore, measures motion on the scale of optical wavelengths, independent of the size of the illuminated volume.

The characterization of chemical reaction kinetics by both fluorescence and resistance methods seems convincing. Results so far, however, have not indicated advantages of these methods over conventional means for homogeneous solutions. However, both fluorescence and resistance methods have been applied to the study of membrane properties with, apparently, more significant adantages. The fluorescence method can probe molecular mobility over small regions of membranes including, presumably, selected regions. The resistance method can monitor the kinetics of ion-channel formation. Number fluctuation measurements by light scattering only occasionally offer advantages over ordinary light scattering. In those instances, which generally involve entire cells rather than single molecules, they do seem to have demonstrated utility, with the alternative being not any of the methods discussed here but such devices as automatic tracking microscopes.

Given the fairly rapid proliferation of schemes for implementing concentration correlation analyses, together with the progress that may be anticipated in instrumentation, the prospects for correlation analysis maturing to widespread utility appear bright. However, one must expect that comparable resources, time, and effort extended to perturbation methods can significantly improve the state of that art, as indeed other chapters of this volume prove. Hence, it seems very likely that concentration correlation methods will be most useful for the study of system where their inherent requirement for small numbers of reactants is a particular advantage. Of all such systems, the most tractable will be those for which a chemical "amplification" precedes experimental detection. We saw one example of this in the membrane permeability study. Similar features might characterize catalytic phenomena, in which changes in a very few active units, enzymes, would be detected not by direct observation but by their effect on the concentrations of substrates. Concentration correlation analysis will prove its full potential if it can move microkinetic studies out of the chemistry laboratory and into the physiology laboratory.

References

ARAGON, S.R., PECORA, R.: Fluorescence correlation spectroscopy and Brownian rotational diffusion. Biopolymers 14, 119-137 (1975)

BERNE, B.J.: Chemical and biological applications of laser light-scattering. Acc. Chem. Res. 6, 318-322 (1973)

BLOOMFIELD, V.A., BENBASAT, J.A.: Inelastic light scattering study of macromolecular reaction kinetics. I. The Reactions $A \rightleftharpoons B$ and $2A \rightleftharpoons A_2$. Macromolecules 4, 609-617 (1971)

CALLEN, H.B., GREENE, R.F.: On a theorem of irreversible thermodynamics. Phys. Rev. 86, 702-710 (1952)

CHEN, Y.: Fluctuations and noise in kinetic systems. II. Open ensembles at equilibrium and steady state. J. Chem. Phys. 59, 5810-5813 (1973)

CHEN, Y., HILL, T.L.: Fluctuations and noise in kinetic systems, application to K^+ channels in the squid axon. Biophys. J. 13, 1276-1295 (1973)

CUMMINS, H.Z., CARLSON, F.D., HERBERT, T.J., WOODS, G.: Translational and rotational diffusion constants of tobacco mosaic virus from Rayleigh linewidths. Biophys. J. 9, 518-546 (1969)

DEGIORGIO, V., LASTOVKA, J.B.: Intensity – correlation spectroscopy. Phys. Rev. A4, 2033-2050 (1971)

EHRENBERG, M., RIGLER, R.: Rotational Brownian motion and fluorescence intensity fluctuations. Chem. Phys. 4, 390-401 (1974)

ELSON, E.L., MAGDE, D.: Fluorescence correlation spectroscopy. I. Conceptual basis and theory. Biopolymers 13, 1-27 (1974)

FEHER, G.: Determination of kinetic parameters from the frequency spectrum of fluctuations. Biophys. Soc. Abstracts 10, 118a (1970)

FEHER, G., WEISSMAN, M.: Fluctuation spectroscopy: determination of chemical reaction kinetics from the frequency spectrum of fluctuations. Proc. Nat. Acad. Sci. (Wash.) 70, 870-875 (1973)

FLEURY, P.A., BOON, J.P.: Laser light scattering in fluid systems. In: Advances in Chemical Physics. PRIGOGINE, I., RICE, S.A. (eds.); Vol. XXIV, pp. 1-93. New York: Interscience 1973 (This very broad survey includes an extensive bibliography.)

GREENE, R.F., CALLEN, H.B.: On the formalism of thermodynamic fluctuation theory. Phys. Rev. 83, 1231-1235 (1951)

GREENE, R.F., CALLEN, H.B.: On a theorem of irreversible thermodynamics. II. Phys. Rev. 88, 1387-1391 (1952)

HERBERT, T.J., WEBB, W.W., ELSON, E.L.: Lateral diffusion in lipid membranes, in preparation (1976)

HUGHES, A.J., JACKEMAN, E., OLIVER, C.J., PIKE, E.R.: Photocorrelation spectroscopy: dependence of linewidth error on normalization, clip level, detector area, sample time, and count rate. J. Phys. A: Math., Nucl., Gen. 6, 1327-1336 (1973)

JOHNSON, D.H., WEBB, W.W.: Measurement of the rate of strain tensor in a turbulent liquid using light scattered from asymmetric particles. Bull. Am. Phys. Soc. 17, 1084 (1972); additional results in preparation.

KELLY, H.C.: A comparison of the information-gathering capacities of photon correlation devices. IEEE J. Quantum Electron. QE-7, 541-550 (1971)

KOPPEL, D.E.: Statistical accuracy in fluorescene correlation spectroscopy. Phys. Rev. A-10, 1938-1945 (1974)

KOPPEL, D.E., SCHAEFER, D.W.: Scaled photocount correlation of non-gaussian scattered light. Appl. Phys. Lett. 22, 36-37 (1973)

MAGDE, D., ELSON, E., WEBB, W.W.: Thermodynamic fluctuations in a reacting system-measurement by fluorescence correlation spectroscopy. Phys. Rev. Lett. 29, 705-708 (1972)

MAGDE, D., ELSON, E.L., WEBB, W.W.: Fluorescence correlation spectroscopy. II. An experimental realization. Biopolymers 13, 29-61 (1974)

PECORA, R.: Quasi-elastic light scattering from macromolecules. Ann. Rev. Biophys. and Bioeng. 1, 257-276 (1972)

SCHAEFER, D.W.: Dynamics of number fluctuations; motile microorganisms. Science 180, 1293-1295 (1973)

SCHAEFER, D.W., BERNE, B.J.: Light scattering from non-gaussian concentration fluctuations. Phys. Rev. Lett. 28, 475-478 (1972)

SCHAEFER, D.W., PUSEY, P.N.: Statistics of non-gaussian light. Phys. Rev. Lett. 29, 843-845 (1972)

SCHMITZ, K.S., SCHURR, J.M.: Rotational relaxation of macromolecules determined by dynamic light scattering. II. Temperature dependence for DNA. Biopolymers 12, 1543-1564 (1973)

SCHURR, J.M., SCHMITZ, K.S.: Rotational relaxation of macromolecules determined by dynamic light scattering. I. Tobacco mosaic virus. Biopolymers 12, 1021-1045 (1973)

SINGER, S.J., NICOLSON, G.L.: The fluid mosaic model of the structure of cell membranes. Science 175, 720-731 (1972)

WEISSMAN, M., FEHER, G.: Observation of energy (thermal) fluctuations in an electrolyte solution. J. Chem. Phys. 63, 586-587 (1975)

YEH, Y.: Experimental study of reaction kinetics by light scattering. II. Helixcoil transition of the copolymer deoxyadenylate-deoxythymidate (dAT). J. Chem. Phys. 52, 6218-6224 (1970)

YEH, Y., KEELER, R.N.: Experimental study of reaction kinetics by light scattering. I. The polarized Rayleigh component. J. Chem. Phys. 51, 1120-1127 (1969a)

YEH, Y., KEELER, R.N.: New Probe for reaction kinetics — The spectrum of scattered light. Quart. Rev. Biophys. 2, 315-349 (1969b)

ZINGSHEIN, H.P., NEHER, E.: The equivalence of fluctuation analysis and chemical relaxation measurements: A kinetic study of ion pore formation in thin lipid membrane. Biophys. Chem. 2, 197-207 (1974)

Dynamics of Substitution at Metal Ions

D. N. HAGUE

I. Introduction

Metal ions are involved in many biochemical reactions. It is well known that all living things have a requirement for the so-called "trace elements", many of which are metals, and it has been estimated that as many as 27% of enzymic reactions either require or are greatly enhanced by the presence of one or more metal ion. But when the list of these reactions is inspected, it is found that the number of metals concerned is small. In Table 1 are listed the metals which are known to be involved in enzymic reactions and, although some allowance must be made for the large number of reactions of this type as yet uncharacterized, the general picture is unlikely to change as our knowledge increases. Four cations figure more prominently than any others: the alkali metals Na^+ and K^+ and the alkaline earths Mg^{2+} and Ca^{2+}. All four share

Table 1. Metal ions in proteins and enzymes

Metal	Systems	Occurrence
Na^+, K^+, Mg^{2+}, Ca^{2+}	Involved in muscle and nerve action, the activation of numerous enzyme reactions, etc.	Very widespread; together constitute some 99% of the total metal content of the human body
Fe	Haem-proteins, oxidases, etc.	>100 known examples
Zn^{2+}	Peptidases, phosphatases, etc.	> 50 known examples
Mn^{2+}	Decarboxylases, dehydrogenases, etc.; can frequently replace Mg^{2+} in vitro	> 50 known examples
Cu	Oxidases, "blue" proteins, etc.	> 50 known examples
Co	B_{12} enzymes; Co^{2+} can frequently replace Zn^{2+} in vitro	~ 20 known examples
Mo	Xanthine oxidase, nitrogen metabolism, etc.	~ 10 known examples
(Li^+, V, Cr, Cd^{2+}, Ni^{2+} and a few others		very few known examples)

with the sixth-placed metal Zn^{2+} the unfortunate property of being, as it were, excessively modest: their lack of suitable spectral and magnetic properties has made it necessary to resort to such techniques as atomic absorption spectroscopy to detect them at all and no reliable method has yet been found for characterising their immediate environment when they are in solution. Much of our knowledge of their role in biological reactions has resulted from such stratagems as replacing them with appriopriate coloured (for example, Co^{2+} for Zn^{2+}) or paramagnetic (for example, Mn^{2+} for Mg^{2+} and tervalent lanthanides for Ca^{2+}) ions and the extent of our ignorance can, perhaps, be appreciated when it is realised that the nature and, indeed the number of the binding groups for the metal is unknown in all but a very few cases. It is not the purpose of the present Chapter to consider the general topic of metals in biological systems (for discussions of which the reader is referred to the admirable reviews of WILLIAMS (1970), MILDVAN (1970), HUGHES (1972), and others) but rather to concentrate on those aspects which have been studied by chemical relaxation techniques. It is, however, important to appreciate at the outset the extent of the problems involved in trying to obtain a detailed kinetic picture of these reactions and to see why, in the following pages, such frequent recourse is had to model systems. The models are not generally designed to mimic, let us say, the behaviour of the metal in an enzyme (since in such an attempt we are almost certain to fail) but rather to try to obtain an indication of the type of behaviour the metal is likely to adopt in the sort of environment found in the living system.

The fundamental property of metal ions which leads to their prevalence in biochemical reactions is that they are Lewis acids, or electrophiles, and can therefore accept lone pairs of electrons from suitable donors, thereby forming complexes. In some cases the prime purpose of the metal seems to be to hold a macromolecule in a particular configuration by combining with basic groups which may be widely separated from one another in the primary structure. (An additional feature may be further to polarise one or more of the bonds in the macromolecule, thereby rendering them more susceptible to nucleophilic attack.) A structure of this type has a long lifetime compared with the half-life of the enzymic reaction and therefore chemical relaxation (or any other kinetic technique) is not relevant in a study of the part played by the metal. We shall be concerned with the other group of systems; those in which the nature of the inner coordination sphere of the metal changes on a time scale comparable to, or shorter than, that of the overall chemical change. Normally, the formation of the metal-macromolecule complex will be followed by at least one other step, such as the transfer of an electron or the reaction of a coordinated ligand, but we shall have little to say about steps other than complex formation and dissociation. This omission is not through prejudice but a lack of knowledge. Redox reactions, for example, tend to have associated with them large free energy changes (ΔG^{o}) and small enthalpy changes (ΔH^{o}), making them unsuitable for study by the temperature-jump technique. Although there is no reason why coupling with another reaction (such as proton transfer) should not be employed so as to make the effective thermodynamic parameters favourable, only very few redox reactions have been studied by chemical relaxation (see, for example, HALPERN et al., 1963).

Complex formation for most univalent and bivalent and for many tervalent metal ions occurs rapidly; it is, however, in many cases markedly slower than a diffusion-controlled process. We shall first outline the picture for 1:1 complexes involving comparatively simple ligands and will then consider how this is modified in the first place when the ligand is large and multidentate (as in the reaction of the alkali metals with macrocyclic ligands) and in the second place when the metal is already bound to a ligand (which does not dissociate during the course of the reaction). In the latter case, the metal may be bound to a ring system, as with the iron in the haem ring of haemoglobin, or to a number of groups, usually through oxygen, nitrogen or sulphur, which may adopt a less symmetrical (although, perhaps, a no less rigid) configuration around the metal ion. Although we shall concentrate on metals which are found in biological situations and reactions which have been studied by the chemical relaxation methods, it must be mentioned that two other techniques, stopped-flow and the kinetic use of n.m.r. spectroscopy, have played a large part in the study of these reactions. Also the majority of the work has been done with a metal ion, Ni^{2+}, whose claim to a major place in a work on molecular biology is, to say the least, scant. We shall restrict our discission to complex formation in water.

Our plan, therefore, is to try and answer such questions as these: What, if any, is the *inherent reactivity* of each biologically important metal towards a potential ligand? How does the presence of groups other than solvent molecules in the inner coordination sphere modify this? What effect does the nature of the incoming ligand have on the picture? Although, as we shall see, it is possible to generalize about the kinetics of complex formation, it is also possible to comment quantitatively on the dissociation process and therefore the lifetime of the complex. In many cases, it is probably the latter possibility which is the more useful in molecular biology.

II. Formation of 1:1 Complexes with Small Ligands

The formation of the complex in aqueous solution between a bivalent metal ion $M(OH_2)_6^{2+}$ and a neutral monodentate ligand L can be represented as follows:

$$M(OH_2)_6^{2+} + L \underset{k_d}{\overset{k_f}{\rightleftharpoons}} [M(OH_2)_5L]^{2+} + H_2O. \qquad (1)$$

Reaction (1), which assumes that the coordination number of the metal remains unchanged, may be discussed mechanistically in terms of the scheme shown in Figure 1. One extreme is the S_N1 (or dissociative) pathway (I) in which the rate-determining step is the removal of the first solvent molecule from the inner hydra-

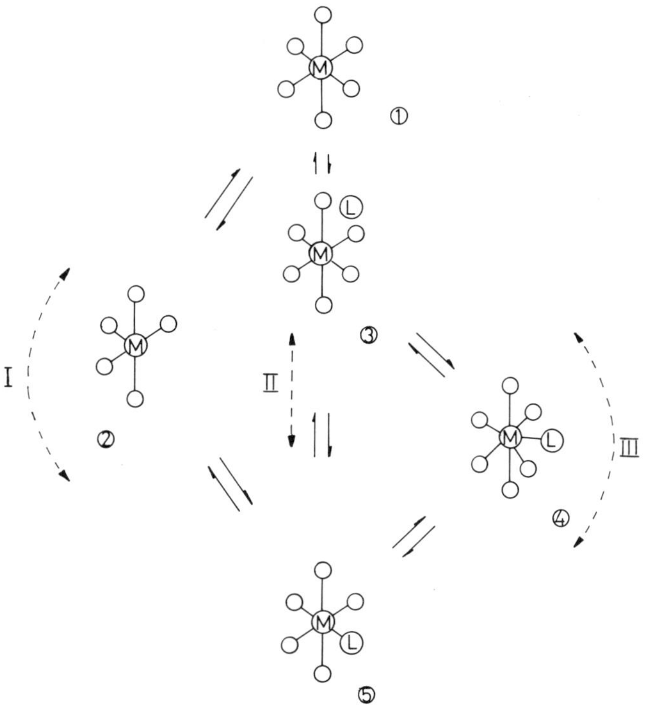

Fig. 1. Schemes representing formation and dissociation of a complex between hydrated metal ion $M(OH_2)_6^{2+}$ and monodentate ligand L. Water molecules represented by open circles

tion sphere of the metal to give an intermediate of reduced coordination number (2). The other extreme is the S_N2 (or associative) pathway (III) in which the bond-making process is more important than the bond-breaking process and the intermediate (4) has an increased coordination number (in this case, seven). Here, the nature of the incoming ligand is rather more important than in the S_N1 pathway. It is beyond the scope of this generally to consider the arguments in any detail, but it is now generally agreed that neither of these extremes best rationalises all the evidence. Instead, substitution at a labile metal ion is generally considered to have a dissociative-interchange mechanism (pathway II), first proposed by EIGEN (EIGEN, 1963; EIGEN and WILKINS, 1965).

The hydrated metal ion and the hydrated ligand diffuse together rapidly to form an outer-sphere complex (3) (sometimes called, more specifically, an ion-pair if the ligand carries a formal negative charge) in which the lone-pairs on the ligand are separated from the metal ion by the inner coordination sphere of the metal. (The water surrounding the ligand, being generally less strongly held than that surrounding the metal, is displaced more readily. This first step can therefore usually be regarded as a composite one comprising diffusion and water-loss at the ligand. However, with some of the larger cations, such as Ca^{2+}, Ba^{2+} and Sr^{2+}, the water molecules in the inner hydration sphere are so weakly held that there is coupling between the diffusion process, water-loss at the ligand and water-loss at the metal (cf. EIGEN and TAMM, 1962)). This is followed by the loss of the water molecule from the inner coordination sphere of the metal, coupled with the formation of the M-L bond. It is assumed that the incoming ligand enters the inner coordination sphere from a position somewhere on one of the faces of the octahedron formed by the inner-sphere water molecules as one of the latter departs, but the exact form of the transition state is uncertain. Indeed, it is probably not even meaningful to try to distinguish between an S_N1- and and S_N2-type process for a reaction occurring on a time scale as short as that of complex formation with most metal ions; the rate-determining step is almost certainly a concerted process in which water and the incoming ligand change places.

Metal ions can be classified into three groups according to the relative rates of steps (1-3) and (3-5). Group A contains those with relatively low charge density at their surface; for example, the heavier alkaline earths and all the alkali metals. As we have already seen, the water molecules are held so weakly that the rate of their loss is comparable with the diffusion-controlled value of around $10^9 - 10^{10}\,s^{-1}$. The complex formation rate constant for an ion in this group is therefore approximately equal to the maximum possible value, viz. $10^9 - 10^{10}$ ℓ/mol/s, unless chelation effects are important (see below).

The second group, group B, contains metals of intermediate surface charge density (for example, Mg^{2+}, Fe^{2+}, Co^{2+} and Ni^{2+}) in which the water molecules are held more strongly. Step (3-5) is now considerably (i.e. > 10 times) slower than step (1-3), and so a fast pre-equilibrium step is followed by rate-limiting loss of a water molecule from the inner coordination sphere. To a first approximation (provided that the concentration of the outer-sphere complex (3) is small compared with that of the uncomplexed metal ion), the observed overall formation rate constant k_f is given by $k_f = K_{os}\, k_{ex}$, where K_{os} is the equilibrium constant for the formation of the outer-sphere complex and k_{ex} is the rate constant for the water-loss process. On this picture, k_{ex} is equivalent to the rate constant for water-exchange at the metal ion and this can frequently be measured independently by n.m.r., especially when the metal ion is paramagnetic. Unfortunately, it has not yet been possible to measure K_{os} directly in reacting systems, so it has not been possible to confirm the mechanism unequivocally. However, the values of k_f for metals in group B do appear to be more or less typical of the metal and comparable to the product of the (pseudo-first order) rate constant for water exchange at the metal and the value of K_{os} calculated on the basis of an equation derived by FUOSS (see below).

The third group, group C, contains metal ions of sufficiently high surface charge density that the water molecules are very strongly held and hydrolysis (i.e. the dissociation of a proton from a bound water molecule) can occur before water-loss. This group includes the small, highly charged ions such as Be^{2+} and Al^{3+} and most of the tetravalent ions; more important from a biochemical point of view, it probably also includes Fe^{3+}. It is the least well understood group since the overall mechanism may involve quite a number of non-separable steps. Substitution at the hydrolysed ion $M(OH)^{(n-1)+}$ is faster than at the non-hydro-

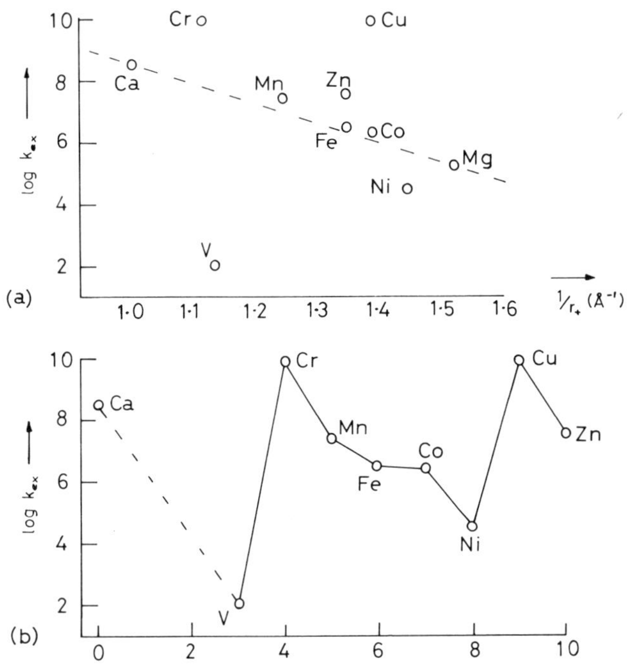

Fig. 2a and b. Rate constants for water exchange (k_{ex}) at divalent metal ions (a) as function of ionic radius, and (b) as function of number of 3d-electrons on M^{2+}. Dashed line in (a) has been drawn through the two $3d^0$ cations Mg^{2+} and Ca^{2+}

lysed ion M^{n+}_{aq} [in fact, $\overset{aq}{M}(OH)^{(n-1)+}_{aq}$ is usually in group B] and so anything which promotes hydrolysis will also promote substitution. In any event, for group C metals the overall rate constant depends to a significant extent on the nature of the incoming ligand (for example, its basicity). This influence is much less apparent with metals of groups A and B, except in so far as the formation of the second and subsequent bonds may become rate-determining, as discussed in greater detail below.

What, now, determines the "typical" rate constant for complex formation at a metal ion or, in effect, the rate constants for water-exchange? The charge density at the surface of the surface of the metal ion ($\propto$ charge : radius) is certainly important and it is found that, for metals in which the interaction with ligands is primarily electrostatic in nature, the logarithmic values of k_f and k_{ex} are approximately inversely proportional to the ionic radius r_+ for metals of a given charge type. However, the presence of partially filled d-orbitals on the metal ion can cause considerable complication as indicated in Figure 2. The bivalent ions of the first few transition elements are not well characterised, but for the remainder of the series there is evidently a deviation from the simple inverse dependence on ionic radius. Although there is a gradual decrease in r_+ as the atomic number increases, this is not sufficient to explain the low values of k_{ex} for V^{2+} and Ni^{2+}, and especially the high values for Cr^{2+} and Cu^{2+} (r_+ for Cu^{2+} = 0.72 Å and for Ni^{2+} = 0.69 Å). For a detailed discussion of the currently accepted views on the transition metals the reader is referred to a standard work on mechanistic inorganic

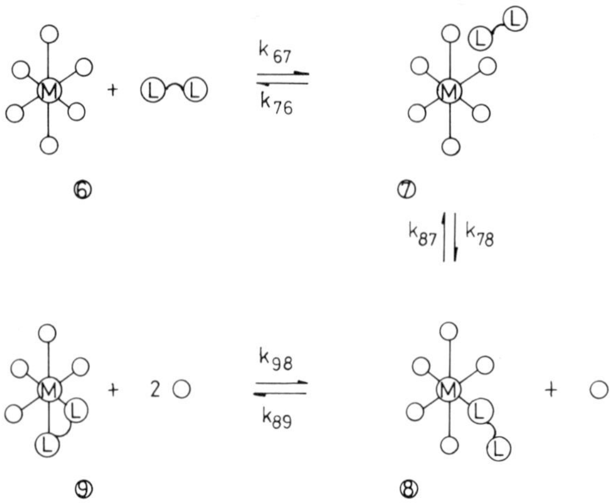

Fig. 3. The formation and dissociation of a complex between $M(OH_2)_6^{2+}$ and bidentate ligand L-L

chemistry (e.g. BASOLO and PEARSON, 1967), but it is worth mentioning here that the high lability of a water molecule bound to Cu^{2+} is almost certainly a function of the Jahn-Teller distortion produced by the interaction between the ligand field and the 3d-orbitals on the metal. As a result of this, the two axial water moelcules are much further from the metal ion than the four in the equatorial plane and are therefore much less strongly bound. Whereas the quoted value of the ionic radius of the metal is based on measurements to tightly-bound (in this case, equatorial) ligands, the exchange rate constant is that for the weakly-held (axial) water.

So far, we have considered complex formation in which the ligand is monodentate. How is the picture modified when the ligand binds through two or more groups? In Figure 3 we represent the formation of the complex of M^{2+} with the neutral bidentate ligand L-L (subsequent steps analogous to (8-9) will be added if the ligand is multidentate). The outer-sphere complex (7) is similar to the corresponding species (3) in Figure 1, but there is an additional intermediate (8) in which only the first metal-ligand bond has been formed. If the steady-state approximation is applied to the formation of the intermediates (7) and (8), the observed rate constants for the formation and dissociation of the final complex are given by, respectively,

$$k_f = \frac{k_{67}\ k_{78}\ k_{89}}{k_{76}\,(k_{87} + k_{89})} \quad \text{and} \quad k_d = \frac{k_{87}\ k_{98}}{(k_{87} + k_{89})} \ .$$

Now, if ring-closure is fast compared with dissociation of the monodentate intermediate (i.e. $k_{89} \gg k_{87}$) then these expressions reduce to $k_f = k_{67}\ k_{78}/k_{76}$ and $k_d = k_{87}\ k_{98}/k_{89}$, and the kinetics of formation of the chelate are identical to those observed with a monodentate ligand (since $k_{67}\ k_{78}/k_{76} = K_{os}\ k_{ex}$). If, on the other hand $k_{87} \gg k_{89}$, the expressions reduce to $k_f = k_{67}\ k_{78}\ k_{89}/k_{76}\ k_{87}$ and $k_d = k_{98}$, and the rate-limiting step becomes ring closure (and opening). There is now reasonable evidence that, with metals of groups B and C, the ring closure step (or steps

Fig. 4. The formation and dissociation of the 2:1 complex of L with $M(OH_2)_6^{2+}$

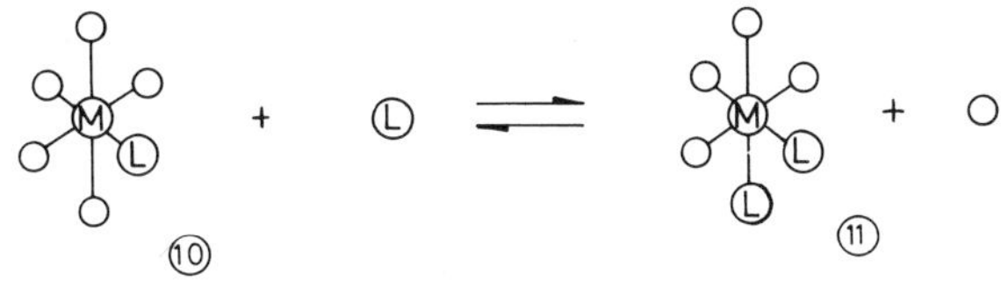

for a multidentate ligand) is (are) fast (non-rate-determining) in most cases. (It is noteworthy that it is unnecessary to invoke a possible labilizing effect on the remaining solvent molecules of the first metal-ligand bond to explain the similarity in overall rate constant k_f and activation enthalpy $\Delta H_f^{\ddagger}$ for reactions of a given metal ion with similar mono- and bidentate ligands. As with the "neighbouring group effects" observed in substitution reactions at carbon, the high rate of step (8-9) compared to that of the combination of a second ligand L to ML^{2+} [(10-11), Figure 4] originates in the statistical advantage of a unimolecular over a bimolecular reaction. In other words, the concentration of the outer-sphere complex is much higher in reaction (8-9) than in reaction (10-11) under analogous conditions, leading automatically to a larger overall rate.)

Several systems have been found in which it is suggested that, for one reason or another, this "normal" picture does not apply. A value of k_f which is lower than expected can be rationalised in terms either of k_{87} being abnormally high or k_{89} being abnormally low. The former occurs when the first metal-ligand bond is relatively weak and it has been proposed that the formation of a six-membered chelate ring (as when the ligand is β-alanine) rather than the more stable five-membered ring (as when it is α-alanine) can cause the latter. (The generality of a reduced k_f for a six-membered ring is, however, by no means established.) Unusually high values of k_f have been explained in terms of an internal conjugate-base mechanism in which hydrogen bonding can occur between the incoming ligand and a bound water molecule while the ligand is in the outer sphere of the metal. Examples of these and other deviations from "normal" behaviour will be found in the literature (see, for example, HAGUE, 1971, 1972, 1974). The majority of these results have been obtained with Ni_{aq}^{2+} and it is the present author's view that, interesting though they undoubtedly are, it is definetely too early to generalise from them; especially if such generalisation includes the majority of biologically important cations which are so much more labile than nickel.

III. Formation of 1:1 Complexes with Large Ligands

The type of problem we touched upon at the end of the last section is obviously of considerable interest to the molecular biologist, who is likely to be faced with the task of deciding whether the formation (or dissociation) of a complex between a metal ion and a macromolecule is likely to be rapid or not. He will want to know: under what circumstances is it reasonable to take k_f as the "typical" value for the particular metal and when should a different value be expected? In the case of a deviation from "typical" behaviour, he will also want to know how large it is likely to be. The short answer to these questions is that we do not yet know enough about the kinetic behaviour of most labile metals to be able to say with any degree of certainty. However, we do have some indications and some of the relevant results will be considered in this section.

In general, the value of k_f is unlikely to be larger than the typical value for the metal and charge type of the ligand concerned (values of which are given at the end of the chapter). The only cases of larger values in the literature have been rationalised in terms of the internal conjugate base mechanism, and this is only applicable for bi- or multidentate ligands in which at least one of the incoming groups is more basic than one of the water molecules bound to the metal. Since the pK_a values of most metal ions are greater than seven, this effectively restricts us to incoming ligands containing amino- or hydroxy-groups. If such a rate enhancement were to be observed, it is unlikely to be larger than one, or at most two, orders of magnitude. It is much more likely that a deviation from typical behaviour should be on the slow side. As we have seen, such a retardation implies that the closure of the first (or a subsequent) chelate ring has become partly or wholly rate-limiting rather than the substitution of the first water molecule in the inner hydration sphere of the metal.

Multidentate ligands can be divided into three categories: large open-chain ligands, flexible ring systems (macrocyclics) and rigid ring systems. Largely as a result of detailed studies with nickel, TURAN and RORABACHER (1972) have made three generalizations concerning the mechanism of complex formation with large multidentate ligands in the first two of these groups. They will be given here because they are likely to be found to apply also to most of the metals listed in Table 1.

1.In linear multidentate ligands, such as the straight-chain polyamines (e.g. Formula I), the terminal (primary) donor atoms are much more reactive than the internal (secondary) donor atoms, to the extent that the reaction path involving first-coordinate bond formation with a terminal donor atom represents the predominant reaction route.

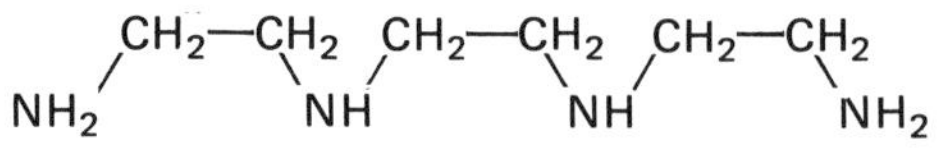

(I)

2. For branched multidentate ligands (such as the ethylenediaminetetraacetate anion Formula II and analogous ligands) involving both primary and tertiary donor atoms, the reaction path involv-

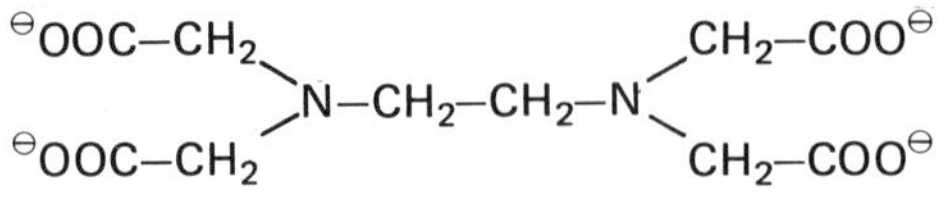

(II)

ing first-coordinate bond formation to a primary (terminal) donor atom predominates to such an extent that it represents the only kinetically significant route to the final complex, even when the terminal donor atoms form such weak bonds with the metal that the rate-determining step is shifted to the point of second-coordinate bond formation (closure of the first chelate ring, i.e. $k_{87} >> k_{89}$, cf. Fig. 3).

3. Extrapolation of these conclusions to ligands involving a high degree of steric hindrance, such as the macrocyclic ligands, suggests that these species will be very slow to react and, considering the large potential energy barriers to internal rotation, will probably involve a shift in the rate-determining step to the point of second-bond formation even when all donor atoms form "strong" bonds with the metal.

An interesting series of macrocyclic ligands of comparatively low (in biochemical terms) molecular weight are the ionophores, such as valinomycin, which are capable of rendering biological and artificial membranes selectively permeable to alkali metal cations. It is not known how they do this but one of the proposed models involves the formation of 1:1 complexes between the "carrier" and the metal in which the metal ion sits inside in a "hole" formed by the ligand. In any event, it has been shown in a number of cases that the macrocycle undergoes a conformational change as it forms the complex. An outline of the evidence is given in, for example, HAYNES et al. (1971); two examples will be quoted here.

IVANOV et al. (1969) have used n.m.r., i.r., o.r.d., and dipole moment techniques to study valinomycin (Formula III) and its

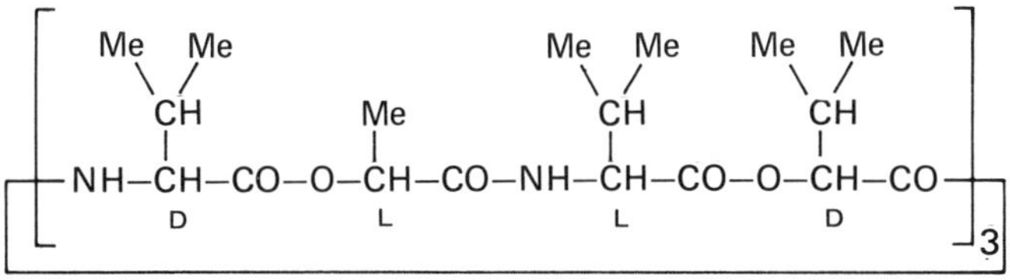

(III)

potassium complex in solvents of varying polarity. They conclude that valinomycin can exist in two forms, in one of which the amide groups form six intramolecular hydrogen bonds and in the other of which only three such bonds are formed; as the solvent polarity increases, so does the proportion of (Formula III) in the latter form. Complexation with K^+ is accompanied by conformational changes in which three formerly outward-facing ester carbonyls become oriented towards the metal ion, which sits in the middle of the ring. Changes in the polarity of the solvent do not affect the shapes of the o.r.d. curve of the complex, suggesting that the latter is quite rigid. A particularly interesting feature of this conformation is that the K^+ ion and the system of hydrogen bonds which helps to maintain the structure are effectively shielded from solvent action by the hydrophobic

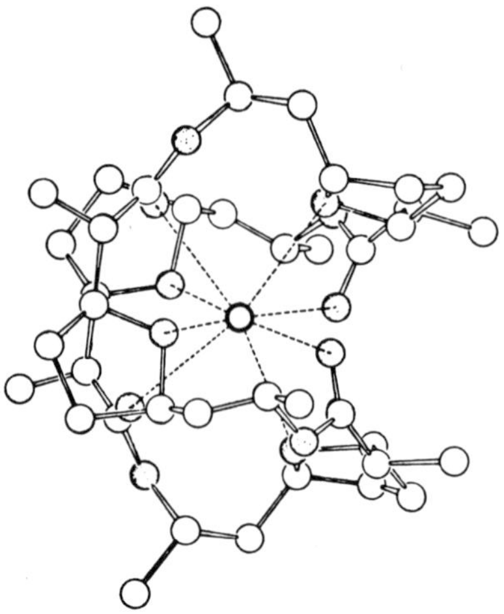

Fig. 5. The nonactin-K^+ complex (see KILBOURN, B.T., DUNITZ, J.D., PIODA, L.A.R., SIMON, W.: J. Mol. Biol. 30, 559 (1967))

branched side-chains at the periphery of the molecule (Fig. 5). Thus, the macrocycle must undergo considerable conformational change in order to form the complex. Similar changes have been shown to occur (DOBLER, 1972) on the binding of nonactin (Formula IV; R = H) to K^+ and it has been confirmed that the metal ion must

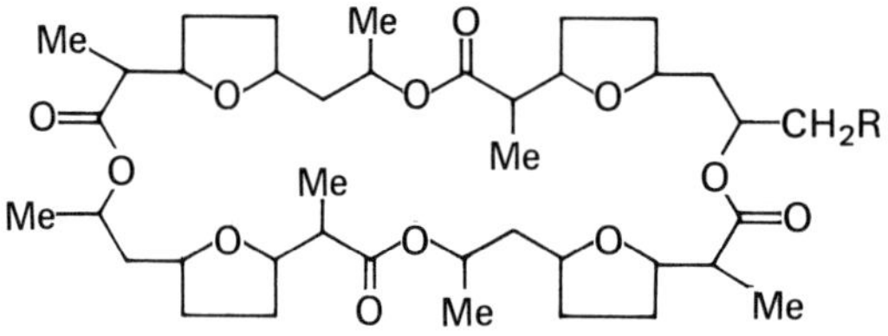

(IV)

be stripped of its hydration shell as it enters the nonactin molecule. The loss of hydration energy (ca. 80 kcal/mol) must evidently be compensated by interactions between the cation and its eight coordinated oxygen atoms in the complex.

Chemical relaxation techniques have been used to measure the kinetics of the binding of univalent cations to molecules of this type and it has been found that the reactions are surprisingly rapid. In Table 2 we list the rate constants for complex formation of Na^+ with several macrocycles in methanol. (The reason for the choice of solvent was technical expediency; rather similar values of k_f are to be expected for the corresponding reactions in water.) It is interesting to note that the limiting value of the formation rate constant k_f expected for a diffusion-controlled reaction with an uncharged species in methanol is ca. $3 \cdot 10^9$ ℓ/mol/s. Thus, for monactin (Formula IV; R = Me), dinactin and trinactin (the latter two molecules are methyl-substituted derivatives of monactin) the reaction rates are only 10 - 50 times lower than for a diffusion-controlled process. It is also interesting to note that the value of k_f for the charged carrier nigericin is, as expected, significantly larger than those for the neutral ligands; it is also considerably nearer to the diffusion-controlled value.

Table 2. Rate constants for the complex formation of Na^+ with carriers in methanol

Carrier	$k_f/\ell/mol/s$	$k_d/1/s$
Nigericin	$\sim 2 \times 10^{10}$	$\sim 2 \times 10^6$
Monactin	$\sim 2 \times 10^8$	4×10^5
Dinactin	6×10^7	5×10^4
Trinactin	7×10^7	4×10^4
Valinomycin	7×10^6	5×10^5

Data from WINKLER, R.: Structure and Bonding 10, 1-24 (1972)

A similar picture has been revealed in the binding of the crown ether dibenzo-30-crown-10 (Formula V) with univalent cations in

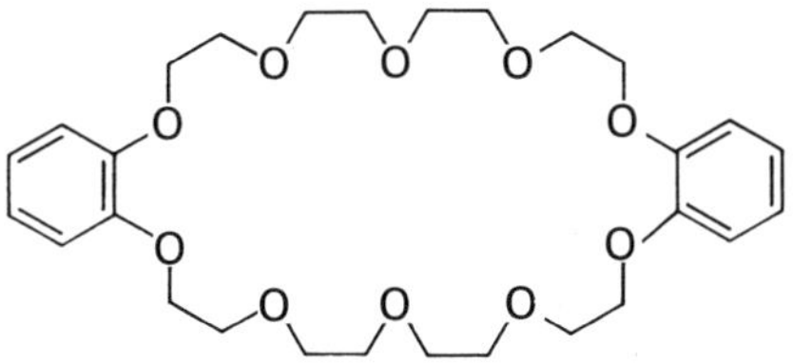

(V)

methanol. The kinetics are consistent with a two-step mechanism

$$(CR)' \xrightleftharpoons{\text{fast}} (CR) + M^+ \underset{k_d}{\overset{k_f}{\rightleftharpoons}} MCR^+$$

where (CR)', (CR) and MCR^+ represent, respectively, an unreactive form of the crown ether, an open configuration of the ether (which is the predominant species in the absence of cations), and the metal complex in which the ether is in a closed configuration. The rate constants k_f for the ions studied are given in Table 3.

It may turn out that the transport mechanism for the passage of the alkali cations through biological membranes does not involve 1:1 carriers of the type we have been discussing. Nonetheless, these kinetic results are still interesting since they indicate how small the retarding effects of a flexible macrocyclic ligand may be in the formation of 1:1 metal complexes.

In an earlier study, CABBINESS and MARGERUM (1970) compared the rates of reaction of Cu^{2+}_{aq} with the somewhat flexible 14-membered macrocyclic ligand (Formula VI; also similar macrocycles), with a comparable open chain polyamine (Formula VII) and with a porphyrin ligand (haematoporphyrin IX - Formula VIII). The macrocycles of type (Formula VI) react more slowly than the open-chain ligand by a factor of $10^3 - 10^4$. This is a rather larger effect than was found with the alkali metals and it could be that the highly non-equivalent nature of the coordination sites in Cu^{2+} referred to above is distorting the pattern in some way.

Table 3. Rate constants for complex formation with dibenzo-30-crown-10 in methanol

M^+	$k_f/\ell/mol/s$	$k_d/1/s$
Na^+	$>1.6 \times 10^7$	$>1.3 \times 10^5$
K^+	$(6 \pm 2) \times 10^8$	$(1.6 \pm 0.5) \times 10^4$
Rb^+	$(8 \pm 2) \times 10^8$	$(1.8 \pm 0.4) \times 10^4$
Cs^+	$(8 \pm 1) \times 10^8$	$(4.7 \pm 0.6) \times 10^4$
NH_4^+	$>3 \times 10^7$	$>1.1 \times 10^5$
Tl^+	$(8 \pm 1) \times 10^8$	$(2.5 \pm 0.3) \times 10^4$

Data from CHOCK, P.B.: Proc. Nat. Acad. Sci. (Wash.) 69, 1939 (1972)

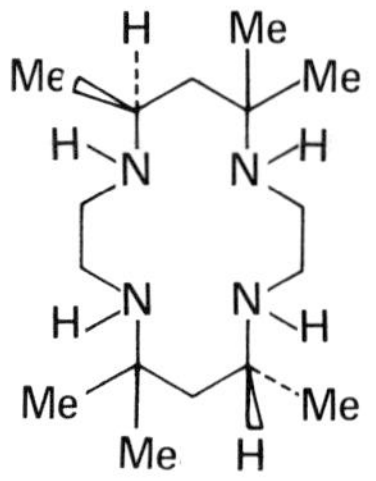

(VI)

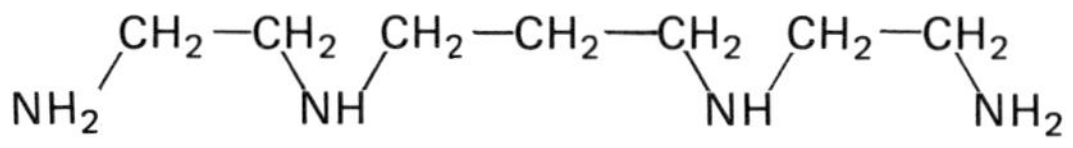

(VII)

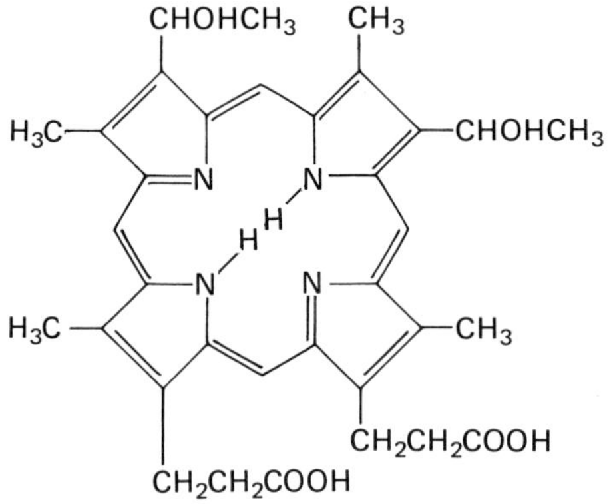

(VIII)

What is interesting, however, is that the more rigid porphyrin is less reactive than the open-chain ligand by a factor of 10^9. (Similar results have been obtained by other workers for the incorporation of other metals into porphyrin and related ring systems – see HAMBRIGHT, 1971).

It appears, then, that the rate constant for complex formation with a multidentate ligand is likely to lie within two orders of magnitude of the "normal" value for the metal (see below). Twisting or folding of the ligand is important for the multiple desolvation of the metal ion, however, and if something interferes drastically with this (rigidity, steric hindrance or internal hydrogen bonding, for example), then a retardation of a factor of rather more than 10^2 is possible.

IV. The Effect of Bound Ligands

A. Non-Ring Systems

In many, though by no means all, enzyme-catalysed reactions which exhibit an absolute requirement for a metal ion, it has been suggested that the metal M forms a bridge or link between the enzyme E and the substrate S and/or product P (see, for example, MILDVAN, 1970). The enzyme is usually a multidentate ligand and the intermediate EMS (or EMP) is pictured as a ternary complex (Formula IX)

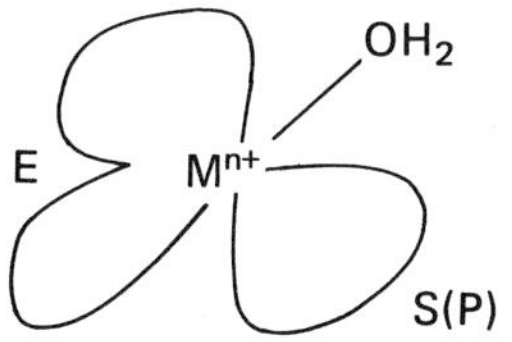

(IX)

which may still contain one or more water molecules in the inner coordination sphere of the metal. Since the enzymic turnover must involve at least one complex formation and dissociation step between a ligand and a metal ion which is already bound in a 1:1 complex, an important question must be: what is the effect of a bound ligand on the kinetics of the subsequent complex formation of the metal? In what ways, if any, are the kinetics and mechanism of the formation of EMS different from those of MS or ME? In view of the complexities of such systems and frequently the lack of knowledge of the nature (and even number) of binding groups of the metals in the actual enzymic systems, most of the work reported to date on this type of system has been on models rather than the enzymes themselves. For technical reasons it is usually convenient to choose systems of the type shown in Figure 6a in which one of the ligands L_1 remains bound to the metal for a time which is long compared to the half time of the binding of the other, L_2. This is generally, though not always, the case if the stability constant of the complex ML_1 is at least two or three orders of magnitude larger than that of ML_2. The kinetic parameters of the reaction of ML_1 with L_2 are then compared with those for the analogous reaction of the unsubstituted metal with L_2 (Fig. 6b).

A priori, we would expect the largest effects, both on k_f and on k_d, to be observed when both ligands carry negative charge. On the assumption that L_2 is bidentate and that the "normal" behaviour (see Sect. II) is observed for $M + L_2$, we might expect the partial replacement of the inner hydration sphere of the

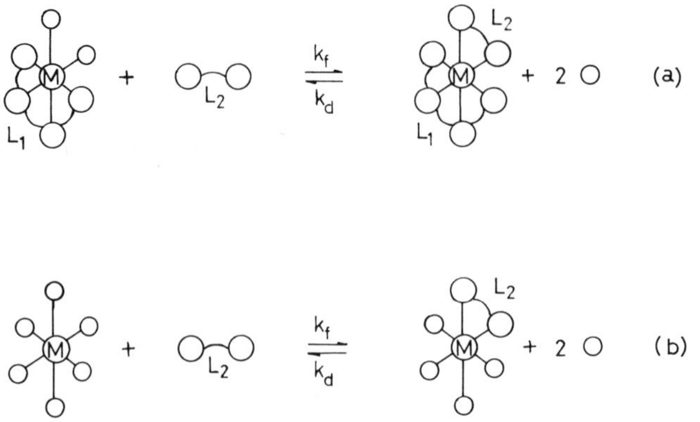

Fig. 6a and b. The formation of the complex between the bidentate ligand L_2 and (a) the complex $[ML_1(OH_2)_2]$, (b) the hydrated metal ion $M(OH_2)_6$

metal by L_1 to affect k_f in one or more of the following ways:

1. Since the net positive charge on the metal is reduced (or even converted into an overall negative charge), K_{os}, and therefore k_f, will be reduced.

2. Since L_1 is held by the metal ion more strongly than the water molecules it has replaced, the remaining water molecules will now be held less strongly than in M^{2+}_{aq}. This will lead to an increase in k_{78} (k_{ex}) and therefore in k_f.

3. k_{89} will be reduced through the steric and electronic interaction between L_1 and L_2, with the result that ring closure now contributes to the rate-determining step, thereby reducing k_f.

Two other effects must also be considered, one trivial, the other, if it occurs, major. The former is a reduction in k_f associated with the fact that in ML_1 the number of replaceable water molecules is less than in M^{2+}_{aq}. This can be corrected for by multiplying the measured k_f in the former case by an appropriate statistical factor. In the case of an octahedrally coordinated metal ion in which n is the number of coordination positions occupied by L_1, this factor is empirically taken as $6/(6-n)$. Such an adjustment has been applied in the formation rate constant k^s_f. The other possible effect of L_1 is to change the coordination number of the metal. Such changes are certainly important in zinc chemistry but our lack of a reliable tool for determining the coordination number of this metal in its complexes in solution seriously inhibits any discussion of their effects on k_f.

In the remainder of this section we shall consider factors (3) in turn and finally comment on the effects of coordination number change. The results on which we shall base our discussion are shown in Figures 7 - 11 and in Figure 12 are listed the ligands and the abbreviations used for them.

1. Outer-Sphere Complex Formation

The Fuoss equation, which is used to calculate the equilibrium constant for the formation of an outer-sphere complex between a metal ion and a charged ligand is given in Eq. (2); N_o is the

$$K_{os} = \frac{4\pi N_o\ a^3}{3000} \exp\ [-U(a)/kT] \qquad (2)$$

Avogadro number, k the Boltzmann constant, T the temperature and a the distance of closest approach of the hydrated metal ion and the ligand, in Ångströms. U(a) is the Coulomb interaction between the two ions and is taken to be equal to $\frac{z_1 z_2 e^2}{aDkT}$, where z_1, z_2 are the electronic charges on the ions, e the electron charge and D the dielectric constant of the medium (assumed to be a continuum). Although this formula is only applicable to charged ligands, similar equations have been derived by different workers for neutral ligands. If a is in the range 3 - 5 Å, values of K_{os} are obtained which, when taken with the water-exchange rate constants k_{ex} for several metals, yield values of k_f for 1:1 complexes which are comparable to those measured independently (see above). They also correctly predict the increase in k_f by a factor of 2 - 3 for each unit of charge-product, with a rather larger increment on going from 0 to -1, as the metal and ligand are varied. (A similar trend is found in proton-transfer reactions – see, for example, EIGEN et al., 1964).

It is clear from Figures 7 - 11 that application of the Fuoss equation, using the net charge on the metal species, to ternary complex formation does not correctly predict the variation in k_f; there is no large decrease in k_f of the type predicted (1 to 2 orders of magnitude) for charged L_2 as the negative charge on ML_1 builds up. The observed increase, or, in some cases, only slight decrease, in k_f could only be explained by a correspondingly large increase in k_{ex}. Unfortunately, the water-exchange data for the partially substituted ions are generally not available, but the lack of any large increase in k_f for neutral L_2 argues against such compensation. (It is noteworthy that in only one of the many cases quoted in Figures 7 - 11 is k_f for a particular ML_1 more for a neutral L_2 than for a negative L_2; that case is $Mn(uda)^-$ + HOx, for which the activation parameters suggest (HAGUE et al., 1972) that some second-sphere interaction of the type reported by EATON (1971) might be involved.)

ML_1 is therefore continuing to behave as a positive centre even though it may carry an overall negative charge; the local charge density on the metal ion is evidently more important in determining k_f than the overall charge on the substituted metal ion. (An analogous argument may be invoked for the similarity in values of k_f for the reaction of Ox^- and nsa^{2-} with magnesium species, Hückel calculations indicating that the difference in the electron densities at the coordinating atoms in the two ligands is considerably less than the one electron suggested by the formal charges. Although the available data with Mn^{2+} support this rationalization in terms of local charge density on incoming ligand rather than overall charge, the possibility does exist that ring-closure is contributing to the rate-determining step with nsa^{2-}).

2. Labilisation of Remaining Water Molecules

In the first paper concerned with the effect of bound ligands on complex formation (HAMMES and STEINFELD, 1962), the temperature-

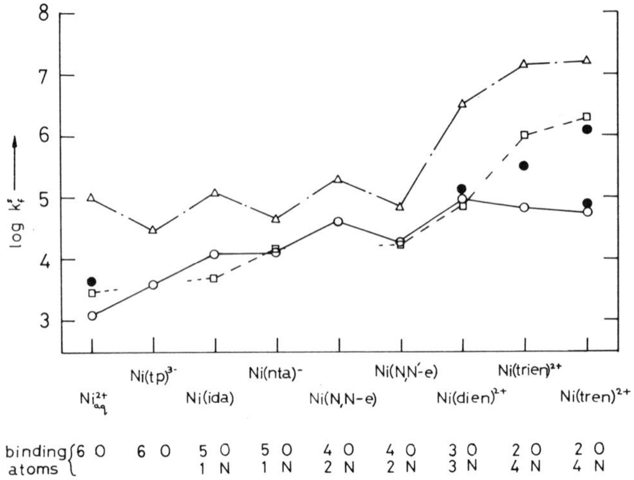

Fig. 7. Statistically adjusted rate constant (log k_f^S, see text) for reaction of nickel (II) species with charged and neutral ligands: nsa^{2-} (Δ), pada (o), NH_3 (□), H_2O (●). In the case of water exchange, first-order rate constants have been converted into second-order rate constants by treating solvent as reagent of concentration 55.5 M to permit direct comparison with other rate constants. Sources of data: COBB, M.A., HAGUE, D.N.: J.C.S. Faraday I 68, 932 (1972); HAGUE, D.N., KINLEY, K.: J.C.S. Dalton 249 (1974); HAGUE, D.N., WHITE, A.R.: unpublished results; HUNT, J.P.: Co-ordination Chem. Rev. 7, 1 (1971); JONES, J.P., BILLO, E.J., MARGERUM, D.W.: J. Am. Chem. Soc. 92, 1875 (1970); MARGERUM, D.W., ROSEN, H.M.: J. Am. Chem. Soc. 89, 1088 (1967)

jump technique was used to study the binding of one, two and three molecules of glycine, diglycine or imidazole to nickel (II) or cobalt (II). For the complexes with glycine, it was found that k_f increased with each successive replacement of the neutral water molecules in the inner coordination sphere by the charged glycinate anion, for both Ni^{2+} and Co^{2+}. No such acceleration was found with imidazole and diglycine and it was concluded, very reasonably, that the presence of the negatively charged carboxylate group in the inner sphere of the metal preduces a "drainage" of the positive charge from the metal ion and a consequent loosening of the remaining water molecules. (The diglycine result was explained by suggesting that bonding between this ligand and the metal takes place through the uncharged portions of the ion rather than the carboxylate.) Unfortunately, this does not appear to be the case. Further work with nickel (II), some of it summarised in Figure 7, suggests that the dominant factor is the number of bound aliphatic N atoms, although clearly some other influence is also important.

Rather less information is available on the other metals but for none of them (Figs. 8 - 11) does it appear that the number of bound aliphatic N atoms is the dominant factor (although the low stability constants of complexes of Mg^{2+} and Mn^{2+} with nitrogen-binding ligands seriously limits the range which can be studied). Without more detailed information on the nature of the bonding

Fig. 8. Statistically adjusted rate constants (log k_f^S) for reaction of magnesium (II) species with charged and neutral ligands: nsa^{2-} (Δ), Ox^- (□), HOx (o). Sources of data: CAYLEY, G.R., HAGUE, D.N.: J.C.S. Faraday I 68, 2259 (1972) and unpublished results; HAGUE, D.N., MARTIN, S. R., ZETTER, M.S.: J.C.S. Faraday I 68, 37 (1972)

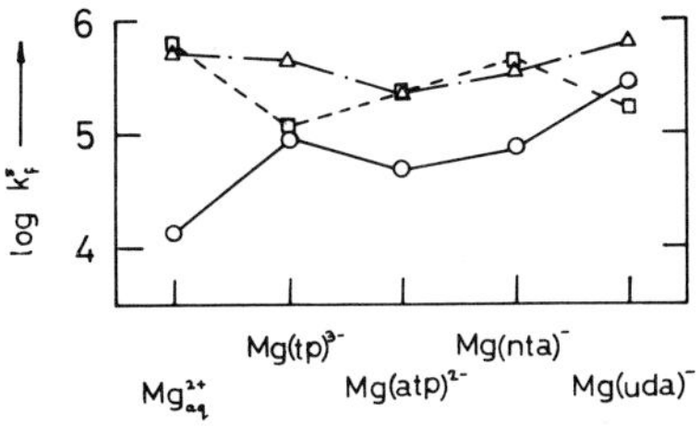

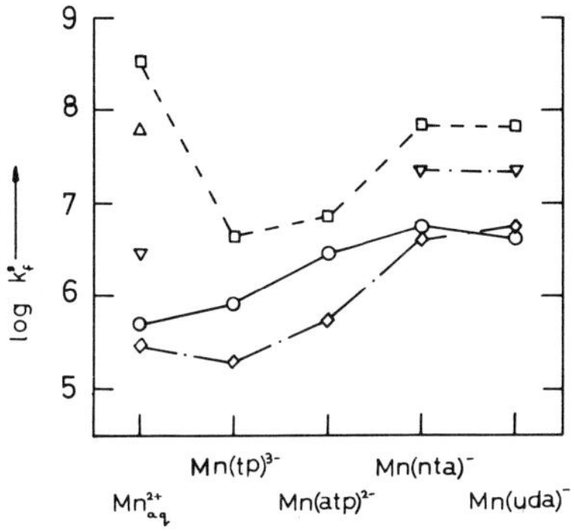

Fig. 9

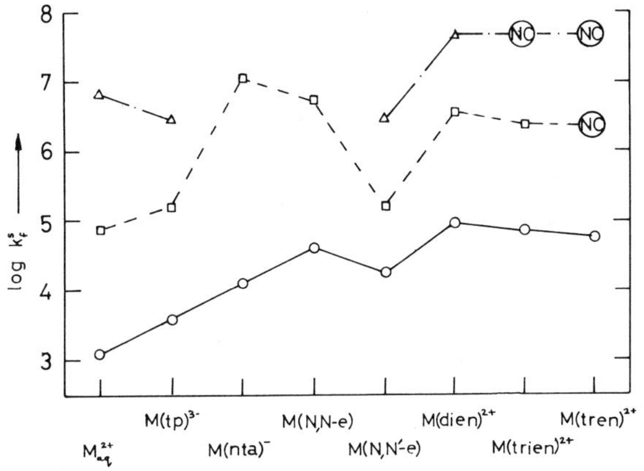

Fig. 10

Fig. 9. Statistically adjusted rate constant (log k_f^S) for reaction of manganese (II) species with charged and neutral ligands: nsa^{2-} (Δ), Ox^- (◇), $acac^-$ (∇), HOx (o), bipy (□). Sources of data: HAGUE, D.N., MARTIN, S.R.: J.C.S. Dalton 254 (1974); HAGUE, D.N., MARTIN, S.R., WILAIRAT, P.: unpublished results; HAGUE, D.N., MARTIN, S.R., ZETTER, M.S.: J.C.S. Faraday I 68, 37 (1972)

Fig. 10. Statistically adjusted rate constant (log k_f^S) for reaction of pada with zinc (II) (Δ), cobalt (II) (□) and nickel (II) (o) species. (NC) indicates that no complex is formed in a particular case. Sources of data on zinc (II) and cobalt (II): CAYLEY, G.R., HAGUE, D.N.: Trans. Faraday Soc. 67, 786 (1971); COBB, M.A., HAGUE, D.N.: Trans. Faraday Soc. 67, 3069 (1971); COBB, M.A., HAGUE, D.N., WHITE, A.R.: unpublished results

in those complexes it is unwise to make too sweeping generalisations, but it appears that in the case of cobalt (II) at least the particular stereochemistry of the bound ligand and the consequent distortion of the ligand field of the metal can have at least as large an effect on the lability of the remaining water molecules as any increase in the metal-ligand bond strength. It is not, on reflection, surprising that the presence of strongly held ligands in the inner coordination sphere of the metal does not inevitably have a profound effect on the lability of the remaining water molecules since in many cases the origin of the high stability constant of ML_1 is entropic rather than enthalpic – and this is particularly true for ligands which carry large negative charges.

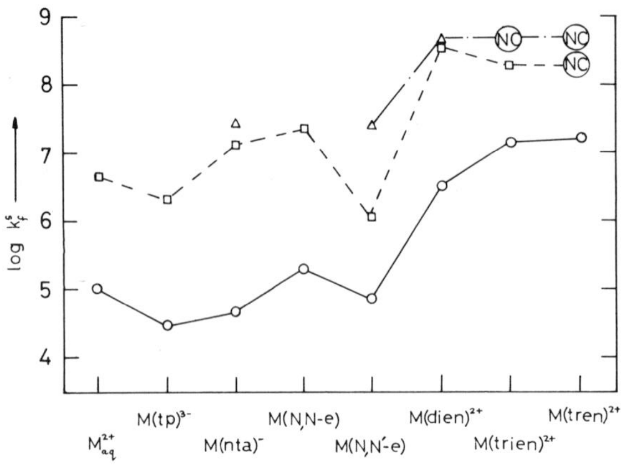

Fig. 11. Statistically adjusted rate constant (log k_f^s) for reaction of nsa^{2-} with zinc (II) (Δ), cobalt (II) (□) and nickel (II) (o) species. (NC) indicates that no complex is formed in a particular case. Sources of data on zinc (II) and cobalt (II): HAGUE, D.N., HENSHAW, J.S., McADAM, M.E., ROANTREE, M.R., WHITE, A.R., WILAIRAT, P.: unpublished results

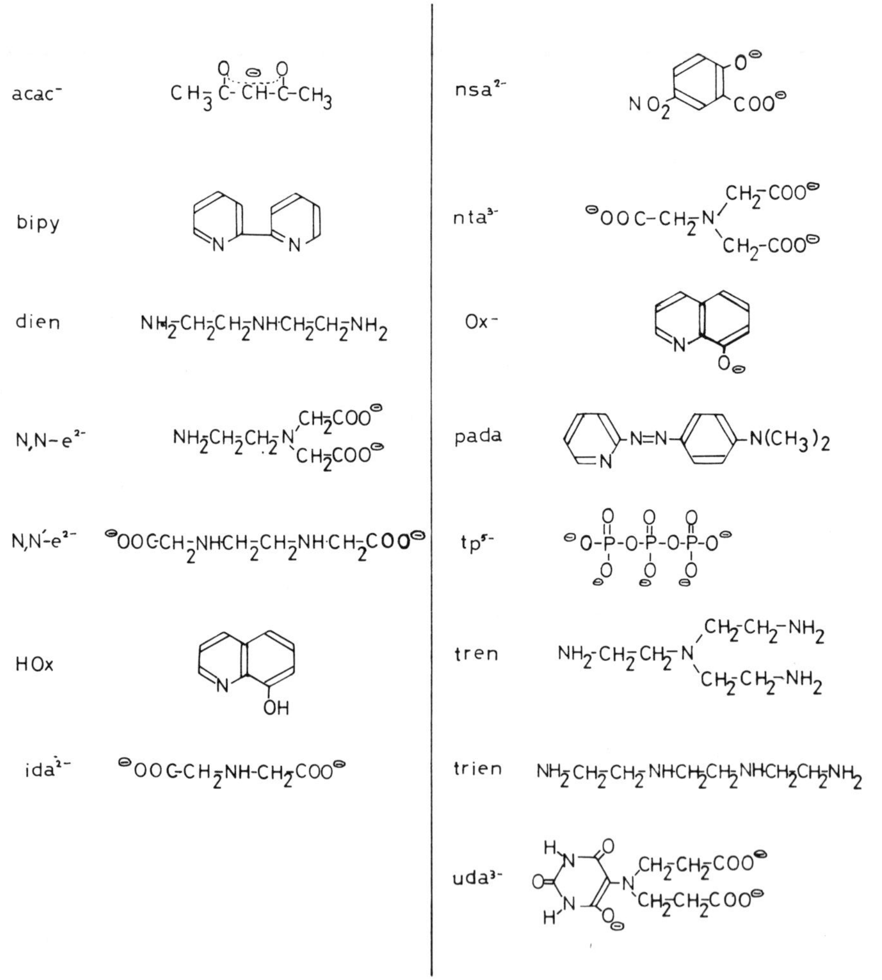

Fig. 12. The ligands, with abbreviations, used in Figs. 7 - 11

3. Steric and Electronic Interaction Between Ligands

This is a much more difficult matter on which to generalise from the comparatively few results at present available. It is probably fair to suggest that, if the incoming ligand is charged, there is a reduction in log k_f of about 0.5 when the bound ligand carries a high (≥ 3) negative charge, but there certainly does not seem to be the large reduction which might intuitively be expected, with the possible exception of $Mn(tp)^{3-}$ and $Mn(atp)^{2-}$ + Ox^-. Similarly, there is no evidence in the present data to suggest large steric interactions. These conclusions are borne out by the comparatively small variation in k_d for complexes of a given metal with a particular L_2. (The large difference in k_f for the reaction with pada and NH_3 of the tetrammines of Ni^{2+} is consistent with rate-limiting ring closure in the case of the bidentate ligand, but this is probably associated with the large number of Ni-N bonds – leading to a high value of k_{87} – rather than a direct interaction between the two ligands.)

4. Coordination Number Change at the Metal

Again, we do not have the data to permit confident generalisation. The only complex ML_1 in which the coordination number of the metal is known to be other than six is $Co(tren)^{2+}$, in which it is five. Interestingly, this is the only complex of cobalt (II) of those studied which did not form a ternary complex with pada or nsa^{2-}. The corresponding ternaries with $Ni(tren)^{2+}$, which contains octahedrally coordinated metal, form with ease.

Although the coordination number of the metal is not known unequivocally in most complexes of zinc, there is considerable evidence that it is less than six in $Zn(dien)^{2+}$, $Zn(trien)^{2+}$ and $Zn(tren)^{2+}$. As with $Co(tren)^{2+}$, no ternary complex is formed with pada or nsa^{2-} by the latter two, but in the case of $Zn(dien)^{2+}$ there seems to be no unusual change in k_f which can be attributed to the lower coordination number of the metal. It seems that the effect of coordination number change is seen in the thermodynamics rather than the kinetics (put another way: in k_d but not in k_f) and that the important factor is the coordination number and nature of the ligands in the potential ternary complex rather than in the initial 1:1 complex.

B. Ring Systems

Although several chemical relaxation studies have been reported on the formation of complexes with metals which are bound in haem and other ring systems, the complex nature of the ligands involved has precluded any systematic work of the type discussed above. For example, THUSIUS (1971) has reported the rate constants for the formation and dissociation of a number of cobalamins CBM-L. The values of k_f (for L = SCN^-, SO_3^{2-}, $S_2O_3^{2-}$, NCO^-, N_3^-, I^- and Br^-) are rather similar (170 - 2300 ℓ/mol/s at 25°C) whereas those of k_d range between $<10^{-5}$ and 590 1/s. The oxidation state of the metal in these complexes is technically +3 and the ring sys-

Table 4. Typical rate constants for complex formation with selected metals

Li^+		Fe^{2+}	2×10^6	La^{3+}	
Na^+		Co^{2+}	2×10^6	Ce^{3+}	
K^+	in the range $10^9 - 10^{10}$	Ni^{2+}	5×10^4	Pr^{3+}	
Rb^+		Cu^{2+}	$1 \times 10^8 - 5 \times 10^9$	Nd^{3+}	1×10^8
Cs^+		Zn^{2+}	5×10^7	Sm^{3+}	
Mg^{2+}	5×10^5	Cd^{2+}	5×10^8	Eu^{3+}	
Ca^{2+}	5×10^8			Gd^{3+}	
Sr^{2+}	1×10^9	Cr^{3+}	1×10^{-5}	Tb^{3+}	
Ba^{2+}	1×10^9	Fe^{3+}	50	Dy^{3+}	
V^{2+}	50	Co^{3+}	1×10^{-5}	Ho^{3+}	2×10^7
Cr^{2+}	$1 \times 10^8 - 5 \times 10^9$	VO^{2+}	1×10^3	Er^{3+}	
Mn^{2+}	5×10^7	$MoO_3(OH)^-$	1×10^8	Tm^{3+}	
				Yb^{3+}	
				Lu^{3+}	

The values quoted are second-order rate constants (ℓ/mol/s) reaction with a charged ligand L^-; appropriate values for neutral ligands are typically 1/50 - 1/10 of these. In many cases only very limited data are available and for some species values differing by two or more orders of magnitude have been reported. These values are therefore to be taken only as an indication of the rate constants to be expected

tem has evidently brought about a very large increase in the lability of its remaining water molecule since k_f for complexes of cobalt (III) is typically ca. 10^{-6} ℓ/mol/s. Other systems which have been studied include those involving haemoglobin and myoglobin (see, for example, SCHUSTER and ILGENFRITZ, 1969) and horse-radish peroxidase (DUNFORD and HASINOFF, 1971); further references to these and other systems are given in HAGUE (1971, 1972, 1974).

V. Summary

The rate constant for the formation of a metal complex k_f is generally determined by the nature of the metal ion and it is simply related to the rate constant for water exchange, k_{ex}. For 1:1 complexes, k_f can be taken, to a first approximation, as equal to K_{os} k_{ex}, where K_{os} can be evaluated from the Fuoss equation (typically, ca. 0.1 ℓ/mol for neutral ligands and 1 ℓ/mol for univalent anions). Values of k_f for biologically important cations are shown in Table 4. (In view of the current use of several of the lanthanides as n.m.r. shift reagents in biological systems, these cations have also been included.) They may be reduced in a particular case if some feature of the ligand, such as the steric factor or the need for the removal of an internally hydrogen-bonded proton, intrudes – an effect likely to be more important the higher the value of k_{ex}. It is also possible, in rather special circumstances, to observe values of k_f which are larger than these "typical" values.

For metal-bridged ternary complexes, k_f may also be affected by the nature of the bound ligand. It is likely to be increased by up to two orders of magnitude, although larger effects are possible, especially if the metal is at the centre of a haem-type ring system; the retarding effect of ligand-ligand charge repulsion generally appears to be rather small. The rate constant for the dissociation of a metal complex, and hence its lifetime, may be estimated simply from the equilibrium constant and k_f; ligand-ligand effects are likely to be larger than in the case of k_f, although it is possible to make fewer generalisations here.

References

BASOLO, F., PEARSON, R.G.: Mechanisms of Inorganic Reactions, 2nd ed., pp. 124-246. New York: Wiley 1967

CABBINESS, D.K., MARGERUM, D.W.: Effect of macrocyclic structures on the rates of formation and dissociation of copper (II) complexes. J. Am. Chem. Soc. 92, 2151-2153 (1970)

DOBLER, M.: The crystal structure of nonactin. Helv. Chim. Acta 55, 1371-1384 (1972)

DUNFORD, H.B., HASINOFF, B.B.: Rapid reaction kinetics involving the iron-porphyrin site of horseradish peroxidase. Adv. Chem. Series (Am. Chem. Soc.) 100, 413-426 (1971)

EATON, D.R.: Structure of the second coordination sphere of metal complexes and its role in catalysis. Adv. Chem. Series (Am. Chem. Soc.) 100, 174-186 (1971)

EIGEN, M.: Fast elementary steps in chemical reaction mechanisms. Pure Appl. Chem. 6, 97-115 (1963)

EIGEN, M., KRUSE, W., MAASS, G., DE MAEYER, L.: Rate constants of protolytic reactions in aqueous solution. Progr. Reaction Kinetics 2, 287-318 (1964)

EIGEN, M., TAMM, K.: Schallabsorption in Elektrolytlösungen als Folge chemischer Relaxation. Ber. Bunsenges. physik. Chem. 66, 93-121 (1962)

EIGEN, M., WILKINS, R.G.: The kinetics and mechanism of formation of metal complexes. Adv. Chem. Series (Am. Chem. Soc.) 49, 55-67 (1965)

HAGUE, D.N.: Complex formation with labile metals and reactions of biochemical interest, in Inorganic Reaction Mechanisms (Specialist Period. Rep., Chem. Soc.) 1, 209-262 (1971); 2, 196-218, 227-243 (1972); 3, 261-296, 329-348 (1974)

HAGUE, D.N., MARTIN, S.R., ZETTER, M.S.: Kinetics of ternary complex formation between magnesium and manganese(II) Species and 8-Hydroxyquinoline. J.C.S. Faraday I 68, 37-46 (1972)

HALPERN, J., LEGARE, R.J., LUMRY, R.: Measurement of the rate of electron transfer between tris-(4,7-dimethyl-1,10-phenanthroline)-iron(II) and hexachloroiridate (IV) by the T-jump method. J. Am. Chem. Soc. 85, 680-682 (1963)

HAMBRIGHT, P.: The coordination chemistry of metalloporphyrins. Co-ordination Chem. Rev. 6, 247-268 (1971)
HAMMES, G.G., STEINFELD, J.I.: Relaxation spectra of some nickel (II) and cobalt (II) complexes. J. Am. Chem. Soc. 84, 4639-4643 (1962)
HAYNES, D.H., PRESSMAN, B.C., KOWALSKY, A.: A nuclear magnetic resonance study of $^{23}Na^{+}$ complexing by ionophores. Biochemistry 10, 852-860 (1971)
HUGHES, M.N.: The Inorganic Chemistry of Biological Processes. London: Wiley 1972
IVANOV, V.T., LAINE, I.A., ABDULAEV, N.D., SENYAVINA, L.B., POPOV, E.M., OVCHINNIKOV, Yu.A., SHEMYAKIN, M.M.: The physico-chemical basis of the functioning of biological membranes: The conformation of valinomycin and its K^{+} complex in solution. Biochem. Biophys. Res. Comms. 34, 803-811 (1969)
MILDVAN, A.S.: Metals in Enzyme Catalysis. In: The Enzymes. BOYER, P.D. (ed.), Vol. II, 3rd ed., pp. 445-536. New York: Academic Press 1970
SCHUSTER, T.M., ILGENFRITZ, G. in ENGSTRÖM, A., STRANDBORG, B. (eds.): Symmetry and Function of Biological Systems at the Macromolecular Level, p. 181. New York: Wiley-Interscience 1969
THUSIUS, D.: Kinetics of cobalamin substitution reactions. J. Am. Chem. Soc. 93, 2629-2635 (1971)
TURAN, T.S., RORABACHER, D.B.: Reactions of Aquonickel (II) ion with N-alkyl-substituted ethylenediamines. Inorg. Chem. 11, 288-295 (1972)
WILLIAMS, R.J.P.: The biochemistry of sodium, potassium, magnesium and calcium. Quart. Rev. 24, 331-365 (1970)

Dynamics of Proton Transfer in Solution

P. SCHUSTER, P. WOLSCHANN, and K. TORTSCHANOFF

I. Introduction

Proton transfer reactions in principle belong to the simplest possible chemical processes, since only a naked proton is exchanged between the two molecules concerned. This formal visualization of course, can be valid at best in the vapor phase and even there at least one stable hydrogen bonded complex is formed as an intermediate in almost every concrete example:

$$\text{A-H+B} \rightleftharpoons \underset{\text{NC}}{[\text{A-H...B}]} \rightleftharpoons \text{A}^- + \text{HB}^+ \qquad (1a)$$

In general this complex will be an aggregate of two neutral molecules (AH..B) as indicated above. There are, however, some examples where a structure close to an associated ion pair ($A^-..HB^+$) is formed (cf. Sect. II.B and II.C). In solution proton transfer becomes much more complex. Stable hydrogen bonded ion pairs appear due to stabilization by solvent molecules (1b). Clusters of three and more molecules are formed. In associated liquids, fi-

$$\text{A-H+B} \rightleftharpoons \underset{\text{NC}}{[\text{A-H...B}]} \rightleftharpoons \underset{\text{IC}}{[\text{A}^-\text{...HB}^+]} \rightleftharpoons \text{A}^-\text{+HB}^+ \qquad (1b)^{1}$$

nally, a whole family of individual rearrangements of molecular complexes accompanies the pure proton transfer reactions and it becomes impossible to give a detailed molecular picture of the entire dynamic processes.

It is a long time since proton-transfer in solution was regarded as a simple process. More and more experimental material was made available by the application of new techniques to study fast chemical reactions and as a result of this increasing knowledge

[1]In this equation and everywhere in this chapter we will not indicate the solvent species by a special notation, since it should be clear from the text if we are dealing with isolated molecules in the vapor phase or with solvated particles in the liquid state. By NC (neutral complex) we will denote the association of two neutral molecules and by IC (ionic complex) the corresponding ion pair which is formed after proton transfer.

interpretation of the data became more and more difficult. Nowadays, proton transfer reactions are by no means a completely explored and well-understood field. In fact very little is known on proton transfer in nonaqueous solutions. It is a kind of paradox that the kinetics of proton transfer in water is much better investigated, although we are far away from a reliable picture of the molecular details of liquid water structure.

Mainly two different kinds of techniques were applied to study the dynamics of proton transfer in solution: nuclear magnetic resonance and chemical relaxation methods. During the past decade both techniques were developed to such a high degree of perfection that the application to systems of biological interest also became possible. Especially in the last two years a great number of papers dealing with dynamic aspects of proton transfer have appeared. An extensive discussion of all of them certainly would go beyond the limits of this article. We decided therefore to restrict ourselves here to some basic examples, which will be treated in some detail and are closely related to our own research work in this field, and to a brief survey of the most recent literature on the application of relaxation methods to proton transfer. Some new developments in experimental techniques will also be mentioned. Several other review articles and summaries dealing with proton transfer entirely or in part have been published in the past. They will be discussed briefly in a separate paragraph.

II. Theoretical Background of Proton Transfer

In this paragraph we intend to present a very brief summary of some theoretical results and thermodynamic data related to proton transfer, which seem to be very useful for a complete understanding of the kinetic data. For this purpose a splitting of thermodynamic quantities into vapor phase and solvation data was found to be very useful.

A. Proton Affinities

Before starting a detailed discussion on proton transfer it seems useful to regard the basic energetic data more closely. Proton affinities, ΔE_{PA}, are the energies of protonation of molecules or ions in the vapor phase extrapolated to absolute zero (O K). They correspond to the process:

$$A + H^{+} \rightleftharpoons [AH]^{+} \qquad (2a)$$

$$\Delta E_{PA}(A) = E([AH]^{+}) - E(A); \quad E(H^{+}) = O \qquad (2b)$$

Proton affinities as such represent the energetic contribution to an absolute acidity scale of molecules, atoms and anions. As it is obvious from Eq. (2), the energy of any proton transfer reaction can be obtained as a difference of two proton affinities, e.g. for the reaction shown in Eq. (1) we find $\Delta E = \Delta E_{PA}(B) - \Delta E_{PA}(A^-)$. Proton affinities for a variety of small molecules and anions are shown in Table 1. Interestingly, protonation of a molecule or atom, in general, is highly exothermic. Provided no better proton acceptor is available, any kind of chemical bond is easily attacked by protons. This result simply reflects the high reactivity of the free proton.

Some years ago, many physicochemists would have been rather suspicious of the usefulness of such a general acidity scale. Due to very extensive studies on protonation equilibria and species occurring in highly acidic superacids such as e.g. "magic acid" ($HF.SbF_5$) and others (OLAH, 1973), however, we know a great deal about the chemistry of weakly solvated protons. The almost equal proton affinities of hydrogen fluoride and methane indicate that we can expect a good deal of interesting chemistry involving carbocations when we succeed in keeping water and compounds or functional groups of similar proton affinities away from the reaction mixture.

For our purpose the table of proton affinities should tell us simply what kind of molecules or functional groups we can expect to be protonated in a certain environment. So far we have not discussed the influence of zeropoint energies and the effect of temperature, particularly entropic effects on protonation equilibria. Since we are mainly interested in proton transfer in solution we will come back to these effects when we discuss the influence of the environment on energies of proton transfer (cf. Sect. II.E).

B. Stability of Hydrogen Bonded Molecular Complexes

Recently, a number of review articles and books on hydrogen bonding in general have been published (see e.g. SCHUSTER et al., 1976b and references therein). Energetic aspects of hydrogen-bonded complexes have been discussed extensively (KOLLMAN and ALLEN, 1972; KOLLMAN et al., 1975; SCHUSTER, 1973, 1976a). We can restrict ourselves here to a short table summarizing most reliable theoretical and experimental data on hydrogen bond strength in the vapor phase. Considering the energies of interaction in Table 2 we can subdivide the hydrogen-bonded complexes into two classes: ordinary hydrogen bonds as occurring between neutral, polar molecules with the essential functional groups and strong hydrogen bonds formed by suitable combinations of ions and polar molecules.

In the first class of complexes the energies of interaction in general lie between $5 < |\Delta E| < 10$ kcal/mol. For a given proton donor molecule we find an increase of $|\Delta E|$ with the proton affinity of the acceptor molecule forming the complex. The inverse is also true: for a given acceptor the hydrogen bond strength increases with the acidity of the proton donor.

Complexes of the second type are usually formed in one of the following reactions:

Table 1. Proton affinities and proton transfer energies in the vapor phase (energies in kcal/mole)

Reaction[a]	ΔE_{calc}[b]	Reference	ΔH_{exp}[b]	ΔW_0^o[b]	Reference
$He + H^+$	- 46.0	WOLNIEWICZ (1965)	⩽ - 41.5		VEDENEYEV et al. (1966)
			- 42.4		CHUPKA and RUSSEL (1968)
$Kr + H^+$			- 85±7		VEDENEYEV et al. (1966)
			⩽ 78.3		CHUPKA and RUSSEL (1968)
$H_2 + H^+$	-102	KUTZELNIGG et al. (1967)	⩽ 91.4		CHUPKA et al. (1968)
$HF + H^+$	-121	LISCHKA (1973)			
$CH_4 + H^+$	-129	DYCZMONS and KUTZELNIGG (1974)	-122±7		VEDENEYEV et al. (1966)
$CO + H^+$	-120	HOPKINSON et al. (1968)	-131	5.0	HOPKINSON et al. (1968)
$C_2H_4 + H^+$	-169	ZURAWSKI et al. (1973)	-157±8		VEDENEYEV et al. (1966)
$H_2CO + H^+$	-172	HOPKINSON et al. (1968)	-161±3		HOPKINSON et al. (1968)
$H_2O + H^+$	-174	LISCHKA (1973)	-169±2	5.3	VEDENEYEV et al. (1966)
	-168	DIERCKSEN et al. (1975)	-164±4		BEAUCHAMP and BUTTRILL (1968)
			-168±3		CHONG et al. (1972), COTTER and KOSKI (1973)
$H_2S + H^+$			-178±2		BEAUCHAMP and BUTTRILL (1968)
$CH_3OH + H^+$			-180±3		VEDENEYEV et al. (1966)
$C_2H_5OH + H^+$			-193±9		VEDENEYEV et al. (1966)
$NH_3 + H^+$	-213	HOPKINSON et al. (1968)	-216		VEDENEYEV et al. (1966)
$e_0^- + H^+$	-314[c]		-314[c]		
$J^- + H^+$			-308	3.2	HERZBERG (1950), SCHUSTER (1976b)
$Br^- + H^+$			-318	3.7	HERZBERG (1950), SCHUSTER (1976b)
$Cl^- + H^+$			-324	4.1	HERZBERG (1950), SCHUSTER (1976b)
$F^- + H^+$	-373	LISCHKA (1973)	-366	5.8	HERZBERG (1950), SCHUSTER (1976b)
$HO^- + H^+$	-396	LISCHKA (1973)	-411	7.5	HOPKINSON et al. (1968)

$H_2N^- + H^+$	-427	HOPKINSON et al. (1968)	-380 (-422)	8.8	HOPKINSON et al. (1968)
$H_3N + HJ$	102	SCHUSTER (1976b)			
$H_3N + HBr$	112	SCHUSTER (1976b)			
$H_3N + HCl$	118	SCHUSTER (1976b)			
$H_2O + HCl$	143	SCHUSTER (1976b)			
$H_3N + HF$	160	SCHUSTER (1976b)			
$H_2O + HF$	185	SCHUSTER (1976b)			
$H_3N + H_3N$	214	HOPKINSON et al. (1968)			
$H_2O + H_2O$	222	LISCHKA (1973)			
$HF + HF$	252	LISCHKA (1973)			

[a] The reactions regarded here are the following: $B + H^+ \rightleftarrows [BH]^+$ and $B + HA \rightleftarrows [BH]^+ + A^-$.

[b] Energies obtained from quantum mechanical calculations (ΔE), reaction enthalpies (ΔH) and zero point enery corrections (ΔW_0^0): $\Delta E \sim \Delta H - \Delta W_0^0$; proton transfer energies were calculated as differences of the most reliable proton affinities available.

[c] Ionization potential of H.

Table 2. Energies of interaction in hydrogen bonded complexes (SCHUSTER, 1976a; DILL et al., 1975; KOLLMAN et al., 1975)

Proton acceptor	Proton donnor	ΔE_{calc}[a] (kcal/mole)	ΔH_{exp} (kcal/mole)
H_2S	HSH	- 0.7	
HCl	HCl	- 1.5	
H_2O	HNH_2	- 2.3	
H_3N	HNH_2	- 2.7	- 4.5
HF	HF	- 3.5	- 6±1
H_2O	HOH	- 3.7	- 5.1
H_2O	HF	- 9.4	
H_3N	HF	-12.2	
H_3N	HCl	-12.6	
Cl^-	HOH	-11.9	-13.1
HO^-	HOH	-24.3	-22.5
F^-	HOH	-24.1	-23.3
H_3N	HNH_3^+	-25.9	-24.8
HF	HFH^+	-30.7	
H_2O	HOH_2^+	-32.2	-31.6
F^-	HF	-44[b]	-60.2

[a]Energy values obtained by the most extended ab initio quantum mechanical calculation available at present. For original references see SCHUSTER (1976a).
[b]KEIL and AHLRICHS (1976).

$$XH^+ + Y \rightleftharpoons [X\text{-}H\text{-}Y]^+ \rightleftharpoons X + YH^+ \qquad (3a)$$

$$X^- + HY \rightleftharpoons [X\text{-}H\text{-}Y]^- \rightleftharpoons XH + Y^- \qquad (3b)$$

The energies of interaction are substantially larger than in the previous examples and lie between $15 < |\Delta E| < 45$kcal/mol. The higher complex stability, essentially, is a result of the large electrostatic contribution (ΔE_{COU}, see e.g. SCHUSTER, 1976a,b) due to ion-dipole attraction. Additionally, highly symmetric structures like $[FHF]^-$ or $[H_2OHOH_2]^+$ represent individual molecular entities rather than simple intermolecular associations, since the geometry of the isolated subsystems is changed drastically on complex formation. The central hydrogen atom is involved here in two bonds of equal covalent character (e.g. SCHUSTER, 1976b). Using the information presented in Tables 1 and 2 we are already in a position to give a very rough estimate on the possible existence and nature of stable complexes formed by two neutral molecules in the vapor phase. All pairs of molecules, which can form ordinary hydrogen bonds, will form stable complexes as well, since the depths of the potential curves at the minima are always large enough to guarantee a sufficient number of bound vibrational levels. The difference in energy between a hypothetical neutral complex and the corresponding ion pair (ΔE_{NI}) can be derived from Eq. (4). ΔE_{NI} consists of two major contributions of opposite

$$AH+B \longrightarrow \underset{NC}{[AH\ldots B]};\ \underset{IC}{[A^{-}\ldots HB^{+}]} \longleftarrow A^{-}+HB^{+}$$

$$\Delta E_{NI} = E_{IC}-E_{NC} = \Delta E_{PA}(B) - \Delta E_{PA}(A^{-}) + \Delta E_{COU}(IC) + \ldots$$

$$\Delta E_{COU}(IC) \sim -\frac{e_0^2}{R_{AB}} \qquad (4)^{2}$$

sign: the proton transfer energy $\Delta E_{PT} = \Delta E_{PA}(B) - \Delta E_{PA}(A^{-})$ corresponding to proton transfer at infinite separation and the electrostatic contribution to the energy of ion pair formation. There are many other energetic terms contributing to ΔE_{NI}, but these are much smaller and cancel each other out to a large extract. In almost all cases two neutral molecules will form a neutral complex (NC). For comparison with highly unpolar media like the interior of some proteins or inside membranes it is interesting to see if there is any chance that aggregates with structures close to ionic complexes (IC) will occur in the vapor phase. We have chosen, therefore, a series of complexes including the most promising candidates for stable associated ion pairs. They are represented, of course, by those pairs of molecules, which show the smallest proton transfer energies in absolute value (ΔE_{PT} in Table 1). Among all examples we have considered here, the series of complexes formed from ammonia and the hydrogen halides best meet the above condition. In this series the association of ammonia and hydrogen fluoride, $NH_3 \cdot HF$, plays an exceptional role (Table 3). Our crude estimation leads to an energy difference ΔE_{NI}=40kcal/mol, thus indicating that the nature of this aggregate comes very close to a neutral complex (NC). For all the other ammonia-hydrogen-halide complexes, however, we obtain ΔE_{NI} values in the order of $-3 < \Delta E_{NI} < 3$ kcal/mol. Consequently, the structures of these associations seem to have an intermediate structure between the extreme examples NC and IC or, accounting for some errors in our estimate, might well resemble an ionic complex. The important conclusion we would like to draw from these results can be summarized as follows: for some pairs of molecules ionic complexes can be formed readily by a proton transfer reaction in the vapor phase even if the ion pair becomes unstable with respect to the neutral complex at larger distances.

---

[2]The estimates presented here are based on a comparison of two hypothetical states, NC and IC, with frozen geometries, which do not necessarily correspond to the actual structure of the complex (for numerical details cf. SCHUSTER, 1976b). In all real examples investigated until now only one complex is formed in the vapor phase, which neglecting entropic effects for the moment, will come close to NC if $\Delta E_{NI} \gg 0$, close to IC if $\Delta E_{NI} \ll 0$ or which will be of intermediate character in case $\Delta E_{NI} \sim 0$.

Table 3. Stability of ionic complexes in associations of ammonia and hydrogen halides

Base	Acid	ΔE_{PT} (kcal/mole)	e_0^2/R_{NX} (kcal/mole)	$\Delta E_{PT} - e_0^2/R_{NX}$ (kcal/mole)
H_3N	HF	160	124	36
H_3N	HCl	118	117	1
H_3N	HBr	112	109	3
H_3N	HJ	102	102	0
H_2O[a]	HCl[a]	143	107	36

[a]For comparison, cf. SCHUSTER (1976b)

C. Potential Curves for Proton Transfer

The existence of one or the other or both intermediate complexes in proton transfer reactions is demonstrated best by potential curves for proton motion along the hydrogen bond. Although various model potentials have been used extensively in the past, direct information on the shape of the actual potential has become available only very recently. Very extensive quantum mechanical calculations have been performed (see e.g. SCHUSTER et al., 1974, and a recent paper STØGÅRD et al., 1975). Additionally, some results were obtained from vapor phase dynamical studies (cf. Sect. II.D).

In ordinary hydrogen-bonded complexes asymetric single minimum potential curves for proton motion (Fig. 1) are obtained at the

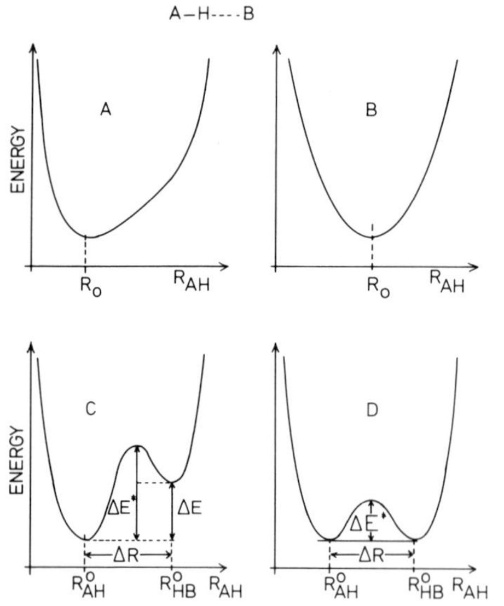

Fig. 1A-D. Potential curves for proton transfer along hydrogen bonds. (A) Asymmetric, (B) symmetric single minimum potential; (C) asymmetric, (D) symmetric double minimum potential

equilibrium distance of the two heavier atoms R^o_{XY}. The position of the proton at the energy minimum corresponds very well to that which we would expect for a neutral complex. The original two subsystems can be easily recognized since the molecules are hardly distorted on complex formation.

Among the complexes studied in detail until now only one exception was found: $H_3N.HCl$ (CLEMENTI and GAYLES, 1967; KOLLMAN et al., 1975). Again a single minimum potential at the equilibrium distance $R_{NCl} = 2.85$ Å was found. The position of the energy minimum ($R^o_{HCl} = 1.67$ Å) however, indicates substantial elongation of the HCl bond compared to the free hydrogen chloride molecule ($R^o_{HCl} = 1.27$ Å). In fact the equilibrium position of the proton is closer to the nitrogen atom $R^o_{NH} = 1.18$ Å, but compared to the ammonium ion (NH_4^+ : $R^o_{NH} = 1.03$ Å), we realize non-negligible elongation of the NH bond too. The complex is a whole, therefore, is characterized best as an intermediate structure between the neutral association (NC) and the hydrogen-bonded ion pair (IC).

In case the two heavy atoms involved in the hydrogen bond are fixed at distances substantially larger than R^o_{XY} asymmetric double minimum potentials have been obtained in all calculations reported until now (Fig. 1).

Complexes with strong hydrogen bonds, on the other hand, show entirely different potential curves for proton motion which are either symmetric or come very close to a symmetric curve (Fig. 1). Depending on the intermolecular distance (R_{XY}) one or two equivalent energy minima were found. The most accurate calculations performed on strong hydrogen bonds until now suggest single minimum potentials for $[FHF]^-$ (STØGARD et al., 1975) $[H_2OHOH_2]^+$ (MEYER et al., 1973) and $[HOHOH]^-$ (STØGARD et al., 1975) and a double minimum potential for $[H_3NHNH_3]^+$ (MERLET et al., 1972) at the equilibrium geometries. In $[FHF]^-$ and even much more pronounced in $[H_2OHOH_2]^+$ very flat potential curves for the motion of the central proton were obtained. Displacement of this proton, consequently, does not require much energy and a number of interesting results can be explained by this fact: the crystal field can shift the equilibrium position of the proton easily and many different geometries of $[H_2OHOH_2]^+$ and $[FHF]^-$ were found in the solid state (LUNDGREN and OLOVSSON, 1976; WILLIAMS and SCHNEEMEYER, 1973). Additionally, structures with hydrogen bonds of the strong symmetric type and flat potentials show unusually high electric polarizabilities (JANOSCHEK et al., 1972; ZUNDEL, 1976).

D. Dynamics of Proton Transfer in the Vapor Phase

Due to new developments in experimental methodology, in particular to the flowing afterglow technique (for a recent review cf. FERGUSON, 1972), dynamic studies on proton transfer reactions in the vapor phase have now become accessible. We will mention here only the recent results obtained by the Canadian research group (BOHME, 1975; HEMSWORTH et al., 1974; BETOWSKI et al., 1975) for reactions of type

$$XH^+ + Y \rightleftharpoons X + HY^+ \qquad (5a)$$

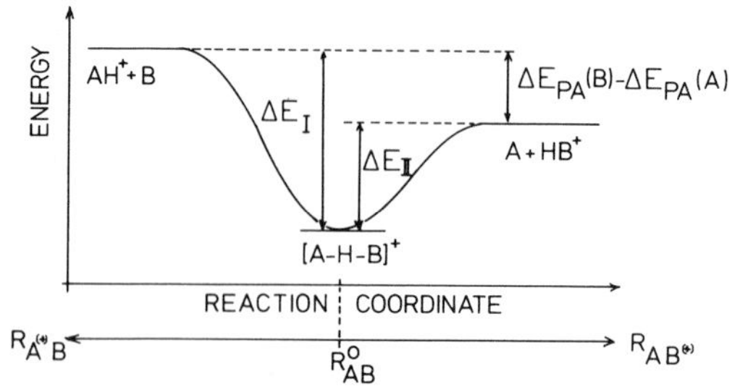

Fig. 2. Potential curves for ion-molecule proton transfer reactions in the vapor phase

and

$$XH + Y^- \rightleftharpoons X^- + HY \qquad (5b)$$

Exothermic proton transfer reactions of both types were found to proceed extremely rapidly at thermal energies, $10^{-10} < k$ (300 K) < 10^{-8} cm^3 $molecule^{-1}$ s^{-1}, without any detectable energy or entropy of activation. Proton transfer thus occurs almost at every collision.

Collision rate coefficients for these proton transfer reactions can be predicted by classical electrostatic theory of ion-molecule reactions provided a correction for preorientation of the molecule in the field of the approaching ion is introduced (average-dipole orientation theory, SU and BOWERS, 1973). BOHME (1975) suggests an energy profile as shown in Figure 2 for both kinds of proton transfer reactions. Clearly, the energy minimum corresponds to the hydrogen-bonded complex discussed previously (cf. Eq. (3)).

Besides their informative value on the nature of proton transfer processes as such, flowing afterglow studies on reactions of type (5) provide excellent access to important thermodynamic data such as proton affinities, heats of formation for anions, ionization potentials and electron affinities (BOHME, 1975).

E. Gas Phase Solvation

High pressure mass spectrometry as developed during the last decade (KEBARLE, 1972) provides an excellent tool for investigations on ionic clustering equilibria in the vapor phase. Since clusters containing up to eight and more molecules surrounding a central ion have been identified and determined quantitatively, the whole series of processes was called gas phase solvation. The data obtained directly from mass spectrometric studies are equilibrium constants for stepwise cluster formation:

$$X^{\pm} + nS \xrightleftharpoons{K_{0,1}} [XS]^{(\pm)} + (n-1)S \xrightleftharpoons{K_{1,2}} [XS_2]^{(\pm)} + (n-2)S \rightleftharpoons \cdots\cdots \rightleftharpoons [XS_{n-1}]^{(\pm)} + S \xrightleftharpoons{K_{n-1,n}} [XS_n]^{(\pm)} \qquad (6)$$

Thermodynamic functions, ΔG^o, ΔH and ΔS^o are derived from equilibrium constants and their dependence on temperature. Liquid phase solvation data, in principle, differ from the results of gas phase studies by two crucial points: in condensed phase the number of molecules approaches infinity ($n \to \infty$) and, moreover, the solvent molecules to interact substantially, forming the liquid solvent ($nS \rightleftarrows S_n$). Nevertheless, vapor phase solvation data present valuable information on the energetic background of inner solvation shells (cf. also SCHUSTER et al., 1975). For our purpose here we are exclusively interested in the data on solvation of protons and hydroxide ions.

Gas phase proton hydration has been studied extensively (KEBARLE et al., 1967; GRIMSRUD and KEBARLE, 1973). One of the results obtained is concerned directly with one of the early questions of proton solvation in water: is there any particular stable structure of the hydrated proton with a fixed number of water molecules such as the highly symmetric $[H_9O_4]^+$ complex postulated by EIGEN (1963)?

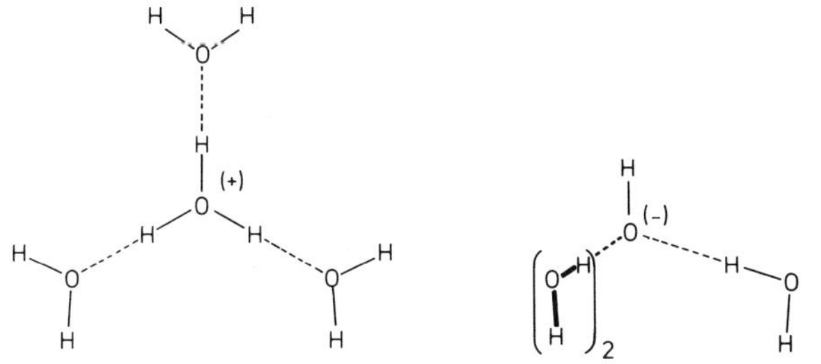

I $[H_9O_4]^+$ II $[H_7O_4]^-$

The results of vapor phase studies are summarized in Figures 3 and 4 and give no hint on a dominating species among the hydrated protons, $H^+(H_2O)_n$. Both functions $[\Delta H_{n-1,n}]$ and $[\Delta G^o_{n-1,n}]$ decrease

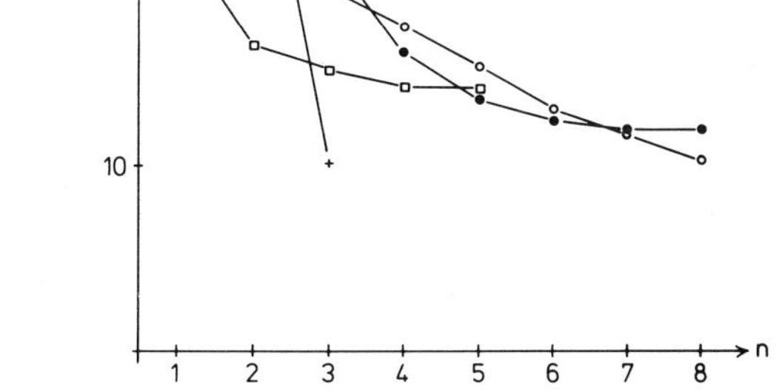

Fig. 3. Proton solvation enthalpies ($\Delta H_{n-1,n}$) in the vapor phase (o ... $H^+(H_2O)_n$, ● ... $H^+(CH_3OH)_n$, + ... $H^+(CH_3OCH_3)_n$ and □ ... $HO^-(H_2O)_n$) (KEBARLE, 1972; GRIMSRUD and KEBARLE, 1973)

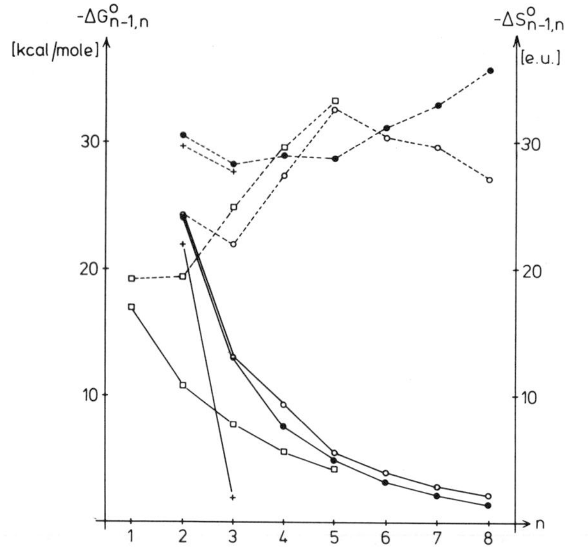

Fig. 4. Free proton solvation enthalpies ($\Delta G^o_{n-1,n}$) and solvation entropies ($\Delta S^o_{n-1,n}$) in the vapor phase (ΔG: ——, ΔS: ---- ; o ... $H^+(H_2O)_n$, ● ... $H^+(CH_3OH)_n$, + ... $H^+(CH_3OCH_3)_n$ and □ ... $HO^-(H_2O)_n$) (KEBARLE, 1972; GRIMSRUD and KEBARLE, 1973)

gradually with increasing n. Very careful inspection of the curves shows a small inflexion or irregularity in slope at n=4 from which one might derive that within the series of decreasing stability $[H_9O_4]^+$ is somewhat more stable than the other complexes.

Compared to the results of proton hydration water molecules are bound less strongly to hydroxide ion (ARSHADI and KEBARLE, 1970). Here, even a small limit on an entranced stability of the anion $[H_7O_4]^-$ is hardly detectable. Although vapor phase data are not conclusive for predicting structure in the liquid state, we have to admit at least that there is no support for a thermodynamic preference of particular hydrated ions like $[H_9O_4]^+$ or $[H_7O_4]^-$ from the results reported here.

The entropy of stepwise hydration, $\Delta S^o_{n-1,n}$, shown in Figure 4 is large and negative ($\Delta S^o \sim -30$ e.u.). Large scattering of the data indicates substantial experimental errors. On the average, however, the $\Delta S^o_{n-1,n}$ values do not change very much with increasing cluster size. The negative sign and the relatively large value of the reaction entropy, clearly result from the immobilization of the water molecule, when it is attached to the complex. Complete fixation of a water molecule resulting in frozen translational and rotational degrees of freedom would lead to $\Delta S^o_{n-1,n} = -45.1$ e.u. at T = 300 K. From this and the observed value we can expect residual mobility of the molecules in the cluster.

Comparing proton solvation data for different solvent molecules – water, methanol and dimethylether – (Figs. 3 and 4) we realize that there is a drastic difference between the first two molecules and the third. Solvation of the proton by methanol is very similar to hydration. $\Delta G^o_{n,n-1}$ values are roughly the same due to compensating changes in ΔH and ΔS^o. In the case of methanol as solvent molecule both ΔH and ΔS^o are somewhat larger in absolute value. Dimethyl ether, however, forms only a first solvation layer around the proton. After this layer is completed (n=3) no further solvent molecule is bound.

These results represent an illustrative example for the difference in structure of solvation shells formed in associated liquids, where solvent molecules act as both hydrogen bond donors and acceptors, and in the other polar solvents lacking specific structure due to hydrogen bonding.

Finally, we would like to mention here also studies on competitive solvation as reported by PAYZANT et al. (1973). In mixtures of water and methanol vapor the proton is preferentially solvated by methanol molecules in small clusters ($n<9$) whereas the inverse becomes true for larger complexes ($n>9$). Proton solvation in ammonia-water mixtures is much more complicated. As expected from proton affinities (Table 1) the proton is almost exclusively bound to ammonia forming a NH_4^+ cation, which then represents the center for further solvation. In the first solvation layer of NH_4^+ ammonia is preferred over water by a factor of about 20. The second layer, however, has an opposite composition: here water molecules dominate. Ionic clusters consisting of different solvent molecules, as we have seen, can have distinct ordered structures.

F. Theoretical Concepts and Mechanisms of Proton Transfer in Solution

In the previous section we tried to give a representative summary of the most important vapor phase data on proton transfer available due to new techniques. Applying this information to proton transfer in solution we have to face enormous complication of different origin, which prevents any straightforward theoretical ab initio treatment of chemical processes in the liquid state. Several model theories have been conceived and were applied more or less successfully to interpret the great variety of experimental data. We cannot present a complete survey of all kinds of approaches in a few lines here, but we will try to point out the most important problems, which seem to be the subject of promising investigations in the future and give some references as entries to more extensive literature.

Double minimum potentials for proton transfer at the equilibrium distances (R^o_{XY}) of stable complexes are rather exceptional in the vapor phase. In solution the situation can be expected to be entirely different. As a consequence of the interaction with the surrounding, intermolecular distances in solvated ionic complexes are remarkably larger than in the vapor phase (e.g. $[H_5O_2]^+$) isolated complex: $R^o_{OO} = 2.39$ A (KRAEMER and DIERCKSEN, 1970); crystal: $R^o_{OO} = 2.61$ A (average distance, LUNDGREN and OLOVSSON, 1976). Therefore double minimum potentials will occur in most examples. Additionally, solvation will diminish the energy differences between neutral and ionic complexes, since the solvation energy of an ion pair is much more negative than that of an aggregate of neutral molecules. Highly polar solvents might even invert the relative stability and the ionic complex will become the favored structure. In solvents with high dielectric constants, finally, the ion pair will dissociate into free ions.

In general we have to distinguish two kinds of potentials for proton motion in a solvated complex: (1) the adiabatic potential, where the whole complex is at equilibrium with the exception of the position of the proton and perhaps more realistic for actual processes (2) the potential curve for proton motion at frozen geometries of the environment. Different kinds of model calculations simulating a solvation shell were performed on $H_3N..HF$ (YAMABE et al., 1973; SCHUSTER et al., 1974) and $[H_2O.H.OH_2]^+$ (SCHUSTER et al., 1974) and confirmed qualitatively by the suggestions made above. Quantitatively reliable results are not available now and cannot be expected for the near future, because all reasonable molecular models are much too large for accurate ab initio calculations. Nevertheless, two additional results could be derived from these model studies: (1) solvation increases the barriers for proton transfer already at fixed intermolecular distance and (2) proton transfer in a neutral complex causes substantial rearrangement in the first solvation shell. Solvent molecules are bound much more tightly in the first solvation shell of the ionic complex. Consequently a net contraction of the solvation shell takes place, when the proton is shifted towards the ionic energy minimum and we observe a negative ΔV on the macroscopic level.

Proton motion in a double well potential has been investigated theoretically by several groups. BRICKMANN (1976) recently presented a review on a quantum mechanical study on proton tunneling. The life times (τ_L) for a given protomeric structure depends largely on three parameters of the double minimum potential: ΔR, $\Delta E^{\ddagger}$ and ΔE (cf. Fig. 1). Ultrafast proton transfer ($\tau_L \sim 10^{-12}$ s, see e.g. GRUNWALD, 1965) can occur in complexes, where the parameters are in the following ranges: $\Delta R < 0.6$ Å, $\Delta E^{\ddagger} < 20$ kcal/mol and $|\Delta E| < 0.06$ kcal/mol. These conditions are fulfilled in structures with moderately strong hydrogen bonds formed between N, O and F atoms.

Two theoretical treatments both starting from the theory of absolute reaction rates were mainly applied to many-step reactions, which are not directly related to biochemistry. We refer here only to the original articles or summaries and a more recent paper in the field: (1) a theoretical relation between rate constants, barrier heights and free enthalpies of reaction (MARCUS, 1968, 1969; KREEVOY and KONASEVICH, 1971) and (2) a correction for proton tunneling (CALDIN, 1969; CALDIN et al., 1975), both applied mainly to slow reactions involving ionic dissociation of CH bonds.

Conventional transition state theory of chemical reactions is based on the assumption of an equilibrium distribution of states at the activated complex. Proton transfer along hydrogen bonds can be extremely fast as we have seen from vapor phase data and consequently the solvent molecules in the environment hardly have a chance to equilibrate. From proton conductivities in water and ice EIGEN (1963) on the contrary, suggested that the proton jumps

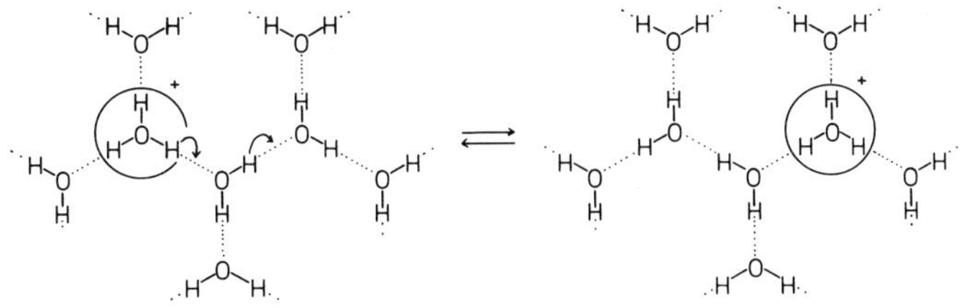

Fig. 5. Two proton jump mechanisms for excess proton migration in water as suggested by FANG et al., 1973

back and forth between the two equilibrium positions until the surrounding solvent molecules undergo an appropriate structural rearrangement and thereby create a new center for the excess proton. Nonadiabatic reactions have been investigated by a number of theoreticians, who have proposed various models to overcome this difficulty in proton transfer reactions (see e.g. FISCHER et al., 1969; CHRISTOV, 1968; KURZ and KURZ, 1972). Further complication arises in associated liquids, where hydrogen bonded networks might extend over a certain length. Strong correlation between the motion of individual protons lead to more or less simultaneous jumps of two or even more protons along the chain of appropriately hydrogen bonds. FANG et al. (1973) performed model calculations on the cluster $[H_{21}O_{10}]^+$ and found evidence for the importance of synchroneous two proton motions (Fig. 5). As we have seen proton motion in a cluster is strongly coupled to reorientation phenomena in the surrounding liquid, which have to be considered in detail and presumably on a molecular level in further developments of the theory of proton transfer in solution.

In contrast to the open problems in theory, there is a great variety of experimental facts, which can be interpreted perfectly by phenomenological kinetic theory and to which we will now turn.

III. Proton Transfer in Aqueous Solution

From previous sections we have to expect that dynamics of proton transfer in associated liquids is much more complicated than in non hydrogen-bonded solvents. Despite the complex nature of the elementary steps on a molecular level the phenomenological kinetics of proton transfer in aqueous solution is very well established nowadays. Rate constants and mechanisms have been confirmed by different methods of investigation. Chemical relaxation techniques, mainly temperature-, pressure- and electric field jumps as well as ultrasound absorption, and NMR methods have been applied. In a way NMR investigations are complementary to relaxation studies because they can be extended easily to symmetric exchange reactions for which ΔH, $\Delta\mu$ and ΔV vanish. From a practical point of view it seems appropriate to split the discussion into two parts, one dealing with inter-, the other with intramolecular proton transfer in water. Furthermore, we decided to make another division: we will concentrate here on the mechanisms and the results for small and medium size molecules, whereas in Sect. V we will come back to investigation on compounds of predominantly biological interest.

A. Intermolecular Proton Transfer

More than ten years ago intermolecular proton transfer in aqueous systems was studied extensively by EIGEN and coworkers and their results have been summarized in a series of articles (EIGEN and DE MAEYER, 1963; EIGEN, 1963; EIGEN et al., 1964). A three-step mechanism (7a), formally identical to that shown already in Eq. (1b), was applied:

$$\underset{(a)}{AH + B} \underset{k_{ba}}{\overset{k_{ab}}{\rightleftharpoons}} \underset{(b)}{AH...B} \underset{k_{cb}}{\overset{k_{bc}}{\rightleftharpoons}} \underset{(c)}{A^-...HB^+} \underset{k_{dc}}{\overset{k_{cd}}{\rightleftharpoons}} \underset{(d)}{A^- + HB^+} \qquad (7a)$$

In aqueous solution proton transfer usually occurs much faster than the diffusion controlled formation of the complexes $\underline{b}$ and $\underline{c}$ ($k_{ba} \ll k_{bc}$ and $k_{cd} \ll k_{cb}$). The complexes need not necessarily be as simple as indicated here. NMR measurements (e.g. GRUNWALD and RALPH, 1975) have shown that proton transfer can take place also in complexes with one or two intermediate water molecules, e.g.:

$$\underset{(b')}{AH...{}_{H}OH...B} \underset{k'_{cb}}{\overset{k'_{bc}}{\rightleftharpoons}} \underset{(c')}{A^-...HO_H...HB^+} \qquad (7a')$$

For stationary concentration of $\delta c_b \sim 0$, $\delta c_c \sim 0$ we obtain a single-step mechanism for proton transfer:

$$AH + B \underset{k_{21}}{\overset{k_{12}}{\rightleftharpoons}} A^- + HB^+$$

$$k_{12} = \frac{k_{ab}}{1+\frac{k_{ba}}{k_{cd}}\frac{k_{bc}}{k_{cb}}}; \qquad k_{21} = \frac{k_{dc}}{1+\frac{k_{cd}}{k_{ba}}\frac{k_{cb}}{k_{bc}}} \qquad (7b)$$

The well-known results of relaxation studies on reactions of this type can be summarized in a few statements:

1. Proton transfer to the amphoteric solvent molecule occurs in two parallel reactions of protolysis and hydrolysis (8a,b), which are of equal importance around neutral pH 7:[3]

$$AH\ (+\ H_2O) \underset{k_R}{\overset{k_D}{\rightleftharpoons}} A^- + H^+ \cdot (H_2O) \qquad (8a)$$

$$AH + OH^- \underset{k'_D}{\overset{k'_R}{\rightleftharpoons}} A^-\ (+\ H_2O) \qquad (8b)$$

[3] In Eq. (8) the solvent molecule directly involved is included in parenthesis in order to demonstrate the complete analogy to Eq. (7). It will be omitted in the following equations.

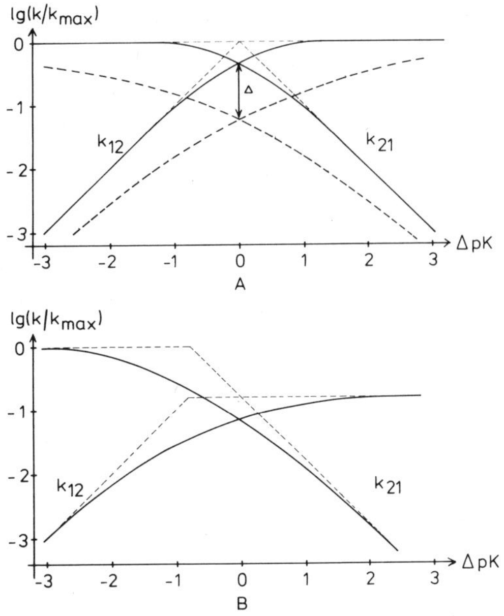

Fig. 6. Rate constants for intermolecular proton exchange in aqueous solution as function of difference in pK-values of two competing acids AH and BH. (A) Equal diffusion controlled limits for k_{12} and k_{21}; —— diffusion controlled rates; ---- slower rates due to insufficient structures of hydration shells, e.g. as found for acidic SH, PH and CH groups. (B) Different diffusion controlled limits for k_{12} and k_{21}, e.g. as a consequence of charge separation, $AH+B \rightleftarrows A^-+HB^+$

2. Rate constants for intermolecular proton transfer follow a generalized BRØNSTED relation:

$$\frac{d\log k_{12}}{d\Delta pK} = \alpha; \quad \frac{d\log k_{21}}{d\Delta pK} = -\beta = -(1-\alpha); \quad 0 \leq \alpha \leq 1 \qquad (9)$$

ΔpK stands here for the difference in pK values between the two acids ($\Delta pK = pK_{HB^+} - pK_{HA}$). Diagrams according to Eq. (9) are shown in Figure 6. In case the complexes b and c are held together by moderately strong hydrogen bonds e.g. those of the type OH...O, OH...N, NH...O or NH...N, one of the two rate constants, k_{12} or k_{21}, reaches the limits of diffusion-controlled processes. The other constant of course is determined by the value of ΔpK itself:

$$k_{21} = k_{12} \cdot K = k_{12} \cdot 10^{-\Delta pK} \qquad (10)$$

At the crossing point of the two curves, $k_{12} = k_{21}$ or $\Delta pK = 0$, the rate constants have to be smaller than the maximum value at least by a factor of two, since the complexes will dissociate in both directions with equal probability. Many concrete series of compounds investigated come very close to the limits of diffusion control (Table 4).

Two kinds of deviations from the ideal curves were reported:
3. The diffusion controlled limits for $\Delta pK \ll 0$ and $\Delta pK \gg 0$ are different. This deviation will be found very often, particularly in the case of proton transfer processes involving a change in net charges like that shown in Eq. (7). The encounter of the oppositely charged ions will be accelerated by electrostatic forces and consequently the diffusion controlled limit for k_{21} is somewhat larger (Fig. 6B). Another common reason leading to two dif-

Table 4. Proton transfer in aqueous solution

Reaction	$k_R[M^{-1}s^{-1}]$	$k_D[s^{-1}]$	Temp (°C)	Method	Reference
$H^+ + OH^- \rightleftarrows H_2O$	$1.4 \cdot 10^{11}$	$5 \cdot 10^{-5}$	25	E	EIGEN and DE MAEYER (1955)
$H^+ + H_2O \rightleftarrows H_3O^+$ [a]	$1.1 \cdot 10^{10}$	-	25	NMR	LOEWENSTEIN and SZÖKE (1962)
$OH^- + H_2O \rightleftarrows H_2O + OH^-$ [a]	$3.8 \cdot 10^{9}$	-	25	NMR	MEIBOOM (1961)
$H^+ + F^- \rightleftarrows HF$	$1.0 \cdot 10^{11}$	$7 \cdot 10^{7}$	25	E-disp.	EIGEN and KUSTIN (1960)
$H^+ + HS^- \rightleftarrows H_2S$	$7.5 \cdot 10^{10}$	$4.3 \cdot 10^{3}$	25	E-disp.	EIGEN and KUSTIN (1960)
$H^+ + NH_3 \rightleftarrows NH_4^+$	$7.7 \cdot 10^{8}$		25	NMR	EMERSON et al. (1960)
$H^+ + (CH_3)_3N \rightleftarrows (CH_3)_3NH^+$	$5.4 \cdot 10^{8}$		25	NMR	GRUNWALD (1967)
$NH_4^+ + H_2O + NH_3 \rightleftarrows NH_3 + H_2O + NH_4^+$	$1.0 \cdot 10^{8}$	-	25	NMR	GRUNWALD et al. (1957), BIANCHIN et al. (1975)
$NH_4^+ + NH_3 \rightleftarrows NH_3 + NH_4^+$	$1.17 \cdot 10^{9}$	-	25	NMR	GRUNWALD et al. (1957), BIANCHIN et al. (1975)
$CH_3NH_3^+ + CH_3NH_2 \rightleftarrows CH_3NH_2 + CH_3NH_3^+$	$4.0 \cdot 10^{8}$	-	25	NMR	GRUNWALD et al. (1957), BIANCHIN et al. (1975)
$CH_3NH_3^+ + H_2O + CH_3NH_2 \rightleftarrows CH_3NH_2 + H_2O + CH_3NH_3^+$	$5.3 \cdot 10^{8}$	-	25	NMR	GRUNWALD et al. (1957), BIANCHIN et al. (1975)
$(CH_3)_2NH_2^+ + (CH_3)_2NH \rightleftarrows (CH_3)_2NH + (CH_3)_2NH_2^+$	$5.0 \cdot 10^{7}$	-	25	NMR	GRUNWALD et al. (1957), BIANCHIN et al. (1975)
$(CH_3)_2NH_2^+ + H_2O + (CH_3)_2NH \rightleftarrows (CH_3)_2NH + H_2O + (CH_3)_2NH_2^+$	$5.6 \cdot 10^{8}$	-	25	NMR	GRUNWALD et al. (1957), BIANCHIN et al. (1975)
$(CH_3)_3NH^+ + (CH_3)_3N \rightleftarrows (CH_3)_3N + (CH_3)_3NH^+$	$3.0 \cdot 10^{7}$	-	25	NMR	GRUNWALD et al. (1957), BIANCHIN et al. (1975)
$(CH_3)_3NH^+ + H_2O + (CH_3)_3N \rightleftarrows (CH_3)_3N + H_2O(CH_3)_3NH^+$	$3.7 \cdot 10^{8}$	-	25	NMR	GRUNWALD et al. (1957), BIANCHIN et al. (1975)
$(C_2H_5)_3N + H_2O \rightleftarrows (C_2H_5)_2NH^+ + OH^-$	$1.7 \cdot 10^{10}$		30	NMR	RALPH and GRUNWALD (1967)
	$2.0 \cdot 10^{10}$		30	sound	BRUNDAGE and KUSTIN (1970)

$H^+ + CH_3COO^- \rightleftarrows CH_3COOH$	$4.5 \cdot 10^{10}$	$7.8 \cdot 10^{5}$	25	E-disp.	EIGEN and SCHOEN (1955)
$H^+ + \Phi COO^- \rightleftarrows \Phi COOH$[b]	$3.5 \cdot 10^{10}$	$2.2 \cdot 10^{6}$	25	E-disp.	EIGEN and EYRING (1962)
$H^+ + pNO_2\Phi O^- \rightleftarrows pNO_2\Phi OH$[b]	$3.6 \cdot 10^{10}$	$2.6 \cdot 10^{3}$	25	E-disp.	EIGEN and KUSTIN (1960)
$H^+ + CH_3CO\bar{C}HCOCH_3 \rightleftarrows CH_3COCH{=}COHCH_3$	$3.0 \cdot 10^{10}$	$1.7 \cdot 10^{2}$	25	T-jump	EIGEN and EYRING (1962)
$H^+ + CH_3CO\bar{C}HCOCH_3 \rightleftarrows CH_3COCH_2COCH_3$	$1.2 \cdot 10^{7}$	$1.4 \cdot 10^{-2}$	25	T-jump	EIGEN and EYRING (1962)
$OH^- + CH_3COCH{=}COHCH_3 \rightleftarrows CH_3CO\bar{C}HCOCH_3$	$1.6 \cdot 10^{7}$	28	25	T-jump	EIGEN and EYRING (1962)
$OH^- + CH_3COCH_2COCH_3 \rightleftarrows CH_3CO\bar{C}HCOCH_3$	$4.0 \cdot 10^{4}$	$3.5 \cdot 10^{-1}$	25	T-jump	EIGEN and EYRING (1962)
$OH^- + CH_3C({=}O)NHCH_3 \rightleftarrows CH_3C({=}O)\bar{N}CH_3 + HOH$	$5.2 \cdot 10^{6}$		21	NMR	BERGER et al. (1959)
$H^+ + CH_3C({=}O)NHCH_3 \rightleftarrows CH_3C({=}O)N^+H_2CH_3$	$3.8 \cdot 10^{2}$		23	NMR	RALPH and GRUNWALD (1969a)
H^+ + imidazole $\rightleftarrows$ [imidazolium]$^+$	$2.4 \cdot 10^{10}$	$1.0 \cdot 10^{3}$	25	NMR	RALPH and GRUNWALD (1969a)
	$1.5 \cdot 10^{10}$	$1.5 \cdot 10^{3}$	13	T-jump	EIGEN et al. (1960)
[imidazolium]$^+$ + H_2O + imidazole $\rightleftarrows$ imidazole + H_2O + [imidazolium]$^+$	$1.07 \cdot 10^{8}$	-	25	NMR	RALPH and GRUNWALD (1969a)
[lutidinium]$^+$ + $2H_2O$ + lutidine $\rightleftarrows$ lutidine + $2H_2O$ + [lutidinium]$^+$[b]	$1.05 \cdot 10^{8}$	-	25	NMR	ROSENTHAL and GRUNWALD (1972)
H^+ + purine $\rightleftarrows$ [purinium]$^+$	$2.4 \cdot 10^{10}$	-	25	NMR	MARSHALL and GRUNWALD (1969)
$OH^- + pNO_2\Phi CH_2CN \rightleftarrows pNO_2\Phi\bar{C}HCN + H_2O$	26		25	back-titr.	HIBBERT and LONG (1972)

	k_{12} $[M^{-1}s^{-1}]$	k_{21} $[M^{-1}s^{-1}]$			
$CH_3COOH + HCOO^- \rightleftarrows CH_3COO^- + HCOOH$	$4.6 \cdot 10^{7}$	$4.6 \cdot 10^{8}$	20	sound	EIGEN and DE MAEYER (1963)
$CH_3COOH + \Phi NH_2 \rightleftarrows CH_3COO^- + \Phi NH_3^+$[b]	$2.4 \cdot 10^{8}$	$3.5 \cdot 10^{8}$	20	sound	EIGEN and DE MAEYER (1963)
CH_3COOH + imidazole $\rightleftarrows CH_3COO^-$ + [imidazolium]$^+$	$1.2 \cdot 10^{9}$	$6.3 \cdot 10^{6}$	20	sound	EIGEN and DE MAEYER (1963)
ΦOH + imidazole $\rightleftarrows \Phi O^-$ + [imidazolium]$^+$[b]	$8.5 \cdot 10^{6}$	$9.9 \cdot 10^{9}$	20	sound	EIGEN and DE MAEYER (1963)

Table 4 (continued)

Reaction	k_{12} $[M^{-1}s^{-1}]$	k_{21} $[M^{-1}s^{-1}]$	Temp (°C)	Method	Reference
$\Phi OH+NH_3 \rightleftarrows \Phi O^- + NH_4^{+b}$	$2.8 \cdot 10^8$	$1.4 \cdot 10^7$	20	sound	EIGEN and DE MAEYER (1963)
ΦOH+piperidine $\rightleftarrows \Phi O^-$+[piperidinium]$^{+b}$	$9.0 \cdot 10^8$	$6.4 \cdot 10^7$	20	sound	EIGEN and DE MAEYER (1963)
$HS(CH_2)_2OH$+imidazole $\rightleftarrows {}^-S(CH_2)_2OH$+[imidazolium]$^+$	$1.4 \cdot 10^6$	$5.0 \cdot 10^8$	20	sound	EIGEN and DE MAEYER (1963)
$HS(CH_2)_2OH+(CH_3)_2NH \rightleftarrows {}^-S(CH_2)_2OH+(CH_3)_2NH_2^+$	$2.4 \cdot 10^7$	$8.6 \cdot 10^5$	20	sound	EIGEN and DE MAEYER (1963)

[a] For the purpose of comparing rates for symmetric proton exchange with bulk of water are given as second order rate constants.

[b] Φ represents the phenylgroup C_6H_5, for structure of lutidine see VI.

ferent maximum values is found in acids, which can form stable intramolecular hydrogen bonds in water. Since the hydrogen bond has to be broken before proton transfer can occur, the corresponding rate constant is prevented from reaching the maximum value. Two typical examples for this effect are represented by maleic acid anion (III) and the whole family of dyes derived from o-hydroxyazobenzene (IV).

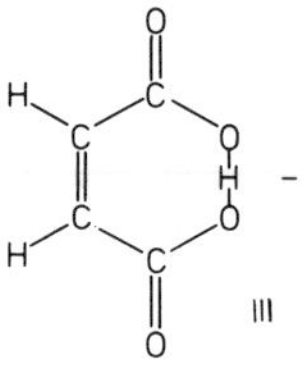

III

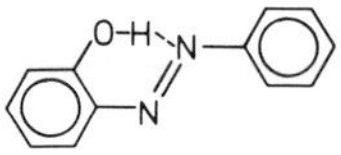

IV

4. Proton transfer is substantially slower than the diffusion-controlled limit in compounds, which can form only weak hydrogen bonds being unstable in aqueous solutions or do not form hydrogen bonds at all, like acids of the type RSH, RR'PH or RR''R'''CH. In this case both curves $k_{12}(\Delta pK)$ and $k_{21}(\Delta pK)$ are shifted to smaller values (Fig. 6A).

Some examples of typical rate constants for dissociation and proton transfer obtained by different methods are presented in Table 4. First of all we realize that excellent agreement between relaxation and NMR results in some cases (GRUNWALD and RALPH, 1975). It is important to verify that both techniques in principle are able to give the same results for simple proton exchange reactions. By NMR methods we measure the rate of exchange for a particular nucleus – in most cases it will be a proton – between two states differing in their magnetic environment. There are cases where different relaxation times can be obtained from chemical relaxation and NMR studies. One example, representative also for others, will be mentioned briefly: the rate constant k_D for the dissociation of acetic acid in aqueous solution has been determined by electric field jump method to be $k_D = 7.8 \cdot 10^5 s^{-1}$ (EIGEN and DE MAEYER, 1963). First-order proton exchange measured by NMR technique turned out to be more than 100 times faster ($\tau^{-1} \sim 1.10^8 s^{-1}$, LUZ and MEIBOOM, 1963).

The relaxation observed by NMR corresponds to a symmetric proton exchange ($\Delta\mu = 0$) involving two water molecules (V) which cannot be detected by relaxation methods.

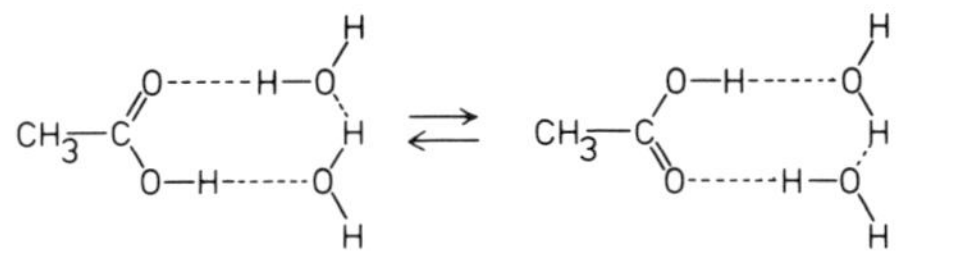

V

The possibility of investigating symmetric proton transfer, in fact makes NMR methods an important complementary tool to relaxation kinetics. Data on proton transfer from an excess proton or hydroxide ion to the water bulk have been presented in the literature (MEIBOOM, 1961; LUZ and MEIBOOM, 1964; LOEWENSTEIN and SZÖKE, 1962). As expected these processes are somewhat slower than the neutralization reaction. From a detailed rate law of symmetric proton exchange between the trimethylamine and trimethylammonium cation GRUNWALD (1963) tried to derive a value for the exchange rate of protons between the hydration layer of the amine and the water bulk: $k_H \sim 1 \cdot 10^{10} s^{-1}$ (see also GRUNWALD and RALPH, 1971, 1975). Comparison of the exchange rates for different nuclei in the same molecule (LOEWENSTEIN and MEIBOOM, 1957; GRUNWALD and RALPH, 1975) makes it possible to draw conclusions on the number of water molecules participating in proton exchange. Symmetric proton transfer with and without interstitial water molecules are competitive and their relative importance is determined by the stereochemistry of the reaction center (BIANCHIN et al., 1975). The series of methyl-substituted ammonium ions represents an excellent example for a change in the predominant reaction on substitution (Table 4). In case of 2,4-lutidine (VI) even two-center molecules can participate in an intermolecular proton transfer process (ROSENTHAL and GRUNWALD, 1972). Proton

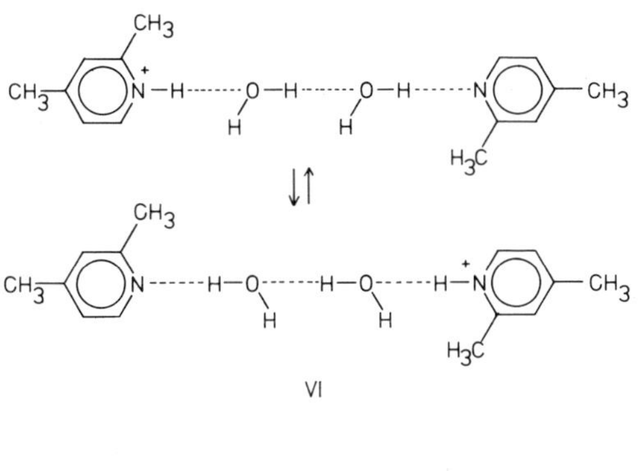

VI

transfer reactions to amines have been studied also by other groups of authors (KRESGE and CAPEN, 1975; LEYDEN and MORGAN, 1969). Interestingly the rate of Walden inversion (VII) could be derived together with the rate of proton exchange of dibenzylmethylammonium ion in aqueous solution (MORGAN and LEYDEN, 1970).

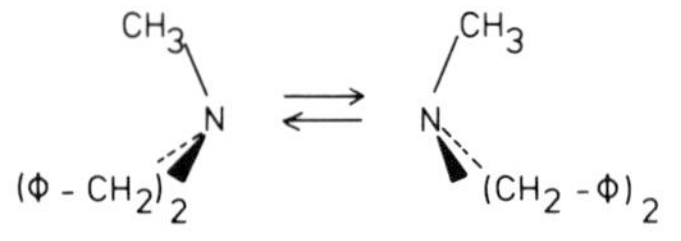

VII

Another investigation of this kind yielded proton exchange rates together with rate constants for nitrogen inversion in acidic aqueous solutions of N,N'-dimethylpiperazine in cis and trans form (SUDMEIER and OCCUPATI, 1968). Some papers dealt with NMR investigations on proton transfer involving various amides (SHEINBLATT, 1970), thiourea (VOLD and CORREA, 1970), and N,N'-dimethylurea (WHIDBY and MORGAN, 1973). Aqueous urea solutions have been studied extensively (HUNSTON and KLOTZ, 1971; VOLD et al., 1970) up to concentrations of 10 M urea. The main idea in these investigations, of course, was to learn more about the role of urea in denaturation processes of proteins. Protonation of urea in aqueous solution occurs at one of its nitrogen atoms. The rate of proton exchange was found to be catalyzed by acid and base:

$$\tau^{-1} = k_H \cdot c_H{}^+ + k_{OH} \cdot c_{H^+} + k_o \cdot c_{H_2O} \qquad (11)$$

The constant obtained from a fit to the data measured in solutions of varying urea concentration ($1.5\ M < c_o < 5\ M$) are $k_H = 3.5 \pm 1.0 . 10^6 M^{-1} s^{-1}$, $k_{OH} = 2.4 \pm 0.8 . 10^6 M^{-1} s^{-1}$ and $k_o < 4 . 10^{-2} M^{-1} s^{-1}$, thus leading to a pH profile almost symmetric to pH=7. At very high concentration aggregation of urea occurs and exchange rates become concentration-dependent.

More recent chemical relaxation studies on proton transfer mainly dealt with deviations from the ideal rate constants derived from diffusion control. In the first group of papers investigations of acids forming strong internal hydrogen bonds are presented. The first class of compounds with relatively slow rates for proton exchange derivatives of o-hydroxyazobenzene (IV) has been mentioned already. New studies in this field applying temperature jump technique were performed by ROSE and STUEHR (1968, 1971, 1972) and by PERLMUTTER-HAYMAN and SHINAR (1975). Three other classes of acids (VIII, IX and X) do form strong internal hydrogen bonds as well and similarly slow deprotonation rates were obtained from temperature jump studies for VIII (ROSE and STUEHR, 1971, 1972), IX (FUENO et al., 1973) and X (HIBBERT, 1973, 1974). Another group of papers deals with proton transfer to and from compounds which form weak hydrogen bonds to the solvent or do not form hydrogen bonds at all in aqueous solution. AHRENS and MAASS (1968) investigated various thiols by ultrasound absorption technique. Temperature jump studies on proton transfer to acids with deprotonable CH bonds — e.g. the keto forms of the tautomeric

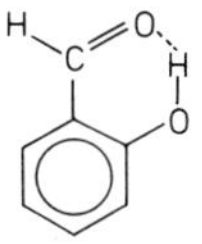

VIII

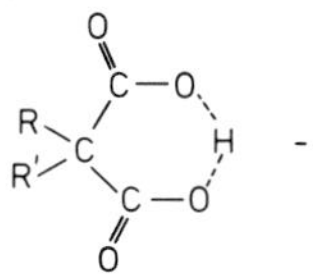

IX

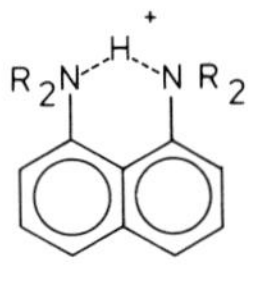

X

acids XI – in aqueous solution were performed on diacetyl acetone (STUEHR, 1967), acetylacetone (AHRENS et al., 1970a) and ethyl acetoacetate and some of its derivatives (BROUILLARD and DUBOIS, 1974). A pressure jump study on barbituric acid and the corresponding compounds with 5-alkyl and phenyl substituents has been reported recently by KOFFER (1975). In all these investigations proton transfer involving the tautomeric enol forms (XI) and the enolate anions was found to be diffusion controlled. Enolic OH groups, therefore, do not show reduced deprotonation rates. Finally we would like to add two other papers to the list wherein temperature jump studies on CH-acids with α-cyano groups (XII and XIII, HIBBERT and LONG, 1972) and on cyanines (XIV, DYNES et al., 1972) are reported.

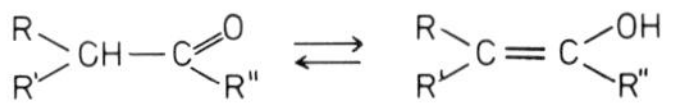

XI

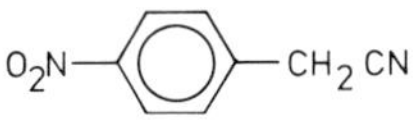

XII

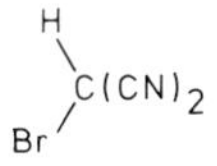

XIII

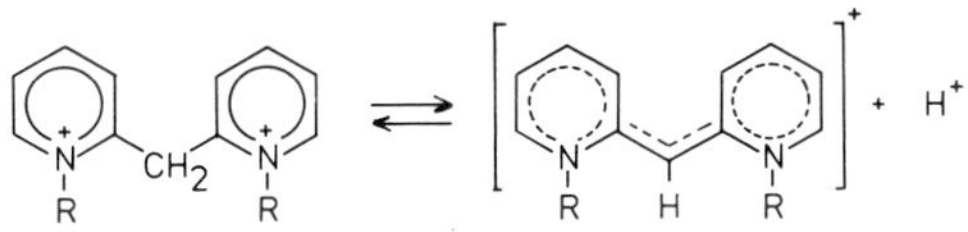

XIV

A number of investigations on intermolecular proton transfer in aqueous solutions of α-amino acids, purine and pyrimidine bases as well as other compounds of biological interest will be discussed in Section V.

B. Intramolecular Proton Transfer

Although intermolecular proton transfer in aqueous solution has been studied extensively and its kinetics is well understood, only a few papers dealing with proton exchange within a molecule are found in the literature. Intramolecular proton transfer on the other hand is of particular importance in many biochemical processes especially in proton catalysis. Therefore, we well present here a more extensive discussion of one particular example and then give a survey of other systems investigated.

Intramolecular proton transfer, evidently, can occur only in molecules with two or more binding sites for protons. Therefore a two basic acid represents the simplest example of a molecule, which can be appropriate to study this kind of reaction. Three equilibrium constants are sufficient to describe the concentrations of the four individual forms of a two basic acid completely (Table 5). Most of the examples discussed here are electrically neutral between the two pK values and we will accordingly call the four species cation ($AH_2^+ = A^+$), neutral molecule ($AH = A_N$), zwitterion ($AH' = A_Z$) and anion (A^-). For our purpose it will be most appropriate to choose the following three constants for characterization:

1. the isoelectric point: $pK_I = \frac{1}{2}(pK_1 + pK_2)$
2. the differences in pK values: $\Delta pK = pK_2 - pK_1$
3. the isomerization constant: $K_T = A_Z/A_N \sim c_Z/c_N$

Various mechanisms for internal proton transfer have been proposed (see e.g. MAASS and PETERS, 1972; WHITE and SLUTSKY, 1972a). They can be classified according to five different types of processes, all leading to a net reaction $A_N \rightleftharpoons A_Z$:

1A. Direct proton transfer within an already existing intramolecular hydrogen bond. As we know from previously discussed examples this process is extremely fast in the preferred direction[4] provided the necessary geometrical requirements for a moderately strong internal hydrogen bond are fulfilled.

1B. Proton transfer within the hydration shell of the molecule along a chain of appropriately oriented water molecule, which reaches from one functional group involved in proton exchange to the other. This process can be expected to occur very fast as well. Previously reported NMR data (Table 4) have shown that rate constants for an intermolecular process of this type with one or two interstitial water molecules are of similar magnitude to the rates for direct proton exchange. One example for internal proton transfer with participation of water molecules from the hydration shell has been mentioned already: the symmetric internal proton exchange in acetic acid (V), where two water molecules are involved ($k = 1 \cdot 10^8 s^{-1}$, LUZ and MEIBOOM, 1963).

For stereochemical reasons processes of type 1A or 1B are mutually exclusive. There are, of course, systems with the appropriate functional groups for proton transfer, but neither the stereochemical requirements for mechanism 1A nor for 1B are fulfilled. In this case either substantial reorganization of the hydrogen shell has to occur *before* proton transfer takes place and consequently the rate for the process could be low, or one of the following mechanisms is realized. These processes are not intramolecular in a very strict sense, but the overall result is the same as that of the internal proton transfer.

---

[4] In highly unsymmetrical systems, $K_T >> 1$ or $K_T << 1$ the unfavored process has an activation energy at least as large as the energy difference between the two protomeric structures and therefore has to be slow.

Table 5. Proton transfer processes of a two basic acid in aqueous solution

Symbols: $A_+ \equiv HAH^+$; $A_N \equiv HA$; $A_Z \equiv AH$ and $A_- \equiv A^-$

$c_H \sim [H^+]$, $c_{OH} \sim [OH^-]$, $c_+ \sim [A_+]$, $c_N \sim [A_N]$, $c_Z \sim [A_Z]$, $c_- \sim [A_-]$.

Monomolecular processes:

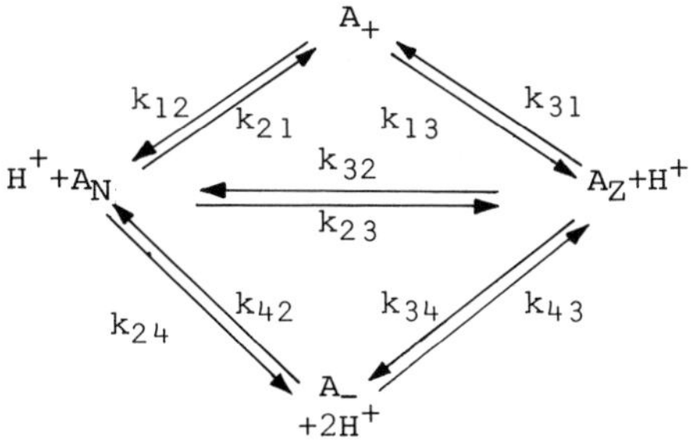

For each reaction step hydrolysis also occurs, e.g.

$$A_+ + OH^- \underset{k'_{21}}{\overset{k'_{12}}{\rightleftharpoons}} A_N$$

(All other hydrolytic steps are defined analogously)

Biomolecular processes:

$$A_N + A_+ \underset{k^+_{32}}{\overset{k^+_{23}}{\rightleftharpoons}} A_+ + A_Z \qquad A_N + A_N \underset{k^0_{32}}{\overset{k^0_{23}}{\rightleftharpoons}} A_Z + A_Z$$

$$A_N + A_- \underset{k^-_{32}}{\overset{k^-_{23}}{\rightleftharpoons}} A_- + A_Z$$

Highly improbable steps not considered explicitely here are e.g.

$$A_+ + A_- \rightleftharpoons A_N + A_N$$

$$A_+ + A_- \rightleftharpoons A_N + A_Z$$

$$A_+ + A_- \rightleftharpoons A_Z + A_Z$$

Equilibrium constants: pK_1, pK_2

$$K_T = \frac{[A_Z]}{[A_N]} \sim \frac{c_Z}{c_N}$$

Microscopic dissociation constants:

$$K_{1N} = \frac{[A_N]\cdot[H^+]}{[A_+]} \sim \frac{c_N \cdot c_H}{c_+} \quad ; \quad K_{1Z} \text{ analogously}$$

$$K_{2N} = \frac{[A_-]\cdot[H^+]}{[A_N]} \sim \frac{c_- \cdot c_H}{c_N} \quad ; \quad K_{2Z} \text{ analogously}$$

$$K = K_{1N} + K_{1Z} \; ; \; \frac{1}{K_2} = \frac{1}{K_{2N}} + \frac{1}{K_{2Z}}$$

2. Proton transfer proceeding in a two-step mechanism involving diffusion-controlled dissociation and recombination (Table 5). There are, of course, two ways for this kind of proton exchange, one involving the cation A_+, the other the anion A_- as intermediate.

3. Transfer of a single proton between two molecules of A in different forms. The more probable processes of this kind are one-step proton exchange with the cation A_+ or the anion A_- acting as a catalyst. In molecules with NH and OH bonds participating in proton transfer these bimolecular processes can be expected to occur diffusion-controlled. Another more complex mechanism might involve the rare encounter of a cation A_+ and an anion A_-. On proton transfer two neutral molecules ($2A_N$), two zwitterions ($2A_Z$) or one of each kind (A_N+A_Z) can be formed.

4. Simultaneous exchange of two protons. Similar to process 1 very restrictive sterical requirements have to be fulfilled so that the rate of transfer is fast enough to compete successfully with the other processes. To give an example, carboxylic acids have the appropriate stereochemistry for that kind of simultaneous proton exchange in the dimer. Finally, also in a process of type 4 one, two or more water molecules might participate.

Provided both concentration and pH dependence of the relaxation time for overall proton transfer $A_N \rightleftharpoons A_Z$ is known in a sufficiently large range we can distinguish between the different mechanisms. Only for a decision between type 1A and 1B is additional information necessary. In general, we will be able to exclude one or both from an inspection of the molecular structure.

The whole reaction scheme shown in Table 5 is characterized by four relaxation times. As long as the net reaction observed corresponds to internal proton transfer ($A_N \rightleftharpoons A_Z$) we can apply stationary conditions for the cation A_+ and the anion A_- (symbols for individual concentrations are defined in Table 5):

$$\frac{dc_+}{dt} = 0, \ \frac{dc_-}{dt} = 0; \quad |\delta c_N| \sim |\delta c_Z| >> \left(|\delta c_H|, |\delta c_+|, |\delta c_{OH}|, |\delta c_-|\right) \quad (12)$$

Furthermore we can expect the autodissociation equilibrium of water to be readily established, $c_H \cdot c_{OH} = K_W$. Under these conditions we are left with a single relaxation process, the relaxation time of which we will simply denote by τ.

The amplitude function Γ for total proton transfer, one of the factors determining the intensity of changes observable in chemical relaxation studies, can be obtained easily:

$$\Gamma = \frac{1}{\frac{1}{c_N} + \frac{1}{c_Z}} = c_o \frac{K_T}{(1+K_T)\left(1+K_T+\frac{c_H}{K_{1N}} + \frac{K_{2N}}{c_H}\right)} \quad (13)$$

The shape of the curve Γ(pH) is shown in Figure 7 together with experimental results on internal proton transfer in 3-hydroxypyridine (3-HP, XV) measured by different techniques. Maximum re-

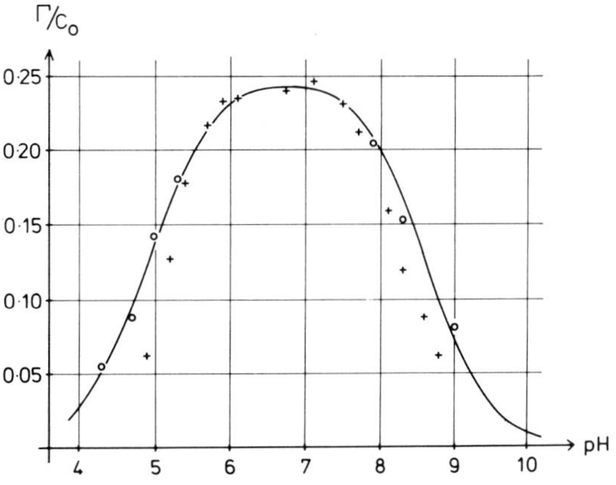

Fig. 7. Amplitude of chemical relaxation process due to internal proton exchange in 3-hydroxypyridine as a function of pH (—— theoretical curve, see Eq. (13); + ... T-jump relaxation, o ... ultrasound absorption)

lative amplitudes $(\Gamma/c_o)_{max}$ are obtained at pH = pK_I and, of course, depend largely on K_T. $(\Gamma/c_o)_{max}$ vanishes and consequently internal proton transfer is not observable for very small and very large K_T values. $(\Gamma/c_o)_{max}$ has the largest value for an equilibrium constant

$$K_T = \left[1 + (2\cdot 10^{-\Delta pK/2})\right]^{1/2} \sim 1;\ \Delta pK>2. \tag{14}$$

For stationary conditions the total reciprocal relaxation time τ^{-1} is obtained simply as a sum of individual contribution from the different processes discussed above:

$$\tau^{-1} = \tau_1^{-1} + \tau_2^{-1} + \tau_3^{-1} + \tau_4^{-1} \tag{15}$$

By standard mathematical techniques we can easily evaluate the various contributions in terms of the rate constants summarized in Table 5:

$$\tau_1^{-1} = k_{23} + k_{32} \tag{15a}$$

$$\tau_2^{-1} = \frac{(k_{12}+k'_{12}c_{OH})(k_{31}c_H+k'_{31})+(k_{21}c_H+k'_{21})(k_{13}+k'_{13}c_{OH})}{k_{12}+k'_{12}c_{OH}+k_{13}+k'_{13}c_{OH}} + \tag{15b}$$

$$+ \frac{(k_{42}c_H+k'_{42})(k_{34}+k'_{34}c_{OH})+(k_{24}+k'_{24}c_{OH})(k_{43}c_H+k'_{43})}{k_{42}c_H+k'_{42}+k_{43}c_H+k'_{43}}$$

$$\tau_3^{-1} = \frac{(k^+_{23}+k^+_{32})\frac{c_H}{K_{1N}} + (k^-_{23}+k^-_{32})\frac{K_{2N}}{c_H}}{1+K_T + \frac{c_H}{K_{1N}} + \frac{K_{2N}}{c_H}} \cdot c_o \tag{15c}$$

and $$\tau_4^{-1} = \frac{4(k^o_{23}+k^o_{32})}{1+K_T + \frac{c_H}{K_{1N}} + \frac{K_{2N}}{c_H}} \cdot c_o \tag{15d}$$

At constant pH we find a simple concentration dependence of the total relaxation time:

$$\tau^{-1} = \alpha + \beta \cdot c_o \tag{16}$$

α and β of course are functions of pH. Interestingly, all four contributions show different pH dependence. τ_1^{-1} is pH independent. τ_2^{-1} becomes proportional to c_H or c_{OH} respectively at the low or high pH limit, (15b') and (15b'')

$$\text{pH} \ll 7: \quad \tau_2^{-1} \sim \frac{k_{12}k_{31}+k_{21}k_{13}}{k_{12}+k_{13}} \cdot c_H \tag{15b'}$$

$$\text{pH} \gg 7: \quad \tau_2^{-1} \sim \frac{k'_{42}k'_{34}+k'_{24}\ k'_{43}}{k'_{42}+k'_{43}} \cdot c_{OH} \tag{15b''}$$

and has a minimum at pH=7. τ_3^{-1}, on the other hand, has roughly the form of a negative Gaussian curve (see Fig. 11). It reaches constant values at low and high pH (15c') and shows a minimum in the neighborhood of pH $\sim$ pK_I.

$$\text{pH} \ll 7: \quad \tau_3^{-1} \sim (k^+_{23}+k^+_{32})c_o; \quad \text{pH} \gg 7: \quad \tau_3^{-1} \sim (k^-_{23}+k^-_{32})c_o \tag{15c'}$$

τ_4^{-1} finally has its maximum value at pH=pK_I and the reaction becomes slow at low and high pH. The actual relaxation time, τ(pH, c_o), is a superposition of all these functions and might look quite complicated. A few characteristic examples are shown in Figure 8A-D. For this illustration we neglected the constant or roughly constant terms τ_1^{-1} and τ_4^{-1} and assumed all intermolecular proton transfer steps to proceed with diffusion controlled rates. From these figures we realize that only in case $pK_I \sim 7$ can we expect to find concentration dependence in both, the acidic and the alkaline branch. For $pK_I < 7$ concentration dependence at low pH becomes important only at very high c_o values. For weaker acids $pK_I > 7$, in contrary, only concentration dependence in the low pH region will be observable. Internal proton transfer simulated by indirect diffusion controlled steps becomes slowest at low concentration and in systems for which $pK_I \sim 7$, $K_T \sim 1$ and ΔpK is large.

Fortunately, the stationary treatment is appropriate at physiological conditions. Coupling of reactions becomes important only at low pH ($c_H \sim c_o$) and high pH ($c_{OH} \sim c_o$). For one particular example we calculated coupling of the process corresponding to τ_2^{-1} and τ_3^{-1} with the protonation reactions of A_+ (Fig. 9). The stationary relaxation time is almost identical with the slower of

---

Fig. 8A-D. Relaxation time for indirect internal proton transfer in a two-basic acid as a function of pH and concentration ($\tau_2^{-1} + \tau_3^{-1}$, see Eqs. (15b,c)) under assumption of diffusion control (rate constants: $k'_{12}=k_{21}=k'_{13}=k_{31}=k'_{24}=k_{42}=k'_{34}=k_{43}= 2\cdot 10^{10}M^{-1}s^{-1}$, $k^+_{23}=k^+_{32}=k^-_{23}=k^-_{32}= 4\cdot 10^8 M^{-1}s^{-1}$. The individual curves correspond to concentrations decreasing by a factor of $\sqrt{10}$, topmost curve: c_o=0.1M, lowest curve: $c_o \rightarrow 0$, $\tau^{-1}= \alpha$(pH). (A) pK_I=7, ΔpK=3, K_T=1; (B) pK_I=5, ΔpK=3, K_T=1; (C) pK_I=7, ΔpK=1, K_T=1; (D) pK_I=7, ΔpK=3, K_T=100)

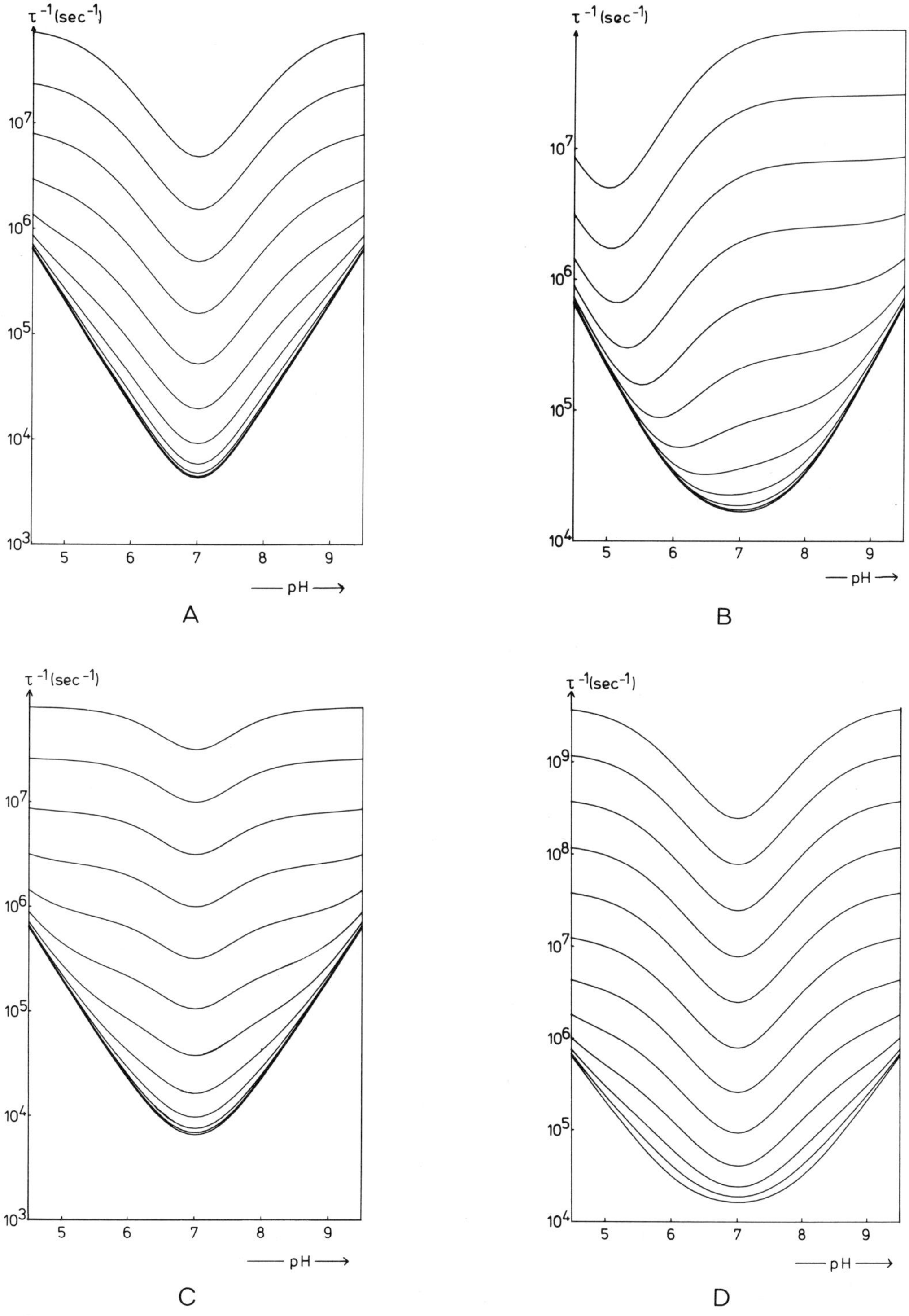

Fig. 8 A-D

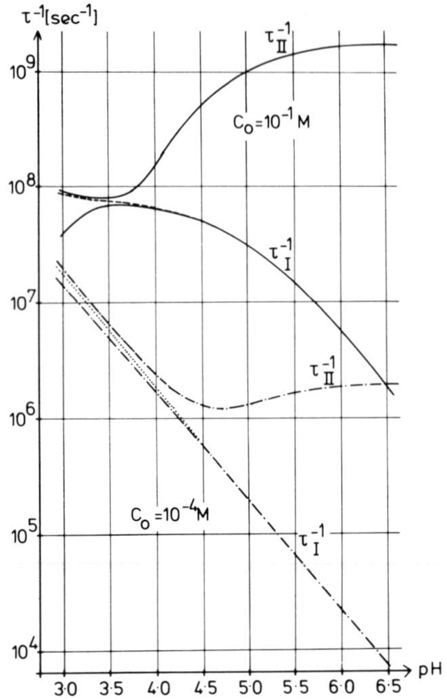

Fig. 9. Validity of stationary state assumption for internal proton transfer in a two basic acid in range $pK_1<pH<pK_2$ (c_o= 0.1M ---- stationary τ^{-1}). —— exact solutions, τ_I^{-1} and τ_{II}^{-1}; c_o=1.10^{-4}M; .... stationary τ^{-1}, -·-·- exact solutions, τ_I^{-1} and τ_{II}^{-1})

the two relaxations τ_I and τ_{II} in the whole region, where internal proton transfer has a big enough amplitude. In cases where direct proton transfer (τ_I^{-1}) is fast, coupling of the two processes is more important (see e.g. MAASS and PETERS, 1972 or PETERS and MAASS, 1974).

Only very recently a complete analysis of an experimental example was presented (SCHUSTER et al., 1976a). The compound chosen, 3-HP (XV) perfectly meets the requirements for slow internal proton transfer derived above (3-HP, aqueous solution, T=25°C: pK_I=6.77, K_T=1.05 and ΔpK=3.71). Ultrasound absorption and temperature jump techniques were applied. At constant pH the concentration dependence Eq. (16) was fulfilled up to concentration as high as 0.1 M. The dependence of relaxation times on pH is shown in Figures 10 and 11. The curves correspond very well to the calculated example

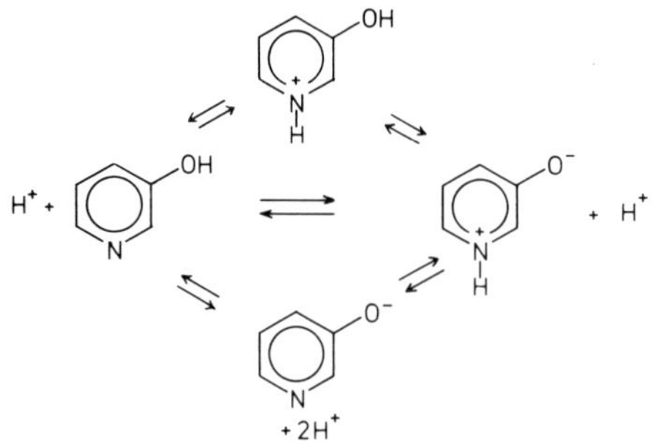

XV

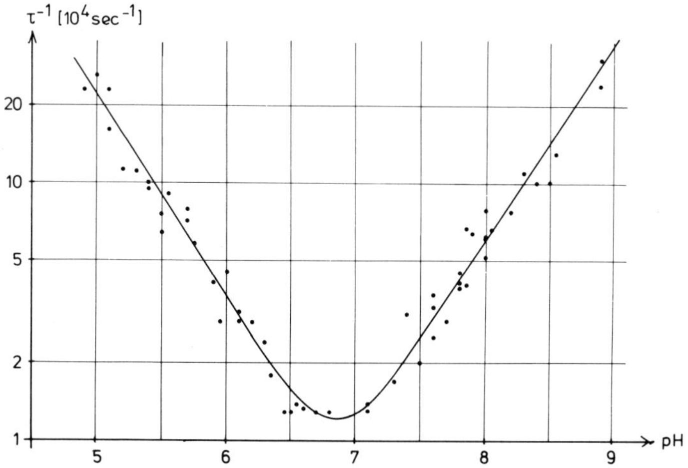

Fig. 10. Relaxation time for internal proton transfer in 3-hydroxypyridine (c_o=2·10^{-4}M, T=25°C, I=0.1; scattering of experimental points comes mainly from pH instability of diluted unbuffered solutions (SCHUSTER et al., 1976a)

shown in Figure 8A. Rate constants for processes 2 and 3 were obtained by a fit to the experimental data. The individual numbers are shown in Table 6 and agree well with those estimated for diffusion controlled proton transfer (see also Table 4). The rates for direct (τ_1^{-1}) and uncatalyzed bimolecular (τ_4^{-1}) proton transfer are too small to be determined ($k_{23} \sim k_{32} < 2 \cdot 10^3 s^{-1}$; $k^o_{23} \sim k^o_{32} < 2.5 \cdot 10^6 M^{-1}s^{-1}$). From these results we can conclude that proton transfer within the hydration shell of 3HP plays only an inferior role if it does occur at all. The predominant reactions are consecutive protonation-deprotonation steps or at higher concentration proton exchange with the catalytically active cation A_+ and anion A_-. Evidently, the structure of the hydration is such, that no appropriate chain of water molecules between the two functional groups does exist or can be formed easily. It might be interesting to

Table 6. Rate constants for proton transfer in aqueous solutions of 3-hydroxypyridine (SCHUSTER et al., 1976a; T=25°C, I=0.1)

k^+_{23}	$3.7 \cdot 10^8 M^{-1}s^{-1}$
k^+_{32}	$3.5 \cdot 10^8 M^{-1}s^{-1}$
k^-_{23}	$4.6 \cdot 10^8 M^{-1}s^{-1}$
k^-_{32}	$4.4 \cdot 10^8 M^{-1}s^{-1}$
$k_H = \dfrac{k_{12} \cdot k_{31} + k_{21} \cdot k_{13}}{k_{12} + k_{13}}$	$2 \cdot 10^{10} M^{-1}s^{-1}$
$k_{OH} = \dfrac{k'_{42} \cdot k'_{34} + k'_{24} \cdot k'_{43}}{k'_{42} + k'_{43}}$	$1.5 \cdot 10^{10} M^{-1}s^{-1}$

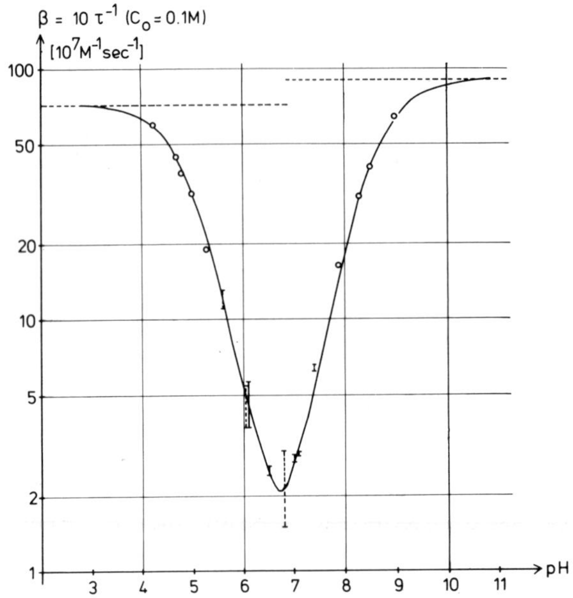

Fig. 11. Concentration dependence of relaxation times for internal proton transfer in 3-Hydroxypyridine (c_0=0.1M, T=25°C, I=0.1, o ... ultrasound absorption, Ⅰ... cabel T-jump relaxation, ⫶... conventional T-jump, values extrapolated from concentration dependence at low concentrations

keep in mind that internal proton transfer can be as slow as $\tau \sim 10^{-4}$ s in aqueous solution of neutral pH, although all steps involved occur diffusion controlled.

Previous studies, using ultrasound absorption technique were more or less restricted to a single pH value and have shown that internal proton transfer in the isomeric aminobenzoic acids (EIGEN and DE MAEYER, 1963; WHITE and SLUTSKY, 1972b), again occurs in an indirect way (processes 2 or 3). The latter authors reported that even for o-aminobenzoic acid (XVI) no evidence for direct proton transfer could be detected[5]. This result is very hard to explain and further investigations on the conformation of o-aminobenzoic acid in water are inevitable.

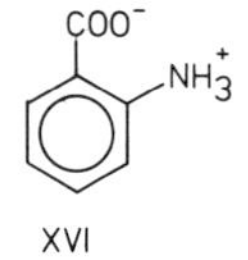

XVI

$HSCH_2CH_2NH_2$

XVII

$HSCH_2CH(COO^-)-NH_3^+$

XVIII

$HSCH_2CH_2CH(COO^-)-NH_3^+$

XIX

Proton transfer in various aminothiols (XVII, XVIII, XIX) has been studied by MAASS and PETERS (1972) again using ultrasound absorption technique. They could show, that in this class of compounds the rate of proton transfer depends on the distance between the two functional groups and is substantially higher than that derived from a diffusion controlled indirect mechanism (pH =

[5] They did find direct proton transfer in XVI in methanol and acetone.

9.5, T=25°C, τ^{-1} (XVII) = $3.0 \cdot 10^7 s^{-1}$, τ^{-1} (XVIII) = $3.6 \cdot 10^8 s^{-1}$, τ^{-1} (XIX) = $2.3 \cdot 10^7 s^{-1}$, compared to $\tau_2^{-1} < 1 \cdot 10^6 s^{-1}$). Proton exchange was found to be concentration-dependent. Since the pH dependence of τ has not been measured, it is difficult to attribute one of the two processes (3 or 4) to the concentration dependence found. From stereochemical considerations we would suggest that proton transfer in 2-aminothiols (XVII or XVIII) occurs either directly or most probably in a structure with one water molecule between the two functional groups (XX).

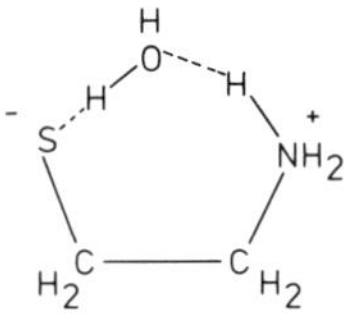

XX

$$H_2NCH_2{-}COOH \underset{k_{32}}{\overset{k_{23}}{\rightleftharpoons}} H_3\overset{+}{N}CH_2{-}COO^-$$

XXI

Another series of candidates for internal proton transfer is presented by α-aminoacids, which predominantly occur as zwitterions in aqueous solution. K_T is very large and consequently, chemical relaxation studies are very difficult to perform. The sensitivity needed seems to be far outside the possibilities of our present equipment. An NMR-study on internal proton transfer in glycine (XXI) was reported by SHEINBLATT and GUTOWSKY (1964) leading to rate constants ($k_{23} = 4.4 \cdot 10^7 s^{-1}$, $k_{32} = 175\ s^{-1}$; $K_T = 2.5 \cdot 10^5$), which again exceeds the protonation-deprotonation two-step mechanism. It is not known whether an interstitial water molecule participates in proton transfer or not.

Some other examples for internal proton transfer in aqueous solution, dealing with adenine and vitamin B_6 derivatives will be mentioned in Section V.

IV. Proton Transfer in Non-Aqueous Solvents

Non-aqueous solvents most appropriately are subdivided into two classes – protic non-aqueous solvents, which behave rather similarly to water and aprotic solvents – and will be discussed in two different subsections here.

A. Protic, Non-Aqueous Solvents

Non-aqueous protic solvents like alcohols, carboxylic acids liquid ammonia etc. are of minor importance in biochemical research and will be treated only briefly here. In principle, most of these solvents contain hydrogen-bonded clusters or chains which are suitable for proton transport. Similar structures as in water can be expected for solvated protons and defect protons. There are, however, several substantial differences: due to the unique structure of the water molecule – almost spherical in shape with

Table 7. Proton transfer in protic solvents

Reaction	Solvent	T (°C)	k	Reference
$H_3O^+ + H_2O$	H_2O	25	$6 \cdot 10^{11} s^{-1}$	MEIBOOM (1961)
$CH_3OH_2^+ + CH_3OH$	CH_3OH	25	$1.7 \cdot 10^{11} s^{-1}$	GRUNDWALD et al. (1962)
$NH_4^+ + NH_3$	NH_3 (liquid)	25	$6 \cdot 10^{9} s^{-1}$	CLUTTER and SWIFT (1968)
$HO^- + H_2O$	H_2O	25	$2.1 \cdot 10^{11} s^{-1}$	MEIBOOM (1961)
$CH_3O^- + CH_3OH$	CH_3OH	25	$1.9 \cdot 10^{10} s^{-1}$	GRUNDWALD et al. (1962)
$NH_2^- + NH_3$	NH_3 (liquid)	25	$8 \cdot 10^{8} s^{-1}$	CLUTTER and SWIFT (1968)
$H_3O^+ + N(CH_3)_3$ [a]	H_2O	25	$3.0 \cdot 10^{10} M^{-1} s^{-1}$	GRUNWALD (1967)
$CH_3OH_2^+ + N(CH_3)_3$ [a]	CH_3OH	25	$6 \cdot 10^{9} M^{-1} s^{-1}$	GRUNWALD (1967)
$CH_3COOH \cdot 2H_2O$ [b]	H_2O	25	$1 \cdot 10^{8} s^{-1}$	LUZ and MEIBOOM (1963)
$C_6H_5COOH \cdot 2CH_3OH$ [b]	CH_3OH	25	$1.3 \cdot 10^{5} s^{-1}$	GRUNWALD et al. (1963)

[a] The rate constant given is that for the energetically favored diffusion controled process.

[b] Intramolecular proton transfer according to V.

two and two proton donor and acceptor sites in an almost tetrahedral arrangement – mobilities and transfer rates reach something like an optimum value in aqueous solutions and are smaller in the other associated liquids. In alkaline or acidic solvents like ammonia or carboxylic acids the centers of acidity scales are shifted towards the extremes known for water: weak acids will either become strong or hardly dissociable at the other extreme. Most non-aqueous solvents have much lower dielectric constants than water and ion association is favored strongly. Additionally internal proton transfer equilibria, $A_N \rightleftarrows A_Z$, are shifted towards the neutral forms. This fact can be interpreted in terms of reduced donicities or bulk donicities (GUTMANN, 1971) in non-aqueous solvents.

Results of NMR studies on comparable proton transfer reactions in water, methanol and liquid ammonia are shown in Table 7. As we can see the rate constants of comparable reactions most probably diffusion controlled in all solvents considered decrease in the following series: water, methanol and ammonia. This decrease is also reflected by the lower mobilities of lyonium and lyate ions in non-aqueous solvents. The only exceptions of reaction rates reported in a recent review (GRUNWALD and RALPH, 1975) concern proton transfer from the lyonium ion to N,N-diethyl-m-toluidine and phenyl-dimethylphosphine, which seem to be larger in methanol than in water.

Table 8. Proton transfer in t-butanol according to the mechanism in Eq. (17) (T=38°C)[a]

Ammonium ion	Anion	$k (M^{-1}s^{-1})$
$CH_3NH_3^+$	Cl^-	$2.6 \cdot 10^7$
	OTs^-	$2.4 \cdot 10^7$
$(CH_3)_3NH^+$	Cl^-	$1.1 \cdot 10^5$
	Br^-	$5.3 \cdot 10^5$
	OTs^-	$7.0 \cdot 10^5$

[a]COCIVERA and GRUNWALD (1965); see also GRUNWALD and RALPH (1975).

Some NMR investigations on proton transfer in t-butanol and acetic acid demonstrate the drastic influence of ion pair formation, which introduces new specificities into the exchange rates. In t-butanol the symmetric reaction has been studied by COCIVERA and GRUNWALD (1965). Comparing the data for methylammonium and tri-

$$RR'R''NH^+..X^- + (CH_3)_3COH + RR'R''N \underset{}{\overset{k}{\rightleftharpoons}} RR'R''N + (CH_3)_3COH + RR'R''NH^+..X^- \tag{17}$$

methylammonium ions bound to various counterions we realize two interesting differences (Table 8). In the latter example the rates are smaller by two orders of magnitude and the anion has a pronounced influence on the rate constants. Both factors can be explained by the most probable structures of the two corresponding ion pairs (XXII and XXIII).

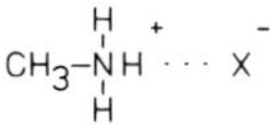

XXII

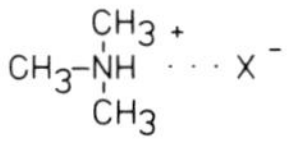

XXIII

In XXII, certainly, a proton from the periphery of the ion pair is transferred, whereas in XXIII the proton involved in the hydrogen bond has to be removed. Clearly, more energy is needed in the second case and additionally the anion X exhibits direct influence on the rate of exchange. Similar studies on proton exchange between ion pairs in acetic acid solution were performed (GRUNWALD and PRICE, 1964; CRAMPTON and GRUNWALD, 1971a,b; GRUNWALD and PUAR, 1967a,b). The rate constants again appeared to be rather high ($k \sim 10^8 M^{-1}s^{-1}$). Specifities were explained by the formation of hydrogen bonded dimers of ion pairs (XXIV). Exchange rates for the mercaptoprotons in mercaptoacetic acid and its methyl ester (XXV) were investigated in acetic acid solutions (WHIDBY and LEYDEN, 1970). The rate constants were found to be very low indeed (T=25°C, $k \sim 1\ s^{-1}$). Addition of metal acetates (cations: K^+, Na^+, Ba^{2+}) increases the relaxation drastically and presents

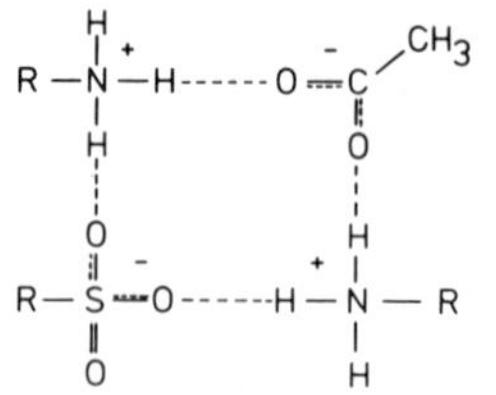
XXIV

XXV

evidence for ion pair catalysis of proton transfer. Catalyzed exchange rates were found to lie in the range of $10^3 M^{-1} s^{-1}$ and to depend on the nature of the cation.

B. Aprotic Solvents

Aprotic solvents, polar and unpolar, are of some interest for biochemical model studies because they simulate to some extent the environment of socalled "hydrophobic pockets" in proteins. Other examples for aprotic environments are represented by the interior of membranes and the central parts of the double helices of nucleic acids.

The major difference between protic and aprotic solvents is the lack of hydrogen bonding in the latter. As we have seen from vapor phase solvation data there is almost no difference between protic and aprotic solvents concerning proton solvation enthalpies in the first solvation layer. In most aprotic solvents, however, we realize a sharp break in the curves $\Delta H_{n-1,n}(n)$ after the first layer is completed. Energies of interaction are substantially smaller or almost zero in the second layer. We have to expect therefore normal mobilities for protons and defect protons in aprotic solvents. On the other hand, aprotic solvents can vary largely in dielectric constants and solvation power, approximately described as donicity or acceptivity (GUTMANN, 1971; MAYER et al., 1975) thus leading to equilibrium constants for ion pair formation or ionization[6] and for ion pair dissociation, which differ extremely from solvent to solvent. Proton transfer processes are formulated appropriately as a sequence of consecutive equilibria similar to that we have discussed already for the vapor phase (see Eq. 1b).

From kinetic data it is necessary to distinguish between an "encounter complex" (EC) and the hydrogen-bonded neutral complex (NC). At the side of ionized species a solvent-separated ion pair (SC) might play a similar role as the encounter complex.

---

[6] In case of molecules MX the formation of ion pairs is usually called "ionization" and represents the step preceding dissoziations into free ions:

$$MX \rightleftharpoons M^+X^- \rightleftharpoons M^+ + X^- \qquad (18)$$

Table 9. Diffusion-controlled association of neutral molecules and oppositely charged ions as a function of dielectric constants (T=25°C, ROBINSON, 1975)

	Neutral molecules				Ions		
Solvent[a]	ε	k_{12} ($M^{-1}s^{-1}$)	k_{21} (s^{-1})	K_{12}[b] (M^{-1})	k_{65} ($M^{-1}s^{-1}$)	k_{56} (s^{-1})	K_{65}[b] (M^{-1})
Benzene	2	$1.5\cdot 10^9$	$4.8\cdot 10^9$	0.31	$83\cdot 10^9$	$5.5\cdot 10^{-14}$	$1.5\cdot 10^{24}$
Chlorobenzene	5	$1.5\cdot 10^9$	$4.8\cdot 10^9$	0.31	$33\cdot 10^9$	27.5	$1.2\cdot 10^9$
Water	80	$1.5\cdot 10^9$	$4.8\cdot 10^9$	0.31	$2.8\cdot 10^9$	$2.25\cdot 10^9$	1.24

[a]"Solvent" represents here only an example of a liquid with the dielectric constant used in the calculation.

[b]$K_{12} = k_{12}/k_{21}$; $K_{65} = k_{65}/k_{56}$.

$$AH+B \underset{k_{21}}{\overset{k_{12}}{\rightleftharpoons}} \underset{EC}{AH\ B} \underset{k_{32}}{\overset{k_{23}}{\rightleftharpoons}} \underset{NC}{AH..B} \underset{k_{43}}{\overset{k_{34}}{\rightleftharpoons}} \underset{IC}{A^-..HB^+} \underset{k_{54}}{\overset{k_{45}}{\rightleftharpoons}} \underset{SC}{A^-HB^+} \underset{k_{65}}{\overset{k_{56}}{\rightleftharpoons}} A^-+HB^+ \qquad (19)$$

The drastic influence of dielectric constants on the sequence of reactions (19) was demonstrated nicely by ROBINSON's (1975) estimate presented in a recent review article. He assumed diffusion control for the first and last step in the series. Rate and equilibrium constants were calculated by DEBYE's (1942) extension of the Einstein-Smoluchowski theory. As we can see from Table 9 the series of reactions simply ends at the solvent separated ion pair in solvents with low dielectric constant, since the probability for dissociation is negligibly small for all practical considerations.

In order to illustrate the thermodynamics of proton transfer in aprotic solvents we want to mention here only two representative examples of phenol-amine interactions in an unpolar and polar solvent – toluene and acetonitrile. NEERINCK et al. (1968) have studied complex formation between various phenols and tri-n-butylamine in toluene as solvent (Table 10).

Depending on the acidity of the phenolic component either the neutral, hydrogen bonded complex (NC) or the ion pair (IC) is formed predominantly. Increasing (absolute) reaction enthalpy ($-\Delta H$) goes parallel to increasing association constants and ion pair preference. In acetonitrile BEIER and SCHUSTER (1977) analyzed complex formation between p-nitrophenol and piperidine. In order to obtain a satisfactory fit to the experimental data $[BHB]^+$ complex formation has to be taken into account as well:

$$B + HB^+ \rightleftharpoons BHB^+ \qquad (20)$$

equilibrium constant: K_{BHB}

Table 10. Complex formation and equilibrium constant $\bar{K}$ for the over all reaction of substituted phenols with tri-n-butylamine in toluene (T=25°C; NEERINCK et al., 1968)

Acid	Nature of the complex		$\bar{K}(M^{-1})$
Phenol	NC		13
4-Chloro-phenol	NC		18.9
4-Nitro-phenol	NC	IC	120
2,4-Dinitrophenol	IC		380
2,4,6-Trinitrophenol	IC		1075

The following equilibrium constants were obtained (T=25°C): $K_{13} = 5.9M^{-1}$, $K_{34} = 10.3$, $K_{46} = 4.5\cdot10^{-5}M$; $K_{BHB} = 26M^{-1}$. The association constant for the ion pair ($K_{64} = 2.2\cdot10^{4}M^{-1}$) fits well into the series of estimated values (Table 9, $\varepsilon(CH_3CN) = 38$).

The kinetics of neutral complex formation has been studied for some special example like cyclolactams and substituted bases of RNA and DNA. Since these investigations are closely related to models for base pairing they will be discussed in Section V.

CALDIN and his coworkers have investigated proton transfer in chlorobenzene and similar solvents by stopped flow, microwave and laser-induced temperature jump method. Their results have been reviewed recently (ROBINSON, 1975; CROOKS, 1975; CALDIN, 1975a) and therefore will be mentioned only very briefly here. As a consequence of unfavorable equilibria and the particular method of detection (UV absorbance) applied, only one relaxation process was absorbed. Appropriately, the proton transfer systems studied are divided in three groups:

1. Proton transfer between acids containing hydroxyl groups — some highly acidic phenols, 2,4-dinitrophenol (XXVI), Bromophenol Blue (XXVII) and Magenta E (XXVIII) have been studied — and strong

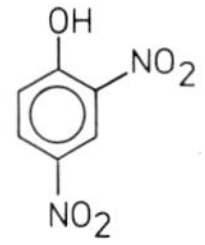

XXVI

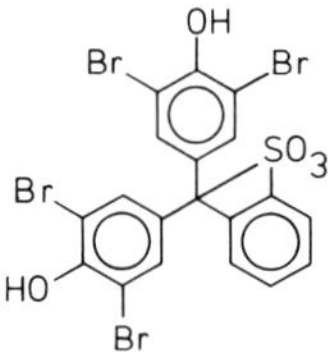

XXVII

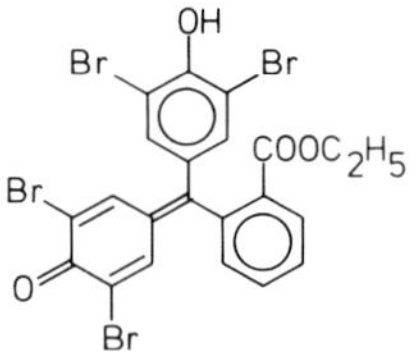

XXVIII

bases mainly tertiary aliphatic amines were found to be very fast. Overall rate constants for ion-pair formation measured by microwave temperature jump were found to lie in the range $10^8 < k_R < 10^9M^{-1}s^{-1}$. The actual proton transfer step (NC $\rightleftharpoons$ IC) is not rate-determining (CALDIN et al., 1973a,b). Solvent effects have been studied (BURFOOT et al., 1974; BURFOOT and CALDIN, 1976) and the fact that the values obtained are somewhat smaller than the values

Table 11. Proton transfer reactions of Bromophenol Blue (XXVII) with pyridine and derivatives in chlorobenzene (T=25°C, ROBINSON, 1975)

Base	K_{34} (M^{-1})	k_{34} (s^{-1})	k_{43} (s^{-1})	k_R [a] ($M^{-1}s^{-1}$)
Pyridine	$3.0 \cdot 10^2$	$1.5 \cdot 10^4$	50	$4.5 \cdot 10^5$
2-Methylpyridine	$4.3 \cdot 10^3$	$1.9 \cdot 10^5$	44	$5.7 \cdot 10^6$
2,6-Dimethylpyridine	$1.9 \cdot 10^4$	$1.3 \cdot 10^6$	70	$3.9 \cdot 10^7$
2,4,6-Trimethylpyridine	$9.1 \cdot 10^4$	$8.9 \cdot 10^6$	98	$2.7 \cdot 10^8$

[a] $k_R = K_{13} \cdot k_{34}$.

for diffusion control is explained by an additional stable complex with a certain life-time, which has to be placed between the encounter complex (EC) and the neutral aggregate (NC) in the sequence of reactions (Eq. (20)). From Table 10 we see that in the systems investigated the ion pair appears to be by far more stable than the neutral complex. Once more, we find that downhill proton transfer in a double minimum potential occurs extremely fast.

2. Proton transfer between highly acidic phenols (Bromophenol Blue (XXVII) and weak bases like pyridine and its methylderivatives has been studied in the millisecond time range by stopped flow and laser temperature jump technique (CROOKS and ROBINSON, 1970, 1971). Again the net reaction in chlorobenzene can be formulated as ion pair formation from free reactants, thus indicating that neutral complexes are only scarcely populated at equilibrium. The kinetic data can be explained by a drastic simplification of the full mechanism shown in Eq. (20) which consists only of a preequilibrium and the rate-determining proton transfer step:

$$AH + B \underset{}{\overset{K_{13}}{\rightleftharpoons}} AH...B \underset{k_{43}}{\overset{k_{34}}{\rightleftharpoons}} A^{-}...HB^{+} \qquad (21)$$

Evidence for the preequilibrium is also obtained from otherwise negative reaction enthalpies (ROBINSON, 1975). The relaxation time then is simply given by the following equation:

$$\tau^{-1} = K_{13}k_{34}(c_{AH}+c_B) + k_{43} \qquad (22)$$

For the discussion of the kinetic data a value for K_{13} was derived from a closely related reaction between the relatively weaker acid Magenta E (XXVIII) and pyridine, where only the neutral complex is formed ($K_{13} \sim 30\ M^{-1}$). Rate and equilibrium constants obtained thereby are summarized in Table 11.

3. As a representative example for proton transfer involving CH bonds the reaction of 4-nitrophenyl-nitromethane (XXIX) with strong bases has been studied in various solvents by stopped

Table 12. Rate constants for proton transfer involving CH bond dissociation (XXIX, T=25°C; CALDIN and MATEO, 1973)

Solvent	$k_R (M^{-1}s^{-1})$	
	Tetramethylguanidine	Triethylamine
Toluene	$2.3 \cdot 10^3$	132
Chlorobenzene	$7.1 \cdot 10^3$	-
Methylenechloride	$5.2 \cdot 10^3$	-
Acetonitrile	$5.9 \cdot 10^3$	12.2

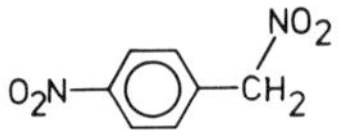

XXIX

flow technique (CALDIN and MATEO, 1973, 1975, 1976). The overall proton transfer reactions are rather slow and correspond directly to the proton transfer step since no stable complex is formed as a result of the extremely weak proton donor properties of CH bonds involved in hydrogen bonding. Effect of deuterium substitution has been studied and evidence for proton tunneling was found for the reaction with the extremely strong base (tetramethylguanidine) in unpolar media (toluene and chlorobenzene as solvents). For the purpose of comparison rate constants are shown in Table 12.

NMR studies on proton transfer in aprotic solvents have been performed mainly for the purpose of comparison with aqueous solutions. As we have seen previously proton transfer rates in water can be entranced by interstitial water molecules (see Table 4). In aprotic solvents, of course, there are no "catalytic proton conductors". One good example demonstrating the importance of a molecular picture including solvent molecules for an understanding of the rate constants measured should be mentioned here: proton transfer between ammonium ion and ammonia occurs almost exclusively without an intermediate water molecule (Table 4). Accordingly, proton exchange rates are almost the same in different solvents: $k(H_2O) = 1.17 \cdot 10^9 M^{-1}s^{-1}$ (GRUNWALD et al., 1957; BIANCHIN et al., 1975) and $k(DMSO) = 1.21 \cdot 10^9 M^{-1}s^{-1}$ (BIANCHIN and DELPUECH, 1972). Tertiary amines, however, due to their bulky structure exchange their proton predominantly via an intermediate water molecule in aqueous solution (Table 4). An illustrative example is presented by the sterically hindered amine N,2,6-trimethylpiperidine (XXX). Isomerization of the corresponding ammonium

XXX

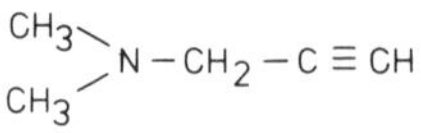

XXXI

ion with the proton in axial position (equatorial methyl group, e) to that with an equatorial proton (axial methyl group, a) in-

Table 13. Rates for deprotonation and inversion at the nitrogen atom of N,2,6-trimethylpiperidine (XXX, DELPUECH and DESCHAMPS, 1967, 1970; BIANCHIN et al., 1975)

Solvent	Rate constants k	
	Deprotonation ($M^{-1}s^{-1}$)	Inversion (s^{-1})
Water	$1.1 \cdot 10^8$	$\sim 10^3$
DMSO	$2.6 \cdot 10^4$	$\sim 10^6$

volves three steps of consecutive changes at the nitrogen atom: deprotonation, inversion and protonation. As expected from our previous discussion deprotonation is slow in DMSO and fast in water. Due to stabilization of a certain lone pair conformation by hydrogen bonding the inverse relation is found for the inversion rates (DELPUECH and DESCHAMPS, 1967, 1970; BIANCHIN and DELPUECH, 1974; BIANCHIN et al., 1975; see Table 13).

Similar NMR investigations comparing rate constants for proton transfer in water and DMSO have been performed on substituted acetylenes (XXXI, BIANCHIN et al., 1975, CHRISMENT and DELPUECH, 1972), benzylalcohol (BIANCHIN et al., 1975; GRUNWALD et al., 1962). Other proton transfer studies on benzylmercaptane in DMSO have been reported (DELPUECH and NICOLE, 1974). Finally, an investigation on proton exchange between di- and trimethylammonium ion and the corresponding amines in acetonitrile (BIANCHIN and DELPUECH, 1972) as well as internal proton transfer studies on pyrazole and imidazole in ethers and ketones as solvents (NESMEYANOV et al., 1975a,b) should be mentioned here.

V. Biochemical Model Studies

In this section we want to present a summary of proton transfer studies on compounds which are somewhat related to biologically important processes. It seems appropriate to subdivide this presentation into four subsections. Proton transfer involving amino acids will be the subject of the first part then we will discuss purines, pyrimidines and related model compounds like cyclic amides or lactams. Coenzymes, especially those of the vitamin B_6 group and various other compounds will be mentioned in the third part and finally a very brief summary on proton transfer studies involving model type macromolecules will be given. This last section is not considered to present an extensive review, which might well fill a whole chapter if all the literature available would have been included, but it will provide an entry to more extended summaries to the interested reader and a discussion of some representative examples.

A. Amino Acids

Apart from functional groups containing acidic protons in the side chains α-amino acids in general represent two basic acids and consequently, the reaction scheme shown in Table 5 can be applied to analyze the relaxation processes observed in aqueous solution. Most of the important α-amino acids, however, are characterized by such values of the three equilibrium constants (pK_1, pK_2 and K_T) which permit a largely simplified treatment. K_T, in general, is so large in α-amino acids, that the presence of the neutral form H_2NCH_RCOOH, can be neglected completely in aqueous solution. NMR studies on glycin in aqueous solution gave some information on the internal proton transfer reaction (SHEINBLATT and GUTOWSKY, 1964; COOPER et al., 1973; LEIPERT and NOGGLE, 1975; CHANG and GRUNWALD, 1976: $K_T = 2.5 \cdot 10^5$; $k_{23} = 4.4 \cdot 10^7 s^{-1}$, $k_{32} = 175\ s^{-1}$, cf. also Sect. III.B). Furthermore, both pK values differ substantially from neutral pH ($pK_1 << 7 << pK_2$). Therefore, proton transfer reactions in α-amino acids can be represented simply by the following mechanism with two single steps occurring separately in different pH ranges. At low pH, the range of the first deprotonation step ($pK_1 \sim 2$) the protolytic reaction dominates by far. Similarly, it is sufficient to consider the hydrolytic step at high pH ($pK_2 \sim 10$). Using the same notation for the individual rate constants as shown in Table 5 we obtain:

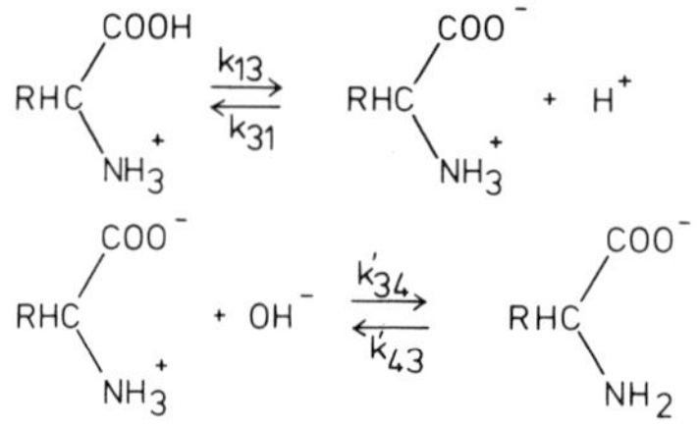

XXXIIa

XXXIIb

A number of studies on the deprotonation of zwitterions of amino acids at alkaline pH (XXXIIb) have been performed by ultrasound absorption technique (BUSSE and MAASS, 1969; APPLEGATE et al., 1968; INOUE, 1970; WHITE et al., 1971; HUSSEY and EDMONDS, 1971a, b; GRIMSHAW et al., 1973). The results obtained are summarized in Table 14.

HUSSEY and EDMONDS (1971a) also observed ultrasound absorption in the low pH range corresponding to the first deprotonation step (XXXIIa) and could determine values for the corresponding rate constants.

In general, the rate constants obtained do not differ appreciably from those we would expect from a theory of diffusion controlled intermolecular proton transfer. GRIMSHAW et al. (1973) introduced a parameter σ into the expression for diffusion controlled rates, which accounts for steric hindrance in the presence of bulky substituents.

Functional groups with acidic protons in the side chains may cause additional complications. Ultrasound absorption in aqueous solutions of lysine and arginine have been studied by HUSSEY and EDMONDS (1971b). In fitting experimental points to the theoretical

Table 14. Intermolecular proton transfer in aqueous solutions of biologically important amino acids[a]

Amino acids	$k_{31} \cdot 10^{-10}$ ($M^{-1}s^{-1}$)	$k'_{34} \cdot 10^{-10}$ ($M^{-1}s^{-1}$)	$k'_{45} \cdot 10^{-10}$ ($M^{-1}s^{-1}$)	Reference
GLY	1.0	4.1	-	HUSSEY and EDMONDS (1971a)
		1.9	-	APPLEGATE et al. (1968)
β-alanine		1.3	-	APPLEGATE et al. (1968)
γ-aminobutric		1.5	-	APPLEGATE et al. (1968)
ASP		1.2	-	APPLEGATE et al. (1968)
GLU		0.7	-	APPLEGATE et al. (1968) BUSSE and MAASS (1969)
SER		2.2	-	WHITE et al. (1971)
THR		2.0	-	WHITE et al. (1971)
LYS		3.5	2.4	HUSSEY and EDMONDS (1971b)
			2.6	BUSSE and MAASS (1969)
ARG		3.0	2.0	HUSSEY and EDMONDS (1971b)
ALA		1.9	-	GRIMSHAW et al. (1973)
VAL		1.4	-	GRIMSHAW et al. (1973)
LEU		1.7	-	GRIMSHAW et al. (1973)
ILE		1.3	-	GRIMSHAW et al. (1973)
PRO		1.5	-	GRIMSHAW et al. (1973)

[a]Only bimolecular rate constants, which are close to diffusion-controlled limit are given here; T=25°C. L-α-amino acids are denoted by biochemical abbreviations.

curves the assumption of a single relaxation process was found to be sufficient. Accordingly, the authors propose a model for more or less independent, subsequent deprotonation steps, e.g. in case of lysine (see also GRUNWALD et al., 1976):

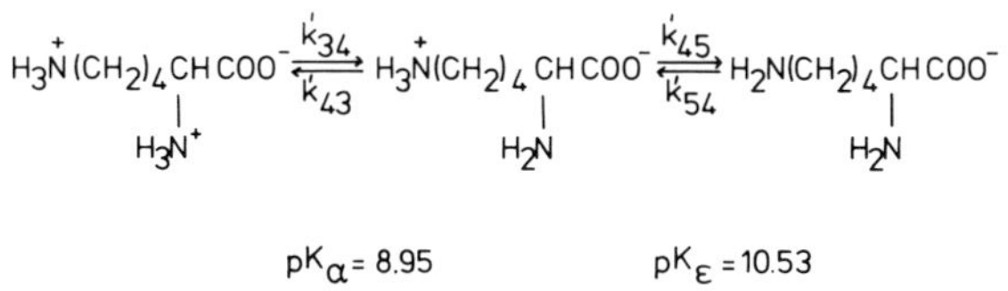

XXXIII

For technical reasons — too high frequencies — aqueous arginine solutions have not been studied up to the completion of the last deprotonation step (pK(guamidogroup) = 12.48). α-Amino acids with sulfhydryl groups in the side chain do not follow such a simple mechanism. As we have mentioned previously (Sect. III.2), intramolecular proton exchange between both functional groups has been observed (MAASS and PETERS, 1972; PETERS and MAASS, 1974; GRUNWALD et al., 1976). Another candidate for strong interaction between two functional groups at high pH is represented by tyrosine (BUSSE and MAASS, 1969). Data on protolytic reactions in this amino acids are rather scarce (INOUE, 1970). Comparing the pK values with those of closely related model compounds we might expect, however, a more or less independent dissociation ($K_T << 1$) of the phenolic hydroxyl proton in contrast to cysteine:

TYR:	$pK_2 = 9.11$	$pK_3 = 10.07$
PHE:	$pK_2 = 9.13$	
phenol:		$pK = 9.98$
CYS:	$pK_2 = 8.33$	$pK_3 = 10.78$
S-methyl-cysteine:	$pK_2 = 8.97$	
3-mercaptopropionic acid:		$pK_2 = 10.05$

In glutamic and aspartic acid we would expect a more complicated proton transfer mechanism at low pH.

Protolytic dissociation reactions of dicarboxylic acids have been studied recently by ultrasound absorption technique (SANO and YASUNAGA, 1973). They have investigated the whole series of homologous dicarboxylic acid up to adipic acid and the two cis-trans isomer, maleic and fumaric acid. A single relaxation process was observed in all cases with only two exceptions: malonic and maleic acid. These two acids are also well-known candidates for the formation of strong hydrogen bonds in the monoanions. In all other dicarboxylic acids a situation analogous to that we have discussed in the case of lysine and arginine arises. The results on succinic and glutaric acid obtained in this investigation can be expected to resemble closely the first two dissociation steps in aspartic and glutamic acid respectively.

Histidine, finally, remains the only candidate among the naturally occurring α-amino acids, which can participate in proton transfer reaction at physiological pH. Unfortunately, no relaxation studies on these systems and under the conditions we are interested in here have been reported in the literature. For the chemical relaxation in histidine some studies on imidazole might serve as an appropritate substitute (Table 4, see also RALPH and GRUNWALD, 1968, 1969a,b).

B. Purines and Pyrimidines

Relaxation studies on purines and pyrimidines have been undertaken mainly for two reasons: base pairing in DNA and RNA involves the formation of at least two hydrogen bonds. Several model compounds, mainly carboxylic acids, cyclic lactames (XXXIV) and α-pyridone (XXXV) have been investigated in order to learn more about the formation reaction of cyclic aggregates. Two kinds of arguments suggest studies in aprotic solvents. Because of competition with hydrogen-bond formation to the solvent most of

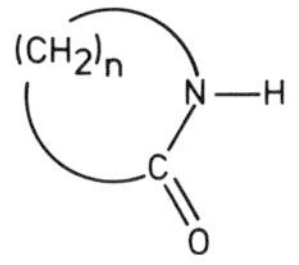

XXXIV a: n = 3 b: n = 4 c: n = 5

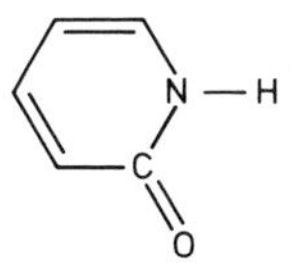

XXXV

the compounds do not form stable aggregates in water or other protic solvents. Secondly, the interior of nucleic acids is expected to be resembled best by an aprotic environment of varying polarity.

Proton transfer studies on purines and pyrimidines in aqueous solution have been performed in order to obtain information on protonation – deprotonation and internal proton exchange reactions of purines, pyrimidines and some closely related bases, which occur in nucleic acids and nucleoside-containing cofactors.

1. Formation of Hydrogen Bonded Complexes

Only two experimental techniques were found appropriate to study association reactions of the type we are interested in here: measurement of the chemical contribution to ultrasound absorption and field induced increments of dielectric loss. The latter method, originally developed and applied to association reactions in unpolar media by BERGMANN et al. (1963); BERGMANN (1963) and DE MAEYER et al. (1968) has been improved substantially during the last few years (HOPMANN, 1970, 1974; HELLEMANS and DE MAEYER, 1975).

Association of benzoic acid in unpolar solvents like hexane, tetrachloromethane, toluene and chloroform was studied some time ago by MAIER (1960) and BORUCKI (1967). They used the ultrasound absorption method and succeeded in interpreting their results by a simple dimerization mechanism. Trimerization was postulated (MAIER, 1960) only to account for some minor differences (Eq. 23).

$$A + A \rightleftharpoons A_2; \qquad A_2 + A \rightleftharpoons A_3 \tag{23}$$

Similar results have been reported for the association of acetic acid in tetrachloromethane (POSCH and KOHLER, 1967; BECKER and KOHLER, 1972). Ultrasound absorption studies on some higher carboxylic acids and halogeno carboxylic acids (propionic acid: TATSUMOTO (1967); n-butyric, isobutyric, n-valeric, isovaleric and n-caproic acid: SANO et al. (1972a); halogeno carboxylic acids: SANO et al. (1972b)) were interpreted successfully by the assumption of open chain dimers with only one hydrogen bond (A⌒A) in these systems. The relaxation process thus was attributed to the formation of the second hydrogen bond leading to a cyclic dimer (Eq. 24).

$$A\frown A \rightleftharpoons A\subset A \tag{24}$$

Ultrasonic relaxation has been measured also in acetic acid/acetone mixtures (CORSARO and ATKINSON, 1971). In this case it turned out that specific solvent-solute interactions leading to defined 1:1 and 1:2 complexes had to be postulated (RASSING, 1972a) in order to give a satisfactory explanation of the available experimental data.

Cyclic lactames (XXXIV), especially ε-caprolactame (n=5), turned out to be among the most appropriate model compounds for kinetic studies on cyclic dimer formation. Additionally, these molecules come closer to the bases of nucleic acids than carbocylic acids because of their rigid cyclic structures.

Both techniques, ultrasound absorption (e.g. EMARA and ATKINSON, 1972) and field-induced dielectric loss (BERGMANN et al., 1963; HOPMANN, 1970, 1974 and HELLEMANS and DE MAEYER, 1975) have been applied to these systems. Mainly three solvents, cyclohexane, tetrachloromethane and benzene, were used. The data obtained in cyclohexane were found to be most difficult to explain. HOPMANN (1974) postulated a trimerization model (Eq. 25) in order to be able to explain spectroscopic and kinetic data, whereas a di-

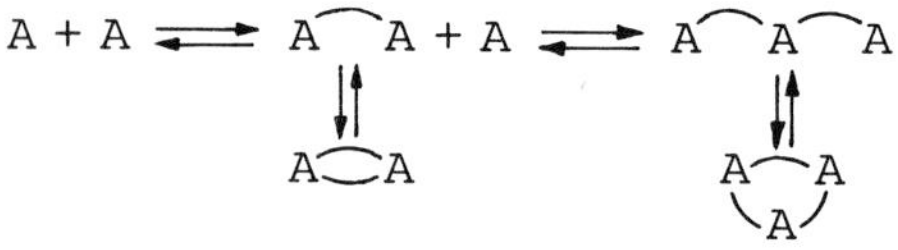

(25)

merization model including open chain (XXXVIa) and cyclic (XXXVIb) dimers was sufficient to interpret the ultrasonic relaxation data (EMARA and ATKINSON, 1972). In the most recent and most advanced

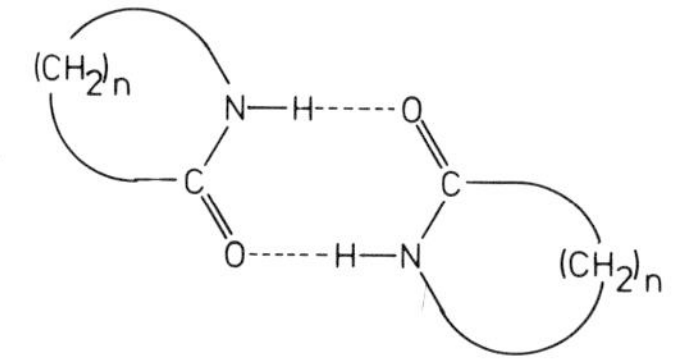

XXXVIa

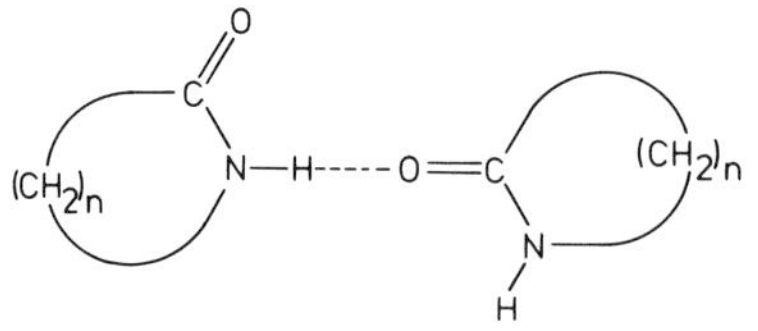

XXXVIb

dielectric relaxation study on the ε-caprolactame/cyclohexane system HELLMANS and DE MAEYER (1975) found a certain discrepancy concerning the heights of the relaxation amplitudes of HOPMANN's measurements, which they propose to correct by the assumption of a different dipole moment for the monomer. Besides that point they could confirm the trimerization mechanism. In benzene solution the situation is much less complicated: a simple dimerization model was found to be sufficient to interpret the experimental data (BERGMANN et al., 1963; HOPMANN, 1970, 1974). From the size of the dielectric increment ($\Delta\mu^2$) HOPMANN (1974) concluded that open chain dimers are formed preferably.

In tetrachloromethane an intermediate situation between cyclohexane and benzene is suggested by the spectroscopic and kinetic data. The bimolecular constants for the formation of the primary association complex were found to lie in the range of $5 \cdot 10^9 M^{-1}s^{-1}$ in all three solvents thus indicating a diffusion controlled process.

Dimerization of α-pyridone has been studied in different solvents and solvent mixtures (HAMMES and SPIVEY, 1966; HAMMES and LILLFORD, 1970). Again a single relaxation process was detected by ultrasound absorption technique. In dioxane a simple dimerization mechanism (Eq. 23) is fulfilled sufficiently well, whereas deviations are observed in more inert solvent mixtures (CCl_4/dioxane). Most probably higher than dimeric aggregates are formed here too. Again the rate constants obtained indicate diffusion-controlled formation of the primary association complexes. Addition of efficient hydrogen bond acceptors like DMSO to inert solvents interestingly, reduces the bimolecular rate constant to some extent.

The reaction, apparently, is no longer diffusion-controlled. HAMMES and LILLFORD (1970) postulate a series of mixed complexes with DMSO molecules bound to the NH groups of α-pyridone in order to explain the results obtained. The structure of these specific complexes involving solvent molecules is similar to those proposed for acetic acid acetone mixtures (RASSING, 1972a).

The kinetics of "backbone free" base pairing has been reported in two papers until now (HAMMES and PARK, 1968; HOPMANN, 1973). Both authors studied the heteroassociation of adenine and uracil derivatives in aprotic unpolar solvents (chloroform and benzene). Adenine and uracil form homoassociates as well and therefore the mechanism of heteroassociation can be expected to be rather complex. Fortunately, the open chain states of the dimers are scarcely populated and only negligible amounts of adenine dimers are formed under the conditions applied. Consequently the whole reaction scheme can be described sufficiently well by Eq. (26).

$$U{\supset}U \rightleftharpoons U + U + A \rightleftharpoons U{\supset}A \qquad (26)$$

The most probable structure of the dimeric complexes formed in aprotic and unpolar solvents are shown in XXXVII and XXXVIII. The so-called Hoogsten associate (XXXVIII) is favored over the Watson-Crick base pair in case of AU.

Rate and equilibrium constants obtained are summarized in Table 15. Despite the fact that different methods of investigation – ultrasonic relaxation: HAMMES and PARK (1968) and dielectric relaxation: HOPMANN (1973) – have been applied to differently substituted model compounds in different solvents, the results on this model reaction for base pairing agree very well. In both examples presented, the equilibrium constant for the formation of the heteroassociate, AU, is larger than that for homoassocia-

Table 15. Model studies on base pairing in aprotic unpolar solvents[a]

Reaction	Compounds	k_R ($M^{-1}s^{-1}$)	k_D (s^{-1})	K (M^{-1})	Solvent	Reference
U+U ⇌ UU	1-cyclohexyluracil	$1.5 \cdot 10^9$	$2.5 \cdot 10^7$	60	$CHCl_3$	HAMMES and PARK (1968)
	1-isobutyl-6-methyl-uracil	$4.4 \cdot 10^9$	$3.8 \cdot 10^7$	120	Benzene	HOPMANN (1973)
A+U ⇌ AU	1-cyclohexyluracil + + 9-ethyladenine	$4.0 \cdot 10^9$	$3.2 \cdot 10^7$	125[b]	$CHCl_3$	HAMMES and PARK (1968)
	1-isobutyl-6-methyl uracil + + 5'-O-acetyl-2,3-O-isopropylidene-adenosine	$10 \cdot 10^9$	$3.6 \cdot 10^7$	278[b]	Benzene	HOPMANN (1973)

[a] T∿25°C. [b] No independent determination.

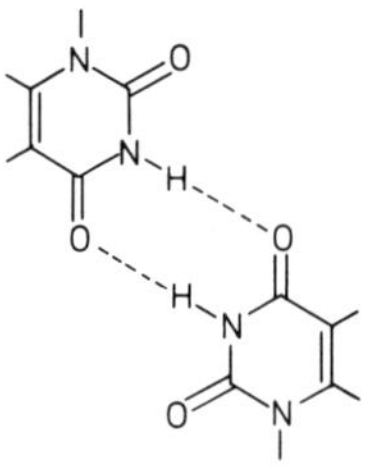

XXXVII

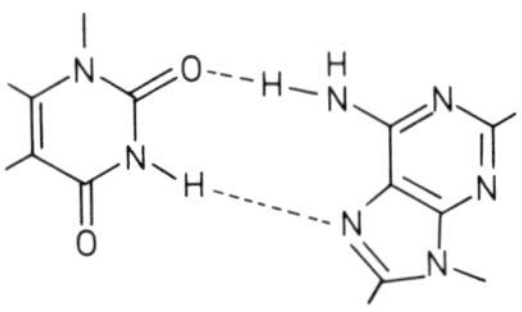

XXXVIII

tion (U_2 or A_2). It is interesting to note here that the preferred heteroassociation has also been observed in case of carboxylic acid dimers. KOHLER et al. (1974) have investigated the system trifluoroacetic acid/acetic acid in tetrachloromethane and found strong preference for the formation of the heterodimer trifluoro-acetic/acetic acid. The kinetics of the base pairing model studies (Table 15) shows, that for both cases homo- and heteroassociation the bimolecular rate constants come close to the diffusion-controlled limits.

Roughly, all the kinetic data, which have become available until now confirm that primary hydrogen bonded complexes are formed under diffusion control provided the solvent molecules are inert enough and do not form specific complexes with the monomers. The rate constants for complex dissociation, of course, are inversely proportional than to the association constants (HAMMES and PARK, 1968).

2. Proton Transfer Reactions on Purines, Pyrimidines and Some Related Heterocyclic Compounds

Proton transfer in aqueous solutions of purines and pyrimidines has been studied by NMR (MARSHALL and GRUNWALD, 1969), ultrasonic relaxation (BUSSE and MAASS, 1969; LANG et al., 1972, 1974; BRENNAN and KUSTIN, 1972) as well as temperature jump techniques. The results on intermolecular proton exchange are summarized in Table 16.

Proton transfer studies on purines in aqueous solutions are complicated by the selfassociation or stacking reactions of these bases (Eq. 27), which may lead to very long stacks at high concentrations. PÖRSCHKE and EGGERS (1972) investigated the thermodynamics and kinetics of base stacking in the case of N^6,N^9-dimethyladenine and evaluated the experimental data by an isodesmic model:

$$A + A \overset{\beta K}{\rightleftharpoons} A_2 + A \overset{K}{\rightleftharpoons} A_3 + A \overset{K}{\rightleftharpoons} \cdot\ \cdot\ \cdot\ \cdot\ \cdot \overset{K}{\rightleftharpoons} A_N \qquad (27)$$

The stacking process gives rise to a single relaxation time. At $T=25^oC$ they obtained: $K \sim 20\ M^{-1}$, nucleation parameter $\beta \sim 2$, $k_R = 9.3 \cdot 10^8 M^{-1}s^{-1}$ and $k_D = 5 \cdot 10^7 s^{-1}$.

In investigations of proton transfer reactions the authors in general tried to avoid complications by choosing conditions under which monomers and dimers predominate. MARSHALL and GRUNWALD

Table 16. Proton transfer reactions on purines and pyrimidines in aqueous solution

Compound	Reaction	$k_R \cdot 10^{10}$ $(M^{-1}s^{-1})$	T^oC	Method	Reference
purine, PH (XXXIX)	$PH+H^+$	2.4[a]	20	NMR	MARSHALL and GRUNWALD (1969)
	$PH+OH_2+P^-$	0.01[a]	20	NMR	MARSHALL and GRUNWALD (1969)
	$PH+OH^-$	1.1	25	US	BRENNAN and KUSTIN (1972)
adenine, AH	$AH+OH^-$	1.0	20	US	BUSSE and MAASS (1969)
uracil, UH	$UH+OH^-$	0.98	20	US	BUSSE and MAASS (1969)
adenosinemonophosphate, AMP	$AMP^{2-}+H^+$	6.5	25	US	LANG et al. (1972, 1974)

[a]Due to stacking reaction of purine in aqueous solution results were found slightly concentration dependent. The values given were obtained in most diluted solutions the authors investigated (c_0=0.04 M). For c_0=0.85 M the authors report values of $k_R=3.1 \cdot 10^{10} M^{-1}s^{-1}$ and $k_R=0.01 \cdot 10^{10} M^{-1}s^{-1}$ respectively.

(1969) found evidence for two kinds of proton exchange processes in their NMR studies on aqueous purine (XXXIX) solutions at acidic or neutral pH. Their major argument, that both rate constants are not influenced very much by the stacking equilibrium is based on the observation of only minor concentration dependence of the rate constants (Table 16). Both constants obtained agree roughly with the assumptions of diffusion control. Ultrasound absorption measurements in alcoholic solutions of purine (XXXIX) present information on the hydrolytic deprotonation rate (BRENNAN and KUSTIN, 1972). The rate constants obtained are somewhat smaller than would be expected for a diffusion controlled process. The authors interpret this deviation by a steric factor which accounts for unsuccessful collisions.

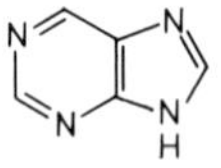

XXXIX

BUSSE and MAASS (1969) reported proton transfer studies on alkaline aqueous solutions of uracil and adenine by means of ultrasonic attenuation. Their rate constants come close to the diffusion controlled limits. Internal proton transfer in aqueous solutions of adenine (XL) has been studied recently by temperature

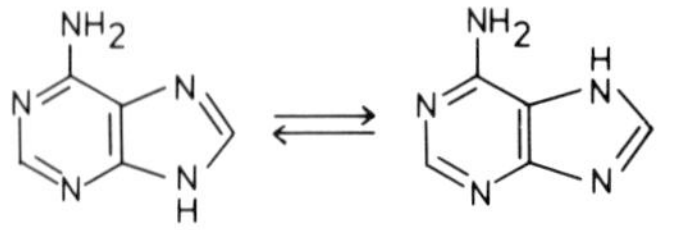

$$K_T = \frac{[N(7)H]}{[N(9)H]}$$

N(9)H N(7)H

XL

jump relaxation technique (DREYFUS et al., 1975). The results obtained can be analyzed easily in terms of reaction mechanism shown in Table 5. In case we identify $A_N \equiv N(9)H$ the results of the authors can be described completely by the following constants for T=10°C (see Tables 5 and 6):

$$k_H = 1.3\cdot 10^{10} M^{-1}s^{-1}, \quad k_{OH} = 0.85\cdot 10^{10} M^{-1}s^{-1}$$

$$k_{23}^{-} = 0.75\cdot 10^{8} M^{-1}s^{-1}, \quad k_{32}^{-} = 2.7\cdot 10^{8} M^{-1}s^{-1}$$

These constants are directly comparable to those reported for 3-hydroxypyridine (SCHUSTER et al., 1976; Table 6). Similarly, DREYFUS et al. (1975) find no conclusive evidence for direct intramolecular proton transfer within the hydration shell of the molecule ($k_{23}+k_{32} < 2\cdot 10^2 s^{-1}$, cf. Sect. III.B). Considering the difference in temperature we find almost no difference between the kinetic data reported for 3-hydroxypyridine and adenine. This observation simply reflects the fact that all reactions in both systems are diffusion controlled and the rate constants are lacking "molecular individuality". Moreover the equilibrium constants of both twobasic acids are not too different

adenine: $K_T \sim 0.3$, $pK_I = 7.0$, $\Delta pK = 5.6$

3-hydroxypyridine: $K_T \sim 1$, $pK_I = 6.8$, $\Delta pK = 3.7$

Ultrasound absorption studies on AMP in aqueous solution have been reported by LANG et al. (1972, 1974). A mechanism accounting for both proton transfer and stacking reactions is used to analyze the kinetic data. The stacks are restricted to dimers. Indirect internal proton transfer from N(1) of the purine ring to the phosphate group is considered too. The rate constants for proton transfer steps are in accord with the assumption of diffusion control. An equilibrium constant $K=7\ M^{-1}$ for the dimerization process is reported. One point, however, seems to need further investigation on certain other model systems. PÖRSCKE and EGGERS (1972) found a negative change in volume for the stacking reaction of N^6,N^9-dimethyladenine ($\Delta V = -6\pm1$ ml/mol), whereas the result of LANG et al. (1972, 1974) is inverse ($\Delta V = 5\pm2$ ml/mol). Of course, there is no need for equal reaction volumes for both stacking processes. A change in sign, nevertheless, is somewhat surprising.

Ending this section, we would like to mention some other recent studies on proton transfer involving the CH bond of the five-membered ring in benzimidazoles, imidazopyridines and purines (ELVIDGE et al., 1973a,b,c, 1974). Tritium exchange of appropriately labeled compounds has been studied in aqueous solution at $T=85^{\circ}C$. The rates were determined by conventional kinetics ($10^{-6} < k_{obs} < 10^{-3}s^{-1}$) and correspond to a mechanism involving several protonation/deprotonation steps. A "bell-shaped" pH-profile indicating slow proton exchange at extreme pH values and relatively fast reactions in the intermediate range was found to be characteristic for the unsubstituted heterocyclic compounds. Typical deviations were observed on blocking certain protonation sites by alkyl substitution.

C. Coenzymes and Other Model Compounds

The kinetics of internal proton transfer in aqueous solutions of pyridoxal-5'-phosphate (PLP, XLI) and related compounds of the vitamin-B_6 group has been investigated by temperature jump

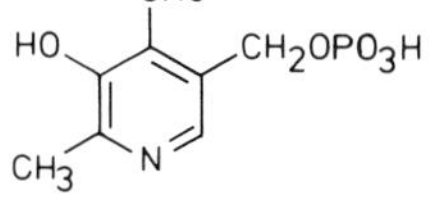

XLI

technique (AHRENS et al., 1969; AHRENS, 1973; WINKLER, TORTSCHANOFF and SCHUSTER, to be published) and ultrasound absorption methods. The general features concerning pH and concentration dependence of the slower relaxation time observed for proton transfer (τ_I in Fig. 9) are roughly the same as those reported for 3-hydroxypyridine (3-HP, SCHUSTER et al., 1976, see also Sect. III.B). At constant total concentration (c_o) we obtain a minimum in the $-\log\tau$/pH plot around neutral pH and a maximum of the relaxation amplitude in a (Γ/c_o)/pH plot nearby. Looking more closely, we observe mainly two kinds of differences: the curves in a $-\log\tau$/pH plot in PLP solutions of low concentration ($c_o = 2\cdot10^{-4}$M, Fig.

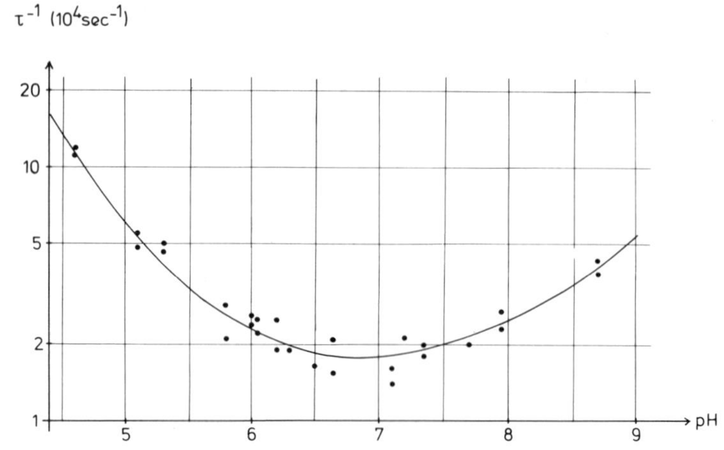

Fig. 12. Relaxation time for internal proton transfer in pyridoxal-5'-phosphate ($c_o = 2 \cdot 10^{-4}$M, T=25°C, I=0.1; scattering of the experimental points mainly comes from pH instability of diluted unbuffered solutions (WINKLER et al., to be published)

12) are much more flat around the minimum than those obtained in case of 3-HP. Especially in the alkaline branch the slope deviates substantially from 1 as has to be expected for a simple proton transfer mechanism (Table 5). There are two reasons observed for this kind of curve broadening:

1. PLP and several other vitamin-B_6 compounds are not present as a single set of species in aqueous solutions. The carbonyl substituent at the C(4) atom gives rise to an hydration equilibrium. Although the hydration reactions are much slower than protonation/deprotonation processes (AHRENS et al., 1970b), we do observe parallel proton exchange reactions for the aldehydes and their hydrates (Fig. 13). There are two arguments against this explanation: Only a single relaxation time has been observed in the studies reported in agreement with the fact that around neutral pH only a few percent of hydrates are present at equilibrium in aqueous solutions of aldehydes with similar structures. The pH

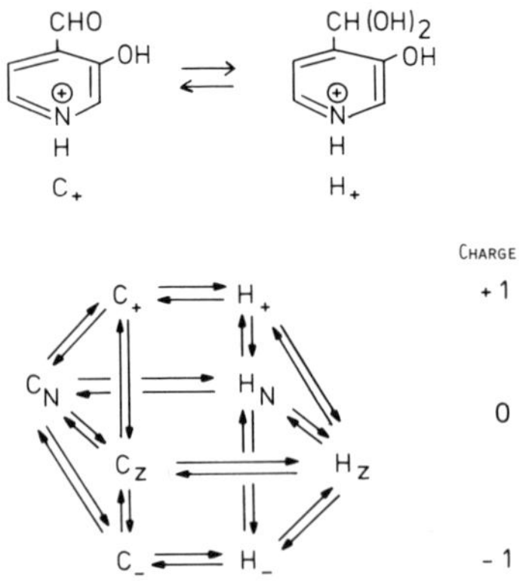

Fig. 13. Internal proton transfer and hydration reactions in 3-hydroxypyridine-4-aldehyde

and concentration dependence of the relaxation time for proton transfer in 5'-deoxypyridoxal (XLII, AHRENS, 1973) can be interpreted well by the mechanism shown in Table 5 although this compound has an aldehyde group (in 5'-deoxypyridoxal the equilibrium constants are the following: $K_T = 4.5$, $pK_I = 6.15$, $\Delta pK = 4.0$, cf. Fig. 8B).

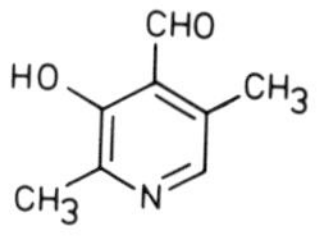

XLII

2. PLP and the other vitamin B_6 compounds with a phosphate group at C(5') represent four basic acids rather than two basic. By conventional spectrophotometric detection we observe mainly proton transfer steps involving the pyridine nitrogen atom and the hydroxy group at C(3). Proton transfer to and from the phosphate group, however, is coupled to the other proton transfer steps thereby leading to a highly complicated reaction scheme. Figure 14 gives an impression of the various processes which have to be considered explicitely. No complete dynamical analysis of various proton transfer reactions in PLP has been performed yet. By a reasonable approximative treatment, however, one can explain the kind of broad curves shown in Figure 12. For a complete kinetic analysis of proton transfer in PLP, the information of relaxation kinetic and electron excitation spectra is probably not sufficient.

The 3-hydroxypyridine unit has been found also in macrocyclic polypeptides like virginamycine, which can act as proton carriers

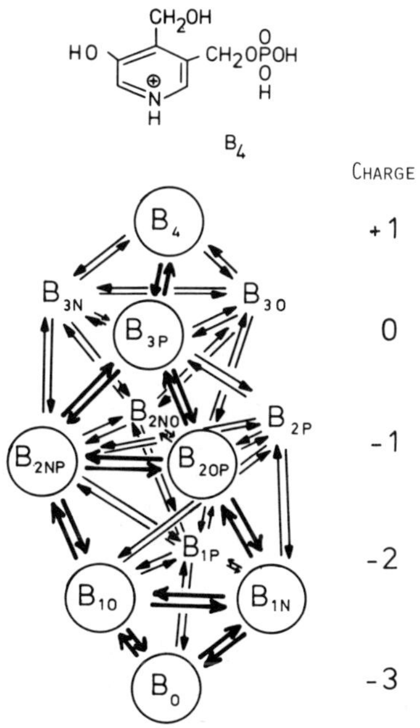

Fig. 14. Internal proton transfer reactions in pyridoxine-5'-phosphate as example of a four basic acid of vitamin B_6 series without hydration reactions; circles: states populated predominantly

in membranes. Since, kinetic studies on this class of compounds will be discussed in another chapter of this volume (GRELL and OBERBÄUMER, 1977) we need not mention any details on this subject here.

HIBBERT and HUNTE (1975) report preliminary studies on proton transfer in meso-tetraphenylporphyrin (XLIII) by temperature jump technique. Because of solubility problems they had to use a mixture of DMSO and water (90/10) as solvent. Interestingly is the fact, no singly protonated species (BH^+) could be detected

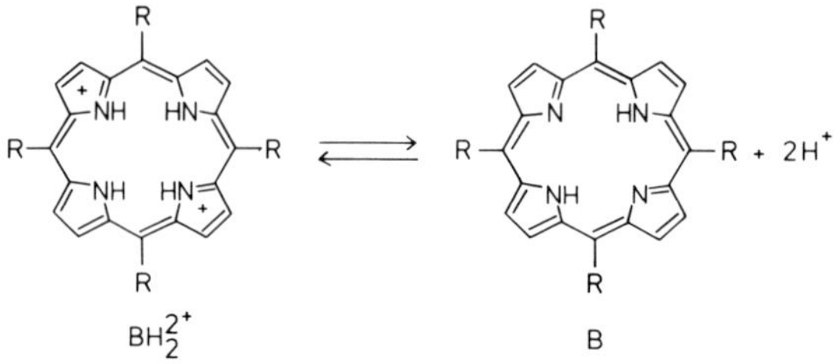

XLIII (R = 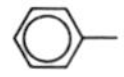)

at equilibrium in agreement with NMR studies on the same compound in $CDCl_3$ as solvent (ABRAHAM et al., 1975). The kinetic data were analyzed by a two step mechanism assuming steady state conditions for BH^+:

$$BH_2^{2+} \underset{k_{21}}{\overset{k_{12}}{\rightleftharpoons}} BH^+ + H^+ \underset{k_{32}}{\overset{k_{23}}{\rightleftharpoons}} B + 2H^+ \qquad (28)$$

From rate and equilibrium constants obtained thereby HIBBERT and HUNTE (1975) conclude that proton transfer in porphyrins occurs substantially slower than expected for diffusion-controlled elementary steps.

D. Macromolecules

Until now we have dealt exclusively with small molecules acting as proton acceptors and donors respectively. Small molecules are characterized by a single or at most a few stable conformations in the vapor phase or in solution. The individual conformations in general correspond to well separated minima of the energy surface. None of these conditions is necessarily fulfilled in case of macromolecules. A priori we cannot expect defined conformations but a distribution of states which do not differ substantially in energy. In this section we will discuss investigations on a few more or less regular synthetic polymers, that do form loose random coils in aqueous solutions. Polymers with peptide linkage will be mentioned in Section VI.

Proton transfer between polyacrylic (XLIVa) and polymethacrylic acid (XLIVb) and phenol red as indicator has been studied in neutral aqueous solution (5<pH<8.5) by WEISS et al. (1971) using

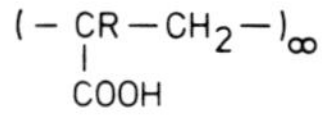

XLIV a: R = H
b: R = CH_3

temperature jump relaxation technique. Interestingly, they observed only a single well-defined relaxation time and no spectrum of relaxation processes. The mutual interaction between two neighboring carboxy groups can be estimated from kinetic studies on glutaric acid which have been mentioned previously (SANO and YASUNAGA, 1973). These authors also found a single relaxation process for only proton transfer reactions in aqueous solution. WEISS et al. (1971) succeeded in explaining concentration and pH dependence of the relaxation time observed by a simple mechanism, which in principle is identical with that proposed by EIGEN (1963) besides the fact that one of the proton exchanging partners is a polyelectrolyte:

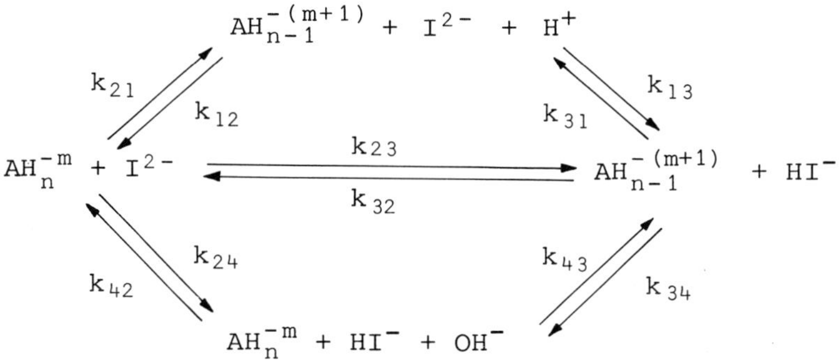

(29)

AH_n^{-m} stands here for the partially deprotonated polyacid, which is usually characterized by the degree of dissociation α: α simply represents the ratio of the number of free protonation sites ($-COO^-$) and the total number of carboxy-groups. IH_2 is the indicator in its fully protonated state. For the slowest relaxation process in the scheme shown in Eq. (29) one obtains under the assumption of diffusion control (c_o being the total concentration of polyacid expressed in terms of monomeric units):

$$\tau^{-1} = k_{23}(1+\frac{K_{HI}}{c_H})(1-\alpha)\ c_o + k_{13}(1+\frac{K_{HI}}{c_H})\ c_H + k_{34}(1+\frac{c_H}{K_{HI}})c_{OH} \qquad (30)$$

The rate constant obtained for proton exchange between the macromolecule and the indicator ($k_{23}=2\pm1\cdot10^7 M^{-1}s^{-1}$ at $T=20^\circ C$) agrees roughly with the value expected from the difference in pKs of the two acids provided we attribute an average pK to the polyacid and use

$$\overline{pK} = -\log\frac{[H^+]\cdot\alpha}{1-\alpha} \sim 6.1 \pm 0.3 \text{ (for XLIVa at } 20^\circ C\text{) and}$$

$pK_{HI} = 7.7$ (cf. Fig. 6). Additionally, the authors point out that we might also expect a certain influence of hydrogen bonding in the partially dissociated acid reducing to some extent the value of the rate constant (Fig. 6b). The final outcome of this in-

Table 17. Hydrolytic deprotonation of low molecular weight polyamines ($T=25^oC$; EMARA et al., 1972)

Base	n^a	k_R ($M^{-1} s^{-1}$)	k_D (s^{-1})
Ethylenediamine	0	$2.9 \cdot 10^{10}$	$1.1 \cdot 10^7$
Diethylenetriamine	1	$2.6 \cdot 10^{10}$	$1.5 \cdot 10^7$
Triethylenetetramine	2	$2.4 \cdot 10^{10}$	$0.4 \cdot 10^7$
Tetraethylenepentamine	3	$2.3 \cdot 10^{10}$	$1.3 \cdot 10^7$

[a] See XLV.

vestigation can be characterized as follows: in the proton transfer reaction the polyacids (XLIVa and b) behave like any other acid in aqueous solution and no evidence was found for a change in conformation coupled to the proton exchange reaction.

Two other papers on dynamical properties of polyacids are dealing with a different frequency rate. MICHELS and ZANA (1969) have studied polyacrylic and polymethacrylic acid (XLIVa,b), ATKINSON et al. (1971) polyacrylic acid and carboxymethylcellulose in aqueous solutions by ultrasound absorption technique. Both groups of authors find additional relaxation processes which they attribute to internal hydrogen bond formation (MICHELS and ZANA, 1969) and association of counterions with the polyanion. A characteristic dependence of the relaxation process on concentration and nature of the counterions but no dependence on the molecular weight of the polyelectrolyte was observed (ATKINSON et al., 1971).

Proton transfer processes in low molecular weight polyamines (XLV, EMARA et al., 1972) and polyethyleneimine (XLVI, BAUMGARTNER et al., 1973) have been studied in aqueous solution by ultrasound absorption technique. The relaxation process in polyamine solutions corresponds to an ordinary proton transfer mechanism. The rate constants obtained (Table 17) agree well with the assumption of diffusion control. In high molecular weight polyethyleneimine the kinetics observed are much more complicated. A spectrum of relaxation times was obtained that the authors interpret as a superposition of proton transfer reactions and counterion polyelectrolyte interactions. Again, no influence of the molecular weight of the polymer on the ultrasonic absorption behavior has been found.

$H_2N\text{-}(CH_2\text{-}CH_2\text{-}NH)_n\text{-}CH_2CH_2NH_2$

XLV

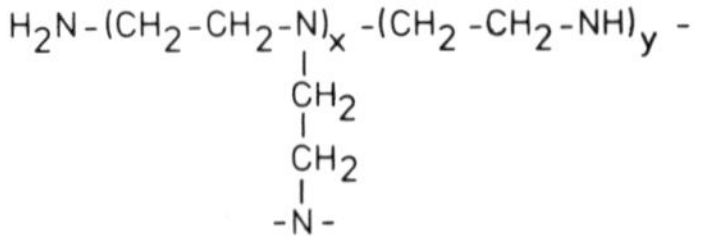

XLVI

Table 18. Hydrolytic deprotonation of oligopeptides ($T=25^{o}C$; APP 1968)

Compound	k_R ($M^{-1}s^{-1}$)	k_D (s^{-1})	pK_2
Glycine	$1.9 \cdot 10^{10}$	$8.0 \cdot 10^{5}$	9.60
Diglycine	$1.6 \cdot 10^{10}$	$0.31 \cdot 10^{5}$	8.25
Triglycine	$2.1 \cdot 10^{10}$	$0.18 \cdot 10^{5}$	7.91

VI. Polypeptides and Proteins

Due to the very different degree of complexity it is appropriate to subdivide the discussion of proton transfer reactions of polypeptides and proteins into two subsections, the first dealing with low molecular weight oligopeptides and a second reserved for high molecular weight compounds.

A. Oligopeptides

A few papers on proton transfer reactions involving small acyclic oligopeptides like diglycine, triglycine (XLVII) and glutathione (XLVIII) have been published during the last few years. NMR- (SWENSON and KOOB, 1970) as well as ultrasound absorption techniques (APPLEGATE et al., 1968; MAASS and PETERS, 1972; PETERS and MAASS, 1974) have been used.

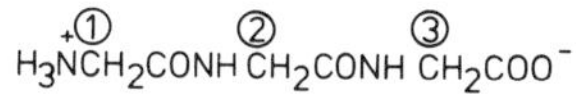

XLVII

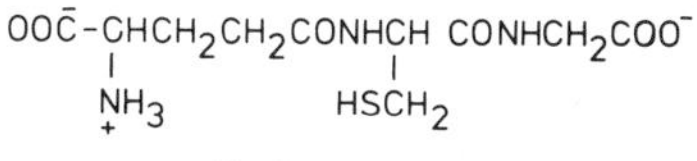

XLVIII

Hydrolytic deprotonation of di- and triglycin occurs diffusion controlled (APPLEGATE et al., 1968). As we can see from Table 18 the differences in rate constants for the reverse reaction goes parallel to changes in pK_2 values. SWENSON and KOOB (1970) have studied another aspect of proton exchange in oligopeptides. They have studied the peptide protons and as expected found base catalysis for this exchange reaction. Of interest, the catalytic constants for the two peptide protons in triglycine (in amino acid (2) and (3) in XLVII) differ by two orders of magnitude and reflect the different influence of ± charged groups in the neighborhood. In concentrated urea solution proton exchange was found to occur at faster or slower rates depending on the individual peptide proton.

Intramolecular proton transfer from the sulfhydryl to the primary aminogroup has been studied in aqueous glutathione (XLVIII) solutions by MAASS and PETERS (1972) and PETERS and MAASS (1974) using ultrasound absorption technique. They found two relaxation processes and attributed the concentration-dependent relaxation time to the internal proton transfer reaction. In glutathione (τ^{-1} = $6.3 \cdot 10^7 s^{-1}$) intramolecular proton transfer is almost one order of magnitude more slow than in cystein (τ^{-1} = $3.6 \cdot 10^8 s^{-1}$, see Sect. III.B), but still somewhat faster than one would expect from diffusion-controlled deprotonation/protonation steps. Internal proton transfer in glutathione, therefore, seems to be a true intramolecular process.

B. High Molecular Weight Polypeptides and Proteins

Proton transfer of course, is a very common elementary step of most kinds of chemical reactions in aqueous solution. Proteins as the highly specific catalysts for all reactions of the living cell, therefore, are frequently involved in various proton transfer steps. Moreover, kinetic investigations giving direct information on proton transfer in proteins are rather rare. Reviewing proton transfer in aqueous protein solutions one could either try to cover all these references and give a representative, comprehensive treatment of all kinds of biochemical reactions or concentrate on a few better known examples, where proton transfer steps have been resolved. We decided here to make the choice in favor of the second possibility. Our treatment therefore will be brief and far from complete with respect to the literature available on proton transfer in high molecular weight polypeptides and proteins.

It seems appropriate to start our brief discussion on proton transfer reactions involving high molecular weight polypeptides and proteins with a few general comments on the theory of conformations of biological macromolecules. Proteins, on one hand have an enormous conformational variety and it is practically impossible to handle them in thermodynamics and chemical kinetics without simplifying model assumptions. A statistical approach based on random walk models or the assumption of ordered structures with nearest neighbor interactions or similar theories seem to be most appropriate. Biological macromolecules, on the other hand, do by no means behave like loose coils under physiological conditions as we have seen in case of the synthetic polymers in Section V.D. Slow, reversible conformational changes strongly dependent on pH are characteristic for most globular proteins (see e.g. POHL, 1972) and agree very well with the assumption of a few well defined stable conformations in aqueous solution, which come close to the known crystal structures. Changes in conformation appear to be highly cooperative. Another complication arises from the limits of experimental information available which often will not be sufficient to discriminate between different model assumptions. Thus, one cannot get rid of the impression that we are leaving the solid ground of more or less simple and unambiguous interpretation of experimental facts in these highly complex systems.

A model theory based on nearest-neighbor interactions, formally identical with the linear Ising model was found to be appropriate for a description of cooperative conformational changes in polymers consisting of a repetitive linear chain with regular or approximately regular side chains. The principles and the mathematical formalisms have been presented many times by various authors. We refer here only to a comprehensive and easy to read summary for the interested reader: ENGEL and SCHWARZ (1970) and SCHWARZ and ENGEL (1972). A formalism for cooperative binding to linear polymers based also on the ideas of the linear Ising model has been developed by SCHWARZ (1972) as well. CARERI et al. (1976) recently pointed out that local conformational changes in proteins and proton transfer processes in aqueous solutions at physiological conditions fall roughly into the same time range and one has to expect complex coupling phenomena between both kinds of reactions. Furthermore, additional complications arise from different behavior of functional groups at the surface of globular proteins and in the interior. Local protonation/deprotonation reactions at the surface will occur readily and have rather limited effects on the stability of a given conformation as a whole, whereas the creation or neutralization of a formal charge on a certain atom or a group of atoms in the inner part strongly disturbs the balance of intermolecular forces and may easily trigger a conformational change of the proton as a whole.

Relaxation studies on aqueous solutions of high molecular weight polypeptides have been performed by a number of authors. We want to mention here only some more recent investigations by ultrasound absorption technique (HUSSEY and EDMONDS, 1971a,b; ZANA and TONDRE, 1972; ZANA, 1972) and closely related older papers (BURKE et al., 1965; PARKER et al., 1968). Based on the theory developed by SCHWARZ (1965) ultrasonic attenuation in aqueous solutions of synthetic polypeptides was attributed to conformational changes, in particular to the helix-coil transition of the macromolecules. More data became available and two other reactions were also discussed as possible candidates for the origin of the ultrasound absorption observed: proton transfer and counterion site binding. ZANA (1972) proposed a set of experiments which seem to be suitable to distinguish between the three mechanisms. Comparing the pH dependence of the chemical contribution to ultrasonic attenuation of the stereochemically pure poly-L-peptide, the poly-D,L-peptide and the salt with unusually large counterions like tetramethylammonium a clear decision should be possible, provided the relaxation observed does not correspond to a complex superposition of different reactions.

Ultrasound absorption of aqueous solutions of poly-L-lysine and poly-D,L-lysine was found to be almost the same over the whole pH range investigated besides a small difference in absolute value (ZANA and TONDRE, 1972). In contrast to the stereochemically defined poly-L-lysine, as it is known from equilibrium studies in aqueous solutions, poly-D,L-lysine does not form a helix and consequently cannot undergo helix-coil transitions. Poly-L-ornithine shows qualitatively the same pH-dependence of ultrasound absorption as found for poly-L-lysine. The absorption curve at constant pH could be fitted with sufficient accuracy by a single relaxation time and does not change substantially when water is substituted by a methanol/water mixture (15/85). From these results the authors (ZANA and TONDRE, 1972) conclude that ultrasound absorption in poly-L-ornithine solutions is caused again by proton transfer and not by a helix-coil transition.

The situation, as it is derived from experiments on poly-L-glutamic acid in aqueous solutions looks different. Comparing the results available with measurements of ultrasonic attenuation in aqueous solutions of poly-D,L-glutamic acid, ZANA (1972) concluded that a helix-coil transition is responsible for the relaxation process observed in poly-L-glutamic acid at pH 5.1, since no corresponding absorption peak has been found in solutions of the stereochemically mixed polymer. HUSSEY and EDMONDS (1971c) proposed a very crude and simple approach to estimate proton transfer contributions to ultrasound absorption of aqueous solutions of polypeptides and proteins. They calculate relaxation times and amplitudes for the various dissociable groups in the macromolecule as simple independent bimolecular reactions making use thereby of the known pK values and the rate constants for diffusion controlled reactions of small molecules with closely related functional groups. Then they assume complete independence of the various groups in the macromolecule and determine the total ultrasonic attenuation simply by adding the individual contributions. By this procedure, which of course ignores all kind of coupling with other processes, they were able to reproduce the ultrasound absorption curve of poly-L-lysine in a qualitatively correct manner and could offer an explanation for the different behavior of poly-L-glutamic acid. In the latter case the absorption maximum due to proton transfer is expected to occur at substantially lower pH ($pH_{max} \sim 3.8$ at $\omega = 40$ MHz). The process observed at pH>5 lies outside the range of detectable proton transfer reactions.

In order to learn more on proton transfer reactions a number of different proteins have been investigated by ultrasound absorption techniques in aqueous solutions. The most frequently studied protein molecules are lysozyme, bovine serum albumin and bovine hemoglobin.

We want to concentrate our discussion here on more recent studies on these three proteins. In their first paper on bovine serum albumin KESSLER and DUNN (1969) report a characteristic pH dependence of ultrasound absorption and velocity. Depending on the frequency of the ultrasound wave and the concentration of the protein a more or less pronounced maximum of ultrasound absorption was obtained around $pH \sim 3.8$ and a sharp increase in the alkaline branch indicated a second maximum at pH>11. KESSLER and DUNN (1969) attributed the ultrasound absorption observed to conformational changes of the protein. ZANA and LANG (1970) have reinvestigated bovine serum albumin solutions and extended their studies to β-lactoglobulin and lysozyme. They observed two strong peaks of ultrasound absorption centered around $pH \sim 3$ and $pH \sim 11.5$ in all protein solutions investigated. The nature of the protein was found to have only little influence on the shape of the "ultrasonic titration curves". Furthermore, very similar pronounced maxima of ultrasound absorption have also been found for diglycine, where no conformational changes are possible. ZANA and LANG (1970), therefore, suggest that proton transfer processes involving side chain carboxy- and aminogroups are responsible for the relaxatio. processes observed:

$$-COOH \rightleftharpoons -COO^- + H^+ \quad (pH_{max} \sim 3) \qquad (31a)$$

$$-NH_3^+ + OH^- \rightleftharpoons -NH_2 \quad (pH_{max} \sim 11) \qquad (31b)$$

Approximate expressions for the pH values at which the maxima in ultrasound absorption appear (pH_{max}) have been derived by APPLEGATE et al. (1968):

$$pH_{max,AH} = \frac{1}{2} (pK_{AH} - \log c_{AH}) \qquad (32a)$$

$$pH_{max,BH^+} = \frac{1}{2} (14 + pK_{BH^+} + \log c_B) \qquad (32b)$$

Here, c_{AH} and c_B represent the concentrations of acidic (COOH) and basic (NH_2) groups which can be derived from the concentration (c_o) and the molecular weight (M_W) of the protein and the number of titratable carboxy and amino groups, n_A or n_B respectively:

$$c_{AH,B} = \frac{10^3 n_{AH,B} \cdot c_o}{M_W} \qquad (33)$$

In principle, DUNN and KESSLER (1970) followed this interpretation in a second paper. The strongest argument in favor of proton transfer as the origin of the relaxation process observed was presented a little later by LANG et al. (1971). They investigated ultrasonic absorption in aqueous protein solutions with urea at 8M concentration. Under these conditions proteins are known to behave like loose coils. Both ultrasound absorption peaks were still present in these solutions. Comparing the data for water and 8M urea intensity changes and pH shifts of the peaks were found. These changes in the ultrasonic titration curves occurred to some extent when 8M urea was added to aqueous solutions of diglycine and acetylglycine. Several reasons may contribute to the shift and intensity changes observed:

1. Changes in pK-values of the dissociable groups.

2. Changes in the numbers, n_{AH} and n_B, of carboxy and amino groups accessible to titration as a consequence of unfolding.

3. Various other changes in protein solvent and protein counterion interactions influencing the low intensity ultrasound absorption.

HUSSEY and EDMONDS (1971c) as mentioned above have proposed a scheme for the calculation of ultrasonic titration curves based on the assumption of independent protonation/deprotonation steps in the polyelectrolyte. They applied this procedure also to a number of proteins in aqueous solutions – bovine serum albumin, human hemoglobin and gelatin – and obtained some kind of qualitative agreement between theoretical and experimental curves. O'BRIEN and DUNN (1972) have analyzed ultrasonic titration curves of bovine hemoglobin by a very similar treatment and proposed proton transfer processes involving the imidazole groups of histidine to be responsible for an additional smaller peak around neutral pH. The neglect of coupling between individual proton

eactions has been criticized by WHITE and SLUTSKY
ese authors could show using bovine hemoglobin as an
t the off diagonal terms are essential for calcula-
of the proton transfer relaxation spectra of polyelectrolytes. Otherwise not even approximately correct ultrasonic titration curves can be expected from the calculation.

Conformational changes in proteins involving proton transfer steps have been investigated by temperature jump relaxation studies on aqueous protein solutions with pH indicators like p-nitrophenol, phenol red, chlorophenol red or bromocresol purple. The reaction mechanisms are expected to be rather complex in this case and in most cases will involve binding processes together with proton transfer and isomerization of the protein. Since these kinds of problems are somewhat on the periphery of our subject here, we will mention only investigations on the three proteins discussed above. One of the first studies in this field has been performed on lysozyme (OWEN et al., 1969). Using phenol red and chlorophenol red as indicators they observed a relaxation process in the millisecond range and attributed it to a conformational change which exposes an ionizable group of pK∿7.3 to the solvent. A similar investigation on bovine serum albumin has been reported recently by HORIUCHI et al. (1975). Systematic studies on binding and isomerization steps involving proteins have shown that even in fairly simple cases a lot more experimental data is necessary before definite conclusions on the mechanisms can be drawn (see e.g. GUTFREUND, 1971, or the studies on NAD binding to glyceraldehyde-3-phosphate dehydrogenase by KIRSCHNER et al., 1971 and KIRSCHNER, 1971). Specific chemical modification of the protein as it has been used in a very recent study on lysozyme with p-nitrophenol as indicator (SEGAWA et al., 1975), allows the identification of aminoacid residues, which participate in the proton transfer reactions observed and may help substantially to clarify mechanistic details. In the concrete example mentioned above the authors could show that the residues GLU 35 and HIS 15 of bovine hemoglobin are involved in proton transfer to p-nitrophenol and that the proton of GLU 35 is kinetically immobilized to a certain extent by the surrounding part of the protein.

Ending this section we would like to refer to another kind of protonation/deprotonation processes involving macromolecules, which is expected to be observable by dielectric relaxation studies. SCHWARZ (1970) has calculated the contribution of proton transfer in zwitterions to the dielectric increment. In case of macromolecules like proteins in a strong electric field the chemical reactions may complete successfully with dipole reorientation by rotational diffusion.

Summarizing the data available we can state that there is clear evidence for proton transfer reactions of acidic and basic side chain groups in proteins. Corresponding relaxation processes have been observed by ultrasound absorption techniques. Relaxation times estimated from these data fit into the picture of protonation-deprotonation reactions involving either the acidic and basic functional groups alone or only their near surroundings on the surface of the macromolecule. Still open to further investigations remains the question how proton transfer reactions are coupled dynamically to conformational changes of proteins.

VII. Experimental Techniques

As we have seen from previous sections, investigations on elementary steps of proton transfer reactions in solution are essentially coupled to the stage of development in methods to study fast chemical reactions.

Flow methods due to the development of special mixing devices can be used, nowadays, for investigations of chemical reactions down to a few milliseconds (CHANCE, 1974). However, the shortest time intervals which are necessary to mix the solutions seem to be reached now and one cannot expect dramatic improvements with respect to time resolution in this field. Most proton transfer processes are still faster and flow-techniques as well as conventional methods to study chemical reactions can be applied to proton transfer reactions only in a very few exceptional cases. These reactions are mainly concerned with the dissociation of CH or other unpolar bonds. A few other investigations by flow techniques are dealing with slow dissociation processes in case of unfavorable preequilibria (CALDIN and HARBRON, 1962; CALDIN et al., 1968; CROOKS, 1975).

During the last two decades relaxation methods became available and turned out to be important and useful techniques for studies on fast reactions. In general the various methods are classified by the external parameter which is disturbed: temperature, pressure or electrical field strength. Moreover single pulse and periodic perturbation methods are distinguished.

Additionally, NMR methods based on line width analysis and NMR pulse technique were also found to be appropriate for studies of fast chemical reactions.

In the following part references concerning recent review articles and new developments in methodology are given:

1. *Relaxation methods* (EIGEN and DE MAEYER, 1963, 1974; HAMMES, 1974a). The theoretical basis of relaxation methods as well as the background has been discussed extensively (EIGEN and DE MAEYER, 1963, 1974; SCHWARZ, 1968; HAMMES, 1974a).

a) Temperature jump techniques (EIGEN and DE MAEYER, 1963, 1974; HAMMES, 1974b; CALDIN, 1975b). Fast increase on temperature is accomplished by different techniques, the most widely used being the discharge of a high voltage capacitor. The major disadvantage comes from its restriction to solutions with sufficient electric conductance. By conventional temperature jump techniques (EIGEN and DE MAEYER, 1963) relaxation times of 0.5 s $> \tau >$ 10 μs can be measured without difficulties. In a further development using the discharge of coaxial cable capacitor heating of the solution might be achieved within 50 ns and even faster (HOFFMAN, 1971, 1972). The relaxation process is followed by spectrophotometric or fluorimetric methods. Microwave absorption can also be used to generate temperature jumps (ERTL and GERISCHER, 1961; CALDIN and CROOKS, 1967). With a suitable microwave pulse generator the solution can be heated through dielectric relaxation within 1 μs. This method is especially convenient for investigation in non-aqueous solvents showing appropriate microwave absorption. An alternative method uses heating by a laser flash (HOFFMANN et al., 1968; CALDIN et al., 1971; CALDIN et al., 1973c; TURNER et al., 1972; GIANNINI, 1975). Necessarily one species of the chemical system or the solvent itself has to absorb at the laser wavelength. Time resolution of this technique can be reduced down to 0.02 μs. Aqueous as well as nonaqueous solutions can be investigated. The major problem is concerned with interference of photochemical decomposition of the reactants. In general the relaxation process can be followed by conductometric, spectrophotometric or fluorimetric detection.

Finally, two other techniques should be mentioned here: stopped flow temperature jump technique (HAMMES, 1974b; VEIL, 1974) and repetitive temperature jump technique (BRUMM et al., 1968).

b) Pressure jump techniques (EIGEN and DE MAEYER, 1963, 1974; KNOCHE, 1974). Different methods using pressure jump techniques have been described and applied over the following range of relaxation times: $20\ \mu s < \tau < 50$ s. Detection was achieved by conductometric methods (JOST, 1966; HOFFMANN and PAULI, 1966; HOFFMANN and YEAGER, 1970; KOFFER, 1975).

c) Electric field jump techniques (EIGEN and DE MAEYER, 1963, 1974; EYRING, 1975; OLSEN et al., 1974).

d) Ultrasonic methods (EIGEN and DE MAEYER, 1963, 1974; RASSING, 1975; STUEHR, 1974). Apart from other phenomena ultrasonic attenuation can be caused by the chemical relaxation. Both reaction enthalpies and volumes contribute to the absorption measured. Various resonance and pulse techniques have been developed during the last years. It is possible to measure ultrasound absorption conveniently between frequencies of $10\ kHz < \omega < 10\ GHz$. The major improvement concerns the necessary volumes of solutions to be investigated. Nowadays, 1 ml is sufficient for the most refined techniques. Because of the large range of relaxation times that can be observed, acoustical methods are widely used for investigations on chemical systems, especially for studies on fast proton transfer reactions (PETERS, 1973; EGGERS, 1967; EGGERS and FUNCK, 1973).

e) Dielectric relaxation (NORTH, 1975; CROSSLEY, 1975; DE MAEYER and PERSOONS, 1974). Dielectric relaxation is roughly the counterpart to ultrasonic attenuation in the domain of the electric field. Dispersion and in some cases also absorption of field-induced dielectric increments are measured. The method is sensitive to changes in molecular electric moments during chemical reactions ($\Delta\mu$). It is applicable mainly to aprotic, unpolar solvents and covers a frequency range of $0.1 < \omega < 10$ GHz. For recent developments see HOPMANN (1970), WAGNER et al. (1975).

2. Nuclear magnetic resonance (JEN, 1974; SWIFT, 1974; JACKMAN and COTTON, 1975). An indirect method for the investigation of fast chemical processes is based on the measurements of nuclear relaxation times (τ_1 and τ_2) in NMR experiments. The transversal (spin-spin)-relaxation time (τ_2) can be measured by continuous wave methods (CW), e.g. by line shape analysis or pulsed NMR methods like the spin echo techniques. Additionally some investigations have been carried out by methods determining the spin lattice or the longitudinal relaxation time (τ_1). Continuous wave, as well as multiple resonance techniques, have been used for this purpose. Chemical relaxation times (τ_1 and τ_2) caused by exchange of nuclei in different magnetic environments provide information on rate constants for the corresponding reactions. The time domain of dynamical NMR spectroscopy is mainly between $10 \text{ ms} < \tau < 1 \text{ s}$. In special cases the time resolution can be extended also to shorter relaxation processes. NMR methods are extremely useful for investigation on "symmetric" exchanges for which reaction enthalpies, reaction volumes etc. vanish and which therefore cannot be studied by relaxation methods. For the application of NMR techniques to proton transfer reactions see JACKMAN and COTTON (1975) and GRUNWALD (1965).

Some other techniques have been applied to proton transfer reactions only occasionally:

3. Electrochemical method (STREHLOW, 1974).

4. Flash and pulse radiolysis (PORTER and WEST, 1974; DORFMAN, 1974; FÖRSTER and VÖLKER, 1975).

5. Raman line broadening (ALBERY, 1967; KREEVOY and MEAD, 1965; IRISH and CHEN, 1970; CHEN and IRISH, 1971).

VIII. Conclusion

In the review presented here we made an attempt to describe the present stage of knowledge on proton transfer in solution. As mentioned already in the introductory remarks it seems impossible to cover all aspects of this widespread and complex field in a single chapter. We concentrated therefore on the kinetics of fast proton transfer reactions as accessible by modern relaxation techniques. Accordingly several kinds of studies have been treated only scarcely here. Among them slow proton exchange reactions investigated mainly by radiochemical methods or conventional kinetics. Additionally, problems which are more concerned with properties of associated liquids than with proton transfer between two solute molecules like dynamics of hydrogen bonded networks, cooperative phenomena in proton motion or proton conductivities etc., have been excluded from this review.

Proton transfer in aqueous solutions has been studied most frequently. Nowadays, the mechanistic aspects of intermolecular proton transfer in water from one small molecule to another are well understood. As long as polar bonds, readily forming hydrogen bonds to water molecules are concerned and no special internal stabilization of the functional groups involved is present, proton exchange is controlled by diffusion. From NMR investigations it became evident that proton transfer between molecules in water can occur either directly or along a chain of a few suitably oriented solvent molecules depending on the stereochemistry of proton acceptor and proton donor. By means of some appropriate examples intramolecular proton transfer in aqueous solution has been studied as well. The true intramolecular process competes with a consecutive series of diffusion-controlled protonation/ deprotonation steps, which involve either free protons or hydroxide ions or correspond to bimolecular proton exchange reactions. The indirect processes dominate in case the hydration shell of the molecule does not fulfil the necessary geometrical requirements, i.e. there is no chain of immobilized water molecules forming a closed cycle of hydrogen bonds between the proton donor and acceptor group. Due to our poor understanding of the structure of hydration layers it is extremely difficult to make reliable predictions on appropriate candidates for true intramolecular proton exchange. In most examples studied a kinetic behavior according to the external many step mechanism was found experimentally.

Mainly for technical reasons proton transfer reactions in nonaqueous aprotic media are much more difficult to investigate. Consequently, experimental data are rare and the mechanistic details are not so well understood as in water. The structure of aprotic liquids, is definitely much simpler than that of water. Nevertheless, the kinetics of proton transfer reactions is more complex. More individual steps can be observed. The reason for this paradoxical situation is explained very simply: as a consequence of strong interaction with solvent molecules, e.g. with water molecules, weak complexes are not formed in aqueous solutions. Even ion pairs dissociate readily in water as a result of both strong solvation power and high dielectric constant. The sequence of processes and the various intermediate complexes in a proton transfer reaction in unpolar solvents seem to be well established nowadays and it is more or less a question of time and technical refinements before extensive quantitative data on various concrete systems will become available.

Proton transfer reactions in biologically important systems have been studied extensively. As far as macromolecules are involved the situation is very complex. Many processes compete, some of them occur in very similar time ranges. Only a few facts are really well established. Fortunately, the evidence for fast protonation-deprotonation reactions involving titrateable functional groups on the surface of proteins is rather good. There are other fast processes, like polyelectrolyte interaction, changes in solvation, local conformational changes for which only little information is available from chemical relaxation studies. The biochemist, however, is mainly interested in the dynamics of proteins on a time scale up to the range of minutes or hours. For a complete understanding we need more information on all processes occurring including binding reactions of substrates and coenzymes, and conformational changes. Moreover the sequence of consecutive events on the time scale represents a center of primary interest. Although most of the individual processes mentioned before have been studied on carefully selected ideal examples and are more or less understood under certain conditions, the reaction dynamics of working enzymes as a whole is rather unknown. Much work has to be done in the future before we can expect a full understanding of these highly complex systems.

Finally, we should comment on the fact that we did not discuss proton transfer involving the other very important class of biological macromolecules, nucleic acids. The reason for that is twofold. First of all only a few relaxation studies have been reported in this field. Two papers on ultrasound absorption of DNA in aqueous solutions of different pH may serve here as examples: LANG and CERF (1969); STURM et al. (1971). Secondly, the interpretation of the data obtained seems to be even more complex than in the case of proteins. Strong conterion dependences characterize all processes involving nucleic acids. Double helix formation is a fast process as long as a few bases are concerned and local conformational changes will couple strongly to protonation/deprotonation reactions. Much work has been done in the field of slow proton exchange studies on polynucleotides and nucleic acids. A very recent paper on this subject using NMR technique is given here as an entry to further references: RORDORF and KEARNS (1975).

The dynamics of both kinds of the most important biological macromolecules, proteins and nucleic acids is so complex that a single experimental technique will never give enough information for an analysis of the individual elementary steps and their dynamic interactions. What we need for a more complete understanding are combinations of results from many different experimental methods of investigation complementary in their way of looking at the macromolecules.

IX. Other Review Articles and Books on Proton Transfer

During the last decade a number of reviews, monographs and collective volumes have been published in this field. We present here a collection of references ordered according to the main subject treated.

Hydrogen bonding and proton transfer in general: CROSSLEY (1970); VINOGRADOV and LINNELL (1971); DAVIES et al. (1973); JOESTEN and SCHAAD (1974); KRESGE (1975); CALDIN and GOLD (1975); SCHUSTER et al. (1976b); BELL (1973).

Proton transfer in aqueous solutions, general acid base catalysis investigated by relaxation methods: ALBERY (1967); EIGEN (1967a); CALDIN and GOLD (1975) by NMR methods: GRUNWALD and RALPH (1975).

Proton transfer and hydrogen bonding in non-aqueous solvents: DE MAEYER (1971); RASSING (1972b); ROBINSON (1975); GRUNWALD and RALPH (1975).

Biochemical aspects, general and especially proton transfer involving proteins: EIGEN and HAMMES (1963); EIGEN (1967b); WANG (1968); HAMMES (1968a,b); HAMMES and SCHIMMEL (1970); SCHLECHTER (1970); GUTFREUND (1971); EIGEN (1974); BENDER and KEZDY (1975), GUTFREUND (1972, 1976); CARERI et al. (1976).

Fastreactions, in general (proton transfer only discussed briefly): HAGUE (1971); HAMMES (1974a); WINKLER (1974); BRADLEY (1975); MATUŠEK (1975); WYN-JONES (1975).

Acknowledgements. This work was financially supported by the Austrian "Fonds zur Förderung der Wissenschaftlichen Forschung" Projekt Nr. 1217 and 2015.

Numerous simulating and helpful discussions with Prof. Dr. Manfred EIGEN are gratefully acknowledged. We want to thank our technical staff, Mrs. J. DURA and Mr. J. SCHUSTER for preparing the manuscript.

References

ABRAHAM, R.J., HAWKES, G.E., HUDSON, M.F., SMITH, K.M.: The nuclear magnetic resonance spectra of porphyrins. Part X. Carbon-13 nuclear magnetic resonance spectra of some meso-tetra-arylporphyrins and their metal chelates. J. Chem. Soc. Perkin II, 204-211 (1975)

AHRENS, M.-L.: Proton transfer kinetics of 5'-deoxypyridoxal. Biochim. Biophys. Acta 320, 86-96 (1973)

AHRENS, M.-L., EIGEN, M., KRUSE, W., MAASS, G.: Relaxationsuntersuchungen der Base-katalysierten Keto-Enol-Umwandlung von Acetylaceton. Ber. Bunsenges. Phys. Chem. 74, 380-385 (1970a)

AHRENS, M.-L., MAASS, G.: Elementarschritte in der Säure-Base-Katalyse. Protonenübertragungsreaktionen in wäßrigen Lösungen. Angew. Chemie 80, 848-849 (1968), Intern. Ed. 7, 818-819 (1968)

AHRENS, M.-L., MAASS, G., SCHUSTER, P., WINKLER, H.: The kinetic behavior of vitamin-B_6 compounds, hydration and proton transfer. FEBS Lett. 5, 327-330 (1969)

AHRENS, M.-L., MAASS, G., SCHUSTER, P., WINKLER, H.: Kinetic study of the hydration mechanism of vitamin B_6 and related compounds. J. Am. Chem. Soc. 92, 6134-6139 (1970b)

ALBERY, W.J.: The role of the solvent in aqueous proton transfers. Progr. React. Kin. 4, 353-398 (1967)

APPLEGATE, K., SLUTSKY, L.J., PARKER, R.C.: Kinetics of proton transfer reactions of amino acids and simple polypeptides. J. Am. Chem. Soc. 90, 6909-6913 (1968)

ARSHADI, M., KEBARLE, P.: Hydration of OH^- and O_2^- in the gas phase. J. Phys. Chem. 74, 1483-1485 (1970)

ATKINSON, G., BAUMGARTNER, E., FERNANDEZ-PRINI, R.: Ultrasonic absorption in aqueous polyelectrolyte solutions.I. J. Am. Chem. Soc. 93, 6436-6443 (1971)

BAUMGARTNER, E., ATKINSON, G., EMARA, M.M.: Ultrasonic absorption in aqueous polyelectrolyte solutions.II. Polyethyleneimine. J. Am. Chem. Soc. 95, 5881-5886 (1973)

BEAUCHAMP, J.L., BUTTRILL, S.E.: Proton affinities of H_2S and H_2O. J. Chem. Phys. 48, 1783-1789 (1968)

BECKER, G., KOHLER, F.: Ultraschalluntersuchungen über die Dimerisation von Essigsäure in Mischungen mit Tetrachlorkohlenstoff. Mh. Chem. 103, 556-570 (1972)

BEIER, G., SCHUSTER, P.: Proton transfer equilibria between phenols and amines. Ber. Bunsenges. Phys. Chem., submitted for publication (1977)

BELL, R.P.: The proton in chemistry. London: Chapman and Hall 1973

BENDER, M.L., KEZDY, F.J.: Proton-transfer in enzymatic catalysis. In: Proton Transfer Reactions. CALDIN, E.F., GOLD, V. (eds.), pp. 385-407. London: Chapman and Hall 1975

BERGER, A., LOEWENSTEIN, A., MEIBOOM, S.: Nuclear magnetic resonance study of the protolysis and ionization of N-methylacetamide. J. Am. Chem. Soc. 81, 62-67 (1959)

BERGMANN, K.: Dielektrische Absorption als Folge chemischer Relaxation. Ber. Bunsenges. Phys. Chem. 67, 826-832 (1963)

BERGMANN, K., EIGEN, M., DE MAEYER, L.: Dielektrische Absorption als Folge chemischer Reaktionen. Ber. Bunsenges. Phys. Chem. 67, 819-826 (1963)

BETOWSKI, D., PAYZANT, J.D., MACKAY, G.I., BOHME, D.K.: Rate coefficients at 297 K for proton transfer reactions with H_2O. Comparisons with classical theories and exothermicity. Chem. Phys. Lett. 31, 321-324 (1975)

BIANCHIN, B., CHRISMENT, J., DELPUECH, J.J., DESCHAMPS, M.N., NICOLE, D., SERRATRICE, G.: A comparison of proton transfer rates in aqueous and non-aqueous solutions. In: Chemical and Biological Applications of Relaxation Spectrometry. WYN-JONES, E. (ed.), pp. 365-373. Dordrecht-Boston: D. Reidel 1975

BIANCHIN, B., DELPUECH, J.J.: Transferts protoniques des sels d'ammonium substitués. IX. Sels de methylammonium dans l'acétonitrile. Bull. Soc. Chim. France 5, 2061-2066 (1972)

BIANCHIN, B., DELPUECH, J.J.: Transferts protoniques de sels d'ammonium substitués XI. Identification des isomères géometriques de l'ion trimethyl 1-cis(2,6)piperidinium par RMN du ^{1}H et du ^{13}C. Tetrahedron 30, 2859-2866 (1974)

BOHME, D.K.: The kinetics and energetics of proton transfer. In: Interactions between Ions and Molecules. AUSLOOS, P. (ed.), pp. 489-504. New York: Plenum Press 1975

BORUCKI, L.: Ultraschallabsorption und Reaktionskinetik der assoziierenden Benzoesäure in verschiedenen Lösungsmitteln. Ber. Bunsenges. Phys. Chem. 71, 504-513 (1967)

BRADLEY, J.N.: Fast Reactions. Oxford: Clarendon Press 1975

BRENNAN, M., KUSTIN, K.: Ultrasonic attenuation in basic purine solutions. J. Phys. Chem. 76, 2838-2841 (1972)

BRICKMANN, J.: Proton motions in hydrogen bonds. In: The Hydrogen Bond - Recent Developments in Theory and Experiments. SCHUSTER, P., ZUNDEL, G., SANDORFY, C. (eds.), vol. I, pp. 217-244. Amsterdam: North Holland Publ. 1976

BROUILLARD, R., DUBOIS, J.-E.: Kinetics of proton transfer reactions in aqueous solution. Alkyl structural effect on CH acid systems. J. Org. Chem. 39, 1137-1142 (1974)

BRUMM, P., KILIAN, F.P., RÜPPEL, H.: Über die Messung chemischer Relaxationsvorgänge durch repetierende Temperatursprunganregung. Ber. Bunsenges. Phys. Chem. 72, 1085-1092 (1968)
BRUNDAGE, R.S., KUSTIN, K.: Ultrasonic attenuation in aqueous triethylamine. J. Phys. Chem. 74, 672-675 (1970)
BURFOOT, G.D., CALDIN, E.F.: Solvent effects on the kinetics and thermodynamics of the fast proton-transfer reactions of trichloracetic acid and picric acid with phenyl diethyl nile blue base in aprotic solvents. J. Chem. Soc., Faraday Trans. I, 72, 963-979 (1976)
BURFOOT, G.D., CALDIN, E.F., GOODMAN, H.: Solvent effects in the kinetics of fast proton-transfer reactions: Diffusion, preliminary complexes, and steric factors. J. Chem. Soc. Faraday Trans. I 70, 105-117 (1974)
BURKE, J.J., HAMMES, G.G., LEWIS, T.B.: Ultrasonic attenuation measurements in poly-L-glutamic acid solutions. J. Chem. Phys. 42, 3520-3525 (1965)
BUSSE, H.G., MAASS, G.: Kinetische Untersuchungen der protolytischen Reaktionen von Aminosäuren und der komplementären Basen Adenin und Uracil. Z. Phys. Chem. N.F. 66, 92-104 (1969)
CALDIN, E.F.: Tunneling in proton-transfer reactions in solution. Chem. Rev. 69, 135-156 (1969)
CALDIN, E.F.: Temperature-jump techniques. Chem. Brit. 11, 4-11 (1975b)
CALDIN, E.F., CROOKS, J.E.: A microwave temperature-jump apparatus for the study of fast reactions in solution. J. Sci. Instr. 44, 449-453 (1967)
CALDIN, E.F., CROOKS, J.E., O'DONNELL, D.: Kinetics and thermodynamics of the proton-transfer reactions of 2,4-dinitrophenol with some amine bases in chlorobenzene studied by a microwave temperature-jump method. J. Chem. Soc. Faraday Trans. I, 69, 993-999 (1973a)
CALDIN, E.F., CROOKS, E.J., O'DONNELL, D.: Kinetics and thermodynamics of hydrogen bond formation and proton transfer in the reactions of tetrabromophenolphthalein ethyl ester with amine bases in chlorobenzene by a microwave temperature-jump method. J. Chem. Soc., Faraday Trans. I, 69, 1000-1015 (1973b)
CALDIN, E.F., CROOKS, J.E., ROBINSON, B.H.: A laser temperature-jump apparatus for the study of fast reactions in solution. J. Phys. E. 4, 165-169 (1971)
CALDIN, E.F., DAWSON, E., HYDE, R.M., QUEEN, A.: Kinetic isotope effects in some reactions of tri-(4-nitrophenyl)-methane with alkoxide bases in various alcoholic media. J. Chem. Soc. Faraday Trans. I, 71, 528-538 (1975a)
CALDIN, E.F., GRANT, M.W., HASINOFF, B.B., TREGLOAN, P.A.: A high-pressure laser temperature-jump apparatus for the study of the rates of fast reactions in solution at pressures up to 3 kbar. J. Phys. E. 6, 349-354 (1973c)
CALDIN, E.F., GOLD, V. (eds.): Proton Transfer Reactions. London: Chapman and Hall 1975
CALDIN, E.F., HARBRON, E.: The kinetics of some fast reactions of 4-nitrobenzyl cyanide in solution at -77°. J. Chem. Soc. 1962, 2314-2319 (1962)
CALDIN, E.F., KASPARIAN, M., TOMALIN, G.: Kinetics of proton-transfer reaction between 4-nitrobenzylcyanide and ethoxide ion in ethanol + ether from -60°-124°C. Trans. Faraday Soc. 64, 2802-2813 (1968)

CALDIN, E.F., MATEO, S.: Kinetic isotope effect in various solvents for the proton-transfer reactions of 4-nitrophenylnitromethane with bases. J. Chem. Soc. Chem. Commun. 854-855 (1973)

CALDIN, E.F., MATEO, S.: Kinetic isotope effects and tunnelling in the proton-transfer reaction between 4-nitrophenylnitromethane and tetramethylguanidine in various aprotic solvents. J. Chem. Soc. Faraday Trans. I, 71, 1876-1904 (1975)

CALDIN, E.F., MATEO, S.: Kinetic isotope effects and tunnelling corrections in the proton-transfer reactions between 4-nitrophenylnitromethane and some tertiary amines in aprotic solvents. J. Chem. Soc. Faraday Trans. I, 72, 112-121 (1976)

CARERI, G., FASELLA, P., GRATTON, E.: Statistical time events in enzymes: A Physical Assessment, in press (1976)

CHANCE, B.: Rapid flow methods. In: Techniques of Chemistry. WEISSBERGER, A. (ed.); Vol. VI, Part II; Investigation of Rates and Mechanisms of Reactions. HAMMES, G.G. (ed.); 3rd ed., pp. 5-62. New York: Wiley Interscience 1974

CHANG, K.C., GRUNWALD, E.: Water participation in proton transfer reactions of glycine and glycine methylester. J. Phys. Chem. 80, 1422-1425 (1976)

CHEN, H., IRISH, D.E.: A Raman spectral study of bisulfate-sulfate systems II. Constitution, equilibria, and ultrafast proton transfer in sulfuric acid. J. Phys. Chem. 75, 2672-2681 (1971)

CHONG, S.L., MYERS, R.A., FRANKLIN, J.L.: Proton affinity of water. J. Chem. Phys. 56, 2427-2430 (1972)

CHRISMENT, J., DELPUECH, J.J.: Acidite et vitèsse de transfert du proton acetylenique de la N,N-dimethylpropargylamine dans le dimethylsulfoxide. Tetrahedron Lett. 6, 521-524 (1972)

CHRISTOV, S.G.: Some aspects of the theory of proton-transfer processes. J.Res.Inst. Catalysis, Hokaido Univ. 16, 169-194 (1968)

CHUPKA, W.A., RUSSELL, M.E.: Photoionization study of ion molecule reactions in mixtures of hydrogen and rare gases. J. Chem. Phys. 49, 5426-5437 (1968)

CHUPKA, W.A., RUSSELL, M.E., REFAEY, K.: Ion molecule and chemiionisation reactions in H_2 by photoionisation. J. Chem. Phys. 48, 1518-1527 (1968)

CLEMENTI, E., GAYLES, J.N.: Study of the electronic structure of molecules VII: inner and outer complex in the NH_4Cl formation from NH_3 and HCl. J. Chem. Phys. 47, 3837-3841 (1967)

CLUTTER, D.R., SWIFT, T.J.: The ammonium-ammonia proton exchange in liquid ammonia. J. Am. Chem. Soc. 90, 601-607 (1968)

COCIVERA, M., GRUNWALD, E.: Hydrogen-bonded structure of ion pairs of ammonium salts as deduced from proton-exchange rates. J. Am. Chem. Soc. 87, 2070-2071 (1965)

COOPER, R.A., LICHTER, R.L., ROBERTS, J.D.: pH-dependent nuclear magnetic resonance spectra of ^{15}N-enriched glycine. Line shape and relaxation studies. J. Am. Chem. Soc. 95, 3724-3729 (1973)

CORSARO, R.D., ATKINSON, G.: Ultrasonic investigation of acetic acid hydrogen bond formation. I. Acetic acid-acetone solutions. J. Chem. Phys. 54, 4090-4095 (1971)

COTTER, R.J., KOSKI, W.S.: Reaction of D_3O^+ with D_2: proton affinity of water. J. Chem. Phys. 59, 784-787 (1973)

CRAMPTON, M.R., GRUNWALD, E.: Kinetics of ion-pair exchange in acetic acid. I. Rate constants deduced from proton exchange of anilinium salts. J. Am. Chem. Soc. 93, 2987-2990 (1971a)

CRAMPTON, M.R., GRUNWALD, E.: Kinetics of ion-pair exchange in acetic acid. II. Effect of the cation on reaction rate. J. Am. Chem. Soc. 93, 2990-2994 (1971b)

CROOKS, J.E.: Fast and slow proton-transfer reactions in solution. In: Proton Transfer Reactions. CALDIN, E.F., GOLD, V. (eds.), pp. 153-177. London: Chapman and Hall 1975

CROOKS, J.E., ROBINSON, B.H.: Reaction between bromphenol blue and aromatic nitrogen bases in chlorobenzene studied with a laser temperature-jump apparatus. Trans. Faraday Soc. 66, 1436-1445 (1970)

CROOKS, J.E., ROBINSON, B.H.: Reaction between bromophenol blue and aromatic nitrogen bases in chlorobenzene studied with a laser temperature-jump apparatus. Trans. Faraday Soc. 67, 1707-1717 (1971)

CROSSLEY, J.: Dielectric relaxation and hydrogen bonding in liquids. Advan. Mol. Relax. Proc. 2, 69-100 (1970)

CROSSLEY, J.: Dielectric relaxation at microwave frequencies. In: Chemical and Biological Applications of Relaxation Spectrometry. WYN-JONES, E. (ed.), pp. 35-40. Dordrecht-Boston: D. Reidel 1975

DAVIES, M.H., ROBINSON, B.H., KEEFFE, J.R.: Physical chemistry, Part I, Kinetics of reactions in solution. Ann. Rep. Chim. Soc. 1973, 123-171 (1973)

DEBYE, P.: Reaction rates in ionic solutions. Trans. Electrochem. Soc. 82, 265-271 (1942)

DELPUECH, J.J., DESCHAMPS, M.N.: Nuclear magnetic resonance spectroscopy: nitrogen inversion rate of 1,2,6-trimethylpiperidine. J. Chem. Soc. Chem. Commun 1967, 1188-1189 (1967)

DELPUECH, J.J., DESCHAMPS, M.N.: Transferts protoniques de sels d'ammonium substitués - III. Spectrographie par résonance magnétique nucléaire de sels de N-methylpiperidinium. Tetrahedron 26, 2723-2738 (1970)

DELPUECH, J.J., NICOLE, D.: Proton transfer to and from sulphur I. Toluene-α-thiol in dimethylsulfoxide. J. Chem. Soc., Perkin II, 1974, 1025-1029 (1974)

DE MAEYER, L.C.M.: Rates and kinetics of hydrogen bonding. Israel J. Chem. 9, 351-376 (1971)

DE MAEYER, L.C.M., EIGEN, M., SUAREZ, J.: Dielectric dispersion and chemical relaxation. J. Am. Chem. Soc. 90, 3157-3161 (1968)

DE MAEYER, L.C.M., PERSOONS, A.: Electric field methods. In: Techniques of Chemistry. WEISSBERGER, A. (ed.), Vol. VI, Part II: Investigation of Rates and Mechanisms of Reactions. HAMMES, G.G. (ed.), 3rd ed., pp. 211-234. New York: Wiley Interscience 1974

DIERCKSEN, G.H.F., KRAEMER, W.P., ROOS, B.O.: SCF-CI studies of correlation effects on hydrogen bonding and ion hydration. Theoret. Chim. Acta (berl.) 36, 249-274 (1975)

DILL, J.D., ALLEN, L.C., TOPP, W.C., POPLE, J.A.: A systematic study of the nine hydrogen bonded dimers involving NH_3,OH_2 and HF. J. Am. Chem. Soc. 97, 7220-7226 (1975)

DORFMAN, L.M.: Pulse radiolysis: In: Techniques of Chemistry. WEISSBERGER, A. (ed.), Vol. VI, Part II: Investigation of Rates and Mechanisms of Reactions. HAMMES, G.G. (ed.), 3rd ed., pp. 463-520. New York: Wiley Interscience 1974

DREYFUS, M., DODIN, G., BENSAUDE, O., DUBOIS, J.E.: Tautomerism of purines. I. N(7)H$\rightleftharpoons$N(9)H equilibrium in adenine. J. Am. Chem. Soc. 97, 2369-2376 (1975)

DUNN, F., KESSLER, L.W.: Further remarks on the ultrasonic properties of bovine serum albumin solutions. J. Phys. Chem. 74, 2736 (1970)

DYCZMONS, V., KUTZELNIGG, W.: Ab initio calculations of small hydrides including electron correlation XII. Theoret. Chim. Acta (Berl.) 33, 239-247 (1974)

DYNES, P.J., CHAPMAN, G.S., KEBEDE, E., SCHNEIDER, F.W.: Kinetics of proton transfer to cyanines. A temperature-jump study. J. Am. Chem. Soc. 94, 6356-6363 (1972)

EGGERS, F.: Eine Resonatormethode zur Bestimmung von Schall-Geschwindigkeit und Dämpfung an geringen Flüssigkeitsmengen. Acustica 19, 323-329 (1967/68)

EGGERS, F., FUNCK, Th.: Ultrasonic measurements with milliliter liquid samples in the 0.5-100 MHz range. Rev. Sci. Instrum. 44, 969-977 (1973)

EIGEN, M.: Protonenübertragung, Säure-Basen-Katalyse und enzymatische Hydrolyse. Teil I: Elementarvorgänge. Angew. Chemie 75, 489-508 (1963), Angew. Chem. Intern. Ed. 3, 1-19 (1964)

EIGEN, M.: Proton transfer and general acid-base catalysis. In: Fast Reactions and Primary Processes in Chemical Kinetics. CLAESSON, S. (ed.), pp. 245-253. New York: Interscience 1967a

EIGEN, M.: Kinetics of reaction control and information transfer in enzymes and nucleic acids. In: Fast Reactions and Primary Processes in Chemical Kinetics. CLAESSON, S. (ed.), pp. 333-369. New York: Wiley Interscience 1967b

EIGEN, M.: Diffusion control in biochemical reactions. In: Studies in the Natural Sciences. MINTZ, St.L., WIDMAYER, S.M. (eds.), Vol. IV: Quantum Statistical Mechanics in the Natural Sciences. KURZUNOGLU, D. (ed.), pp.37-61. New York: Plenum Press 1974

EIGEN, M., DE MAEYER, L.C.M.: Untersuchungen über die Kinetik der Neutralisation. I. Z. Elektrochemie 59, 986-993 (1955)

EIGEN, M., DE MAEYER, L.C.M.: Relaxation methods. In: Techniques of Organic Chemistry. WEISSBERGER, A. (ed.), Vol. VIII/2, 2nd ed., pp. 895-1054. New York: Wiley Interscience 1963

EIGEN, M., DE MAEYER, L.C.M.: Theoretical basis of relaxation spectrometry. In: Techniques of Chemistry. WEISSBERGER, A. (ed.), Vol. VI, Part II: Investigation of Rates of Mechanisms of reactions. HAMMES, G.G. (ed.), 3rd ed., pp. 63-146. New York: Wiley Interscience 1974

EIGEN, M., EYRING, E.M.: Fast protolytic reactions in aqueous solutions of aminobenzoic acids. J. Am. Chem. Soc. 84, 3254-3256 (1962)

EIGEN, M., HAMMES, G.G.: Elementary steps in enzyme reactions. In: Advances in Enzymology and Related Subjects of Biochemistry. NORD, F.F. (ed.). New York: Wiley Interscience 1963

EIGEN, M., HAMMES, G.G., KUSTIN, K.: Fast reactions of imidazole studied with relaxation spectrometry. J. Am. Chem. Soc. 82, 3482-3483 (1960)

EIGEN, M., KRUSE, W., MAASS, G., DE MAEYER, L.C.M.: Rate constants of protolytic reactions in aqueous solution. Prog. React. Kin. 2, 287-318 (1964)

EIGEN, M., KUSTIN, K.: The influence of steric factors in fast protolytic reactions as studied with HF, H_2S and substituted phenols. J. Am. Chem. Soc. 82, 5952-5953 (1960)

EIGEN, M., SCHOEN, J.: Stoßspannungsverfahren zur Untersuchung sehr schnell verlaufender Ionenreaktionen in wäßriger Lösung. Z. Elektrochem. 59, 483-494 (1955)

ELVIDGE, J.A., JONES, J.R., O'BRIEN, C., EVANS, E.A., TURNER, J.C.: Proton transfer from heterocyclic compounds. Part I. Some benzimidazoles. J. Chem. Soc., Perkin II, 1973, 432-435 (1973a)

ELVIDGE, J.A., JONES, J.R., O'BRIEN, C., EVANS, E.A., SHEPPARD, H.C.: Proton transfer from heterocyclic compounds. Part II. Purine, 9-alkylpurines and imidazo[4,5-b]pyridine. J. Chem. Soc., Perkin II 1973, 1889-1893 (1973b)

ELVIDGE, J.A., JONES, J.R., O'BRIEN, C., EVANS, E.A., SHEPPARD, H.C.: Proton transfer from heterocyclic compounds. Part III. Adenine and adenosine. J. Chem. Soc., Perkin II, 1973, 2138-2141 (1973c)

ELVIDGE, J.A., JONES, J.R., O'BRIEN, C., EVANS, E.A., SHEPPARD, H.C.: Proton transfer from heterocyclic compounds. Part IV. Guanine guanosine, hypoxanthines and inosine. J. Chem. Soc., Perkin II, 1974, 174-176 (1974)

EMARA, M.M., ATKINSON, G.: The kinetics of self-association of ε-caprolactam in cyclohexane. In: Advances in Molecular Relaxation Processes. ORVILLE-THOMAS, W.J., BÖTTCHER, C.J.F., MEIXNER,J., REDSHAW, M. (eds.), Vol. IV, pp. 203-216. Amsterdam: Elsevier 1972

EMARA, M.M., ATKINSON, G., BAUMGARTNER, E.: Ultrasonic absorption in aqueous solutions of low-molecular-weight polyamines. J. Phys. Chem. 76, 334-338 (1972)

EMERSON, M.T., GRUNWALD, E., KROMHOUT, R.A.: Proton transfer studies by nuclear magnetic resonance. I. Diffusion control in the reaction of ammonium ion in aqueous acid. J. Chem. Phys. 33, 547-555 (1960)

ENGEL, J., SCHWARZ, G.: Kooperative Konformationsumwandlungen von linearen Biopolymeren. Angew. Chem. 82, 468-479, Intern. Ed. 9, 389-400 (1970)

ERTL, G., GERISCHER, H.: Eine Temperatursprungmethode zur Untersuchung schneller Dissoziationsreaktionen mit Hilfe eines Mikrowellen-Impulses. Z. Elektrochem. 65, 629-633 (1961)

EYRING, E.M.: The electric field jump relaxation method. In: Chemical and Biological Applications of Relaxation Spectrometry. WYN-JONES, E. (ed.), pp. 85-90. Dordrecht-Boston: D. Reidel 1975

FANG, J.K., GODZIK, K., HOFACKER, G.L.: The role of proton-proton interactions in proton transfer processes in ice and water. Ber. Bunsenges. Phys. Chem. 77, 980-990 (1973)

FERGUSON, E.E.: Flowing afterglow studies. In: Ion-Molecule Reactions. FRANKLIN, J.L. (ed.), Vol. II, pp. 363-393. London: Butterworths 1972

FISCHER, F.S., HOFACKER, G.L., SEILER, R.: Model approach to non-adiabatic reaction process. J. Chem. Phys. 51, 3951-3966 (1969)

FÖRSTER, Th., VÖLKER, S.: Laserblitzspektroskopische Untersuchung rascher Protonübertragungsprozesse. Z. Phys. Chem., N.F. 97, 79-96 (1975)

FUENO, T., KAJIMOTO, O., NISHIGAKI, Y., YOSHIOKA, T.: Temperature jump study of the kinetics of reactions of arylethylmalonate monoanions with hydroxide ion. J. Chem. Soc., Perkin II, 1973, 738-741 (1973)

GIANNINI, I.: A laser temperature jump and photolysis apparatus with repetitive tunable excitation. In: Chemical and Biological Applications of Relaxation Spectrometry. WYN-JONES, E. (ed.), pp. 53-59. Dordrecht-Boston: D. Reidel 1975
GRELL, E., OBERBÄUMER, I.: Dynamic aspects of carrier mediated cation transport through membranes. In: Molecular Biology, Biochemistry and Biophysics, Vol. 24 = this volume. Berlin-Heidelberg-New York: Springer 1977
GRIMSHAW, D., HEYWOOD, P.J., WYN-JONES, E.: Proton transfer in some amino-acids studied by the ultrasonic method. J. Chem. Soc. Faraday Trans. II, 69, 756-762 (1973)
GRIMSRUD, E.P., KEBARLE, P.: Gas phase ion equilibria studies of the solvation of the hydrogen ion by methanol, dimethyl ether and water. J. Am. Chem. Soc. 95, 7939-7943 (1973)
GRUNWALD, E.: Rate constants and activation parameters for breaking the $(CH_3)_3N...HOH$ hydrogen bond in aqueous solution. J. Phys. Chem. 67, 2211-2214 (1963)
GRUNWALD, E.: Ultra-fast proton transfer reactions. Prog. Phys. Org. Chem. 3, 317-358 (1965)
GRUNWALD, E.: A nuclear magnetic resonance study of acid dissociation and proton exchange of trimethylammonium ion in methanol. J. Phys. Chem. 71, 1846-1850 (1967)
GRUNWALD, E., CHANG, K.C., SKIPPER, P.L., ANDERSON, V.K.: Kinetics of bifunctional proton transfer. 2. Lysine and cystein in aqueous solutions. J. Phys. Chem. 80, 1425-1431 (1976)
GRUNWALD, E., JUMPER, C.F., MEIBOOM, S.: Kinetics of proton transfer in methanol and the mechanism of the abnormal conductance of the hydrogen ion. J. Am. Chem. Soc. 84, 4664-4671 (1962)
GRUNWALD, E., JUMPER, C.F., MEIBOOM, S.: Kinetics of proton transfer in methanol solutions of benzoic acid and sodium benzoate. J. Am. Chem. Soc. 85, 522-528 (1963)
GRUNWALD, E., LOEWENSTEIN, A., MEIBOOM, S.: Rates and Mechanisms of protolysis of methylammonium ion in aqueous solution studied by proton magnetic resonance. J. Chem. Phys. 27, 630-642 (1957)
GRUNWALD, E., PRICE, E.: Ionization and proton exchange of amines in acetic acid. II. Effect of structure on reactivity and the reaction mechanism. J. Am. Chem. Soc. 86, 2970-2977 (1964)
GRUNWALD, E., PUAR, M.S.: Proton exchange and ion-pair-exchange of N,N-dialkylanilinium salt and basicity of N,N-dialkylanilines in acetic acids. J. Am. Chem. Soc. 89, 6842-6848 (1967a)
GRUNWALD, E., PUAR, M.S.: Proton exchange of phenol in aqueous acid. J. Phys. Chem. 71, 1842-1845 (1967b)
GRUNWALD, E., RALPH, E.K.: Kinetic studies of hydrogen bonded solvation complexes of amines in water and hydroxylic solvents. Acc. Chem. Res. 4, 107-113 (1971)
GRUNWALD, E., RALPH, E.K.: Proton transfer processes. In: Dynamic Nuclear Magnetic Resonance Spectroscopy. JACKMAN, L.M., COTTON, F.A. (eds.), pp. 621-647. New York: Academic Press 1975
GUTFREUND, H.: Transients and relaxation kinetics of enzyme reactions. Ann. Rev. Biochem. 40, 315-344 (1971)
GUTFREUND, H.: Enyzmes: Physical Principles. London: Wiley Interscience 1972, 1st ed.; 1976, 2nd ed.
GUTMANN, V.: Chemische Funktionslehre. Vienna: Springer 1971
HAGUE, D.N.: Fast Reactions. London: Wiley Interscience 1971.
HAMMES, G.G.: Relaxation spectrometry of enzymatic reactions. Acc. Chem. Res. 1, 321-329 (1968a)

HAMMES, G.G.: Relaxation spectrometry of biological systems. In: Advances in Protein Chemistry. ANFINSEN, G.B., EDSALL, J.T., ANSON, M.L., RICHARDS, F.M. (eds.), Vol. XXIII, pp. 1-57. New York: Academic Press 1968b

HAMMES, G.G.: Investigation of Rates and Mechanisms of Reactions. In: Techniques of Chemistry. WEISSBERGER, A. (ed.), Vol. VI, Part II, 3rd ed. New York: Wiley Interscience 1974a

HAMMES, G.G.: Temperature jump methods. In: Techniques of Chemistry. WEISSBERGER, A. (ed.), Vol. VI, Part II: Investigation of Rates and Mechanisms of Reactions. HAMMES, G.G. (ed.), 3rd ed., pp. 147-185. New York: Wiley Interscience 1974b

HAMMES, G.G., LILLFORD, P.J.: A kinetic and equilibrium study of the hydrogen bond dimerization of 2-pyridone in hydrogen bonding solvents. J. Am. Chem. Soc. 92, 7578-7585 (1970)

HAMMES, G.G., PARK, A.C.: Kinetic studies of hydrogen bonding. 1-Cyclohexyluracil and 9-Ethyladenine. J. Am. Chem. Soc. 90, 4151-4157 (1968)

HAMMES, G.G., SCHIMMEL, R.P.: Rapid reactions and transient states. In: The Enzymes. BOYER, P.D. (ed.), Vol. II, pp. 67-114. New York: Academic Press 1970

HAMMES, G.G., SPIVEY, H.O.: A kinetic study of the hydrogen bond dimerisation of 2-pyridone. J. Am. Chem. Soc. 88, 1621-1625 (1966)

HELLEMANS, L., DE MAEYER, L.: Absorption and dispersion of the field induced dielectric increment in caprolactam-cyclohexane solutions. J. Chem. Phys. 63, 3490-3498 (1975)

HEMSWORTH, R.S., PAYZANT, J.D., SCHIFF, H.I., BOHME, D.K.: Rate constants at 297 K for proton transfer reactions with NH_3. Comparison with classical theories and exothermicity. Chem. Phys. Lett. 26, 417-421 (1974)

HERZBERG, G.: Molecular spectra and molecular structure. In: Spectra of Diatomic Molecules, pp. 501-581. New York: Van Nostrand 1950

HIBBERT, F.: Slow proton transfer involving amino-groups. J. Chem. Soc. Chem. Commun. 1973, 463 (1973)

HIBBERT, F.: Temperature jump study of proton transfer from protonated 1,8-bis(dialkylamino)naphthalenes to hydroxide ion in water and aqueous dioxan. J. Chem. Soc., Perkin II, 1974, 1862-1866 (1974)

HIBBERT, F., HUNTE, K.P.P.: Rates of proton transfer for meso-tetraphenylporphyrin in 90% dimethyl sulphoxide-water (v/v). J. Chem. Soc. Chem. Comm. 1975, 728-729 (1975)

HIBBERT, F., LONG, F.A.: Proton transfer from cyanocarbon acids IV. Kinetic ionisation behavior of p-nitrobenzyl cyanide and bromomalonenitrile. J. Am. Chem. Soc. 94, 2647-2651 (1972)

HOFFMAN, G.W.: A nanosecond temperature jump apparatus. Rev. Sci. Instr. 42, 1643-1647 (1971)

HOFFMAN, G.W.: Entwicklung einer schnellen Temperatur-Sprung-Anlage und ihre Anwendung auf kooperative Basenpaarung. Thesis, Braunschweig, 1972

HOFFMANN, H., PAULI, K.: Eine Druckwellenapparatur zur Messung von Relaxationsvorgängen in Lösungen. Ber. Bunsenges. Phys. Chem. 70, 1052-1056 (1966)

HOFFMANN, H., YEAGER, E.: The effect of various inner coordination sphere ligands on the dissociation rate of nickel malonate complexes. Ber. Bunsenges. Phys. Chem. 74, 641-649 (1970)

HOFFMANN, H., YEAGER, E., STUEHR, J.: Laser temperature jump apparatus for relaxation studies in electrolytic solutions. Rev. Sci. Instr. 39, 649-653 (1968)
HOPKINSON, A.C., HOLBROOK, N.K., YATES, K., CSIZMADIA, I.G.: Theoretical study on the proton affinity of small molecules using Gaussian basis sets in the LCAO-MO-SCF-framework. J. Chem. Phys. 49, 3596-3601 (1968)
HOPMANN, R.: Dielektrische Absorption und chemische Relaxation im Megahertz-Bereich: Kinetik und Thermodynamik bivalenter Wasserstoffbrückenassoziate. Ber. Bunsenges. Phys. Chem. 74, 935-936 (1970)
HOPMANN, R.: A chemical relaxation study of base pairing through self-association of uracil and its hetero-association with adenine. Ber. Bunsenges. Phys. Chem. 77, 52-60 (1973)
HOPMANN, R.: Chemical relaxation as a mechanistic probe of hydrogen bonding. Thermodynamics and kinetics of lactam isoassociation in nonpolar solvents. J. Phys. Chem. 78, 2341-2348 (1974)
HORIUCHI, Z., NAKAO, M., ABE, S., TAKAHASHI, K., YASUNAGA, T.: Kinetic study on the isomerization of bovine plasma albumin in aqueous solution by means of temperature jump relaxation method. J. Sci. of the Hiroshima Univ. Ser. A. 39, 37-44 (1975)
HUNSTON, D.L., KLOTZ, I.M.: Proton exchange in aqueous urea solution. J. Phys. Chem. 75, 2123-2127 (1971)
HUSSEY, M., EDMONDS, P.D.: Ultrasonic examination of proton-transfer reactions in aqueous solutions of glycine. J. Acoust. Soc. Am. 49, 1309-1316 (1971a)
HUSSEY, M., EDMONDS, P.D.: Ultrasonic examination of proton-transfer reactions at the α-amino and side-chain groups of arginine and lysine in aqueous solution. J. Acoust. Soc. Am. 49, 1907-1908 (1971b)
HUSSEY, M., EDMONDS, P.D.: Proton transfer reactions. A mechanism for the absorption of ultrasound in aqueous solutions of proteins. J. Phys. Chem. 75, 4012-4019 (1971c)
INOUE, H.: Kinetic studies of proton transfer reactions in amino acid solutions by means of ultrasonic absorption measurement. J. Sci. Hiroshima Univ., Ser. A-2, 34, 17-35 (1970)
IRISH, D.E., CHEN, H.: Equilibria and proton transfer in the bisulfate-sulfate system. J. Phys. Chem. 74, 3796-3801 (1970)
JACKMAN, L.M., COTTON, F.A. (eds.): Dynamic Nuclear Magnetic Resonance Spectroscopy. New York: Academic Press 1975
JANOSCHEK, R., WEIDEMANN, E.G., PFEIFFER, H., ZUNDEL, G.: Extremely high polarizability of hydrogen bonds. J. Am. Chem. Soc. 94, 2387-2396 (1972)
JEN, J.: Chemical exchange and NMR-T_2 relaxation. Advan. Mol. Rel. Proc. 6, 171-183 (1974)
JOESTEN, M.D., SCHAAD, L.J.: Hydrogen Bonding. New York: Dekker 1974
JOST, A.: Stoßwellen in Flüssigkeiten zur Untersuchung von Lösungsreaktionen. Ber. Bunsenges. Phys. Chem. 70, 1057-1060 (1966)
KEBARLE, P.: Higher-order reactions - ion clusters and ion solvation. In: Ion-Molecule Reactions. FRANKLIN, J.L. (ed.), Vol. II, pp. 315-362. London: Butterworths 1972
KEBARLE, P., SEARLES, S.K., ZOLLA, A., SCARBOROUGH, J., ARSHADI, H.: The solvation of the hydrogen ion by water molecules in the gas phase. J. Am. Chem. Soc. 89, 6393-6399 (1967)

KEIL, F., AHLRICHS, R.: Theoretical study of S_N2 reactions. Ab initio computations on HF and CI level. J. Am. Chem. Soc. 98, 4787-4793 (1976)
KESSLER, L.W., DUNN, F.: Ultrasonic investigation of the conformal changes of bovine serum albumin in aqueous solution. J. Phys. Chem. 73, 4256-4263 (1969)
KIRSCHNER, K.: Cooperative binding of nicotinamide-adenine dinucleotide to yeast glyceraldehyde-3-phosphate dehydrogenase. II. Stopped-flow studies at pH 8.5 and 40°C. J. Mol. Biol. 58, 51-68 (1971)
KIRSCHNER, K., GALLEGO, E., SCHUSTER, I., GOODALL, D.: Cooperative binding of nicotinamide-adenine dinucleotide to yeast glyceraldehyde-3-phosphate dehydrogenase. I. Equilibrium and temperature-jump studies at pH 8.5 and 40°C. J. Mol. Biol. 58, 29-50 (1971)
KNOCHE, W.: Pressure-jump methods. In: Techniques of Chemistry. WEISSBERGER, A. (ed.), Vol. VI, Part II: Investigation of Rates and Mechanisms of Reactions. HAMMES, G.G. (ed.), 3rd ed., pp. 187-210. New York: Wiley Interscience 1974
KOFFER, H.: Kinetics of protolytic and keto-enol reactions of some 5-monosubstituted barbituric acids. J. Chem. Soc., Perkin II 1975, 819-821 (1975)
KOHLER, F., FINDENEGG, G.H., BOBIK, M.: Mixtures of trifluoroacetic acid with acetic acid and carbon tetrachloride. J. Phys. Chem. 78, 1709-1714 (1974)
KOLLMAN, P., ALLEN, L.C.: The theory of the hydrogen bond. Chem. Rev. 72, 283-303 (1972)
KOLLMAN, P., McKELVEY, J., JOHANSSON, A., ROTHENBERG, S.: Theoretical studies of hydrogen bonded dimers. J. Am. Chem. Soc. 97, 955-965 (1975)
KRAEMER, W.P., DIERCKSEN, G.H.F.: SCM MO LCGO studies on hydrogen bonding. The system $(H_2OHOH_2)^+$. Chem. Phys. Lett. 5, 463-465 (1970)
KREEVOY, M.M., KONASEVICH, D.E.: The Brønsted α and the primary hydrogen isotope effect. A test of the Marcus theory. Advan. Chem. Phys. 21, 243-252 (1971)
KREEVOY, M.M., MEAD, C.A.: Estimation of very fast reaction rates from the broadening of vibrational spectral lines. Disc. Faraday Soc. 39, 166-171 (1965)
KRESGE, A.J.: What makes proton transfer fast? Acc. Chem. Res. 8, 354-360 (1975)
KRESGE, A.J., CAPEN, G.L.: Deprotonation of the N,N-dimethylanilinium ion. Search for slow proton transfer to nitrogen. J. Am. Chem. Soc. 97, 1795-1797 (1975)
KURZ, J.L., KURZ, L.C.: On the mechanism of proton transfer in solution. Factor determining whether the activated complex has an equilibrated environment. J. Am. Chem. Soc. 94, 4451-4461 (1972)
KUTZELNIGG, W., AHLRICHS, R., LABIB-ISKANDER, I., BINGEL, W.A.: The Hartree-Fock and the correlation energies of the H_3^+ ion and their dependence on the nuclear configuration. Chem. Phys. Lett. 1, 447-450 (1967)
LANG, J., CERF, R.: Absorption ultrasonore dans des solutions d'acide désoxyribonucléique: Étude de la dénaturation alcaline, J. Chim. Phys. 66, 81-87 (1969)
LANG, J., STURM, J., ZANA, R.: Cinétique d'échange de protons dans les solutions aqueuses d'adénosine 5'-monophosphate. C.R. Acad. Sci. Paris, C 275, 597-600 (1972)

LANG, J., STURM, J., ZANA, R.: Ultrasonic absorption in aqueous solutions of nucleotides and nucleosides. II. Kinetics of proton exchange in adenosine-5'-monophosphate. J. Phys. Chem. 78, 80-86 (1974)
LANG, J., TONDRE, C., ZANA, R.: Effect of urea and other organic substances on the ultrasonic absorption of protein solutions. J. Phys. Chem. 75, 374-379 (1971)
LEIPERT, T.K., NOGGLE, J.H.: The nitrogen-15 magnetic resonance of glycine. J. Am. Chem. Soc. 97, 269-272 (1975)
LEYDEN, D.E., MORGAN, W.R.: Proton exchange mechanisms of some tertiary benzylamines. J. Phys. Chem. 73, 2924-2929 (1969)
LISCHKA, H.: Ab initio calculations on small hydrides including electron correlation. IX. Theoret. Chim. Acta (Berl.) 31, 39-48 (1973)
LOEWENSTEIN, A., MEIBOOM, S.: Rates and mechanisms of protolysis of di- and trimethylammonium ions studied by proton magnetic resonance. J. Chem. Phys. 27, 1067-1071 (1957)
LOEWENSTEIN, A., SZÖKE, A.: The activation energies of proton transfer reactions in water. J. Am. Chem. Soc. 84, 1151-1154 (1962)
LUNDGREN, J.O., OLOVSSON, I.: The hydrated proton in solids. In: The Hydrogen Bond - Recent Developments in Theory and Experiments. SCHUSTER, P., ZUNDEL, G., SANDORFY, C. (eds.), Vol. II, pp. 471-526. Amsterdam: North Holland 1976
LUZ, Z., MEIBOOM, S.: Kinetics of proton exchange in aqueous solutions of acetate buffer. J. Am. Chem. Soc. 85, 3923-3925 (1963)
LUZ, Z., MEIBOOM, S.: The activation energies of proton transfer reactions in water. J. Am. Chem. Soc. 86, 4768-4769 (1964)
MAASS, G., PETERS, F.: Relaxationskinetische Untersuchungen intramolekularer Protonenübertragungen an Cystein und verwandten Verbindungen. Angew. Chem. 84, 430-432, Intern. Ed. 11, 428-429 (1972)
MAIER, W.: Ultraschallabsorptionsmessungen zur Kinetik von Assoziationsreaktionen in verdünnten Lösungen. Z. f. Elektrochem. 64, 132-135 (1960)
MARCUS, R.A.: Theoretical relations among rate constants, barriers and bronsted slopes of chemical reactions. J. Phys. Chem. 72, 891-899 (1968)
MARCUS, R.A.: Unusual slopes of free energy plots in kinetics. J. Am. Chem. Soc. 91, 7224-7225 (1969)
MARSHALL, T.H., GRUNWALD, E.: NH-proton exchange of purine in aqueous solution. Effect of molecular complexing on reaction rate. J. Am. Chem. Soc. 91, 4541-4544 (1969)
MATUŠEK, M.: Relaxation methods in the investigation of reaction kinetics in solution. Chemické listy 69, 673-702 (1975)
MAYER, U., GUTMANN, V., GERGER, W.: The acceptor number - a quantitative empirical parameter for the electrophilic properties of solvents. Mh. Chem. 106, 1235-1257 (1975)
MEIBOOM, S.: Nuclear magnetic resonance study of the proton transfer in water. J. Chem. Phys. 34, 375-388 (1961)
MERLET, P., PEYERIMHOFF, S.D., BUENKER, R.J.: Ab initio study of the hydrogen bond in $[H_3N\text{-}H...NH_3]^+$. J. Am. Chem. Soc. 94, 8301-8308 (1972)
MEYER, W., JAKUBETZ, W., SCHUSTER, P.: Correlation effects on energy curves for proton transfer. The cation $[H_5O_2]^+$. Chem. Phys. Lett. 21, 97-102 (1973)

MICHELS, B., ZANA, R.: Absorption ultrasonore de solutions aqueuses d'acides polyacrylique et polyméthacrylique non neutralisées. Kolloid-Z. u. Z. Polymere 234, 1008-1015 (1969)

MORGAN, W.R., LEYDEN, D.E.: Nitrogen inversion of tertiary amines in aqueous acid. J. Am. Chem. Soc. 92, 4527-4531 (1970)

NEERINCK, D., VAN AUNDENHAEGE, A., LAMBERTS, L., HUYSKENS, P.: Calorimetric determination of the equilibrium constant and the heat of formation of hydrogen bonded complexes. Nature (Lond.) 218, 461-462 (1968)

NESMEYANOV, A.N., ZAVELOVICH, E.B., BABIN, V.N., KOCHETKOVA, N.S., FEDIN, E.I.: ^1H and ^{13}C NMR study of tautomerism in azoles I. Proton transfer in the absence of solvation. Tetrehedron Lett. 31, 1461-1462 (1975a)

NESMEYANOV, A.N., ZAVELOVICH, E.B., BABIN, V.N., KOCHETKOVA, N.S., FEDIN, E.I.: ^1H and ^{13}C NMR study of tautomerism in azoles II. Proton transfer in ketone solvents. Tetrahedron Lett. 31, 1463-1464 (1975b)

NORTH, A.M.: Dielectric relaxation. In: Chemical and Biological Applications of Relaxation Spectrometry. WYN-JONES, E. (ed.), pp. 17-34. Dordrecht-Boston: D. Reidel 1975

O'BRIEN, W.D., DUNN, F.: Ultrasonic absorption mechanisms in aqueous solutions of bovine hemoglobin. J. Phys. Chem. 76, 528-533 (1972)

OLAH, G.A.: Carbokationen und elektrophile Reaktionen. Angew. Chem. 85, 183-225 (1973), Intern. Ed. 12, 1273-215 (1973)

OLSEN, S.L., HOLMES, L.P., EYRING, E.M.: Pulsed spectrophotometric chemical relaxation measuremtents. Rev. Sci. Instrum. 45, 859-861 (1974)

OWEN, J.D., EYRING, E.M., COLE, D.L.: A kinetic investigation of an isomerization of aqueous lysozyme near neutral pH. J. Phys. Chem. 73, 3918-3921 (1969)

PARKER, R.C., SLUTSKY, L.J., APPLEGATE, K.R.: Ultrasonic absorption and the kinetics of conformational change in poly-L-lysine. J. Phys. Chem. 72, 3177-3186 (1968)

PAYZANT, J.D., CUNNINGHAM, A.J., KEBARLE, P.: Gas phase solvation of the ammonium ion by NH_3 and H_2O and stabilities of mixed clusters $NH_4^+(NH_3)_n(H_2O)_w$. Can. J. Chem. 51, 3242-3249 (1973)

PERLMUTTER-HAYMAN, B., SHINAR, R.: The kinetics of deprotonation of Tropaeolin O. Intern. J. Chem. Kinetics 7, 453-462 (1975)

PETERS, F.: Verbesserte Doppelstrahl-Pulsmethode zur Messung der Ultraschallabsorption für chemische Relaxationsuntersuchungen. Acustica 28, 159-165 (1973)

PETERS, F., MAASS, G.: On the rates of intramolecular proton transfer in amino-thiol compounds. Proc. 16th conf. German Soc. Biol. Chem., Tübingen, 1973, Stuttgart: Thieme 1974

POHL, F.: Kooperative Konformationsänderungen von globulären Proteinen. Angew. Chem. 84, 931-944, Intern. Ed. 11, 894-906 (1972)

PÖRSCHKE, D., EGGERS, F.: Thermodynamics and kinetics of base-stacking interactions. Europ. J. Biochem. 26, 490-498 (1972)

PORTER, G., WEST, M.A.: Flash photolysis. In: Techniques of Chemistry. WEISSBERGER, A. (ed.), Vol. VI, Part II: Investigation of Rates and Mechanisms of Reactions. HAMMES, G.G. (ed.), 3rd ed., pp. 367-462. New York: Wiley Interscience 1974

POSCH, H., KOHLER, F.: Ultraschalluntersuchungen über die Dimerisation von Essigsäure in Mischungen mit Tetrachlorkohlenstoff. Mh. Chem. 98, 1451-1459 (1967)

RALPH, E.K., GRUNWALD, E.: Proton exchange of triethylammonium ion in aqueous solution. J. Am. Chem. Soc. 89, 2963-2967 (1967)
RALPH, E.K., GRUNWALD, E.: Higher-than-termolecular proton transfer in aqueous solutions of imidazole. J. Am. Chem. Soc. 90, 517 (1968)
RALPH, E.K., GRUNWALD, E.: Kinetics of proton exchange in the ionization and acid dissociation of imidazole in aqueous acid. J. Am. Chem. Soc. 91, 2422-2425 (1969a)
RALPH, E.K., GRUNWALD, E.: Proton exchange and solvation of imidazole in t-butyl alcohol water mixtures. J. Am. Chem. Soc. 91, 2429-2432 (1969b)
RASSING, J.E.: Hydrogen bonding in acetic acid/acetone mixtures. J. Chem. Phys. 56, 5225-5228 (1972a)
RASSING, J.E.: Hydrogen bond kinetics and relaxation spectrometry. Advan. Mol. Relax. Proc. 4, 55-86 (1972b)
RASSING, J.E.: Ultrasonic relaxation spectrometry. In: Chemical and Biological Applications of Relaxation Spectrometry. WYN-JONES, E. (ed.), pp. 1-16. Dordrecht-Boston: D. Reidel 1975
ROBINSON, B.H.: Hydrogen-bonding and proton transfer reactions in aprotic solvents. In: Proton-Transfer Reactions. CALDIN, E., GOLD, V. (eds), pp. 121-152. London: Chapman and Hall 1975
RORDORF, B.F., KEARNS, D.R.: NMR investigation of proton exchange in transfer RNA by high resolution NMR. Biochem. Biophys. Res. Commun. 65, 857-862 (1975)
ROSE, M.C., STUEHR, J.: Rates of internally hydrogen-bonded systems. J. Am. Chem. Soc. 90, 7205-7209 (1968)
ROSE, M.C., STUEHR, J.: The Brønsted slope for internally hydrogen-bonded weak acids. J. Am. Chem. Soc. 93, 4350-4354 (1971)
ROSE, M.C., STUEHR, J.: A linear free energy, enthalpy and entropy system. J. Am. Chem. Soc. 94, 5532-5538 (1972)
ROSENTHAL, D., GRUNWALD, E.: Quatermolecular proton transfer of 2,4-lutidine and its conjugate acid in aqueous solution. J. Am. Chem. Soc. 94, 5956-5961 (1972)
SANO, T., TATSUMOTO, N., MENDE, Y., YASUNAGA, T.: Kinetic studies of intermolecular hydrogen bonding in carboxylic acids by means of ultrasonic absorption measurement. II. Halogen carboxylic acids. Bull. Chem. Soc. Japan 45, 2673-2677 (1972b)
SANO, T., TATSUMOTO, N., NIVA, T., YASUNAGA, T.: Kinetic studies of intermolecular hydrogen bonding in carboxylic acids by means of ultrasonic absorption measurement. I. Fatty Acids. Bull. Chem. Soc. Japan 45, 2669-2673 (1972a)
SANO, T., YASUNAGA, T.: Kinetic studies of dissociation and recombination reaction in aqueous solutions of dicarboxylic acids by means of ultrasonic absorption measurements. J. Phys. Chem. 77, 2031-2038 (1973)
SCHLECHTER, A.N.: Measurement of fast biochemical reactions. Science 170, 273-280 (1970)
SCHUSTER, P.: Neuere Ergebnisse aus der Theorie der Wasserstoffbrücken und Protonenübergänge. Z. Chem. 13, 41-55 (1973)
SCHUSTER, P.: Energy surfaces for hydrogen bonded systems. In: The Hydrogen Bond - Recent Developments in Theory and Experiments. SCHUSTER, P., ZUNDEL, G., SANDORFY, C. (eds.), Vol. II, pp. 25-163. Amsterdam: North Holland 1976a
SCHUSTER, P.: The fine structure of the hydrogen bond. In: Intermolecular Interactions from Diatomics to Biopolymers. PULLMAN, B. (ed.). New York: Wiley Interscience 1976b

SCHUSTER, P., JAKUBETZ, W., BEIER, G., MEYER, W., RODE, B.M.: Potential curves for proton transfer along hydrogen bonds. In: Chemical and Biochemical Reactivity. BERGMANN, E., PULLMAN, B. (eds.), pp. 257-282. Jerusalem: The Israel Academy of Sciences and Humanities 1974

SCHUSTER, P., JAKUBETZ, W., MARIUS, W.: Molecular models for the solvation of small ions and polar molecules. Top. Curr. Chem. 60, 1-107 (1975)

SCHUSTER, P., TORTSCHANOFF, K., WINKLER, H.: Protonenübertragungsreactionen zweibasischer Säuren in wässriger Lösung: 3-Hydroxypyridin. Z. Naturforsch. 31c, 219-224 (1976a)

SCHUSTER, P., ZUNDEL, G., SANDORFY, C. (eds.): The Hydrogen Bond - Recent Developments in Theory and Experiments, Vol. III. Amsterdam: North Holland 1976b

SCHWARZ, G.: On the kinetics of the helix-coil transition of polypeptides in solution. J. Mol. Biol. 11, 64-77 (1965)

SCHWARZ, G.: Kinetic analysis by chemical relaxation methods. Rev. Mod. Phys. 40, 206-218 (1968)

SCHWARZ, G.: Acid-base catalysis of dielectric relaxation of zwitterions. J. Phys. Che. 74, 654-658 (1970)

SCHWARZ, G.: Chemical relaxation of a basic cooperative binding mechanism for linear polymers. Ber. Bunsenges. Phys. Chem. 76, 373-378 (1972)

SCHWARZ, G., ENGEL, J.: Kinetik kooperativer Konformationsumwandlungen von linearen Biopolymeren. Angew. Chem. 84, 615-623, Intern. Ed. 11, 568-575 (1972)

SEGAWA, S., HUSIMI, Y., WADA, A.: Kinetics of proton transfer reactions of lysozyme with p-nitrophenol near neutral pH - A study of dynamic properties of Glu 35 and His 15. Biopol. 14, 2007-2024 (1975)

SHEINBLATT, M.: Determination of an acidity scale from peptide hydrogens from nuclear magnetic resonance kinetic studies. J. Am. Chem. Soc. 92, 2505-2509 (1970)

SHEINBLATT, M., GUTOWSKY, S.: A nuclear magnetic resonance study of the protolysis kinetics of glycine. J. Am. Chem. Soc. 86, 4814-4820 (1964)

STØGÅRD, A., STRICH, A., ALMLÖF, J., ROOS, B.: Correlation effects on hydrogen-bond potentials. SCF CI calculations for the systems HF_2^- and $H_3O_2^-$. Chem. Phys. 8, 405-411 (1975)

STREHLOW, H.: Electrochemical methods for the investigation of fast chemical reactions. In: Techniques of Chemistry. WEISSBERGER, A. (ed.), Vol. VI, Part II: Investigation of Rates and Mechanisms of Reactions. HAMMES, G.G. (ed.), 3rd ed., pp. 285-341. New York: Wiley Interscience 1974

STUEHR, J.: Kinetics of proton transfer and tautomerism in aqueous diacetylactone. J. Am. Chem. Soc. 89, 2826-2832 (1967)

STUEHR, J.: Ultrasonic methods. In: Techniques of Chemistry. WEISSBERGER, A. (ed.), Vol. VI, Part II: Investigation of Rates and Mechanisms of Reactions. HAMMES, G.G. (ed.), 3rd ed., pp. 237-283. New York: Wiley Interscience 1974

STURM, J., LANG, J., ZANA, R.: Ultrasonic absorption of DNA solutions: influence of pH. Biopol. 10, 2639-2643 (1971)

SU, T., BOWERS, M.T.: Ion-polar molecule collisions: the effect of ion size on ion-polar molecule rate constants; the parameterization of the average-dipole-orientation theory. Intern. J. Mass. Spectrom. Ion Phys. 12, 347-356 (1973)

SUDMEIER, J.L., OCCUPATI, G.: Nuclear magnetic resonance studies of proton exchange rate, nitrogen inversion rate and relative "size" of methyl groups and $-NH^+$ protons in N,N-dimethylpiperazine hydrochlorides. J. Am. Chem. Soc. 90, 154-159 (1968)
SWENSON, C.A., KOOB, L.: Nuclear magnetic resonance study of the exchange rates of the peptide protons of glycylglycine and triglycine in water and aqueous urea. J. Phys. Chem. 74, 3376-3380 (1970)
SWIFT, T.J.: Nuclear magnetic resonance. In: Techniques of Chemistry. WEISSBERGER, A. (ed.), Vol. VI, Part II: Investigation of Rates and Mechanisms of Reactions. HAMMES, G.G. (ed.), 3rd ed., pp. 521-563. New York: Wiley Interscience 1974
TATSUMOTO, N.: Ultrasonic absorption in propionic acid. J. Chem. Phys. 47, 4561-4570 (1967)
TURNER, D.H., FLYNN, G.W., SUTIN, N., BEITZ, J.V.: Laser raman temperature jump study of the kinetics of the triiodide equilibrium. Relaxation times in the 10^{-8}-10^{-7} second range. J. Am. Chem. Soc. 94, 1554-1559 (1972)
VEDENEYEV, V.I., GURVICH, L.V., KONDRAT'YEV, V.N., MEDVEDEV, V.A., FRANKEVICH, Ye.L.: Bond energies ionization potentials and electron affinities, 199-200. Scripta Technica Ltd. London: Arnold 1966
VEIL, L.B.: Aufbau einer Stopped Flow Temperatursprung Apparatur und ihre Anwendung auf die Untersuchung biologischer Systeme. Thesis, Braunschweig 1974
VINOGRADOV, S.N., LINNELL, R.H.: Hydrogen Bonding. New York: Van Nostrand Reinhold Co. 1971
VOLD, R.L., CORREA, A.: Nuclear magnetic resonance measurements of proton exchange in aqueous thiourea. J. Phys. Chem. 74, 2674-2680 (1970)
VOLD, R.L., DANIEL, E.S., CHAN, S.O.: Magnetic resonance measurements of proton exchange in aqueous urea. J. Am. Chem. Soc. 92, 6771-6776 (1970)
WAGNER, K., RUDAKOFF, G., FRÖLICH, P.: Das dielektrische Absorptionsverfahren - eine Relaxationsmethode zur Untersuchung der Kinetik schneller Reaktionen mit polaren Komponenten. Z. Chem. 15, 204-206 (1975)
WANG, J.H.: Facilated proton transfer in enzyme catalysis. Science 161, 322-338 (1968)
WEISS, S., DIEBLER, H., MICHAELI, I.: Kinetics of proton transfer reactions of polyacrylic and polymethyacrylic acids with an indicator. J. Phys. Chem. 75, 267-271 (1971)
WHIDBY, J.F., LEYDEN, D.E.: Rates of mercapto proton exchange of mercaptoacetic acid in acetic acid. J. Phys. Chem. 74, 202-205 (1970)
WHIDBY, J.F., MORGAN, W.R.: Investigation of specific acid catalysis by substituted acetic acids on the proton exchange of N,N-dimethylurea by NMR. J. Phys. Chem. 77, 2999-3002 (1973)
WHITE, R.D., SLUTSKY, L.J.: Ultrasonic absorption and relaxation spectra in aqueous bovine hemoglobin. Biopolymers 11, 1973-1984 (1972a)
WHITE, R.D., SLUTSKY, L.J.: Proton-transfer kinetics in the aminobenzoic acids. J. Phys. Chem. 76, 1327-1333 (1972b)
WHITE, R.D., SLUTSKY, L.J., PATTISON, S.: Kinetics of the proton-transfer reactions of serine and threonine. J. Phys. Chem. 75, 161-163 (1971)

WILLIAMS, J.M., SCHNEEMEYER, L.F.: An asymmetric single-minimum-hydrogen bond in the bifluoride ion. A neutron diffraction study of p-toluidinium bifluoride. J. Am. Chem. Soc. 95, 5780-5781 (1973)

WINKLER, H.: Recent progress in the study of fast chemical reactions by relaxation techniques. Endeavour 119, 73-79 (1974)

WOLNIEWICZ, L.: Variational treatment of the HeH^+ ion and the β-decay in HT. J. Chem. Phys. 43, 1087-1091 (1965)

WYN-JONES, E. (ed.): Chemical and Biochemical Applications of Relaxation Spectrometry. Dordrecht-Boston: D. Reidel 1975

YAMABE, S., KATO, S., FUJIMOTO, H., FUKUI, K.: An MO-theoretical calculation of solvent effect upon the $NH_3+HF=NH_4F$ reaction. Theoret. Chim. Acta 30, 327-337 (1973)

ZANA, R.: On the detection of the helix-coil transition of polypeptides by ultrasonic absorption. J. Am. Chem. Soc. 94, 3646-3647 (1972)

ZANA, R., LANG, J.: Effect of pH on the ultrasonic absorption of aqueous solutions of proteins. J. Phys. Chem. 74, 2734-2736 (1970)

ZANA, R., TONDRE, C.: Ultrasonic studies of proton transfers in solutions of poyl(lysine) and poly(ornithine). Implications for the kinetics of the helix-coil transition of polypeptides and for the ultrasonic absorption of proteins. J. Phys. Chem. 76, 1735-1743 (1972)

ZUNDEL, G.: Easily polarizable hydrogen bonds their interactions with the environments. In: The Hydrogen-Bond - Recent Developments in Theory and Experiments. SCHUSTER, P., ZUNDEL, G., SANDORFY, C. (eds.), Vol. II, pp.683-766. Amsterdam: North Holland 1976

ZURAWSKI, B., AHLRICHS, R., KUTZELNIGG, W.: Have the ions $C_2H_3^+$ and $C_2H_5^+$ classical or non-classical structure? Chem. Phys. Lett 21, 309-313 (1973)

Fig. 3. View of the poly(A) single helix (from SAENGER, RIECKE and SUCK, 1975) derived from x-ray data obtained for an ApApA-fragment (9-fold helix with a pitch of 25.4 Å, vertical distance between adenine planes 3.4 Å)

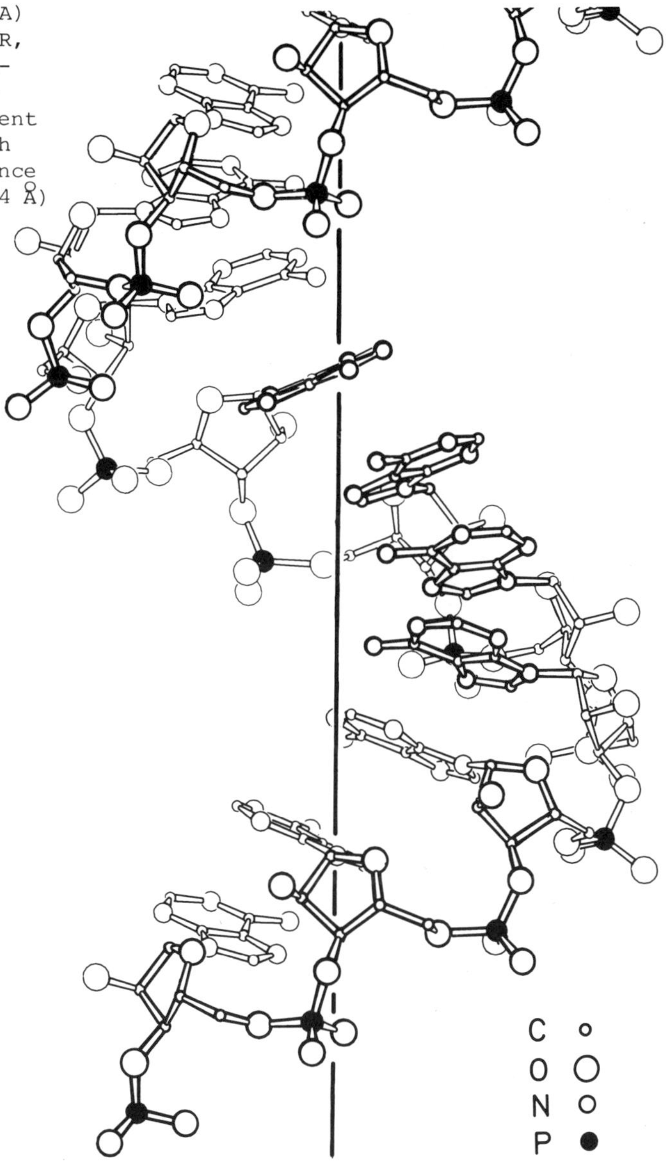

go(A) and poly(A) showed that there are two different relaxation processes (PÖRSCHKE, 1973). Thus the two state model cannot be correct, because it should be characterized by a single relaxation process. However, the second model is not correct either. This can be judged from the time constant observed for single strand relaxation, which is in the range of 0.1 to 1 μs. Thus the rate of the conformation change is surprisingly low, especially when compared to the high stacking rate of monomer bases. This indicates the existence of relatively high activation barriers separating the conformation states of the single strand chain. The existence of such activation barriers is not consistent with an oscillating dimer model.

This example shows that the application of kinetic methods also allows conclusions about the equilibrium states of systems under investigation. The relaxation data indicate the existence of at least three independent conformational states which are already present in the dinucleotide ApA. A model of the poly(A) single-stranded helix according to SAENGER et al. (1975) is given in Figure 3. This may correspond to one of the conformations of the poly(A) chain in aqueous solution. The second conformational state is the random coil form. The third state may be another stacked form of the polymer with a stacking pattern different from that in Figure 3, probably with an altered sugar conformation.

Recently more relaxation data on single strand helix-coil transitions have been obtained using an improved version (PÖRSCHKE, 1976a) of the cable temperature jump apparatus constructed by HOFFMANN (1971). These measurements demonstrated the existence of two relaxation processes associated with the helix-coil transition in oligo(dA) and poly(dA) (PÖRSCHKE, Biopolymers, in press). The absence of "slow" relaxation processes in the model compound 9,9'trimethylenbisadenine (with two adenine bases linked by a trimethylen-bridge) demonstrates, that the slow processes observed in ApA and dApdA are due to the coupling of base stacking with the folding of the sugar phosphate chain.

In contrast to the results obtained for various oligo- and polyadenylates the single strand helix-coil transition of CpC and poly(C) is associated with a single "slow" relaxation process only (PÖRSCHKE, 1976b). Thus there is a remarkable variability in the folding reactions of polynucleotide chains.

III. Ion Condensation to Polynucleotides

The conformation of polynucleotides is strongly influenced by electrostatic interactions between the negatively charges phosphates. These interactions can be regulated by the association of ions to the polyelectrolyte chain. Until recently it has not been possible to analyse the dynamics of these reactions. The development of a technique for selective observation of chemical reactions in the presence of orientation effects has allowed a detailed investigation of ion association to polynucleotides by the field jump method (PÖRSCHKE, 1976c).

Application of high electric fields leads to the dissociation of ions from the polynucleotide chain, which then undergoes a conformation change due to increased electrostatic repulsion. This effect has been used to study the dynamics of ion binding to single stranded polynucleotides. The relaxation observed in field jump experiments provides information about the coupling of the intramolecular helix-coil transition to the intermolecular process of ion binding to the polymer. In the presence of monovalent ions the ion association is fast compared to the helix-coil transition. Evaluation of the relaxation time constants, measured as a function of the ion concentration, leads to the assignment of preequilibrium constants for ion binding and of rate constants for the helix-coil transition. The equilibrium constants for ion association are almost independent of the polynucleotide and are found to be around 60 M^{-1}. The rates found for the helix-coil transition depend upon the type of polymer: helix formation is characterised by rate constants in the range of $2.1 \cdot 10^7$ to $5 \cdot 10^6$ s^{-1}, whereas the rate constants associated with helix dissociation are $5 \cdot 10^5$ to 10^6 s^{-1}.

Field jump experiments performed on polynucleotide solutions containing bivalent ions such as Mg^{++} or Ca^{++}, demonstrate the existence of a separate relaxation process which reflects the association of these ions with the polymer. The time constants observed for this process show a characteristic dependence upon the ion concentration. The experimental data were evaluated according to a model in which a bimolecular step of ion binding to the polymer is followed by an intramolecular stacking reaction. It is found that the rate constants for Mg^{++} association to poly(A) and poly (C) are $2 \cdot 10^{10}$ M^{-1} s^{1} and $1.3 \cdot 10^{10}$ M^{-1} s^{-1} respectively. These rate constants, indicating a diffusion-controlled reaction, are clearly different from rate constants previously observed for Mg^{++} binding to a great variety of ligands with typical rate constants of about 10^5 M^{-1} s^{-1}. The relatively low rates associated with the formation of normal Mg^{++}-complexes are due to "inner sphere" complexation, where the exchange of H_2O-molecules bound to Mg^{++} in the inner hydration sphere is the rate determining step. Thus the high rates observed for Mg^{++}-association to polynucleotides indicate the formation of "outer sphere" complexes without attachment of the Mg^{++} to a specific site.

Examination of the polynucleotide structure shows that the distance between adjacent phosphate residues (7 to 8 A°) is too large for simultaneous site binding of one Mg^{++}, while "inner sphere" complexation at a single phosphate site appears to be too weak. The kinetic data support the view that ions like Mg^{++} surround the polynucleotides in the form of a mobile ion atmosphere. Electrostatic interactions coupled with site binding of ligands to polynucleotides may be reserved for specific proteins with a charge spacing adapted to that of the polynucleotides.

IV. Recognition of Monomer Bases on a Polymer Template

The most important step in replication and transcription is the recognition of complementary nucleotide monomers by the polymer template. The biochemical recognition process is extremely accurate due to the catalysis by well adapted enzymes. In order to understand this process as well as its possible prebiotic precursor, the recognition step has been studied in a simple model system, a monomer-polymer double helix. The desired 1:1 polymer: monomer stochiometry could only be obtained by using a substituted monomer derivative, N6,N9-dimethyladenine, which forms a double helical complex with poly(U) (PÖRSCHKE et al., 1973). The kinetics of the complex formation was investigated by the fast cable temperature jump technique (HOFFMAN, 1971). Relaxation time con-

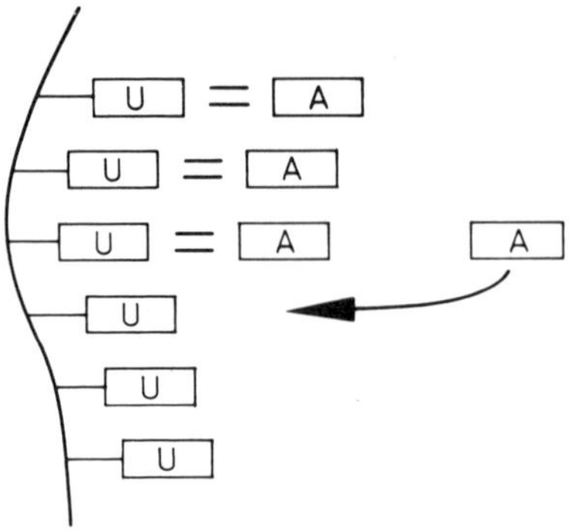

Fig. 4. Scheme of monomer base recognition on a polymer template

stants were observed in the 1 to 10 μs time range (HOFFMAN and PÖRSCHKE, 1973), compared to the 1 to 10 ns relaxation observed for the stacking reaction of N6,N9-dimethyladenine in the absence of a template (PÖRSCHKE and EGGERS, 1972). Thus the specific base stacking directed by hydrogen bonds to a polymer template is much slower than the unspecific base stacking in free aqueous solution.

Evaluation of the relaxation data according to an ISING model led to a rate constant for binding of a monomer next to an already bound monomer on the polymer of $(7.5 \pm 1) \times 10^{7}\ M^{-1}\ s^{-1}$. This rate constant is clearly lower than expected for a diffusion controlled process. There are two possible reasons for the reduction of the rate compared to that observed for the "free" stacking reaction. (1) A steric factor may be partly responsible since the bases can approach their binding site only on a restricted reaction pathway; (2) Hydrogen donor and acceptor sites of the unpaired bases will be hydrogen bonded to water molecules; these complexes must be dissociated, before complementary bases may form hydrogen bonds to each other, and thus will slow down the recombination process.

The influence of steric effects is clearly seen when the data obtained for the 1:1 monomer-polymer complex are compared to measurements of a 1:2 monomer-polymer complex formed from adenosine and poly(U) (DAVIES, 1972). The relaxation times observed for the triple helical complex are in the time range of 1 to 10 ms. The corresponding time constants for the double helical complex, discussed above, are in the range of 1 to 10 μs, whereas the stacking of free bases is characterized by 1 to 10 ns relaxation times. The decreasing time constants of the above series very clearly reflect the decreasing stereochemical complexity associated with the binding of the monomer residue.

V. Helix-Coil Transition of Oligo(A) · Oligo(U)

A. Equilibrium Parameters According to the Cooperative Reaction Model

The formation and dissociation of nucleic acid helices can be observed conveniently by the changes of UV absorbance that result

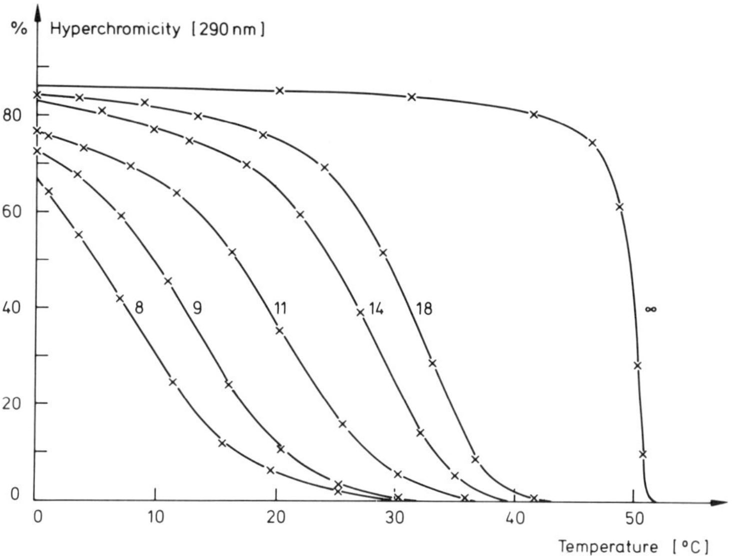

Fig. 5. Hyperchromicity — temperature profiles for A:U oligomer double helices of various chain lengths (Hyperchromicity $100x(E_T-E_C)E_C$, where E_T absorbance at temperature T, E_C absorbance at 100% coil form; $c_t(8)$ = 7.0 mM, $c_t(9)$ = 8.2 mM, $c_t(11)$ = 9.25 mM, $c_t(14)$ = 11.3 mM, $c_t(18)$ = 10.5 mM, $c_t(\infty)$ = 5.0 mM in 50 mM sodium cacodylate pH 6.9) (PÖRSCHKE, 1968, 1971)

from the incorporation of the base chromophores into the ordered helix structures. The relative change of the absorbance given in Figure 5 for oligo(A) • oligo(U) of different chain lengths indicates the formation of helices at low temperatures and their dissociation at high temperatures. The temperature range of the helix-coil transition is shifted to higher temperatures with increasing length of the nucleotide chain. The stability of the double helix reaches a maximum for the "infinitely" long polymer. The slope of the conformational transition also increases with increasing chain length and reaches a maximum for the "infinitely" long chain.

The experimental data may be described by a cooperative model, having at least two model parameters (cf. Fig. 6): (1) a stability constant for the addition of base pairs to a preexisting helix, usually called the chain growth parameter s; (2) a stability constant for the formation of a single isolated base pair, usually defined by βs, where β is the nucleation parameter. Using these

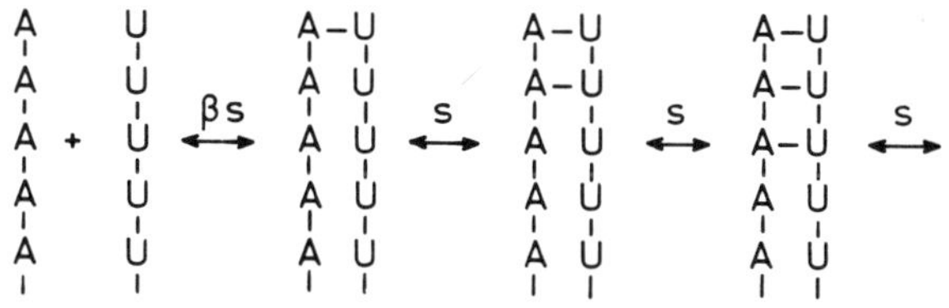

Fig. 6. Scheme of helix formation with chain growth parameter s and nucleation parameter β

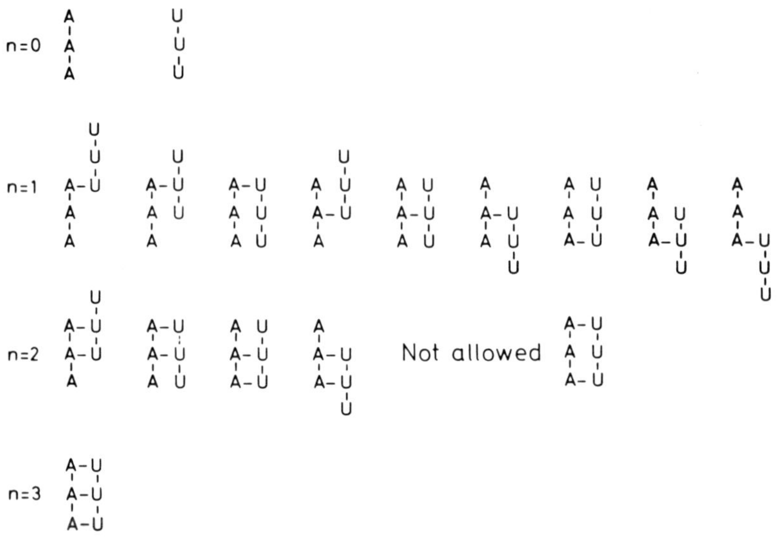

Fig. 7. Helix species formed from triadenylate and triuridylate according to the staggering zipper model

two constants the statistical weight of any helix species may be calculated. The various possibilities of helix formation are shown in Figure 7 for complementary trinucleotides according to a "staggering zipper model". Using the methods of statistical thermodynamics one may sum up the contributions of all the different helix species and calculate theoretical melting curves (cf. APPLEQUIST and DAMLE, 1965). By comparison of the experimental with the calculated melting curves model parameters can be evaluated.

In the oligo(A) · oligo(U) system the nucleation parameter is about 10^{-3} M^{-1}. The chain growth parameter is about 10 around 0°C and decreases to 1 at the melting temperature of the polymer. The difference in stability of the first base pair and the base pairs adjacent to preexisting helices is the basis for cooperativity. This difference is due to the fact that a single isolated base pair is stabilized by hydrogen-bonding, which is a very weak interaction in aqueous solution, whereas following base pairs are stabilized by stacking interactions in addition to hydrogen bonds. The formation of a single base pair is characterized by a rather low negative enthalpy change, whereas the formation of succeeding base pairs is characterized by much higher negative enthalpy changes. Quantitative information about these thermodynamic parameters has been obtained from a detailed analysis of oligomer melting curves (PÖRSCHKE, 1969, 1971).

B. Relaxation Data and Their Interpretation According to an "All or None" Model

Information about the kinetics of the oligomer helix-coil transition can be easily obtained by the temperature jump method

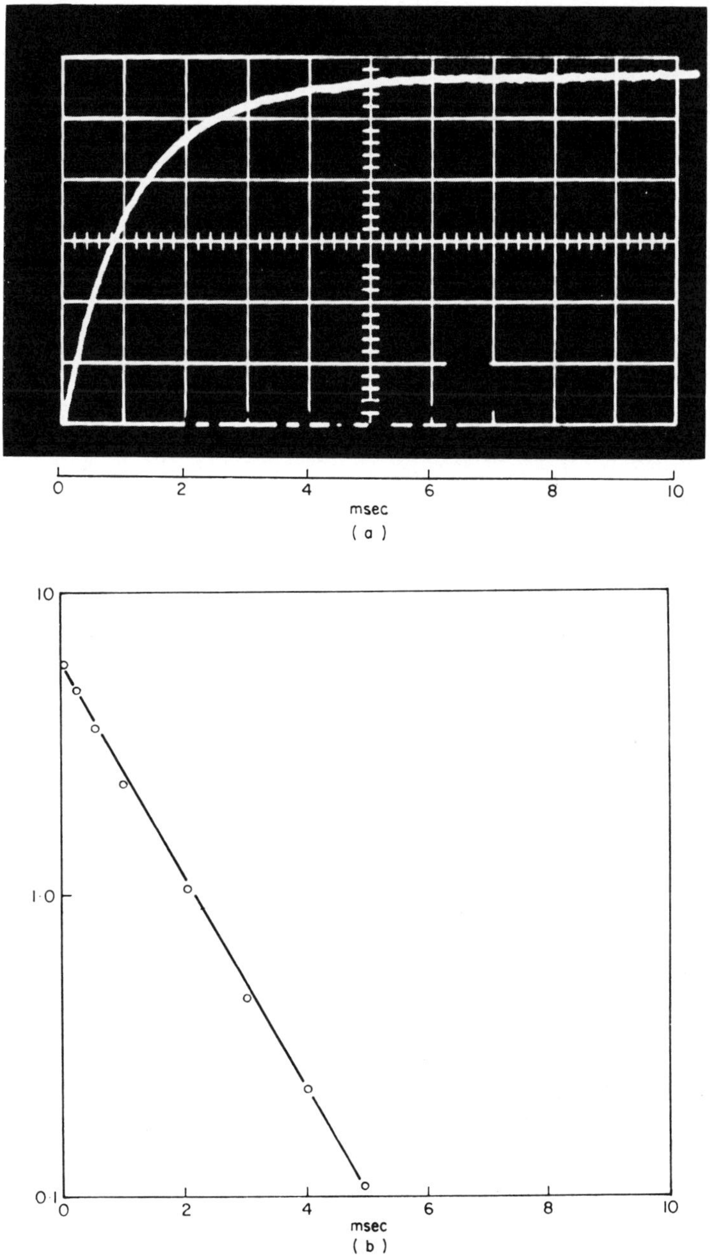

Fig. 8. (a) Change of transmission after a temperature jump with $A(pA)_8$ + $U(pU)_8$. Ratio of A to U is 1/1 in 50 mM sodium cacodylate pH 6.9, T = 3.4°C, C_t = 2.15 mM. (b) Logarithm of the transmission change (arbitrary) units) versus time. (From PÖRSCHKE and EIGEN, 1971)

(PÖRSCHKE, 1968; PÖRSCHKE and EIGEN, 1971). An example of a relaxation signal obtained for an oligo(A) · oligo(U) helix-coil transition is given in Figure 8. A fast relaxation process associated with a relatively low amplitude precedes a big relaxation signal with a time constant usually in the ms time range. A logarithmic plot of the signal change against the time yields a straight line, indicating that this process can be described

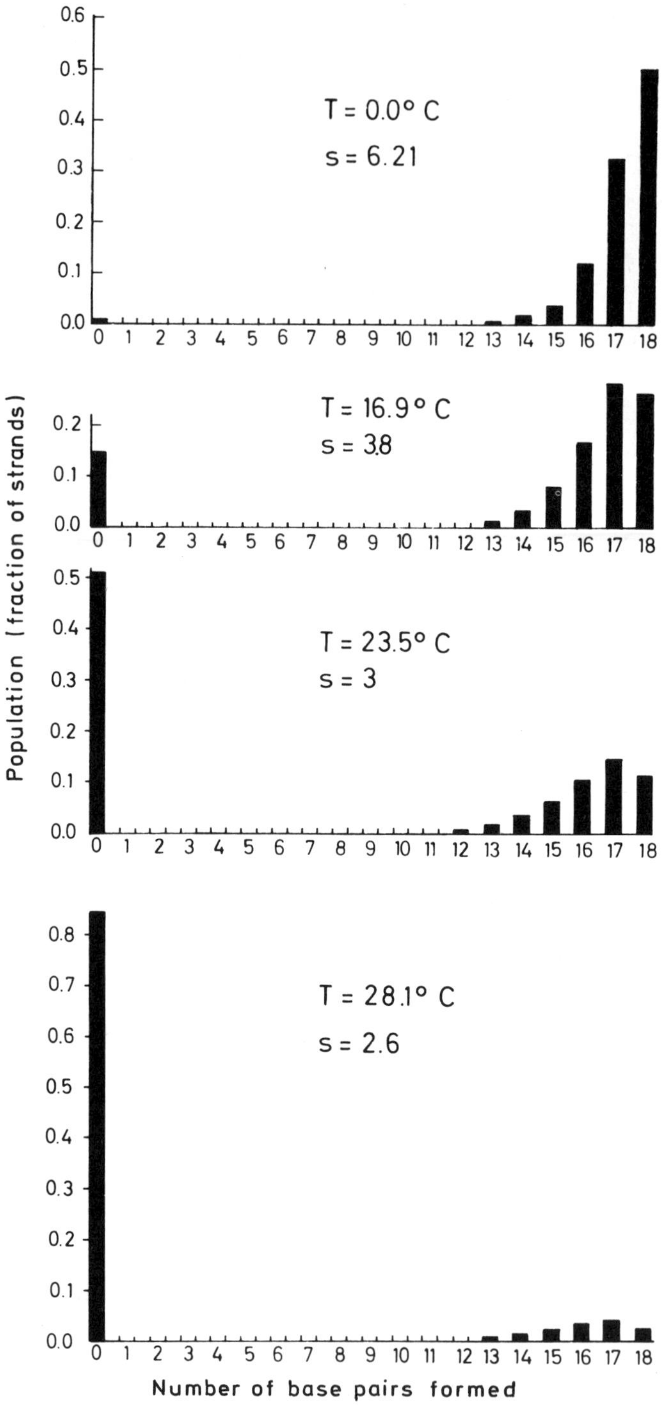

Fig. 9. Population analysis for $A(pA)_{17} + U(pU)_{17}$ at different temperatures ($\beta \cdot c_t = 10^{-7}$). (From PÖRSCHKE and EIGEN, 1971)

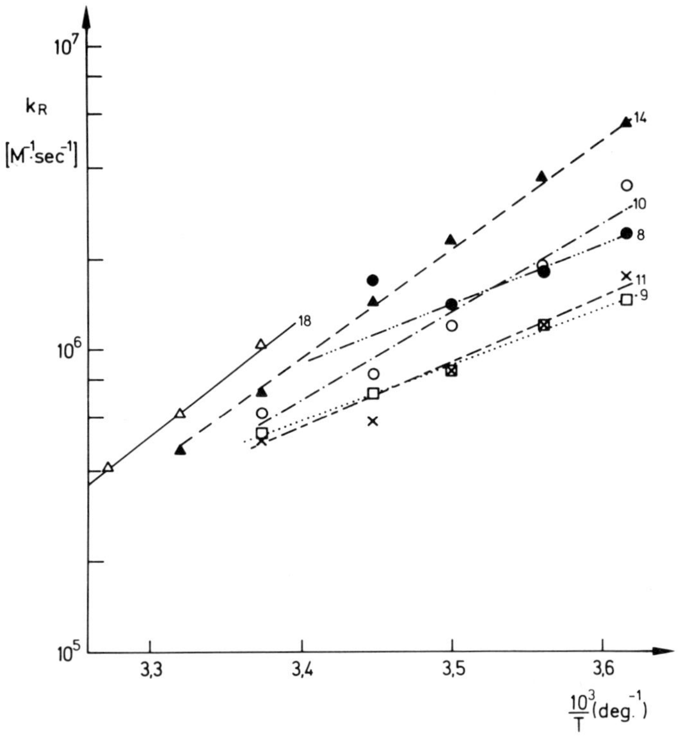

Fig. 10. Rate constants of recombination k_R (logarithmic scale) for different chain lengths N as a function of the reciprocal absolute temperature (50 mM Na-cacodylate pH 6.9); N = 8 ●, N = 9 □, N = 10 O, N = 11 X, N = 14 ▲, N = 18 Δ. (Data from PÖRSCHKE and EIGEN, 1971)

by a single relaxation time. A detailed discussion of this result will be given in Section 4d on the basis of a quantitative model.

For a first explanation it is useful to look at the population of different helix species. The population analysis given in Fiure 9 illustrates that the major states present in solution are the completely base paired double helix and the unpaired single stranded nucleotides. The concentration of the intermediate states is rather low and may be neglected in a first approximation. Thus a first simple description of the helix-coil transition may be given in terms of an all or none model.

If the reciprocal relaxation time is plotted in the usual way as a function of the free nucleotide concentration, a linear relationship is obtained, which allows one to extract rate constants of helix recombination k_R and dissociation k_D according to the simple "all or none" reaction scheme

$$A + U \underset{k_D}{\overset{k_R}{\rightleftharpoons}} H$$

The rates of helix recombination k_R are of the same order of magnitude for all the different oligomer chain lengths (cf. Fig. 10). This result suggests that the rate of recombination is determined

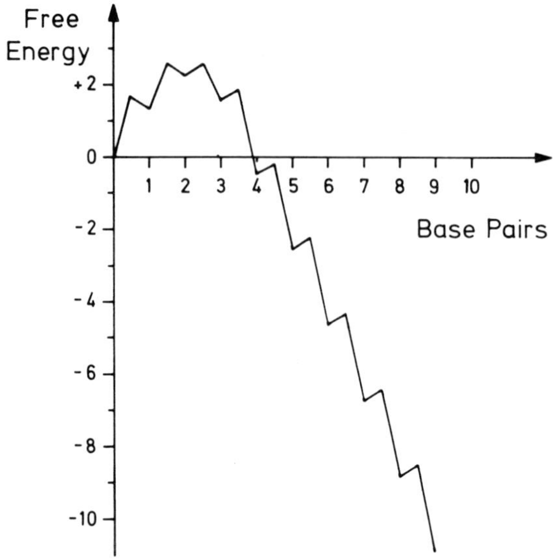

Fig. 11. Relative free energy (arbitrary units) of helix formation as a function of the number of base pairs formed

by the same process for all the different chain lengths. Since we know that the stability of the first base pairs is much lower than that of succeeding base pairs, the nucleation of the helix is expected to be the rate determining step. This is illustrated in Figure 11 which shows the relative free energy as a function of the number of base pairs formed. As soon as the activation barrier, i.e. the formation of the first few base pairs, is overcome, it does not matter how many base pairs are formed subsequently, since the addition of base pairs to a nucleus is a very fast reaction favored by a relatively high decrease in free energy.

Detailed information about the nucleation may be obtained from the value of the activation enthalpy. The experimental data show that helix recombination is associated with a negative enthalpy of activation ($AE_R \approx -9$kcal/mol). This result tells us directly that helix nucleation is not a simple elementary process, since elementary processes are characterized by positive enthalpies of activation. The formation of a single base pair, for instance, corresponds to such an elementary process and should be associated with a positive activation enthalpy. Thus the formation of the first base pair cannot be the rate determining step in AU helix formation. The next step may be to try whether two base pairs are sufficient for helix nucleation. If the formation of the second base pair is the rate determining step (i.e. $k_{12} > k_{10}$ cf. Fig. 12), the rate constant of helix formation k_R is given by

$$k_R = \frac{k_{01}}{k_{10}} \cdot k_{12} = \beta \cdot s \cdot k_{12}$$

In this case the enthalpy of activation AE_R corresponds to the sum of the reaction enthalpy associated with the formation of the first base pair $\Delta H_{\beta s}$ and the enthalpy of activation associated with the formation of the second base pair AE_f

Fig. 12. Reaction scheme of helix formation

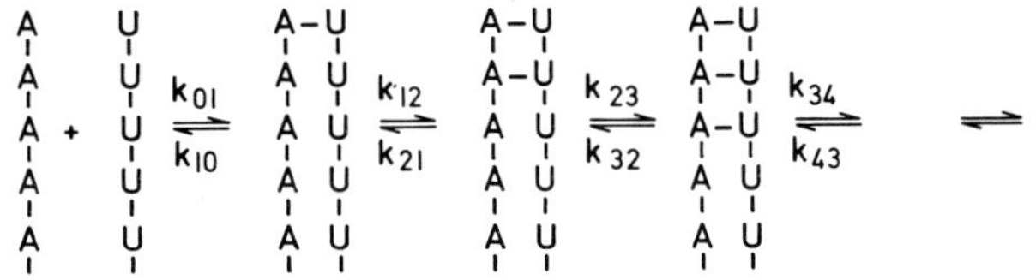

$$AE_R = \Delta H_{\beta s} + AE_f$$

$\Delta H_{\beta s}$ has been determined from oligomer melting curves (PÖRSCHKE, 1971) to be -4 kcal/mol and AE_f is estimated to be 0 to 2 kcal/mol. Thus the sum is negative, but definetely less than the experimental value.

If we go one step further and assume that the formation of the third base pair is rate determining, the following relations[2] apply

$$k_R = \frac{k_{01}}{k_{10}} \cdot \frac{k_{12}}{k_{21}} \cdot k_{23} = \beta \cdot s \cdot s \cdot k_{23} \quad \text{and}$$

$$AE_R = \Delta H_{\beta s} + \Delta H_s + AE_f$$

Using a value of ΔH_s = -6 kcal/mol (PÖRSCHKE, 1971), the calculated activation enthalpy AE_R is very close to the experimental value. If one base pair more would be necessary for nucleation, the calculated activation enthalpy would be higher than the experimental value. Thus we may conclude that formation of the third base pair is the rate determining step in the formation of the AU helices investigated. Model calculations suggest that the free energy profile may look similar to that given in Figure 11. When a third base pair is formed the probability for formation of further base pairs is higher than the probability for the dissociation of the third base pair.

The free energy scheme given in Figure 11 may also be used to discuss the rates of helix dissociation. With an increasing number of base pairs in the helix, the number of activation barriers increases and the rate of dissociation decreases. Even if the rate of dissociation of an individual base pair is rather high, the rate of dissociation of the complete helix may be very low if many base pairs have to dissociate one after another. The expected parameters are actually observed for the oligo(A) · oligo(U) double helices (cf. Fig. 13). With increasing chain length the rates of dissociation become rather low. The average lifetime of a helix containing 14 AU base pairs is about 1 s at room temperature, compared to a lifetime of 0.5 ms for a helix with 8 AU base pairs under corresponding conditions. The increasing number of activation barriers with increasing chain length is also reflected

---

[2]Based upon a simple all or none model; detailed quantitative calculations are discussed in Sections V-C and V-D.

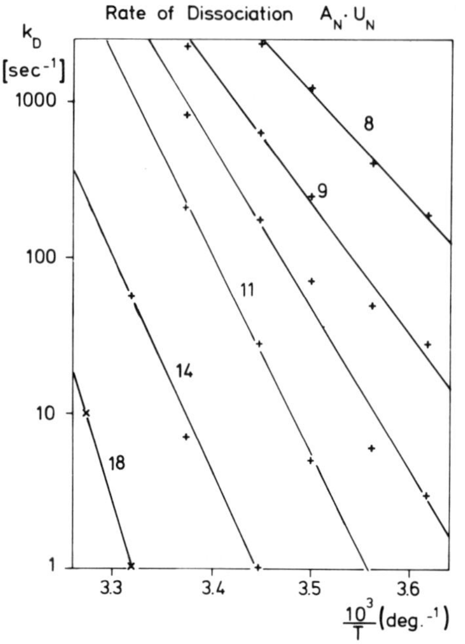

Fig. 13. Rate constants of dissociation k_D (logarithmic scale) for different chain lengths as a function of the reciprocal absolute temperature (50 mM Na-cacodylate pH 6.9). (Data from PÖRSCHKE and EIGEN, 1971)

in the activation enthalpy, which strongly increases with increasing chain length. Similar data on the all or none transition of self-complementary (AU)-oligomers were obtained by CRAIG et al. (1971). The formation of hybrid double helices from riboadenylates and deoxyribothymidylates was analysed by HOGGETT and MAASS (1971). The results are similar to those discussed above.

C. Unzippering at Helix Ends

The discussion in the previous section is concerned with the formation and dissociation of entire helices. From the observed chain length dependence of the rate of dissociation it can be extrapolated that the time for self-unzippering of entire genome double helices is extremely high. In contrast to this *overall* dissociation process, the dissociation of *single base pairs at helix ends* is a rather fast reaction. Direct information about the equilibration of base pairs at helix ends may be obtained from the first part of oligomer relaxation spectra. For an explanation it will be useful to give a short summary of a detailed theory of oligomer helix-coil kinetics.

The process of helix formation and dissociation consists of a series of equivalent reaction steps. Thus most of the steps in the helix-coil transition (cf. Fig. 14) may be described by uniform rate constants. Only the first steps of nucleation have to be described by individual rate constants. The values of these rate constants may first be estimated according to a quasistationary model and then be refined by comparison of computed relaxation spectra to the experimental data. In order to reduce

Fig. 14. Schematic representation of helix formation with intermediates according to a zipper model

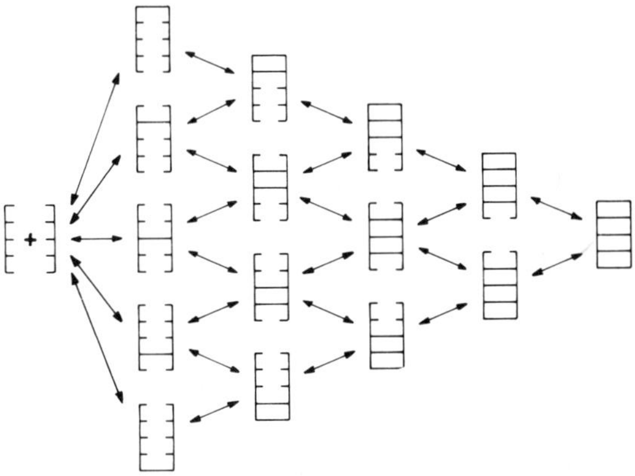

$$A + U \underset{k_{10}}{\overset{k_{01}}{\rightleftharpoons}} H_1 \underset{k_{21}}{\overset{k_{12}}{\rightleftharpoons}} H_2 \underset{k_{32}}{\overset{k_{23}}{\rightleftharpoons}} H_3 \underset{k_{43}}{\overset{k_{34}}{\rightleftharpoons}} H_4 \underset{k_{54}}{\overset{k_{45}}{\rightleftharpoons}} H_5$$

the number of intermediates, helices with the same number of base pairs may be counted as one species. The multiplicity of reaction pathways can be considered by statistical factors (PÖRSCHKE, 1974a).

Examples of relaxation spectra calculated for such a model are given in Table 1. There are as many relaxation times as independent steps. For all the different chain lengths there is one slow relaxation process, clearly separated on the time scale from a spectrum of fast processes. The time constant of this process depends upon the nucleotide concentration. Furthermore, this process is characterized by a relatively high fraction of the total amplitude. These facts indicate that the slow process is associated with the overall helix-coil transition. The rest of the relaxation spectrum reflects the various unzippering reactions at helix ends. The unzippering amplitude increases with increasing chain length of the oligomers. A graphical representation of a theoretical calculation is given in Figure 15, showing the relaxation spectrum for a helix of chain length 18 (PÖRSCHKE, 1974a, b). There are two clearly separated relaxation processes: (1) the faster one, in the μs time range, is the sum of the various unzippering reactions (2) the slower one, with a time constant of about 2 s, corresponds to the relaxation effect described in Section V-B and is associated with the helix-coil transition of the entire helix.

The fast process could not be resolved by the conventional temperature jump technique. However, the development of a fast cable temperature jump apparatus (HOFFMANN, 1971) allowed a measurement of this relaxation process (PÖRSCHKE, 1974b). The observed time constant of about 0.2 μs is in the time range expected from the model calculations. A simulation of the measured spectra yielded a rate constant of the formation of a base pair at a helix end of $8 \times 10^6\ s^{-1}$ and a corresponding constant for dissociation of $2 \cdot 10^6\ s^{-1}$ (at 25°C).

Table 1. Relaxation times τ_i and amplitudes A_i according to a zipper model

N = 8		N = 14	
τ_i	A_i (%)	τ_i	A_i (%)
3.61 ms	98.95	12.66 ms	75.30
0.37 μs	0.84	0.49 μs	13.00
0.23 μs	0.18	0.34 μs	7.02
0.16 μs	0.02	0.23 μs	3.07
0.12 μs	<0.01	0.17 μs	1.12
0.10 μs	<0.01	0.13 μs	0.36
0.07 μs	<0.01	0.10 μs	0.11
0.03 μs	<0.01	0.09 μs	0.03
		0.07 μs	<0.01
		0.07 μs	<0.01
		0.06 μs	<0.01
		0.06 μs	<0.01
		0.04 μs	<0.01
		0.02 μs	<0.01

Chain length N = 8 (14), $c_t = 5 \times 10^{-5}$ M (10^{-3} M)
Temperature jump from 0 → 3.4°C (9.2 → 16.9°C),
rate of chain growth $k_f = 3.84 \cdot 10^6\ s^{-1}$ ($4.55 \cdot 10^6\ s^{-1}$),
s = 8.54 (4.43), fraction of nucleotides involved in
base pairing Θ = 0.211 (0.917). (From PÖRSCHKE, 1974a)

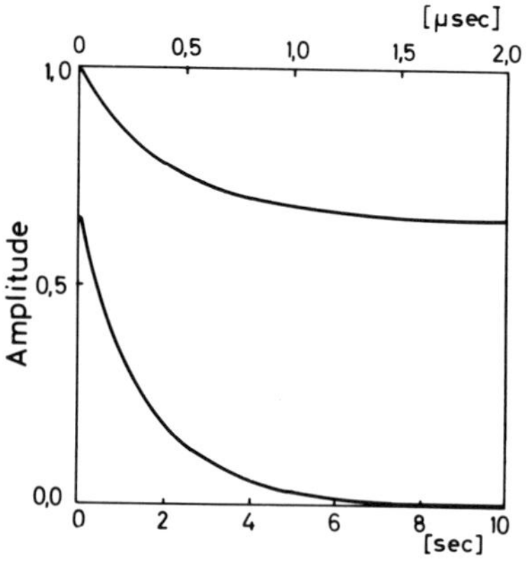

Fig. 15. Relaxation curve calculated according to a zipper model for N = 18 (14°C, c_t = 0.36 mM, $k_f = 4.4 \times 10^6 s^{-1}$). Upper curve reflects the unzippering reaction (upper time scale), whereas the lower curve reflects the "overall" helix-coil transition (lower time scale); signal is in relative normalized units. (From PÖRSCHKE (1974b), cf. for details of calculation)

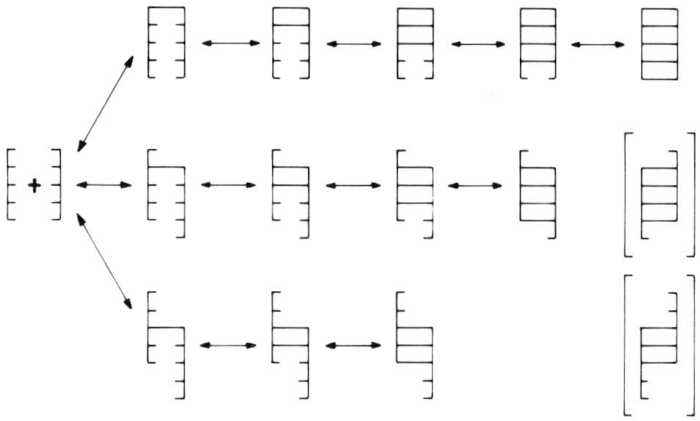

Fig. 16. Scheme of helix formation according to a staggering zipper model with two series of staggered helices. The helix species shown in brackets indicate the two different kinds of helices within each series of staggered species [subspecies with different positions of base pairs (cf. Fig.14) are not shown]. (From PÖRSCHKE, 1974a)

D. Chain Sliding

The reaction model discussed in the previous section did not include "staggered" helix series with the nucleotide chains shifted against each other (cf. Fig. 16). If these helices are also considered in the model calculation, more than one slow relaxation process is obtained (PÖRSCHKE, 1974a). Each of the slow processes reflects the overall helix-coil transition of one separate helix series, which reacts independently. However, the relaxation spectra calculated according to the "non sliding" staggering zipper model are not consistent with the experimental data. The slow part of the experimental relaxation curves can be represented by a single exponential, even when well separated slow effects should be expected according to the model calculations. Thus the model has to be modified. The calculated and experimental results become consistent when a fast coupling reaction between the different helix series is considered. The coupling can be achieved by a chain sliding mechanism illustrated in Figure 17. The chain sliding is initiated by the introduction of a helix defect in the form of a bulge loop at one of the helix ends. When such a bulge loop is present, it may walk in either direction of the double helix. If a bulge loop arives at the other end of the helix, the migration of the loop results in a shift of the nucleotide strands against each other, i.e., a helix of one series is converted to a helix of a different series. Using the estimated thermodynamic stability of bulge loops and the measured rates of base pair equilibration one may calculate that the chain sliding reaction indeed may be very fast. For a helix containing 10 base pairs, the relaxation time constant is expected to be in the μs

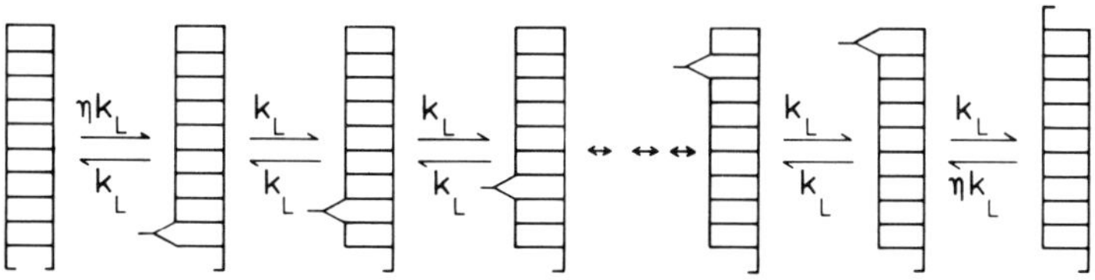

Fig. 17. Scheme of bulge loop diffusion resulting in chain sliding. (From PÖRSCHKE, 1974a)

time range (PÖRSCHKE, 1974a). Chain sliding reactions will occur in helices formed from homopolymers with particularly high rates, but will also be quite important in the annealing process of heteropolymer helices.

VI. The Influence of GC Base Pairs

It is well known that GC base pairs have a much higher thermodynamic stability than AU base pairs. This difference in stability is very clearly reflected in kinetic parameters obtained for helices containing GC base pairs (PODDER, 1971; COUTTS, 1971; PÖRSCHKE et al., 1973). In the recombination process usually 1 or 2 GC base pairs are sufficient for nucleus formation. The dissociation rates observed for helices containing GC base pairs are much lower than those obtained for AU helices of the same chain length. In mixed sequences containing both AU and GC base pairs the rate of helix formation depends upon the position of the GC base pairs (PÖRSCHKE et al., 1973). When the GC base pairs are at a helix end, the rate of helix formation is higher than in the case of helices with GC base pairs in the middle of the sequence. Apparently this effect is due to some steric hindrance and probably also due to electrostatic repulsion. A comparison of the recombination rates of homo-oligomers (e.g. oligo(A) + oligo(U)) with those of hetero-oligomers (containing mixed AU/GC sequences) has led to the suggestion of a nucleate + slip + zip mechanism for helix formation from homo-oligomers (PÖRSCHKE, 1976d). According to this mechanism an oligo(A) · oligo(U) double helix with maximal base pairing may not only be formed by nucleation from completely aligned single strands, but also via staggered helix species. These staggered species may be converted to helices with maximal base pairing by chain sliding and chain growth reactions. This mechanism results in a large increase of the number of nucleation sites with increasing chain length and seems to compensate steric hindrance and electrostatic repulsion effects in the helix formation from homo-oligomers.

VII. Specific Effects in Helix Loops

Loops are very common and important elements of secondary structure in natural nucleic acids. Some experimental data on helix structures containing loops, in particular hairpin helices, indicate that loops may influence the parameters of the adjacent helices considerably (UHLENBECK et al., 1973; GRALLA and CROTHERS, 1973). An example of a rather specific influence is discussed in the following. Under appropriate conditions (low temperature, high salt concentration, low nucleotide concentration) the oligomer $(Ap)_6$-$(Cp)_6$-$(Up)_5$-U forms a double helix containing 6 A-U base pairs and a C_6-loop. The formation of this helix is an intramolecular, first order reaction and thus the rate parameters

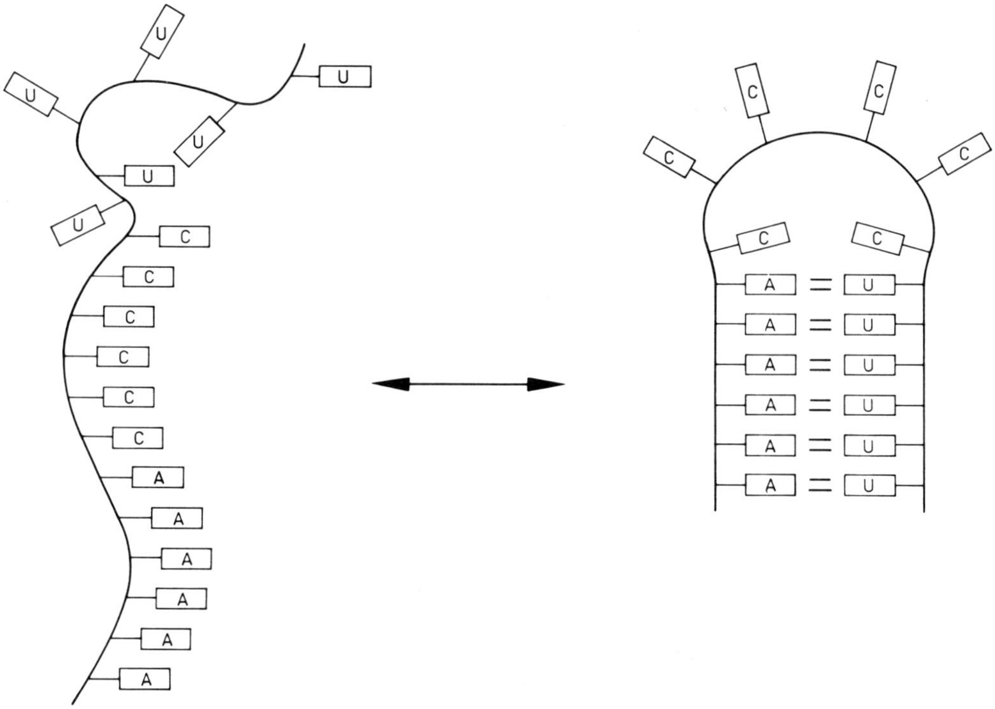

Fig. 18. Scheme of helix formation in $(Ap)_6(Cp)_6(Up)_5U$. Stacking in the single strands reduces the probability of hairpin helix formation

of hairpin formation are not equivalent to parameters obtained for the second order formation of a helix from $A(pA)_5$ and $U(pU)_5$. Nevertheless a comparison of the rate constants is quite instructive. The formation of the hairpin helix is characterized by a rate constant of about $2 \cdot 10^4$ s^{-1} (PÖRSCHKE, 1974c), whereas the rate constant for the formation of a helix from $A(pA)_5$ and $U(pU)_5$ is around 10^7 M^{-1} s^{-1} (extrapolated value from data of PÖRSCHKE and EIGEN, 1971).

In spite of the expected high local concentration of the reaction partners in $(Ap)_6(Cp)_6(Up)_5U$ the observed rate of helix formation is rather low. A simple explanation of this effect can be found when the structure of the single strands is considered. Both adenine and cytosine are characterised by a relatively high stacking tendency, leading to the formation of a single strand helix. This single strand structure decreases the probability of an encounter between the complementary A and U sequences resulting in a decrease of the reaction rate (cf. Fig. 18). The influence of the conformation is specifically reflected in the rates of recombination, whereas the rates of dissociation are very similar for both helices (k_D in the range of 10^4 to 10^5 sec^{-1}). This example demonstrates that kinetic data can also be used as a specific indicator of the influence of molecular conformations.

VIII. Dynamics of Polymer Helix-Coil Transitions

Two different aspects of the helix-coil dynamics of polynucleotides may be discussed separately. The less complex one is the process of helix formation from separated, complementary polymer strands. For many nucleotide systems ranging from synthetic homopolymers (ROSS and STURTEVANT, 1960) to natural polynucleotides (WETMUR and DAVIDSON, 1968) it has been demonstrated that most of the recombination process can be described by a second order rate equation, with a corresponding increase of the rate at increasing single-strand concentration. This part of the reaction is directly analogous to the recombination process observed for oligomer helices. The formation of a helix nucleus is the rate determining step followed by a fast "zipping up" process. However, when the length of the nucleotide chain is increased, the "zipping up" process requires an increasing amount of time. This may be reflected by a deviation from the second order rate law towards a first order relationship. A more important contribution to this deviation apparently comes from the annealing reaction of partly mismatched helix regions. These regions arise during the statistical process of nucleation. The conversion into matched helices often requires the dissociation of helix regions for initiation and hence is a relatively slow reaction. Kinetic data obtained for helix recombination are now widely used for the detection of redundant DNA sequences (WARING and BRITTEN, 1966; HEARST and BOTCHAN, 1973).

The dissociation of long nucleic acid double helices is a more complex reaction. Usually this reaction is characterised by a very broad spectrum of relaxation processes. This is already true for the relatively "simple" helices formed from homopolymers. Helix regions may dissociate at different positions along the double helix resulting in loops of different sizes separated by helix sequences of different lengths. This multiplicity of species

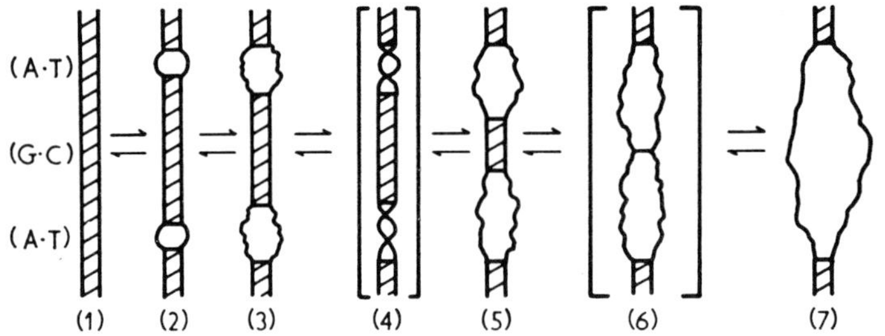

Fig. 19. Scheme of processes involved in DNA denaturation. Upon a first perturbation AT rich helix regions are melted, yielding a distribution of states like 2 and 3. A further perturbation results in an "instantaneous" effect by melting of a few base pairs at the ends of the helical sections, with the twist taken up in adjacent coil regions (4). This step is followed by unwinding with melting of further base pairs 4 → 5. The unwinding reaction is limited by viscous resistance. A further step is the melting of helix regions separating two loops 5 → [6] → 7. (From SPATZ and CROTHERS, 1969)

HOFFMAN, G.W.: A nanosecond temperature-jump apparatus. Rev. Sci. Instr. 42, 1643-1647 (1971)
HOFFMAN, G.W., PÖRSCHKE, D.: Cooperative nonenzymic base recognition. Kinetics of the binding of a base monomer to a complementary polynucleotide template. Biopolymers 12, 1625-1638 (1973)
HOGGETT, J.G., MAASS, G.: Thermodynamics and kinetics of the formation of a hybrid double helix of oligomers of riboadenylic and deoxyribothymidylic acids of definite chain lengths. Ber. Bunsenges. 75, 45-54 (1971)
KASARDA, D.D.: Dilution volume changes of some purine and pyrimidine compounds. Biochem. Biophys. Acta 217, 535-538 (1970)
KAUZMANN, W.: Some factors in the interpretation of protein denaturation. Adv. Prot. Chem. 14, 1-63 (1959)
NEMETHY, G.: Hydrophobe Wechselwirkungen. Angew. Chem. 79, 260-271 (1967)
NEMETHY, G., SCHERAGA, H.A.: The structure of water and hydrophobic bonding in proteins. J. Phys. Chem. 66, 1773-1789 (1962)
PODDER, S.U.: Cooperative non-enzymic base recognition. A kinetic study of interaction between GpGpGpC and GpCpCpC and of self-association of GpGpGpC. Eur. J. Biochem. 22, 467 (1971)
PÖRSCHKE, D.: Thermodynamik und Kinetik der Helix-Knäuel Umwandlung von Oligonucleotiden. Dissertation Braunschweig (1968)
PÖRSCHKE, D.: Cooperative nonenzymic base recognition II. Thermodynamics of the helix coil transition of oligoadenylic + oligouridylic acids. Biopolymers 10, 1989-2013 (1971)
PÖRSCHKE, D.: The dynamics of nucleic acid single-strand-conformation changes. Oligo- and polyriboadenylic acids. Eur. J. Biochem. 39, 117-126 (1973)
PÖRSCHKE, D.: Model calculations on the kinetics of oligonucleotide double helix-coil transitions. Evidence for a fast chain sliding reaction. Biophys. Chem. 2, 83-96 (1974a)
PÖRSCHKE, D.: A direct measurement of the unzippering rate of a nucleic acid double helix. Biophys. Chem. 2, 97-101 (1974b)
PÖRSCHKE, D.: Thermodynamic and kinetic parameters of an oligonucleotide hairpin helix. Biophys. Chem. 1, 381-386 (1974c)
PÖRSCHKE, D.: Cable temperature jump with improved sensitivity and time resolution. Rev. Sci. Instrum. 47, 1363-1365 (1976a)
PÖRSCHKE, D.: The nature of stacking interactions in polynucleotides. Molecular states in oligo- and polyribocytidylic acids by relaxation analysis. Biochemistry 15, 1495-1499 (1976b)
PÖRSCHKE, D.: Thermodynamic and kinetic parameters of ion condensation to polynucleotides. Outer sphere complex formed by Mg^{++} ions. Biophys. Chem. 4, 383-394 (1976c)
PÖRSCHKE, D.: Dynamic properties of short helices. Studia biophys. 57, 35-46 (1976d)
PÖRSCHKE, D., EGGERS, F.: Thermodynamics and kinetics of base-stacking interactions. Eur. J. Biochem. 26, 490-498 (1972)
PÖRSCHKE, D., EIGEN, M.: Cooperative nonenzymic base recognition III. Kinetics of the helix-coil transition of the oligoribouridylic oligoriboadenylic acid system and of oligoriboadenylic acid alone at acidic pH. J. Mol. Biol. 62, 361-381 (1971)
PÖRSCHKE, D., HOFFMAN, G.W., SENEAR, A.: Double helical complex formed from a polynucleotide and a complementary monomer. Nature N.B. 242, 45-46 (1973)

PÖRSCHKE, D., UHLENBECK, O.C., MARTIN, F.H.: Thermodynamics and kinetics of the helix coil transition of oligomers containing GC base pairs. Biopolymers 12, 1313-1335 (1973)
RECORD, M.T., ZIMM, B.H.: Kinetics of the helix coil transition in DNA. Biopolymers 11, 1435-1484 (1972)
RIESNER, D., RÖMER, R., in: Physicochemical Properties of Nucleic Acids. DUCHESNE, J. (ed.), Vol. 2, pp. 237-318. London-New York: Academic Press 1973
ROSS, P.D., STURTEVANT, J.M.: The kinetics of double helix formation from polyriboadenylic acid and polyribouridylic acid. Proc. Nat. Acad. Sci. (Wash.) 46, 1360-1365 (1960)
SAENGER, W., RIECKE, J., SUCK, D.: A structural model for the polyadenylic acid single helix. J. Mol. Biol. 93, 529-534 (1975)
SPATZ, H.C., CROTHERS, D.M.: The rate of DNA unwinding. J. Mol. Biol. 42, 191-219 (1969)
TS'O, P.O.P.: Basic Principles in Nucleic Acid Chemistry, Vol. I, pp. 453-584. New York-London: Academic Press 1974
UHLENBECK, O.C., BORER, P.N., DENGLER, B., TINOCO, I.Jr.: Stability of RNA hairpin loops: A_6-C_m-U_6. J. Mol. Biol. 73, 483-496 (1973)
WARING, M., BRITTEN, R.J.: Nucleotide sequence repetition: A rapidly reassociating fraction of mouse DNA. Science 154, 791-794 (1966)
WETMUR, J.G., DAVIDSON, N.: Kinetics of renaturation of DNA. J. Mol. Biol. 31, 349-370 (1968)

Structural Dynamics of tRNA
A Fluorescence Relaxation Study of $tRNA^{Phe}_{yeast}$

R. RIGLER, M. EHRENBERG, and W. WINTERMEYER

I. Introduction

From the X-ray diffraction studies of KIM et al. (1974) and ROBERTUS et al. (1974) a detailed picture of the crystal structure of phenylalanine-tRNA from yeast ($tRNA^{Phe}_{yeast}$) has emerged. The finding that most of the hydrogen bonds between loop and stem regions which stabilize the tertiary structure of $tRNA^{Phe}$ involve bases of the invariable parts of the primary sequences of tRNA suggests a similarity of the tertiary structure of all tRNAs. It is generally assumed that the crystal structure is predominant also in solution. The support for this assumption is mainly based on the accessibility of nucleotides to chemical modification, and has been reviewed by RICH and RAJBHANDARY (1976). Experiments in which the binding of oligonucleotides to tRNA was investigated, however, indicate interaction between oligonucleotides and complementary sequences particularly in the D-loop (UHLENBECK, 1972; CAMERON and UHLENBECK, 1973) and the anticodon-loop (UHLENBECK, 1972; EISINGER and SPAHR, 1973) which are not accessible in the crystal lattice. Studies of the solution structure of tRNA applying spectroscopic techniques such as low-angle X-ray diffraction (KRIGBAUM and GOODWIN, 1966; CONNORS et al., 1969; PILZ et al., 1970) and NMR (LIGHTFOOT et al., 1973; SHULMAN et al., 1973; DANIEL and COHN, 1975; REID et al., 1975; REID and ROBILLARD, 1975; KEARNS, 1976) were up to now not able to decide whether conformations differing from the crystal structure are present in solution.

In order to make the detailed information contained in the X-ray picture intelligible, one has to know to what extent the conformation of tRNA observed in the crystalline state is involved in the reaction sequences of the translation process. Of particular interest is the relevance of the tertiary structure of tRNA for its recognition by the aminoacyl-tRNA synthetase, by the elongation factor Tu and by the programmed ribosome. We demonstrate in this contribution how information about conformational states of a molecule like tRNA and their chemical transition rates can be obtained by combining the analysis of the excited states of a tRNA, labeled with fluorescent probes at a single site, with the analysis of the chemical kinetics of the tRNA structure.

II. Fluorescent Probes for the Structure of tRNA

The reason for having a fluorescent probe coupled to a unique site in tRNA is two-fold: (1) the quantum yield of the probe when

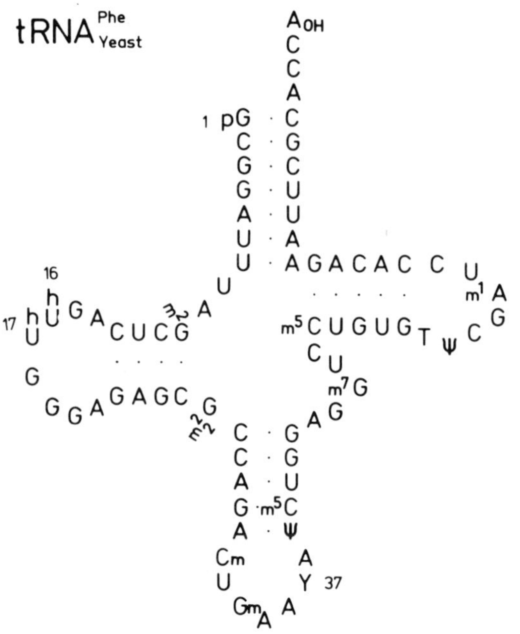

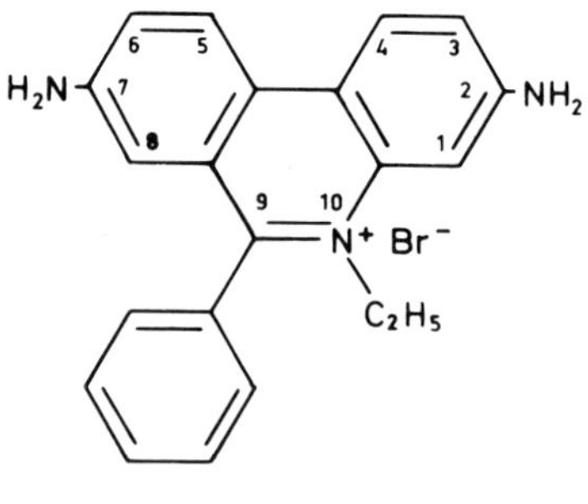

Ethidium bromide

Fig. 1. Primary sequence of $tRNA^{Phe}_{yeast}$ (RAJBHANDARY et al., 1967)

inserted between nucleotide bases is dependent on the degree of intercalation and the accessibility to solvent molecules. The position of the probe with respect to its immediate vicinity will be characterized by the lifetime of the excited state of the probe. Thus changes in structure leading to an altered position of the probe can be detected. To assess if the structure at regions distant from the probe has changed as well , the probe has to be placed at different sites. (2) The probe is used as a fluorescent label for measuring the rotational Brownian motion of the tRNA. For this purpose the lifetime of the excited state of the probe must be comparable with the rotational relaxation time of its carrier (EHRENBERG and RIGLER, 1976; RIGLER and EHRENBERG, 1976).

$tRNA^{Phe}_{yeast}$ (Fig. 1) was labeled at position 16/17 or at position 37 by replacing the corresponding nucleotides by ethidium bromide (WINTERMEYER and ZACHAU, 1971, 1974). From the ethidium derivative at position 37 two fractions can be separated by column chromatography (WINTERMEYER et al., 1976). They have been designated tentatively as $tRNA^{Phe}_{EB37B}$ and $tRNA^{Phe}_{EB37C}$ and probably represent isomers with respect to the amino group of ethidium which has reacted with the ribose. All derivatives can be acylated by aminoacyl-tRNA synthetase and they are capable of transferring phenylalanine to the growing peptide chain on the ribosome (WINTERMEYER and ZACHAU, 1975). Thus two important parts of the tRNA molecule, the D-loop and the anticodon-loop can be probed by this label. For comparison a natural fluorophore, i.e. the Y-base in position 37, can be used. The advantage of EB instead of the Y-base as a probe is its considerably longer lifetime, which facilitates the analysis of the rotational motion of tRNA.

III. Pulsed Fluorescence Measurements

A. The Lifetime of Excited States of the Fluorescent Probe and the Distribution of Conformational States

Placed at a strategic position in the structure, the probe is able to report changes in the structure of the carrier molecule by changes of the lifetime of the excited state of the probe. This parameter allows to separate on a time scale the various conformational states influencing the probe. When the isomers T_1 and T_2 of structure T affect the lifetime of the probe differently, one can detect the presence of these two conformations by a pulsed relaxation experiment. Exciting the fluorescence of the probe by a short pulse of light, the emission of the probe will decay exponentially to the ground state after the ligth pulse has vanished, for each conformation with a different time constant. For the equilibrium

$$T_1 \rightleftharpoons T_2$$

the observed total fluorescence intensity $I_m(t)$ will decay as

$$I_m(t) = \sum_{i=1}^{2} a_i \cdot e^{-t/\tau_i} . \qquad (1)$$

For the situation that T is labeled at a single site the amplitude a_i is related to the concentration of the isomer T_i by

$$a_i = \text{const } [T_i] \varepsilon_i k_i^e \quad (i=1,2) \qquad (2)$$

where ε_i is the absorption coefficient and k_i^e the emission rate in the absence of a quenching process (RIGLER and EHRENBERG, 1973, 1976). When the ratio $(\varepsilon_2 k_2^e)/(\varepsilon_1 k_1^e)$ is known, the equilibrium constant $K=[T_2]/[T_1]$ can be calculated from the ratio of the amplitudes a_2/a_1.

B. Rotational Brownian Motion and Time-Dependent Fluorescence Anisotropy

When the sample is excited with a short pulse of polarized light, an anisotropic distribution of particles in their excited state is created by photoselection. As the particles undergo Brownian rotational motion during the time of emission the original anisotropic distribution of the emission vector of the fluorescent label is randomized as times goes on. By measuring the polarized components of emission as a function of time this randomization process, determined by the rotational diffusion of the probe, linked to this carrier, can be followed.

The component polarized parallel with the polarization vector of the exciting beam ($I_{||}$) decays for a body of arbitrary shape with one lifetime of emission according to:

$$I_{||}(t) = I_m e^{-t/\tau} (1 + \sum_{i=1}^{5} a_i \cdot e^{-t/\tau_{ri}}) \qquad (3)$$

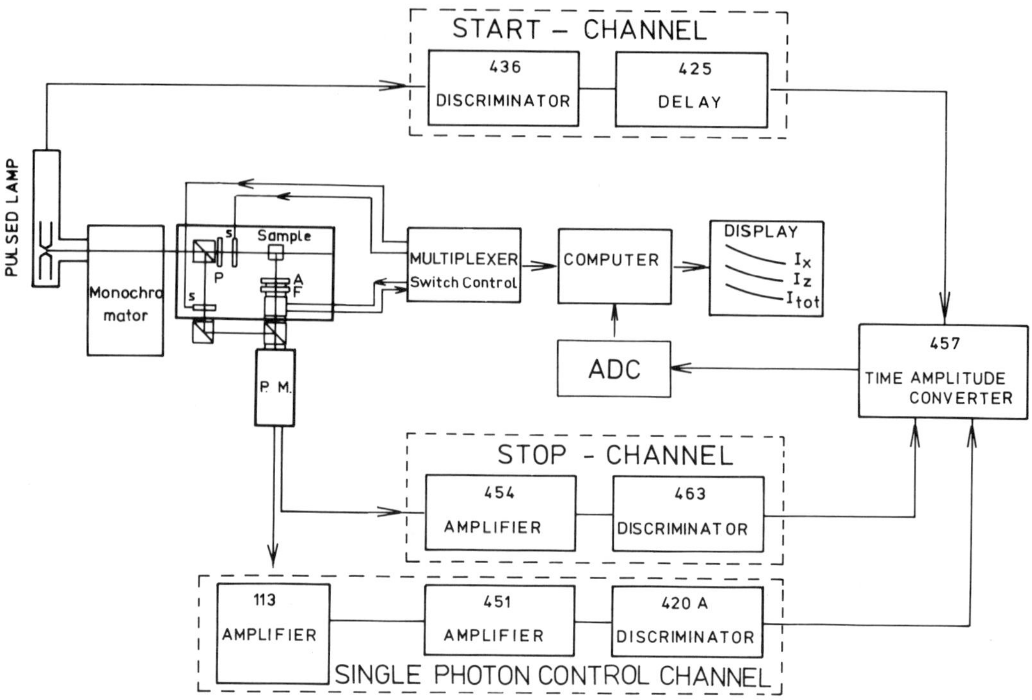

Fig. 2. Scheme of pulsed fluorescence relaxation spectrometer

where τ is the lifetime of emission and τ_{ri} are the rotational relaxation times. They are functions of the three rotational diffusion constants D_i (i=1,2,3). The weight factors a_i are related to the direction cosines determining the orientation of the absorption and emission vector of the probe in the molecular coordinate system of the carrier molecule (CHUANG and EISENTHAL, 1972; EHRENBERG and RIGLER, 1972). For molecules with rotational symmetry, the number of exponentials reduces to three with $\tau_{r_1} = 5D_1+D_2$, $\tau_{r_2} = 2D_2+4D_3$, $\tau_{r_3} = 6D_1$, for a spherical body, to one with $\tau_r = 6D$. Since the rotational diffusion constants are directly related to the molecular volume and the shape of the molecule (MEMMING, 1961), changes in the structure of the rotating unit will be reflected with great sensitivity by its rotational relaxation times. This is in contrast to translational Brownian motion, where the diffusion constant essentially reflects the molecular radius and thus is less sensitive to conformational changes.

C. Instrumentation

For pulsed relaxation experiments a single-photon decay spectrometer constructed in this laboratory was used (Fig. 2). In this instrument the delay between excitation and detection of a single photon and its distribution on the time axis is determined by a multichannel analyzer. The time delay is measured by the combination of a time amplitude converter and an AD-converter which finally addresses the multichannel analyzer. The collection of single photon processes is monitored by a special control channel (single channel analyzer) which prevents all events exceeding the single photon energy level from being measured. To relate the amplitudes of the decay processes to the excitation energy, a split beam arrangement allowing to measure lamp pulse and sample decay in an alternating sequence is used. Details of the set-up have been published (RIGLER and EHRENBERG, 1976).

D. Results

The influence of divalent cations like Mg^{2+} on the tertiary structure of tRNA has been established in several investigations examining the hydrodynamic behavior (FRESCO et al., 1966; RÖMER and HACH, 1975) as well as the optical properties of tRNA such as UV-absorption (FRESCO et al., 1966; RÖMER et al., 1970; RIESNER et al., 1973; LYNCH and SCHIMMEL, 1974), circular dichroism (RÖMER and HACH, 1975; EISINGER and LAMOLA, 1971), the fluorescence of the Y-base (BEARDSLEY et al., 1970; EISINGER et al., 1970; EISINGER and LAMOLA, 1971; ROBINSON and ZIMMERMANN, 1971). These studies however were only able to give a rather general picture of conformational changes under the influence of Mg^{2+} without a detailed relationship between possible conformational states. We have therefore investigated the effect of Mg^{2+} on the structure of $tRNA^{Phe}_{yeast}$ labeled by the fluorescence probe ethidium bromide using the potentialities of time-resolved fluorescence spectroscopy in combination with the analysis of chemical kinetics. We have concentrated our efforts on Mg^{2+} concentrations and temperatures which are known to be of relevance for the biological activity of tRNA and which are required for maintaining the tertiary folding of the tRNA. For all experiments described, a buffer containing 10 mM Tris, pH 7.2, 0.1 M KCl, 0.1 mM EDTA with various concentrations of $MgCl_2$ was used.

When exciting ethidium-labeled $tRNA^{Phe}$ with a short pulse of polarized light, the intensity of the polarized component ($I_{||}$) and of the nonpolarized component (I_m) decays as shown in Figure 3. The decay curves were evaluated by fitting appropriate models using nonlinear regression analysis (MARQUARDT, 1963; MEETER, 1964). Discrimination between different models satisfying the observed decay curves are made with a χ^2-test. The detailed procedure of data evaluation taking into account the finite duration of the lamp pulse has been described previously (GRINVALD and STEINBERG, 1974; RIGLER and EHRENBERG, 1976).

For $tRNA^{Phe}_{EB37C}$ the decay of I_m has to be described by at least two exponential terms:

$$I_m(t) = a_1 \cdot e^{-t/\tau_1} + a_2 \cdot e^{-t/\tau_2}. \quad (4)$$

The parameters were estimated as $\tau_1 = 33$ ns, $\tau_2 = 11$ ns, $a_1 = 4.64 \cdot 10^{-4}$, $a_2 = 4.18 \cdot 10^{-4}$. The decay analysis shown for a single concentration of $MgCl_2$ was carried out over a concentration range between 0 - 20 mM at 10°C. At all concentrations the same two lifetimes τ_1, τ_2 were found (Table 1), while the amplitudes corresponding to τ_1 and τ_2 varied. The amplitude ratio a_2/a_1 increased from

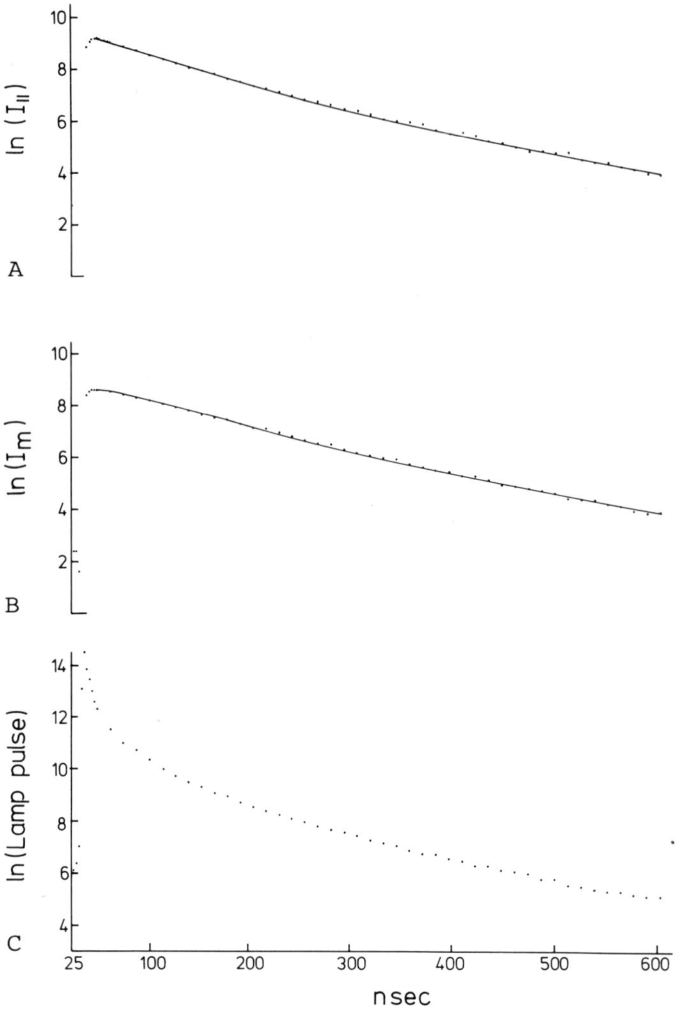

Fig. 3A-C. Decay curves of components $I_{||}$ (A) and I_m (B) of $tRNA^{Phe}_{EB37C}$ ($2 \cdot 10^{-6}$ M) in Tris buffer (0.01 M Tris, 0.1 M KCl, 0.1 mM EDTA, pH 7) with 16 mM $MgCl_2$ as well as of the lamp pulse (C). Sample excited at 510 nm and fluorescence measured through a OG 590 cut-off filter (Schott and Gen., Mainz, W-Germany)

0.17 to about 1 when the $MgCl_2$ concentration was increased to about 16 mM (Fig. 4a). At higher temperatures (25° and 35°C) the situation became more complex and a third exponential term τ_3 was required to describe the decay of $I_m(t)$. The data analysis yielded for τ_3 20 ns while for τ_1 and τ_2 the same values as at 10°C were obtained. The variation of the amplitudes a_1 and a_2 with increasing concentrations of $MgCl_2$ was similar to that found at 10°C, the third amplitude a_3 vanishes as the $MgCl_2$ concentration is inreased (Fig. 4b).

Table 1. Lifetimes for $tRNA^{Phe}_{EB37C}$ at $10^{o}C$. Evaluation for two conformational states. (A) The parameters are determined for each decay curve at various Mg^{2+} concentrations of a titration experiment.
The number of degreees of freedom were 415.
Lifetimes (τ) and rotational relaxation times (τ_r) for $tRNA^{Phe}_{EB37C}$ at $10^{o}C$. (B) Evaluation for two conformational states. The parameters are determined from simultaneous evaluation of all decay curves of a Mg^{2+} titration experiment. The numbers of degrees of freedom at $10^{o}C$ were 1240

	$MgCl_2$ mM	τ_1 (ns)	σ (ns)	τ_2 (ns)	σ (ns)	χ^2
(A)	0.0	34.73	0.47	12.09	1.83	0.99
	0.5	33.75	0.30	8.42	1.46	1.12
	1.0	34.73	0.50	11.91	1.52	0.95
	2.0	34.67	0.59	12.07	1.29	1.16
	4.0	-	-	-	-	-
	8.0	33.81	0.43	10.20	0.57	1.03
	15.0	33.47	0.44	10.62	0.45	1.05

	τ_1 (ns)	σ (ns)	τ_2 (ns)	σ (ns)	χ^2_n
(B)	34.34	0.16	11.15	0.24	1.02
	τ_{r1} (ns)	σ (ns)	τ_{r2} (ns)	σ (ns)	
	28.94	0.79	13.68	1.68	1.09

Provided that the spectroscopic parameters $\varepsilon_i k_i$ (i=1,2,3) are the same, the amplitudes a_1, a_2, a_3 are proportional to the concentration of the three excited states of ethidium. The amplitude ratios a_2/a_1 and a_3/a_1 thus correspond directly to the constants describing the equilibrium between three conformational states of tRNA as reflected by the probe. A proof for the assumption that $\varepsilon_i k_i$ is the same at least for the amplitudes a_1 and a_2 is given later on.

The data from the pulsed decay experiments can be interpreted best by the existence of three conformational states of $tRNA^{Phe}$, one of which is populated to a significant degree only at elevated temperatures (>10°C) and low Mg^{2+} concentrations (<1 mM). The picture described up to now refers to the $tRNA^{Phe}$ labeled with EB at position 37 ($tRNA^{Phe}_{EB37C}$). Essentially the same picture was found for $tRNA^{Phe}$ labeled at position 16/17 (EHRENBERG et al., 1976).

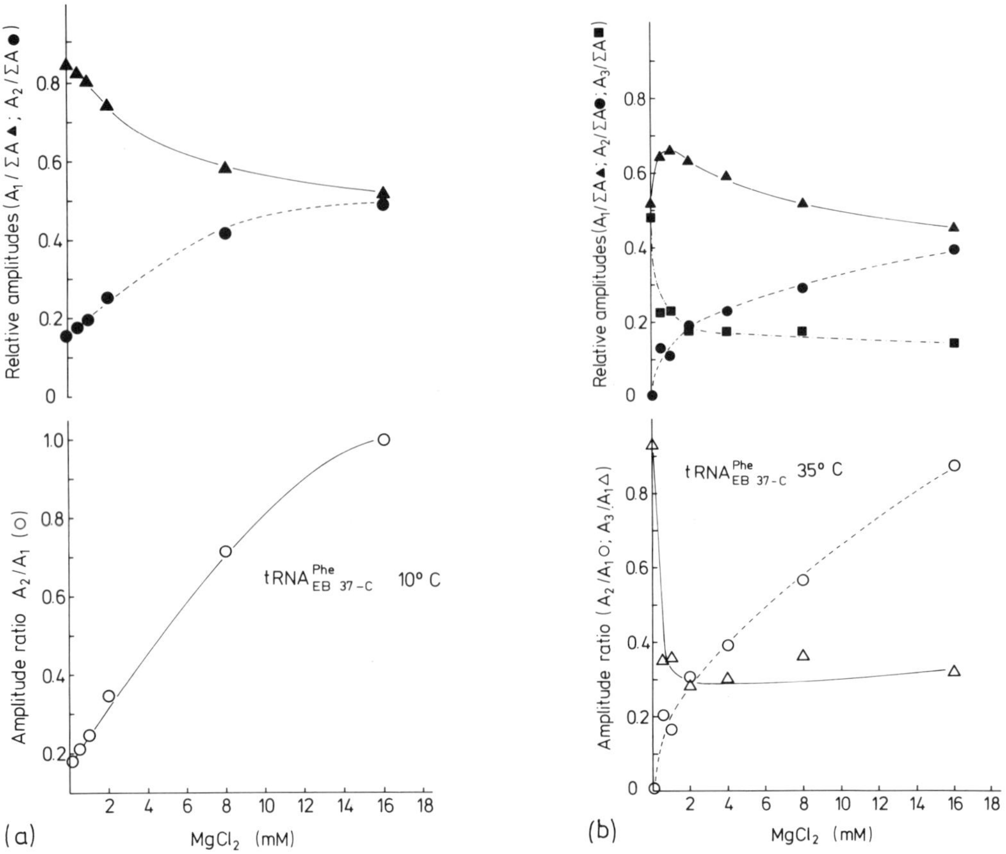

Fig. 4. (a) Dependence of the amplitudes a_1, a_2 and of the amplitude ratio a_2/a_1 of $tRNA^{Phe}_{EB37C}$ ($2 \cdot 10^{-6}$ M) on the concentration of $MgCl_2$ at 10°C, in Tris buffer, $\tau_1 = 34$ ns, $\tau_2 = 11$ ns, $\tau_{r1} = 29$ ns, $\tau_{r2} = 14$ ns. (b) Dependence of the amplitudes a_1, a_2, a_3 and of the amplitude ratio a_2/a_1 and a_3/a_1 of $tRNA^{Phe}_{EB37C}$ ($2 \cdot 10^{-6}$ M) on the concentration of $MgCl_2$ at 35°C in Tris buffer, $\tau_1 = 33$ ns, $\tau_2 = 11$ ns, $\tau_3 = 20$ ns

The existence of two different configurations of $tRNA^{Phe}_{yeast}$ is also indicated by the analysis of the polarized component $I_{||}(t)$ (Fig. 3). To simplify the analysis we have used, as models for tRNA, rotating spheres each of which is described by one rotational relaxation time. For fitting $I_{||}(t)$ at least two rotational relaxation times were required:

$$I_{||}(t) = a_1 \cdot e^{-t/\tau_1}(1+r_1 \cdot e^{-t/\tau_{r1}}) + a_2 \cdot e^{-t/\tau_2}(1+r_2 \cdot e^{-t/\tau_{r2}}) \quad (5)$$

In this model τ_{r1} is the rotational relaxation time of structure T_1 with a lifetime τ_1 and τ_{r2} is the rotational relaxation time of structure T_2 with lifetime τ_2. For τ_{r1} and τ_{r2} the values 29 and 14 ns were obtained, for the corresponding amplitudes r_1 and r_2 the values 0.633 and 0.680 respectively.

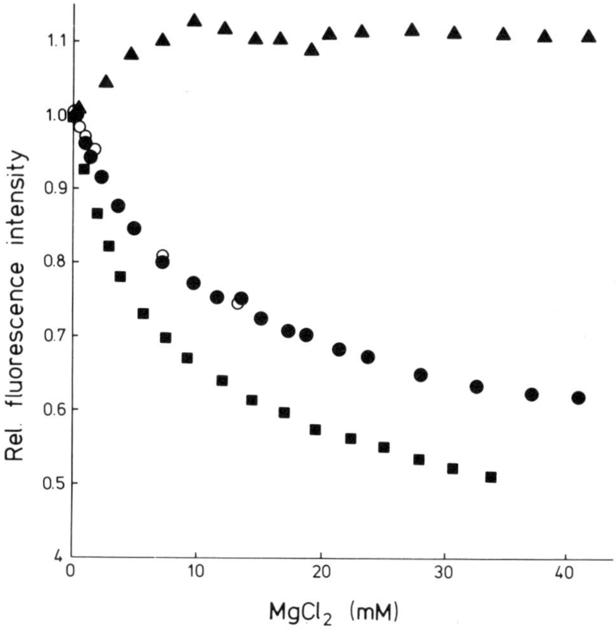

Fig. 5. Dependence of the fluorescence intensity of $tRNA^{Phe}_{EB37C}$ (●), the dodeka fragment 31-42 of Phe_{EB37C} (▲) and of $tRNA^{Phe}_{EB16/17}$ (■) on the concentration of $MgCl_2$ at $10^{o}C$ in Tris buffer. The points denoted by O were calculated from amplitudes and time constants obtained from pulsed experiments. Sample excited at 510 nm and fluorescence measured at 590 nm

The presence of two significantly different rotational relaxation times supports the interpretation that we are dealing with at least two conformational states of tRNA at low temperature. The large difference in τ_{r1} and τ_{r2} suggests a substantial change in the configuration of tRNA. We are at present not able to decide to what extent a changed rotational freedom of the probe itself could influence the observed rotational relaxation times. However, motions of this kind should be expected to occur at shorter times than observed here.

IV. Measurements Under Stationary Excitation

While in a pulsed fluorescence experiment the presence of different excited states can be resolved on a time axis, this information vanishes under stationary excitation. This condition corresponds to the time integral of $I_m(t)$:

$$\bar{I}_m = \int_0^\infty \sum_i a_l\, e^{-t/\tau_i} = \sum_i a_i\, \tau_i . \tag{6}$$

Thus the fluorescence intensity measured under stationary excitation is given by the sum of the lifetimes τ_i weighted by their amplitudes a_i. For the situation exemplified in Figure 4a, where τ_1 was found to be larger than τ_2 and a_2 to increase on the expense of a_1, $\bar{I}_m$ should decrease as the Mg^{2+} concentrations is increased. This behavior is also observed in the experiment where $\bar{I}_m$ is measured directly (Fig. 5). The change in I_m in the experiment with stationary excitation thus reflects the continuous shift from a structure with a long lifetime of its probe to a structure with a short lifetime. A comparison between the fluorescence intensity directly measured ($\bar{I}_m$) and calculated from the pulsed experiments (I'_m) is shown in Figure 5. I'_m has been calculated from the relative amplitudes $a_1/(a_1+a_2)$, $a_2/(a_1+a_2)$ and the lifetimes τ_1, τ_2 according to:

$$I'_m = \frac{a_1\tau_1 + a_2\tau_1}{a_1+a_2} \tag{7}$$

and thus differs from $\bar{I}_m$ by the factor $a_1 + a_2$. From the correspondence between the normalized intensities $\bar{I}_m$ and $\bar{I}'_m$ it follows that $a_1 + a_2$ must be constant and $\varepsilon_1 k_1 = \varepsilon_2 k_2$ since $[T_1]+[T_2]$ is constant. No change in $\bar{I}_m$ was observed for the dodeka fragment 31 - 42 containing the nucleotide bases 31 until 42 as well as EB in position 37. This indicates that Mg^{2+} is unable to induce a particular configuration of this loop section when released from its connection with other parts of the tRNA structure.

The ethidium derivative at position 16/17 ($tRNA^{Phe}_{EB16/17}$) exhibits the same dependence of I_m on increasing concentration of $MgCl_2$ as does $tRNA^{Phe}_{EB37-C}$ (Fig. 5). Since with 0.1 M KCl the Cl^- concentration varies up to 80% at the highest $MgCl_2$ concentrations used, the same titrations were performed also at 0.3 M KCl with the same result. Thus we conclude that the change in the fluorescence parameters of the ethidium-labeled tRNAs is due to the influence of Mg^{2+} ions.

V. Measurements of Chemical Rates

Strong support for the existence of different conformational states of $tRNA^{Phe}$ can be given by the analysis of the rate constants for the isomerization between the conformers. We have therefore investigated the chemical relaxation spectrum of tRNA under various Mg^{2+} concentrations by temperature jump experiments.

A. Instrumentation and Data Evaluation

The experiments were performed in a temperature jump apparatus described previously (RIGLER et al., 1974) which has been complemented by a data evaluation facility. The signals obtained after a temperature jump were stored after A/D conversion in an 8 bit-1K memory (Biomation 802, Palo Alto, California, USA). The stored signal curves were then played back on an oscilloscope together with the simulated exponential progress curves. The simulator (Fig. 6) allows the generation of up to three exponential functions with calibrated amplitudes and time constants. For minimization of the difference ΔY between simulated ($\hat{Y}$) and observed data points (Y) a special electronic control was developed: (1) the difference $\Delta Y(t) = \hat{Y}(t) - Y(t)$ is displayed on the oscilloscope and yields a straight line for an optimal fit;

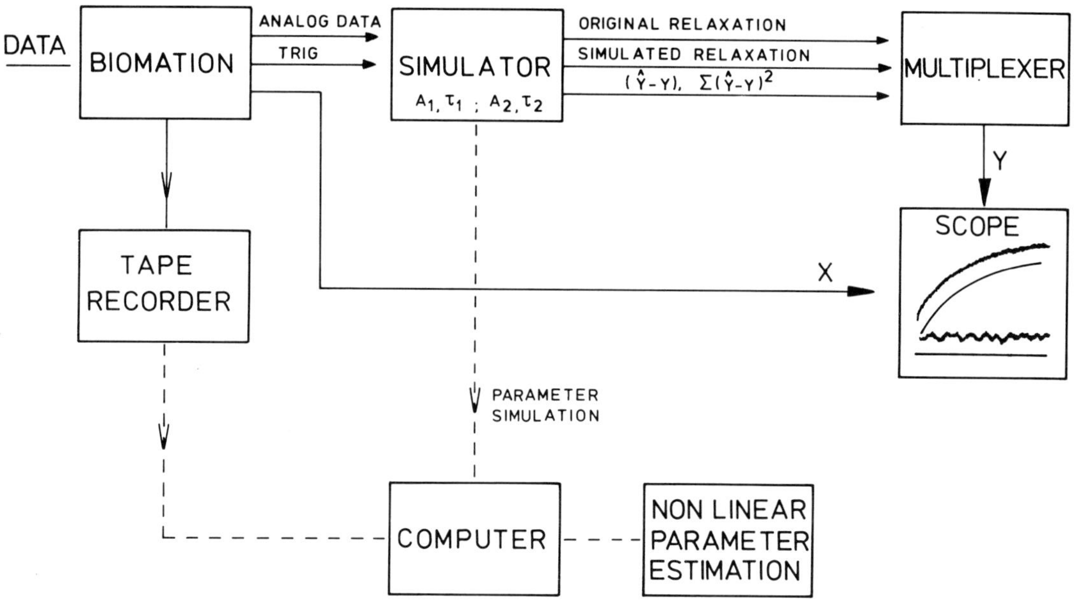

Fig. 6. Scheme of the simulator for chemical relaxation times and amplitudes (LOMAKKA, ISSLER and RIGLER, unpublished)

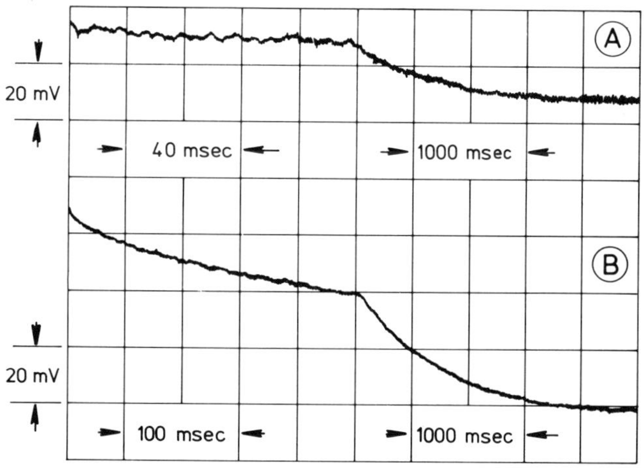

Fig. 7A and B. Reaction progress curves after a temperature jump from 3° to 10^oC. (A) $tRNA^{Phe}_{EB37C}$ ($1 \cdot 10^{-6}$ M) at 8 mM $MgCl_2$, Tris buffer, $\tau_{ch1} = 613$ ms. (B) $tRNA^{Phe}_{EB16/17}$ ($1 \cdot 10^{-6}$ M) at 16 mM $MgCl_2$, Tris buffer, $\tau_{ch1} = 714$ ms, $\tau_{ch2} = 47.4$ ms. Double time axis. Rise time filter in both experiments 1 ms. The sample was excited at 546 nm, the fluorescence was measured through a OG 590 cut-off filter

(2) the least square sum $R = \Sigma(\hat{Y}(T) - Y(t))^2$ is calculated and displayed on a digital volt meter. The estimated amplitudes and time constants were then used as start parameters for nonlinear

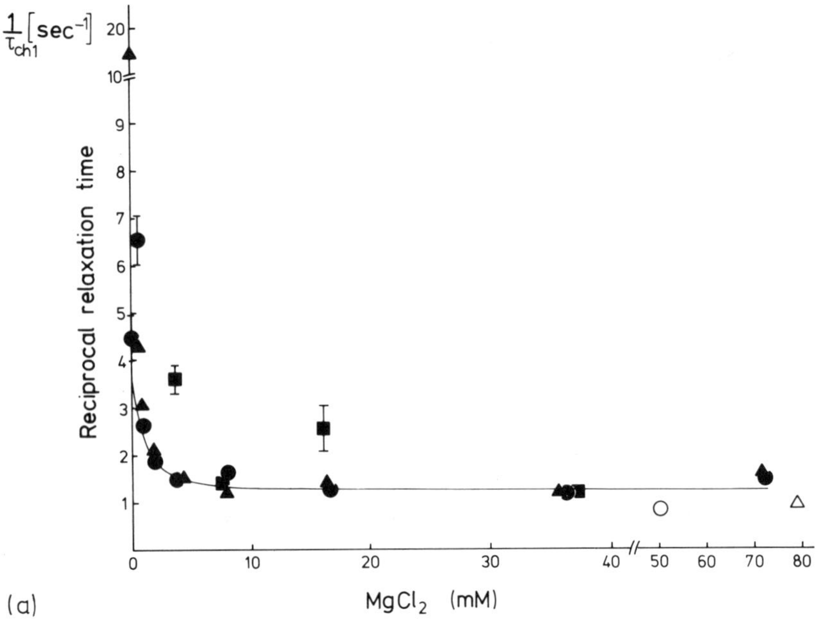

1/τ_ch1 [sec⁻¹]
Reciprocal relaxation time
20
10
9
8
7
6
5
4
3
2
1
0
10
20
30
40
50
60
70
80
MgCl2 (mM)

(a)

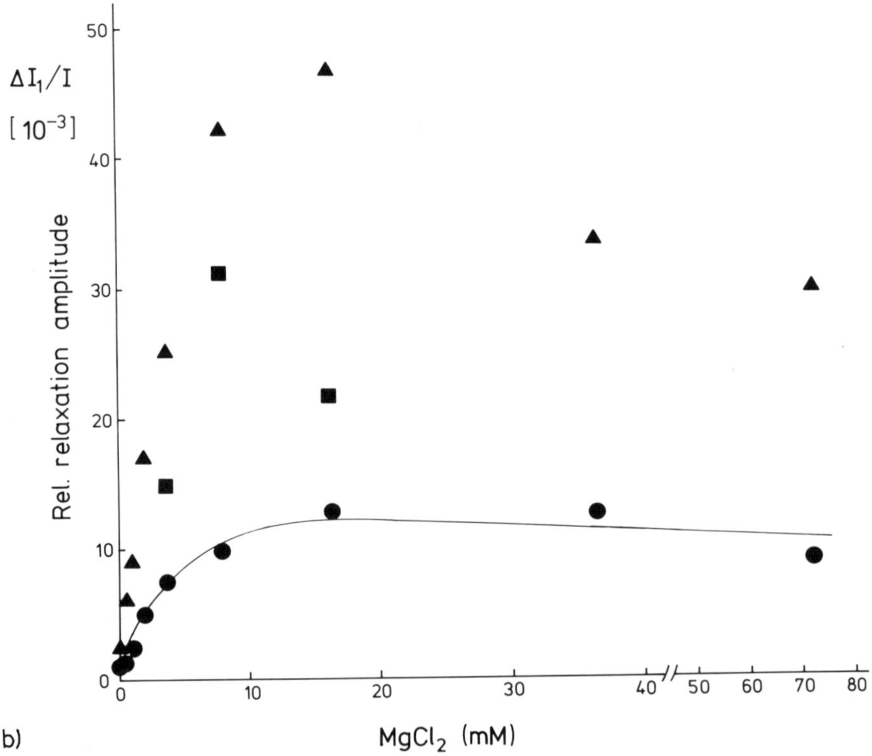

ΔI1/I
[10⁻³]
Rel. relaxation amplitude
50
40
30
20
10
0
0
10
20
30
40
50
60
70
80
MgCl2 (mM)

(b)

regression analysis (MARQUARDT, 1963; MEETER, 1964). These programs use local linearization and successive minimization starting from initial parameter guesses. The accuracy of the parameters is obtained from the linearized estimation problem in close vicinity of the minimum R. Discrimination between different models satisfying the observed relaxation curves are made with a χ^2-test.

B. Results

The time course of the fluorescence intensity of $tRNA^{Phe}_{EB37-C}$ and $tRNA^{Phe}_{EB16/17}$ observed after a temperature jump from 3° to 10°C is displayed in Figure 7. One slow relaxation process of about 600 ms was observed for $tRNA^{Phe}_{EB37C}$ while for $tRNA^{Phe}_{EB16/17}$ in addition a faster process of about 50 ms was found. The dependence of $1/\tau_{ch}$ and the relative relaxation amplitudes ($\Delta I/I$) are shown in Figure 8a,b and Figure 9a,b. The important result is that the reciprocal relaxation time of the fast as well as of the slow process decreases with increasing $MgCl_2$ concentrations and reaches a constant value for high $MgCl_2$ concentrations. The corresponding relaxation amplitude exhibited for the slow process a maximum around 20 mM Mg^{2+} ($tRNA^{Phe}_{EB37C}$, $tRNA^{Phe}_{EB16/17}$) while for the fast process ($tRNA^{Phe}_{EB16/17}$) the maximum was found below 10 mM Mg^{2+}. Though there were indications for the existence of the fast relaxation time (τ_{ch2}) also for $tRNA^{Phe}_{EB37-C}$ at 10°C and 20 mM Mg^{2+}, where the amplitude for τ_{ch1} is maximal, the signals were too low to be discriminated from those of the slow relaxation time (τ_{ch1}) at other Mg^{2+} concentrations. However, at higher temperatures (24° and 36°C, not shown) the fast relaxation process is clearly visible also for $tRNA^{Phe}_{EB37C}$. The behavior of the isomer $tRNA^{Phe}_{EB37B}$ with respect of the dependence of τ_{ch1} and τ_{ch2} and the respective amplitudes resembled closely that of $tRNA^{Phe}_{EB16/17}$ (not shown).

In order to ascertain that τ_{ch} observed at high $MgCl_2$ concentration was not affected by a decrease of the temperature level within the temperature jump cell, usually observed 1 - 2 s after a temperature jump, relaxation experiments were performed also by stopped flow mixing of $tRNA^{Phe}_{EB37-C}$ and $tRNA^{Phe}_{EB16/17}$ with $MgCl_2$ at 10°C. Since one is dealing with an excess of $MgCl_2$ over tRNA, all rate equations can be linearized and the relaxation times obtained from the stopped flow experiment are directly comparable with those obtained from the T-jump experiment. Only a small deviation between the relaxation times observed by both techniques was observed (Fig. 8a), indicating that the values for $1/\tau_{ch}$ obtained from T-jump experiments were slightly overestimated at high Mg^{2+} concentrations.

Fig. 8a and b. Dependence of $1/\tau_{ch1}$ (a) and the relative amplitude $\Delta I_1/I$ (b) on the concentration of $MgCl_2$ for $tRNA^{Phe}_{EB37-C}$ (●), $tRNA^{Phe}_{EB16/17}$ (▲) and $tRNA^{Phe}$ (■) at 10°C. Scale $\Delta I_1/I$ for $tRNA^{Phe}$ enlarged 10 times. Data points with open symbols were determined by stopped flow measurements. S.E. indicated by error bars in those cases where S.E. is bigger than the size of the symbols. Curves: simulations for $tRNA^{Phe}_{EB37C}$ according to Eqs. (12) and (14) with the parameters given in Table 2

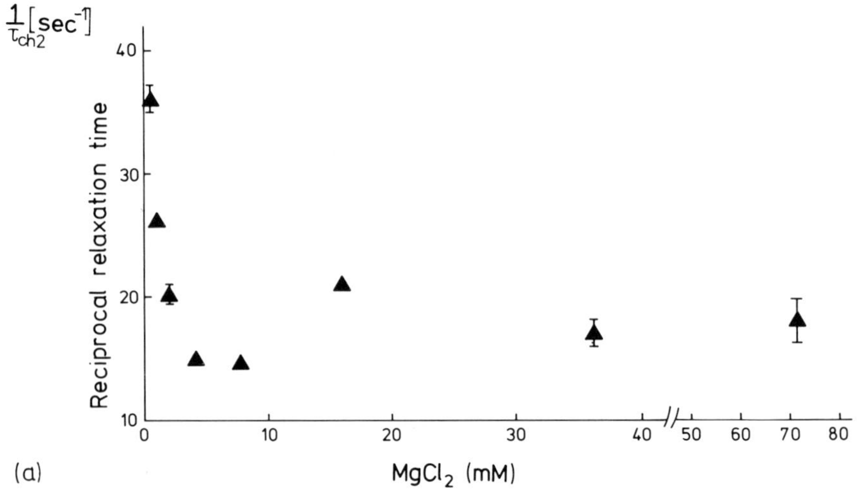

(a)

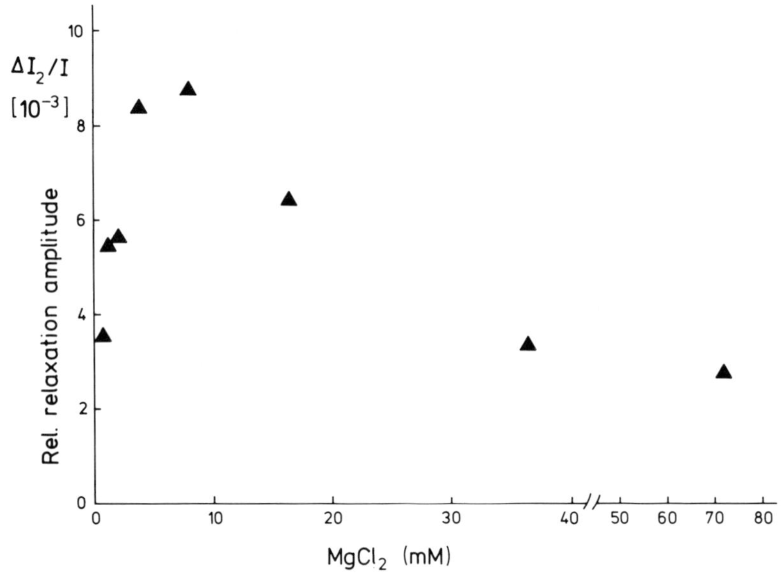

(b)

Fig. 9a and b. Dependence of $1/\tau_{ch2}$ (1) and the relative amplitude $\Delta I_2/I$ (b) of $tRNA^{Phe}_{EB16/17}$ on the concentration of $MgCl_2$ at 10°C

We have not been able at present to resolve fast relaxation processes (below 1 ms) which could be ascribed to the direct binding of Mg^{2+} to tRNA. With the high concentrations of Mg^{2+} used and the binding constants estimated (cf. evaluation of parameters), one expects this relaxation to occur within a few microseconds, assuming that the reaction is diffusion controlled.

For comparison the fluorescence of the Y-base in $tRNA^{Phe}$ was also investigated. Due to the small amplitudes observed (Fig. 8a) the estimates for $1/\tau$ spread considerably more than those for the ethidium derivatives. Within the experimental accuracy $1/\tau$ exhibits the same behavior as observed for $tRNA^{Phe}_{EB37C}$ and $tRNA^{Phe}_{EB16/17}$.

The dependence of the relaxation times on the concentration of $MgCl_2$ indicates the presence of at least two conformational states for $tRNA^{Phe}_{EB37C}$ (as well as for $tRNA^{Phe}$) at 10°C and of at least three conformational states at higher temperatures, in correspondence with the results obtained by pulsed fluorescence decay experiment (Fig. 4a,b). For $tRNA^{Phe}_{EB16/17}$ and $tRNA^{Phe}_{EB37B}$ the third conformational state is visible already at low temperatures. The increase in the observed relaxation time with increasing $MgCl_2$ concentration can be explained by stabilizing conformations with greatly reduced transition rates under the influence of $MgCl_2$.

VI. A Model for Allosteric Conformations of tRNA

As has already been discussed qualitatively all data obtained point towards the existence of at least two conformational states of $tRNA^{Phe}$, the distribution of which is dependent on the presence of Mg^{2+}. We have therefore investigated more closely the simplest model in which the data indicating the presence of only two conformations can be described. It is based on the following experimental data:

(1) The pulsed decay experiment show the presence of two conformations of tRNA, designated T_1 and T_2. The equilibrium is shifted towards T_2 as the $MgCl_2$ concentration is raised,

(2) the chemical relaxation experiments show that the transition rates for the two states decreases with increasing $MgCl_2$ concentrations, suggesting that structure T_2 is trapped with increasing efficiency at high $MgCl_2$ concentrations. Given this basis we propose the following reaction scheme:

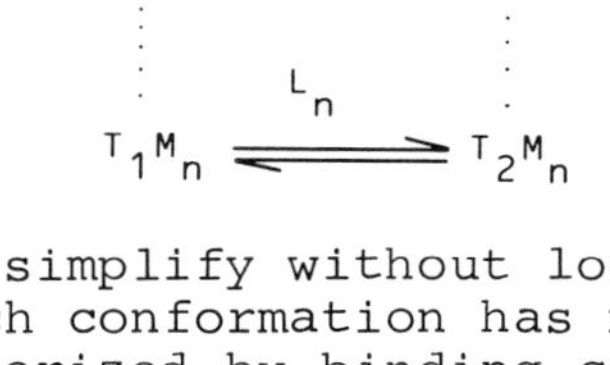

To simplify without losing essential features we assume that each conformation has n identical binding sites for Mg^{2+} characterized by binding constants K_1 and K_2. With the definitions:

$$L_i = \frac{l_i}{l_{-i}} = \frac{[T_2M_i]}{[T_1M_i]}, \qquad i=1,2....n \qquad (8)$$

$$K_j = \frac{n-i+1}{i} = \frac{[T_jM_i]}{[M]\ [T_jM_{i-1}]}, \qquad j=1,2,\ i=1,2...n, \text{ where} \qquad (9)$$

n = number of binding sites for Mg^{2+} denoted by M, we can describe the observables a_2/a_1 in the pulsed decay experiment and $\Delta I/I$ and $1/\tau_{ch}$ in the chemical relaxation experiment as follows:

Using Eq. (2) the amplitude a_j observed in the pulsed experiment corresponds to the sum of all tRNA molecules in state T_j and is given by:

$$a_j = T_j + \sum_{i=1}^{n} [T_jM_i]\ \varepsilon_j k_j^e, \qquad j = 1,2 \qquad (10)$$

and

$$\frac{a_2}{a_1} = L_o \frac{(1+[M]K_2)^n}{(1+[M]K_1)^n} \text{ with } \varepsilon_1 k_1^e = \varepsilon_1 k_2^e \text{ and } L_o = \frac{[T_2]}{[T_1]} \qquad (11)$$

$\frac{a_2}{a_1}$ increases from L in the absence of Mg^{2+} to $L \cdot (\frac{K_2}{K_1})^n$ at high Mg^{2+} concentrations if $K_2 > K_1$.

For the conditions applied, i.e. the concentration of Mg^{2+} in large excess over that of tRNA, we assume equilibration of the binding of Mg to T_1 and T_2 to occur very rapidly compared to the transition between T_1 and T_2. Moreover the binding of Mg^{2+} to one form is not influenced by its binding to the other (large excess of Mg^{2+}). With these simplifications we obtain for $1/\tau_{ch}$:

$$1/\tau_{ch} = \left(\frac{(1+p[M]K_1)^n}{(1+[M]K_2)^n} + \frac{(1+p[M]K_1)^n}{L_o(1+[M]K_2)^n} \right) l_o, \qquad (12)$$

where we have expressed $l_i = l_o \cdot p^i$, i=0,1....n. If p is set equal to 1 the model assumes that Mg^{2+} only influences the back rates l_{-i} but leaves the forward rate l_i unaffected and Eq. (12) reduces to:

$$1/\tau = \left(1 + \frac{(1+[M]K_1)^n}{L_o(1+[M]K_2)^n} \right) l_o. \qquad (13)$$

This expression is identical with that derived by EIGEN (1967, 1968) for the chemical relaxation of the two allosteric states of glyceraldehyde-3-phosphate-dehydrogenase (KIRSCHNER et al., 1966). According to Eq. (13) $1/\tau_{ch}$ decreases from $l_o + l_{-o}$ in the absence of Mg^{2+} and reaches a limiting value $l_o + l_{-o}\left(\frac{K_1}{K_2}\right)^n$ at high Mg^{2+} concentrations.

For the relative amplitude $\Delta I/I$ of the exponential relaxation process with time constant τ_{ch} after a temperature change δT one obtains:

$$\frac{\Delta I}{I} = \frac{\delta T}{RT^2} \cdot \frac{L_o(1+[M]K_1)^n \cdot (1+[M]K_2)^n \cdot (\beta-1)}{[(1+[M]K_1)^n + L_o(1+[M]K_2)^n]\ [(1+[M]K_1)^n + \beta \cdot L_o(1+[M]K_2)^n]}\ X,$$

$$\text{where } X = n\left(\frac{[M]K_1}{1+[M]K_1}\Delta H_1 - \frac{[M]K_2}{1+[M]K_2}\Delta H_2\right) + \Delta H_L, \tag{14}$$

$$\Delta H_1 = \left(\frac{\delta \ln K_1}{\delta T}\right)_P RT^2, \quad \Delta H_2 = \left(\frac{\delta \ln K_2}{\delta T}\right)_P RT^2, \quad \Delta H = \left(\frac{\delta \ln L}{\delta T}\right)_P RT^2$$

and $\beta = \frac{\tau_2}{\tau_1}$ · τ_1 and τ_2 are the fluorescence lifetimes of structure T_2 and T_2 respectively.

$$\frac{\Delta I}{I} \cdot \frac{RT^2}{\delta T} \text{ goes from } \frac{L_o(\beta-1)}{(1+L_o)(1+\beta L_o)} \Delta H_L$$

in the absence of Mg^{2+} over a maximum to

$$\frac{L_n(\beta-1)}{(1+L_n)(1+\beta L_n)}\left(n(\Delta H_1-\Delta H_2)+\Delta H_L\right) \text{ with } L_n = L_o\left(\frac{K_2}{K_1}\right)^n$$

at high concentrations of Mg^{2+}.

A. Evaluation of the Parameters

From the parameters describing the dependence of a_2/a_1, $1/\tau_{ch1}$ and $\Delta I_1/I$ of $tRNA^{Phe}_{EB37C}$ on the concentration of Mg^{2+} L_o and β can be determined directly from pulsed decay experiments. For l_o a good initial guess can be obtained from comparing $1/\tau_{ch}$ (Eq. 12) with a_2/a_1(Eq. 11) at 0 and high Mg^{2+} concentration assuming that p of Eq. (12) is equal to one. Since independent information on the parameters K_1, K_2 and n is not available, we have fitted all parameters of the two-state allosteric model keeping only β fixed. For this purpose all experimental information available was used and the Mg^{2+} dependence of a_2/a_1, $1/\tau_{ch1}$ and $\Delta I_1/I$ was simulated with parameters K_1, K_2, n, L_o, l_o, ΔH_1, ΔH_2, ΔH_L and p. The values given in Table 2 are those for which a minimum of the least square deviations between observed and simulated data points was reached. In this fitting procedure all observables were weighted by their respective standard deviations as evaluated from nonlinear regression analysis of the fluroescence decay and chemical relaxation curves.

As seen from Table 2 accepta›le statistics have been obtained for the parameters K_1, K_2, L_o, l_o, ΔH_L and p while the error for ΔH_1 and ΔH_2 is considerable. Reliable error estimates for n could not be obtained.

As predicted from the experimental observations K_2 is larger than $K_1 \cdot L_o$ (0.118) is smaller than the observed value (0.17). For this experimental point the $MgCl_2$ concentration is not well determined. It is however below 10^{-5} M as estimated from atomic absorption measurements. The ΔH values, of which ΔH_L is safest, indicate that the Mg-induced conformation change is driven by a positive entropy change with regard to the binding of Mg^{2+} as well as to the structural transitions. In our case p is found close to one

Table 2. Parameters (with S.E.) for Model I describing two Mg^{2+}-dependent conformations of $tRNA^{Phe}_{EB37-C}$ at $10^{o}C$

K_1	$= 258 \pm 118\ M^{-1}$
K_2	$= 295 \pm 118\ M^{-1}$
n	$= 21$
L_o	$= 0.118 \pm 0.04$
l_o	$= 0.32 \pm 0.13\ s^{-1}$
ΔH_1	$= -1.5 \pm 2.8$ kcal/mol; $\Delta S_1 = 5.7$ cal/mol deg[a]; $\Delta G_1 = -3.1$ kcal/mol[b]
ΔH_2	$= -1.7 \pm 3.2$ kcal/mol; $\Delta S_2 = 5.3$ cal/mol deg[a]; $\Delta G_2 = -3.2$ kcal/mol[b]
ΔH_L	$= 0.67 \pm 0.43$ kcal/mol; $\Delta S_L = 1.9$ cal/mol deg[a]; $\Delta G_L = 1.2$ kcal/mol[b]
p	$= 1.04 \pm 0.03$

[a]Determined from $\Delta S = (\Delta G - \Delta H)/T$

[b]Determined from $\Delta G = -RT\ln K$.

and is in line with the assumptions previously made by EIGEN (1967) regarding the allosteric behavior of glyceraldehyde-3-phosphate-dehydrogenase.

The model investigated (Model I) is incomplete in the sense that it does not account for the existence of a third conformational state which is evident from the pulsed decay measurements and chemical relaxation experiments at higher temperatures (25° and $35^{o}C$). Here a state with a fluorescence lifetime of 20 ns is observed for $tRNA^{Phe}_{EB37C}$ and a second chemical relaxation time (τ_{ch2}) is found which is an order of magnitude faster than first chemical relaxation time (τ_{ch1}). $1/\tau_{ch2}$ depends on the concentration of $MgCl_2$ in the same manner as does $1/\tau_{ch1}$ and suggests a transition to a new allosteric state. The present two-state model has to be enlarged by an additional conformational state T_3:

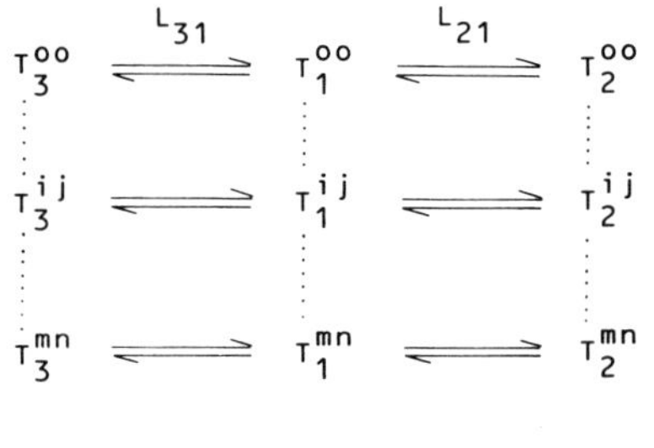

In this scheme the possible transition $T_3 \rightleftharpoons T_2$ described by L_{32} is not indicated. Here each structure state T_k (k=1,2,3) binds n molecules of Mg^{2+} with a binding constant K_k and m molecules of Mg^{2+} with a binding constant Q_k. For this model the amplitude ratio a_k/a_1 (k=2,3) as observed in the pulsed decay experiment as a function of the Mg^{2+} concentration (Fig. 4b) is given by:

$$\frac{a_k}{a_1} = L_{k1} \frac{(1+[M]K_k)^n \ (1+[M]Q_k)^m}{(1+[M]K_1)^n \ (1+[M]Q_1)^m}, \qquad k=2,3 \qquad (15)$$

Thus the expression given for a_2/a_1 in Eq. (11) is a special case of Eq. (15). Relations for $1/\tau_{ch}$ and $\Delta I/I$ can be derived in analogy to those given for the two-state model. The relevance of this three-state model to account for the chemical relaxation spectrum at higher temperatures of $tRNA^{Phe}_{EB37C}$ and for the ethidium derivatives $tRNA^{Phe}_{EB37B}$ and $tRNA^{Phe}_{EB16/17}$ is being investigated.

In summary we can conclude that the two-state model is able to describe within the error limits given the data for $tRNA^{Phe}_{EB37-C}$ observed at low temperature by pulsed decay and chemical relaxation experiments. Its strongest simplification is probably the assumption of n identical binding sites of tRNA for Mg^{2+} with regard to the complex structure of tRNA. As shown in Eq. (15) the existence of a second set of binding sites can be accounted for and a refinement of the two-state model is possible, given sufficient precision of the estimates.

VII. Conformational States of tRNA with Regard to the Biological Role of tRNA

The data analysis which as been performed rigorously only for one ethidium derivative suggests that the configuration of the anticodon-loop in vicinity of the ethidium-labeled can exist in at least two states. From the qualitative behavior of the ethidium derivative in position 16/17 however we are able to conclude that also the D-loop must exist in two different conformations which are closely coupled to those in the anticodon-loop. This in line with the observation that excision of the Y-base in $tRNA^{Phe}$ alters the binding of oligonucleotides to the D-loop (CAMERON and UHLENBECK, 1973). This close coupling probably is the reason why mutants which effect the D-stem of $tRNA^{Trp}$ convert this tRNA, normally decoding the triplet UCG, into a suppressor of the UGA nonsense codon (HIRSH, 1971; HIRSH and GOLD, 1971).

The finding that all ethidium derivatives are biologically active suggests that the derivatized tRNA structure serves as a relevant analog of the structure of the unmodified $tRNA^{Phe}_{yeast}$. Under conditions where a direct comparison between these two types of $tRNA^{Phe}$ was possible no significant difference could be found. It is thus likely that the two conformations described by spectroscopic and kinetic parameters reflect a change of the tRNA structure at least in the D-loop and anticodon regions, but probably also within a larger part of the tertiary structure.

Though we are not able to make definite statements about the significance of the rotational relaxation times in terms of profound changes of the tertiary structure, it is worthwile pointing out that the rotational relaxation time τ_{r2} found for structure T_2 (14 ns) is compatible with a completely folded tRNA structure, as has been suggested by OLSON et al. (1976) on the basis of data from dynamic laser light-scattering experiments performed with yeast $tRNA^{Phe}$. We calculate, using the equivalent sphere approximation, ∿20 Å for the molecular radius which would be consistent with a tRNA structure in which acceptor stem and anticodon stem are folded together. The low ΔH values for both the binding Mg^{2+} to the two conformational states as well as for the transitions between these states are compatible with data obtained from calorimetric measurements on yeast $tRNA^{Phe}$ (RIALDI et al., 1972). They indicate that the reaction is mainly driven by a positive entropy change which for the binding of Mg^{2+} has been related to the loss of the water coordinated to the fully solvated Mg^{2+} ions. The ΔH value for the structural transition $T_1 \rightleftharpoons T_2$ thus suggests a similar loss of the water structure of tRNA, a situation which would be expected for a highly compact form of tRNA.

The cooperative interaction of Mg^{2+} with the tRNA structure stabilizes structure T_2. For this interaction we determine binding constants K_1 and K_2 which are rather small compared to the values which were obtained for the binding of Mg^{2+} to $tRNA^{Phe}$ by calorimetric (RIALDI et al., 1972) and fluorometric (RÖMER and HACH, 1975) measurements. These authors report the presence of about 4 - 6 strong binding sites with a K between 10^5 and 10^6 M^{-1} and of about 20 weak binding sites with a K of $\sim 10^4$ M^{-1}. A comparison with our data is not straightforward since they were obtained at much higher concentrations of neutral salt.

Moreover, we have adopted a detailed model for evaluating our binding constants compared to the conventional Scatchard analysis used by these authors. However, based on own and reported data (SANDER and TS'O, 1971), RIALDI et al. (1972) propose the existence of a third type of binding site with n = 14 and K of order 10^2 M^{-1}, which is within the range of our estimates.

The functional relevance of the two conformations of tRNA is emphasized by the fact that both structures are populated at about equal amounts at conditions which are usually applied in the acylation reaction. We conclude that neither of these conformations can be explained by the denatured forms of tRNA as described by LINDAHL et al. (1966) and by GARTLAND and SUEOKA (1966) which are usually obtained by heating of native tRNAs in the presence of EDTA. The conformational states described by us resemble the structural states of tRNA proposed by COLE et al. (1972). Based on melting studies on several tRNAs including $tRNA^{Phe}$ from *E.coli* these authors propose three structural forms: an "extended" form at low temperature and salt concentrations, a "native" form at low temperature and high salt concentrations, and a "clover leaf" or "close variant" at high temperature and high salt concentrations. Our structural states however all refer to the high salt conditions used by COLE et al. (1972).

In this context experiments are of interest supporting the existence of different forms of tRNA which cannot be connected with the denatured tRNA structure in Fresco's sense. It was found that preincubation of native yeast $tRNA^{Phe}$ with large amounts of phenylalanyl-tRNA synthetase increased the acceptor activity for phenylalanine in the standard incubation assay significantly

(RENAUD et al., 1971). It was suggested that native $tRNA^{Phe}$ contains a certain fraction of inactive molecules which can be converted into active ones by complex formation with the synthetase. The existence of two conformations of yeast $tRNA^{Phe}$ which must differ at or near the 5' terminus of the molecules was demonstrated by HÄNGGI et al. (1970) on the basis of phosphorilation experiments with polynucleotide kinase. Similar effects were observed in experiments on the degradation of tRNA from its 3' end with polynucleotide phosphorylase (THANG et al., 1967, 1971). A detailed review of these results has been given by ZACHAU et al. (1972).

The existence of conformational changes of tRNA and/or the aminoacyl-tRNA synthetase in the tRNA synthetase complex has been postulated from kinetic data (KRAUSS et al., 1976; RIESNER et al., 1976; RIGLER et al., 1976). We are presently performing experiments to see if phenylalanyl-tRNA synthetase is able to influence the equilibrium between the observed conformational states. Since with this kind of experiment the influence of a variety of tRNA-specific structures such as the elongation factor Tu or the programmed ribosome on the conformation of tRNA can be investigated, more detailed information on the role of different states of the tRNA structure can be expected.

It is reasonable to assume that one of the states might represent the crystal structure, although this is by no means certain, since the crystallization procedure might create a minimum in free energy not to be found for tRNA in solution. It is however remarkable that for crystallization of both the orthorombic (KIM et al., 1971) and monoclinic (LADNER et al., 1972) form of tRNA, spermine had to be present. Spermine apparently stabilized a certain conformation of yeast $tRNA^{Phe}$ (ROBINSON and ZIMMERMAN, 1971) which is required for crystallization. From this viewpoint it is not surprising that no differences in fluorescence parameters of the Y-base (KIM et al., 1975) as well as the vibrational modes (CHEN et al., 1975) of $tRNA^{Phe}$ are found when tRNA in the mother liquid (solution structure) and in the crystal (crystal structure) are compared. Further studies of the influence of spermine on the structure of tRNA in solution are required in order to relate the conformations described by us to the X-ray structure.

The existence of different conformations of tRNA and their importance for the function of the molecule at various steps of the translation process have been discussed by several authors (LUCAS-LENART and LIPMANN, 1971; KURLAND, 1975). Evidence is accumulating that tRNA when binding to the programmed ribosome is undergoing a substantial conformational change in the TΨCG-loop (RICHTER et al., 1973; SCHWARZ et al., 1974, 1976). This conformational change has been proposed as a mechanism for recognition of tRNA on the ribosome (KURLAND et al., 1975). In this hypothesis the cognate message acts as an effector for an allosteric selection of the cognate tRNA by the ribosome. An allosteric network composed by the tRNA, the message, and the ribosome is proposed bearing the essential features of the three-state model (II). Similarly discrimination of acylated from nonacylated tRNA by the elongation factor Tu might be based on different structures of tRNA (LUCAS-LENARD and LIPMAN, 1971). Indeed there exists some evidence for a different conformational state of tRNA when aminoacylated (WATANABE and IMAHORI, 1971; TAKASAKI and IMAHORI, 1973; KAN et al., 1976).

A characterization of various possible conformational states of tRNA in terms of their structural, kinetic and thermodynamic parameters at various steps of the translation process seems to be of utmost value for an understanding of the tRNA molecule in its adaptor function between the nucleic acid and the aminoacid code.

Acknowledgement. This study was supported by funds from the K. and A. Wallenberg foundation, the Swedish Cancer Research foundation and the Swedish Nature Science Research foundation.

References

BEARDSLEY, D., TAO, T., CANTOR, Ch.R.: Studies on the conformation of the anticodon loop of phenylalanine transfer ribonucleic acid Effect of environment on the fluorescence of the Y-base. Biochemistry 9, 3524 (1970)

CAMERON, V., UHLENBECK, O.C.: Removal of Y-37 from $tRNA^{Phe}_{yeast}$ alters oligomer binding to two loops. Biochem. Biophys. Res. Comm. 50, 635 (1973)

CHEN, M.C., GIEGÉ, R., LORD, R.C., RICH, A.: Raman spectra and structure of yeast phenylalanine transfer RNA in the crystalline state and in solution. Biochemistry 14, 4385 (1975)

CHUANG, T.J., EISENTHAL, K.B.: Theory of fluorescence depolarization by anisotropic rotational diffusion. J. Chem. Phys. 57, 5094 (1972)

COLE, P.E., YANG, S.K., CROTHERS, D.M.: Conformational changes of transfer ribonucleic acid. Equilibrium phase diagrams. Biochemistry 11, 4358 (1972)

CONNORS, R.G., LABANAUSKAS, M., BEEMAN, W.W.: Structural studies on transfer RNA: The molecular conformation in solution. Science 166, 1528 (1969)

DANIEL, Jr. W.E., COHN, M.: Proton nuclear magnetic resonance of spin-labelled Escherichia coli $tRNA^{Met}_{f1}$. Proc. Nat. Acad. Sci. (Wash.) 72, 2582 (1975)

EHRENBERG, M., RIGLER, R.: Polarized fluorescence and rotational Brownian motion. Chem. Phys. Lett. 14, 539 (1972)

EHRENBERG, M., RIGLER, R.: Fluorescence correlation spectroscopy applied to rotational diffusion of macromolecules. Quart. Rev. Biophys. 9, 69 (1976)

EHRENBERG, M., RIGLER, R., WINTERMEYER, W.: On the structure and conformational dynamics of fluorescence labelled $tRNA^{Phe}_{yeast}$ in solution. In preparation (1977)

EIGEN, M.: Kinetics of reaction control and information transfer in enzymes and nucleic acids. In: Fast Reactions and Primary Process in Chemical Kinetics. CLAESSON, S. (ed.), p. 333. Stockholm: Almquist and Wiksell 1967

EIGEN, M.: New looks and outlooks on physical enzymology. Quart. Rev. Biophys. 1, 3 (1968)

EISINGER, J., FEUER, B., YAMANE, T.: Luminiscence and binding studies on $tRNA^{Phe}$. Proc. Nat. Acad. Sci. (Wash.) 65, 638 (1970)

EISINGER, J., LAMOLA, A.A.: In: Excited States of Protein and Nucleic Acids. STEINER, R.F., WEINRYB, I. (eds.), p. 189. New York: Plenum Press 1971

EISINGER, J., SPAHR, P.F.: Binding of complementary pentanucleotides to the anticodon loop of transfer RNA. J. Mol. Biol. 73, 131 (1973)

FRESCO, J.A., ADAMS, A., ASCIONE, R., HENLEY, D., LINDAHL, T.: Tertiary structure in transfer ribonucleic acids. Cold Spring Harbor Symp. Quant. Biol. 31, 527 (1966)

GARTLAND, W.W., SUEOKA, N.: Two interconvertible forms of tryptophanyl sRNA in E. coli Proc. Nat. Acad. Sci. (Wash.) 55, 948 (1966)

GRINVALD, A., STEINBERG, I.Z.: On the analysis of fluorescence decay kinetics by the method of last-squares. Analyt. Biochem. 59, 583 (1974)

HÄNGGI, U.J., STREECK, R.E., VOIGT, H.P., ZACHAU, H.G.: Phosphorylation od dephosphorylated tRNA and oligonucleotides by polynucleotide kinease. Biochem. Biophys. Acta 217, 287 (1970)

HIRSH, D.: Tryptophan transfer RNA as the UGA suppressor. J. Mol. Biol. 58, 439 (1971)

HIRSH, D., GOLD, L.: Translation of the UGA triplet in vitro by tryptophan transfer RNAs. J. Mol. Biol. 58, 459 (1971)

KAN, L.S., TS'O, P.O.P., SPRINZL, M., Van der HAAR, F., CRAMER, F.: Biophys. J. 16, 11a (1976)

KEARNS, D.R.: High-resolution nuclear magnetic resonance investigations of the structure of tRNA in solution: In: Progress in Nucleic Acid Research and Molecular Biology. COHN, W.E. (ed.), Vol. 18, p. 91. New York: Academic Press 1976

KIM, S.H., LANGLOIS, R., CANTOR, Ch.: Conformation of anticodon loop of yeast $tRNA^{Phe}$ in solution and in crystals as monitored by the Y-base fluorescence. Abstracts, EMBO Workshop on tRNA structure and function, Israel: Nof Ginossar 1975

KIM, S.H., QUIGLEY, G., SUDDATH, F.L., RICH, A.: High-resolution X-ray diffraction patterns of crystalline transfer RNA that show helical regions. Proc. Nat. Acad. Sci. (Wash.) 68, 841 (1971)

KIM, S.H., SUDDATH, F.L., QUIGLEY, G.J., McPHERSON, A., SUSSMAN, J.L., WANG, A.H.J., SEEMAN, N.C., RICH, A.: Three-dimensional tertiary structure of yeast phenylalanine transfer RNA. Science, N.Y. 185, 435 (1974)

KIRSCHNER, K., EIGEN, M., BITTMANN, R., VOIGT, B.: The binding of NAD to yeast D-glyceraldehyde-3-phosphate dehydrogenase: Temperature jump relaxation studies on the mechanism of an allosteric enzyme. Proc. Nat. Acad. Sci. (Wash.) 56, 1661 (1966)

KRAUSS, G., RIESNER, D., MAASS, G.: Mechanism of discrimination between cognate and non-cognate tRNAs by phenylalanyl-tRNA synthetase from yeast. Europ. J. Biochem. 68, 81 (1976)

KRIGBAUM, W.R., GOODWIN, R.W.: Small-angle X-ray study of transfer RNA. Science 154, 423 (1966)

KURLAND, Ch.: Prospects of ribosome structure and function. In: Protein Synthesis. WEISSBACH, H., PESTKA, S. (eds.). New York: Academic Press 1975

KURLAND, C.G., RIGLER, R., EHRENBERG, M., BLOMBERG, C.: Allosteric mechanism for codon-dependent tRNA selection on ribosomes. Proc. Nat. Acad. Sci. (Wash.) 72, 4248 (1975)

LADNER, J.E., FINCH, J.T., KLUG, A., CLARK, B.F.C.: High-resolution X-ray diffraction studies on a pure species of transfer RNA. J. Mol. Biol. 72, 99 (1972)

LIGHTFOOT, D.R., WONG, K.L., KEARNS, D.R., REID, B.R., SHULMAN, R.G.: Assignment of the low field proton nuclear magnetic resonance spectrum of yeast phenylalanine transfer RNA to specific base pairs. J. Mol. Biol. 78, 71 (1973)

LINDAHL, T., ADAMS, A., FRESCO, J.R.: Renaturation of transfer ribonucleic acids through site binding of magnesium. Proc. Nat. Acad. Sci. (Wash.) 55, 941 (1966)

LUCAS-LENARD, J., LIPMANN, F.: Protein biosynthesis. In: Annual Review of Biochemistry. SNELL, E.E., BOYER, P.D., SINSHEIMER, R.L. (eds.). Vol. 40, p. 409. Palo Alto, Calif.: Annual Reviews Inc. 1971

LYNCH, D.C., SCHIMMEL, P.R.: Cooperative binding of magnesium to transfer ribonucleic acid studied by a fluorescent probe. Biochemistry 13, 1841 (1974)

MARQUARDT, D.W.: An algorithm for least-squares estimation of non-linear parameters. J. Soc. Ind. Appl. Math. 11, 431 (1963)

MEETER, D.A.: Non-Linear Least-Squares (Gaushaus). Univ. Wisconsin Computing Center, 1964

MEMMING, R.: Theorie der Fluoreszenzpolarization für nicht kugelsummetrische Moleküle. Z. Physik. Chem. N.F. (Frankfurt) 28, 168 (1961)

OLSON, T., FOURNIER, M.J., LANGLEY, K.H., FORD, N.C.: Detection of a major conformational change in transfer ribonucleic acid by laser light scattering. J. Mol. Biol. 102, 193 (1976)

PILZ, I., KRATKY, O., CRAMER, R., van der HAAR, F., SCHLIMME, E.: On the conformation of phenylalanine specific transfer RNA. Studies on size and shape of the molecule by X-ray small angle scattering. Europ. J. Biochem. 15, 401 (1970)

RAJBHANDARY, U.L., CHANG, S.H., STUART, A., FAULKNER, R.D., HOSKINSON, R.M., KHORANA, H.G.: Studies on polynucleotides, LXVIII. The primary structure of yeast phenylalanine transfer RNA. Proc. Nat. Acad. Sci. (Wash.) 57, 751 (1967)

REID, B.R., RIBEIRO, N.S., GOULD, G., ROBILLARD, G., HILBERS, C.W., SHULMAN, R.G.: Tertiary hydrogen bonds in the solution structure of transfer RNA. Proc. Nat. Acad. Sci. (Wash.) 72, 2049 (1975)

REID, B.R., ROBILLARD, G.T.: Demonstration and origin of six tertiary base pair resonances in the NMR spectrum of E. coli $tRNA^{Val}_1$. Nature (London) 257, 287 (1975)

RENAUD, M., BOLLACK, C., BEFORT, N., EBEL, J.P.: (1971) cited by ZACHAU et al. (1972)

RIALDI, G., LEVY, J., BILTONEN, R.: Thermodynamic studies of transfer ribonucleic acids. I. Magnesium binding to yeast phenylalanine transfer ribonucleic acid. Biochemistry 11, 2472 (1972)

RICH, A., RAJBHANDARY, U.L.: Transfer RNAß Molecular structure, sequence and properties. In: Annual Review of Biochemistry. SNELL, E.E., BOYER, P.D., MEISTER, A., SINSHEIMER, R.L. (eds.). Vol. 45, p. 805. Palo Alto, Calif.: Annual Reviews Inc. 1976

RICHTER, D.L., ERDMANN, V.A., SPRINZL, M.: Specific recognition of G-T-Ψ-C loop (loop IV) of tRNA by 50S ribosomale subunits from E.coli. Nature (New Biol.) 246, 132 (1973)
RIESNER, D., MAASS, G., THIEBE, R., PHILIPPSEN, P., ZACHAU, H.G.: The conformational transitions in yeast $tRNA^{Phe}$ as studied with $tRNA^{Phe}$ fragments. Europ. J. Biochem. 36, 76 (1973)
RIESNER, D., PINGOUD, A., BOEHME, D., PETERS, F., MAASS, G.: Distinct steps in the specific binding of tRNA to aminoacyl-tRNA synthetase. Temperature-jump studies on the serine-specific system from yeast and the tyrosine-specific system form Escherichia coli. Europ. J. Biochem. 68, 71 (1976)
RIGLER, R., EHRENBERG, M.: Molecular interactions and structure as analysed by fluorescence relaxation spectroscopy. Quart. Rev. Biophys. 6, 139 (1973)
RIGLER, R., EHRENBERG, M.: Fluorescence relaxation spectroscopy in the analysis of macromolecular structure and motion. Quart. Rev. Biophys. 9, 1 (1976)
RIGLER, R., PACHMANN, U., HIRSCH, R., ZACHAU, H.G.: On the interaction of seryl-tRNA synthetase with $tRNA^{Ser}$. A contribution to the problem of synthetase-tRNA recognition. Europ. J. Biochem. 65, 307 (1976)
RIGLER, R., RABL, C.R., JOVIN, T.M.: A temperature jump apparatus for fluorescence measurements. Rev. Sci. Instrumen. 45, 580 (1974)
ROBERTUS, J.D., LADNER, J.E., FINCH, F.T., RHODES, D., BROWN, R.S., CLARK, B.F., KLUG, A.: Structure of yeast phenyl-alanine tRNA at 3 Å resolution. Nature (London) 250, 546 (1974)
ROBISON, B., ZIMMERMANN, T.P.: Cation dependence of the transfer reaction catalyzed by phenylalanine transfer ribonucleic acid synthetase from Baker's yeast. J. Biol. Chem. 246, 4664 (1971)
ROBISON, B., ZIMMERMANN, T.P.: A conformational study of yeast phenylalanine transfer ribonucleic acid. J. Biol. Chem. 246, 110 (1971)
RÖMER, R., HACH, R.: tRNA conformation and magnesium binding. A study of yeast phenylalanine-specific tRNA by a fluorescent indicator and differential melting curves. Europ. J. Biochem. 55, 271 (1975)
RÖMER, R., RIESNER, D., MAASS, G.: Resolution of five conformational transitions in phenylalanine specific tRNA from yeast. FEBS Lett. 10, 352 (1970)
SANDER, C., TS'O, P.O.: Interaction of nucleic acids VIII. Binding of magnesium ions by nucleic acids. J. Mol. Biol. 55, 1 (1971)
SCHWARZ, U., LÜHRMANN, R., GASSEN, H.G.: On the mRNA-induced conformational change of AA-tRNA exposing the T-Ψ-C-G sequence for binding to the 50S ribosomal subunit. Biophys. Biochem. Res. Comm. 56, 807 (1974)
SCHWARZ, U., MENZEL, H.M., GASSEN, H.G.: Codon-dependent rearrangement of the three-dimensional structure of phenylalanine tRNA, exposing the T-Ψ-C-G sequence for binding to the 50S ribosomal subunit. Biochemistry 15, 1484 (1976)
SHULMAN, R.G., HILBERS, C.W., KEARNS, D.R., REID, B.R., WONG, Y.P.: Ring-current shifts in the 300 MHz nuclear magnetic resonance spectra of six purified transfer RNA molecules. J. Mol. Biol. 78, 57 (1973)

TAKASAKI, Y., IMAHORI, K.: CD and fluorescence studies of $tRNA^{Phe}$ from Baker's yeast. J. Biochem. 74, 513 (1973)
THANG, M.N., BELTCHEV, B., GRUNBERG-MANAGO, M.: Phosphorolysis of tRNA. Multiple conformational states of tRNA in solution. Europ. J. Biochem. 19, 184 (1971)
THANG, M.N., GUSCHLBAUER, W., ZACHAU, H.G., GRUNBERG-MANAGO, M.: Degradation of transfer ribonucleic acid by polynucleotide phosphorylase. I. Mechanism of phosphorolysis and structure of tRNA. J. Mol. Biol. 26, 403 (1967)
UHLENBECK, O.C.: Complementary oligonucleotide binding to transfer RNA. J. Mol. Biol. 65, 25 (1972)
WATANABE, K., IMAHORI, K.: The conformation difference between $tRNA_f^{met}$ from E. coli. Biochem. Biophys. Res. Comm. 45, 488 (1971)
WINTERMEYER, W., ZACHAU, H.G.: Replacement of the Y base, dihydrouracil and 7-methyl-guanine in tRNA by artificial odd bases. FEBS Lett. 18, 214 (1971)
WINTERMEYER, W., ZACHAU, H.G.: Replacement of odd bases in tRNA by fluorescent dyes. In: Methods of Enzymology. COLOWICK, S.P., KAPLAN, N.O. (eds.). Vol. XXIX, pp. 667-673. New York: Academic Press 1974
WINTERMEYER, W., ZACHAU, H.G.: Characterization of fluorescent derivatives of yeast tRNA in the ribosomal system. Mol. Biol. (Russian) 9, 63; (English) 9, 49 (1975)
WINTERMEYER, W., ROBERTSON, J.M., ZACHAU, H.G.: Fluorescence studies on tRNA conformation and ribosome interaction. Proc. Conf Synthesis Structure and Chemistry of tRNAs and Their Components. WIEWIOROWSKI, M. (ed.). Poland: Poznan 1976
ZACHAU, H.G., STREECK, R.E., HÄNGGI, U.J.: Conformational states of transfer ribonucleic acids. In: Gene Expression and Its Regulation. KENNEY, F.T., HAMKALO, B.A., FAVELUKES, G., AUGUST, J.T. (eds.). New York: Plenum Publishing Corp. 1972

Chemical Relaxation Kinetic Studies of *E. coli* RNA Polymerase Binding to Poly [*d*(*A*—*T*)] Using Ethidium Bromide as a Fluorescence Probe

T. M. JOVIN and G. STRIKER

I. Introduction

Many and perhaps most of the biosynthetic and control mechanisms of a cell are mediated through protein-nucleic acid interactions. Such phenomena are necessarily complex in nature since they involve multisite associations between macromolecules and are subject to a multitude of thermodynamic and environmental influences (temperature, pH, ionic strength, ligands etc.). Unfortunately, detailed information about these processes is available for very few systems to date, due in part to the use of indirect experimental methods (activity measurements, filter binding, ultracentrifugal analysis, etc.). Kinetic data, in particular, often have been obtained under extreme conditions which may not typify the physiological state.

A recent review of recognition mechanisms utilized by DNA-specific enzymes (JOVIN, 1976) emphasizes the conformational flexibility of DNA, manifested in part as local structural fluctuations which modulate to a greater or lesser degree the interactions with proteins. Not yet known, however, are the specific mechanisms by which defined primary nucleotide sequences are recognized, as, for example, in the expression of repression or initiation of transcription.

In this connection one can postulate as distinct and distinguishable events the "recognitory" as opposed to the "functional" activation of a specific site (JOVIN, 1976). That is, does the recognition of a particular sequence involve concerted conformational changes in either the DNA and/or enzyme, and if so, are these followed by additional structural modifications which poise the system for function, e.g. initiation of transcription?

In the case of the *Escherichia coli* RNA polymerase (see CHAMBERLIN and LOSICK, 1976 for a review of this enzyme), considerable in-

---

Abbreviations

1. Hepes: N-2-hydroxyethylpiperazine-N'-2-ethanesulfonic acid.
2. Ethidium bromide: 2,7-diamino-10-ethyl-9-phenylphenanthridium bromide.
3. DTT: dithiothreitol.
4. Poly[d(A-T)]: double-helical alternating copolymer of deoxyribothymidylate and deoxyriboadenylate.

formation is available with respect to these points. The holoenzyme is capable of recognizing promoter sequences which include a binding site with an apparent size of about 42 basepairs. The latter have been isolated by DNase digestion of the unprotected DNA and several sequences are known (PRIBNOW, 1975b and references therein). Initiation of transcription occurs at the midpoint of these regions (HEYDEN et al., 1975; PRIBNOW, 1975a) and 7 basepairs to the left is the start of an "A-T" rich heptanucleotide "binding" sequence common to all known promoters (PRIBNOW, 1975a, b). Another feature in current promoter models (DICKSON et al., 1975; PRIBNOW, 1975b) is the existence of recognition elements located about 35 basepairs "upstream" from the initiation site. Additional evidence for interactions which extend beyond the protected region is provided by the stimulatory action of the catabolite gene activator protein, CAP (ZUBAY et al., 1970) which binds to the left of RNA polymerase (DICKSON et al., 1975) and which ceases to be required in the case of the lac UV5 mutation which affects the heptanucleotide binding sequence (data of J. GRALLA, see PRIBNOW, 1975a,b).

A number of possible mechanisms can be invoked to explain the translation of RNA polymerase from a recognition site to the initiation site (JOVIN, 1976; PRIBNOW, 1975b) but there are few direct experimental data bearing on this point. The "functional" activation of the binary polymerase-promoter complex, however, is better understood. A transition from an inactive "closed" form to a highly stable "open" complex occurs as a consequence of opening about ten basepairs of DNA (SAUCIER and WANG, 1972) at high temperature (WALTER et al., 1967; SAUCIER and WANG, 1972; CHAMBERLIN, 1974a,b; MANGEL and CHAMBERLIN, 1974b) and low ionic strength (FUCHS et al., 1967; HINKLE and CHAMBERLIN, 1972; SAUCIER and WANG, 1972; CHAMBERLIN, 1974a; MANGEL and CHAMBERLIN, 1974a; DOMINGO et al., 1975). Other properties of the enzyme-DNA complex are altered during promoter activation and, of course, during the subsequent steps of initiation and propagation (CHAMBERLIN, 1974b; KRAKOW et al., 1976).

The study which forms the substance of this report had as its aim the elucidation of additional information about the perturbations of DNA structure upon the binding of RNA polymerase. In particular, we wished to know whether effects upon regions of the DNA outside of the immediate binding site could be detected. This consideration and the complexities inherent in studies with natural DNAs (see above) dictated the choice of polynucleotide and experimental approach.

The synthetic alternating copolymer poly[d(A-T)] was selected because it possesses a well-defined uniform sequence unit, its structural and dynamic properties in solution and in the crystalline (fiber) state have been extensively investigated (DAVIES and BALDWIN, 1963; ARNOTT et al., 1974) as has the binding of RNA polymerase (see below), and because the role of A-T rich regions in transcription has been the subject of intensive discussion (BRAM and TOUGARD, 1972; ARNOTT et al., 1974; PILET et al., 1975). As stated above, such sequences have been shown recently to actually exist as important features of the primary binding site for polymerase.

In addition, poly[d(A-T)] is the only synthetic polynucleotide of many tested as templates for RNA polymerase (G. RHODES, personal communication) with which rapid initiation of synthesis is observed in a rifampicin-challenge experiment (MANGEL and CHAMBERLIN, 1974a), a property otherwise characteristic of open promoter complexes[1].

The properties of the stable complexes of RNA polymerase with poly[d(A-T)] have been investigated extensively by filtration (retention on nitrocellulose filters: JONES and BERG, 1966; STEAD and JONES, 1967; ISHIHAMA and HURWITZ, 1969; KRAKOW and v.d. HELM, 1970; SO and DOWNEY, 1970; KRAKOW, 1971; KRAKOW, 1974; STRNISTE and SMITH, 1974), polyacrylamide gel electrophoresis (KRAKOW and v.d. HELM, 1970; KRAKOW, 1971), and by numerous spectroscopic techniques. The binding capacity of the DNA for the enzyme is exceedingly high. In two reports, binding stoichiometries (in units of basepairs/enzyme) of about 200 (KRAKOW, 1971) and 60 (STRNISTE and SMITH, 1974) were reported. The complexes are stable at ionic strengths of about 0.1 or less (SO and DOWNEY, 1970; STRNISTE and SMITH, 1974) and in the absence or presence of a divalent cation (KRAKOW, 1974; STRNISTE and SMITH, 1974). Upon binding, the dimeric holoenzyme dissociates into functionally equivalent monomers (STRNISTE and SMITH, 1974) and if synthesis is not initiated the regulatory σ subunit is retained (KRAKOW and v.d. HELM, 1970). The former property (monomerization) is observed with other DNA-enzyme complexes (SMITH et al., 1967; ANTHONY and GOLDTHWAIT, 1970; BEABEALASHVILLY and SAVOTCHKINA, 1973). Covalent linkages between bound enzyme and residues of poly[d(A-T)] (STRNISTE and SMITH, 1974) or the analogue poly[d(A-S^4T)] (FRISCHAUF and SCHEIT, 1973) are inducible by UV irradiation.

Due to the absence of inherent spectroscopic signals sensitive enough for our purposes, an appropriate reporter molecule was required which would (a) bind to the DNA nonspecifically but not to the enzyme, (b) possess physical properties sensitive to the presence of polymerase, and (c) perturb the system minimally. In addition, the binding had to be noncovalent and susceptible to analysis by relaxation kinetics in order to exploit the capacity of the latter technique for resolving elementary steps in reaction sequences.

The trypanocide ethidium bromide (EB) fulfilled all of our requirements. This planar molecule binds to double helical nucleic acids by "intercalation", a mechanism originally postulated by LERMAN (1961) and FULLER and WARING (1964) and most recently supported by direct X-ray diffraction data (TSAI et al., 1975) and related conformational analyses (ALDEN and ARNOTT, 1975). The fluorescent properties of the free and bound dye have been characterized extensively (LePECQ and PAOLETTI, 1967); for our purposes, the most remarkable and useful effect is the up to 50-fold

---

[1]A potential disadvantage of poly[d(A-T)], shared with all the self-complementary alternating copolymers, is the propensity to form internal loops (hairpins) under certain conditions with a consequent increase in the complexity of equilibrium and kinetic properties (BALDWIN, 1971).

enhancement of the fluorescence upon binding to poly[d(A-T)] (SCHREIBER and DAUNE, 1974). There is an extensive literature concerning other physical properties of ethidium bromide complexes with polynucleotides and oligonucleotides. For example, recent studies demonstrate that the binding of the dye leads to elongation and stiffening of the DNA (FREIFELDER, 1971; LANG, 1971; REINERT, 1973), to unwinding of the helix by an angle per bound molecule estimated at between -17° (ALDEN and ARNOTT, 1975; TICHADOU et al., 1975) and -26° (WANG, 1974; KELLER, 1975; PULLEY-BLANK and MORGAN, 1975), and possibly to other conformational changes as revealed by circular dischroism measurements (AKTIPIS and KINDELIS, 1973; WILLIAMS and SELIGY, 1974).

A number of kinetic investigations have probed the dynamics of ethidium bromide-nucleic acid interactions including the binding to tRNA (stopped-flow and temperature-jump studies, BITTMAN, 1969; TRITTON and MOHR, 1971; TRITTON and MOHR, 1973), the induction of an allosteric transition in poly[d(G-C)] (POHL et al., 1972), and the binding to DNA by fluorescence correlation spectroscopy (MAGDE et al., 1974) and by absorption (BRESLOFF and CROTHERS, 1975) and fluorescence (JOVIN, 1975) temperature-jump relaxation kinetics.

While complex kinetic behavior was observed in many of the above studies, the binding of ethidium bromide to poly[d(A-T)] under certain conditions (low concentrations, sensitive fluorescence detection) is characterized by one relaxation process and has been described formally in terms of a single step reaction (JOVIN, 1975). However, two important points had to be considered carefully before proceeding with the present study. In the first place, ethidium bromide is a known inhibitor of *E.coli* RNA polymerase (WARING, 1964; RICHARDSON, 1973; AKTIPIS and PANAYOTATOS, 1976), apparently due to blocking the formation of some active promoter complexes. Secondly, dye binding is complicated by the phenomenon of site exclusion, i.e. the effect of a bound molecule upon the probability of additional interactions in its vicinity. The first problem was circumvented by utilizing dye concentrations lower than the levels required for perceptible inhibition. In this way, the perturbations of DNA structure were minimized as well. The exclusion effect was treated in detail and constitutes a major topic of this paper.

Our experimental results support the following conclusions: (a) the thermodynamic and kinetic parameters describing the binding of ethidium bromide to poly[d(A-T)] are meaningful only if derived with appropriate consideration of the excluded site effect. We present the formalism for the relaxation kinetic analysis of such a system; and (b) the interaction of enzyme and DNA affects the binding of ethidium bromide in complex ways. The experimental data are compatible with a model according to which each bound enzyme molecule excludes dye binding completely within a small region but modifies the nature of dye-DNA interactions over a larger polynucleotide segment extending beyond the confines of the binding site per se.

II. Experimental Procedures and Data Analysis

A. Materials

Poly[d(A-T)] was synthesized with *Escherichia coli* DNA polymerase I, deproteinized by successive extractions with 0.2% sodium dodecylsulfate in 1 M NaCl and with chloroform-isoamylalcohol (25:1), and dialyzed against 5 mM NaCl, 10 mM EDTA, 2 mM Hepes, pH 7.3. Analytical band sedimentation analysis at neutral pH yielded an $s_{20,w} = 14$ S, corresponding to an average chainlength for the duplex molecule of $3.5 \cdot 10^3$ (STUDIER, 1965). Concentrations of poly[d(A-T)] solutions in nucleotide units were calculated using an $\varepsilon_{260} = 6650\ M^{-1}\ cm^{-1}$ (INMAN and BALDWIN, 1962). DNA concentrations are expressed throghout in terms of nucleotide residues except in discussions of lattice or binding sites, in which case the basepair is taken to be the appropriate unit.

Ethidium bromide was purchased from Serva (Heidelberg) and shown to be pure by thin layer chromatography. Concentrations were determined with the values $\varepsilon_{287} = 5.39 \cdot 10^4$ and $\varepsilon_{460} = 4220\ M^{-1}\ cm^{-1}$ (HSIEH and WANG, 1975).

RNA polymerase was isolated from *E.coli* cells by the method of ARNDT-JOVIN et al. (1975). The holoenzyme was very pure as judged by gel electrophoresis and from the specific activities ($2.4 \cdot 10^4$ and $1.2 \cdot 10^3$ units/mg with poly[d(A-T)] and phage T7 DNA templates, respectively). These values are in good agreement with those previously reported (BERG et al., 1971; BÄHR et al., 1976). Enzyme concentrations were determined by absorbance using the value $E_{1\ cm}^{0.1\%} = 0.65$ at 280 nm (RICHARDSON, 1966) and assuming a molecular weight of $4.9 \cdot 10^5$ (BERG and CHAMBERLIN, 1970). We are indebted to Dr. G. RHODES for the supply of enzyme.

All solutions were prepared and measurements made in a standard buffer: 0.1 M KCl, 10 mM Hepes, 0.1 mM DTT, pH 7.5.

B. Fluorescence Temperature-Jump Measurements. Instrumentation and Conditions

The apparatus for performing relaxation kinetic measurements using fluorescence detection has been described extensively in previous publications (RIGLER and EHRENBERG, 1973; RIGLER et al., 1974; JOVIN, 1975). The standard "Cyclops" cell (JOVIN, 1975) with a heated volume of 0.7 ml and an optical pathlength of 7 mm was used. The excitation wavelength of 313 nm was selected with the low blaze grating of a 1/4-m Jarrel Ash monochromator and 4.5 mm slits. Due to the fact that the observed fluorescence derives virtually from the complexed ethidium bromide alone, the emission anisotropy is independent of the fractional saturation and thus no polarizing elements were used either in the excitation or in the emission paths (see discussion in JOVIN, 1975). The emitted fluorescence was isolated with OG 550-1 low-pass cutoff filters (Schott, Mainz) which have 50% transmission at 550 nm.

The initial temperature within the cell was 21.5°C or, where indicated, 22.5°C. A rapid jump of 3.8° to a final temperature of 25.3°C was achieved by a 20 KV discharge from a 52 nF capacitor. Thermal requilibration was complete within 3 min and jumps could be repeated at 3 - 5 min intervals.

In the case of experiments performed in the presence of enzyme, the solutions containing different concentrations of DNA and ethidium bromide were prepared beforehand and RNA polymerase was added a few minutes before use. Incorporation assays demonstrated no loss of activity in solutions of enzyme (25 nM) and poly[d(A-T)] (3 μM) brought from 0°C to 22°C and "jumped" either 0, 1,2 or 5 times. Thus the enzyme was stable under the selected experimental conditions.

C. On-Line Computer Acquisition of Relaxation Data

The analog signal conditioning unit of the relaxation kinetic apparatus is connected to a PDP 11/20 minicomputer (Digital Equipment Corp., Maynard) for real-time data acquisition (JOVIN, 1975). The system is under software control by MIDAS (Modular Interrogative Data Acquisition System) which consists of a system of input/output data handling routines, an interrogative monitor ASK, and sets of conventions for storing and displaying data.

The actual measured signal (in these studies) is the quotient of the detected fluorescence signal to a reference derived from the excitation beam. Fluctuations in lamp intensity thus are compensated to a considerable degree.

There are two signal paths to the computer. The slow path, under control of ASK, is used to measure the "total signal" (fluorescence) and other static signals. The fast data path is triggered by the high voltage discharge after a selectable delay during which the DC component from temperature-dependent physical parameters such as quantum yield is suppressed. The resultant relaxation signal is transmitted by a coaxial line driver to an analog-digital converter connected to a direct-memory interface. The bandwidth of the recorded signal is limited by a single stage low pass filter. The rise time in these studies was set usually to 50 μs or in some instances to 0.5 ms.

The time course of data acquisition for each individual temperature jump is programmed so as to maximize the information content distributed over the 5 - 6 orders of magnitude of time between the discharge and the onset of themal cooling within the cell (0.5 - 1 s). Five "tiers" of data each consisting of 2K (2048) stored points are collected. The first 4K are successive measurements which follow the initial trigger at a frequency selected by an external pulse generator (usually set to 0.5 MHz) and are stored in direct memory access (DMA) mode. The next three tiers record data over successive periods of 0.041, 0.164, and 1.31 sec, respectively. These data are acquired under program (i.e. processor) control. Individual measurements are made continuously at a base frequency of 50 KHz, i.e. at intervals of 20 μs, but the numbers actually stored in memory consist of averages of n measurements where n in the three lines is 1,4 and 32. In this manner, the degree of data compression and digital filtering increases throughout the course of the experiment. The total signal and excitation reference are recorded automatically at two times, the first at the moment of opening the shutter admitting light to the cell and the second coincident with the discharge initiating the temperature-jump. A short sequence of measurements immediately prior to the jump is recorded as well.

The data set, together with numerous parameters specifying the experimental conditions, is stored on a suitable medium (magnetic tape, disc) and analyzed either with the same computer or (in the present instance) with a UNIVAC 1108 system and procedures described below.

D. Analysis of Relaxation Curves by the Method of Modulating Functions

The analysis of relaxation data involves the resolution of a spectrum generally consisting of a discrete set of negative exponentials. For n components, there exist 2n + 1 parameters to be derived: n relaxation times, n amplitudes, and one "baseline" (the value of the signal at final equilibrium, which is time-invariant in the case of a step function temperature perturbation).

In addition to the problem of noise, which is inherently present in electronic measurements of light intensities, the difficulty in analyzing data of this kind is the notorious nonorthogonality of exponential functions. It is this latter problem that has led many authors, including the present, to avoid the (non-linear) least-squares methods that are generally used to resolve experimental data.

Many experimental techniques other than relaxation kinetics give rise to exponential decays. We shall not review the vast literature on the subject but briefly describe the method used here, which was developed initially for the analysis of fluorescence decay data, and will be published shortly in that connection (STRIKER, in preparation).

LOEB and CAHEN (1963) discuss a general approach for determining the parameters of a system described by a set of differential equations from measurements on that system. The method requires a sufficient number of data points so as to allow integration, but only in a finite time range (as against transform methods, which generally require data to infinity). Due to the process of integration, the curves are permitted to be quite noisy, provided the noise is of a random nature and of sufficiently high frequency so as not to severely distort the shape of the curves.

In the case of perfectly smooth data corresponding to a process described by a set of linear differential equations, it is in principle possible to perform numerical differentiations and solve the resulting equations by ordinary least squares analysis. It is generally impossible, however, to differentiate noisy (real) data. LOEB and CAHEN avoid this problem by multiplying the differential equation term by term by an analytical "modulating function" and integrating the result. Integration by parts transfers the differentiation from one function (the measured variable) to another (the analytical variable). By choosing the latter so that it and a sufficient number of derivatives vanish at the two extremes of the data set, the constant of integration also vanishes. Iteration of this process leaves an equation involving nothing but integrals of the measured data multiplied by some derivative of the modulating function (which is chosen so as to be easily differentiated). A sufficient set of such equations yields the solution for the parameters (the coefficients of the equations). Such a set can be obtained by choosing enough modulating functions.

This method was first applied theoretically to the analysis of multiexponential decay of fluorescence by VALEUR and MOIREZ (1973). The differential equation for exponential decay convoluted with a forcing function was treated, and using noise-free simulated data surprisingly good analyses were shown for curves containing up to three components. The method has been in use in our laboratory for some time, but a number of modifications have been found necessary in dealing with actual experimental data. A major problem is that the matrix representing the derived equations tends to be close to singular and its inversion can produce random solutions. In order to obtain meaningful results, at least for multiexponential data, analysis of the matrix for singularity and computation of pseudoinverses have been found necessary. When the data permit, it is also possible to overdetermine the linear system so as to get a better definition of the solution space. The additional modification of the original method required in order to account for a finite baseline is straightforward and has the effect of introducing an additional coefficient in the linear system.

For the data structure given by the MIDAS system described above, it is expedient to set up a system of equations for each of the tiers to be used in the analysis. (The fastest tier was generally excluded, because of convolution difficulties, and the slowest had no relaxation information but showed, rather, a strong cooling effect. A great deal of baseline information was contained in the fourth tier.) The resulting matrices are transformed so as to correspond to a unique time base. This overdetermined system is solved using singular value analysis.

The solution of the linear equations yields the coefficients of a polynomial whose roots correspond to the relaxation times. The amplitudes, which are not parameters of the differential equations in the nonconvoluted case, can then be determined by simple linear analysis. The results are plotted, along with magnified plots of the deviations of the solution from the experimental values, and the autocorrelation of these deviations are also calculated and plotted (see later in Fig. 3). The autocorrelations play an important role in the acceptance or rejectance of an analysis since an incorrect analysis can still have a low root mean square error due to the nonorthogonality of the exponential decay functions (see GRINVALD and STEINBERG, 1974).

III. Excluded Site Binding of Ethidium Bromide to Poly[d(A-T)]

A. Theory for the Equilibrium State

The theoretical analysis of systems (at equilibrium) comprised of large ligands binding cooperatively or noncooperatively to one-dimensional homogeneous lattices was reviewed and extended by McGHEE and VON HIPPEL (1974), with particular emphasis on the relevance to protein-nucleic acid interactions. Other more recent publications also deal with these questions (SCHELLMAN, 1974, 1975; WYMAN, 1975).

The binding isotherms observed with complexes of ethidium bromide and DNA have been interpreted quantitatively by the neighbor exclusion model (BAUER and VINOGRAD, 1970; BRESLOFF and CROTHERS, 1975). In the case of an intercalating dye interacting with a homogeneous synthetic polynucleotide like poly[d(A-T)], every basepair constitutes a potential binding site and all bound states can be considered to be thermodynamically equivalent. The neighbor exclusion concept states that the act of binding alters the polynucleotide structure in the vicinity of the complex so as to prevent binding to what otherwise would be additional potential sites. The equilibrium distribution of the system is given by

$$r/L = K \cdot f(r) \qquad (1)$$

where r is the fractional saturation of total potential sites (the latter quantity being defined for the naked lattice in the absence of binding), K is the equilibrium association constant for complex formation at an individual site, L is the concentration of free ligand, and f(r), a function, denotes the fraction of total potential sites available for additional binding. Thus by definition and by the application of the conservation of mass condition for the ligand,

$$r = X/S_o \qquad f(r) = S/S_o \qquad L = L_o - X \qquad (2)$$

where X, L_o, and S_o are the concentrations of complex, total ligand, and total potential lattice sites, respectively.

The function f(r) has the following general form (McGHEE and VON HIPPEL, 1974)

$$f(r) = \frac{[1 - nr]^n}{[1 - (n-1)r]^{n-1}} \qquad (3)$$

$$f(r) = 1-r \qquad \text{for } n=1 \qquad (3a)$$

$$f(r) = (1-2r)^2/(1-r) \qquad \text{for } n=2 \qquad (3b)$$

where n is the number of consecutive lattice sites directly blocked by the ligand. It can be shown that the number of potential binding sites which are blocked by the binding of an isolated ligand molecule is equal to 2n-1 (see discussion in McGHEE

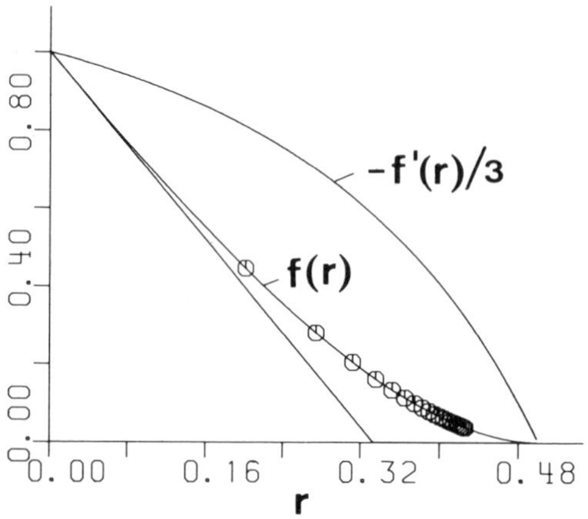

Fig. 1. Theoretical plots of functions f(r) and f'(r) for excluded binding with n=2. Abscissa r is the fractional saturation of lattice sites. Straight unlabeled line: f(r) for nonexcluded binding (n=1) to adjacent nonoverlapping sites consisting of three lattice units, a case approximated by excluded binding (n=2) at low saturation. Symbols on the middle curve: sample calculations using total ligand concentrations of 1-20 μM in equal unitary increments. Total lattice concentration and dissociation constant are 0.5 and 1.7 μM, respectively

and VON HIPPEL, 1974). The function for n=1 [Eq. (3a)] taken together with Eq. (1) constitutes the usual Scatchard representation of binding to independent, equivalent, noninteracting sites (SCATCHARD, 1949).

Not incorporated in Eq. (3) is an additional parameter which can account for cooperativity (positive or negative) in the ligand-ligand interactions. Ethidium bromide does not demonstrate cooperativity in binding to DNA and thus this element of the model is omitted in the present discussion (the general equations are given by McGHEE and VON HIPPEL, 1974).

In the case of ethidium bromide, the parameter n in Eq. (3) appears to be 2 (BAUER and VINOGRAD, 1970; BRESLOFF and CROTHERS, 1975). Thus the quantity -(2n-1) has the value -3 which implies that both basepairs adjacent to a bound molecule are blocked as binding sites for additional ligand molecules. It also follows that the value of r achieved at saturation (infinite concentration of free ligand) is given by 1/n, or 0.5 for ethidium bromide.

These features of Eqs. (1) and (3) are depicted graphically in Figure 1. It is seen that the effect of n>1 is to produce a curved binding isotherm which can be (and often is) misinterpreted as providing evidence for heterogeneous binding, i.e. the presence of "strong" and "weak" binding sites (BLAKE and PEACOCKE, 1968). The limiting values of f(r) for r = 0 and 1/n are 1 and 0, respectively. The derivative of f(r) is given by

$$f'(r) = \frac{[nr(n-1)-2n+1]\ (1-nr)^{n-1}}{[1-(n-1)r]^{n}} \tag{4}$$

$$f'(r) \equiv -1 \qquad \text{for } n=1 \tag{4a}$$

$$f'(r) = \frac{(2r-3)(1-2r)}{(1-r)^{2}} \qquad \text{for } n=2 \tag{4b}$$

The value of $f'(r)$ at $r = 0$ is the initial slope of the binding isotherm (Fig. 1) and equals $-(2n-1)$, i.e. -3 for $n=2$, as already stated above. If the initial linear portion of $f(r)$ is extrapolated to the abscissa (r), the intercept is $1/(2n-1)$ or 0.33, a value typically observed in ethidium bromide binding.

The curvature observed in the functions $f(r)$ and $f'(r)$ (Fig. 1) reflects the increasing difficulty of binding more ligand as the degree of saturation increases. This effect is due to the creation of "gaps" (stretches of lattice between bound ligands) which are smaller than the minimum size required for the ligand, namely n residues. Only through the energetically unfavorable redistribution of the complexes can potential binding sites be "rescued" by amalgamation of small gaps into larger ones. Thus, each binding event reduces the number of potential binding sites by a number $(-f'(r))$ dependent upon the current degree of saturation of the system, r. The limiting values (for $n=2$) are 3 for a totally naked lattice and 0 for one virtually filled.

B. Theory for Relaxation Kinetic Behavior

We postulate that a single pair of rate constants represent the kinetics of binding to a given potential binding site. That is, all recombination and dissociation events are held to be kinetically as well as thermodynamically equivalent. Thus if binding (in our case, intercalation) is represented as a single step reaction, the appropriate differential equation is

$$\dot{X} = k_1 L\, S - k_{-1} X \qquad K = k_1/k_{-1} \tag{5}$$

where k_1 and k_{-1} are the forward and reverse rate constants, respectively, and S is the instantaneous concentration of free potential binding sites on the lattice.

Relaxation experiments are conducted under the condition of small perturbation from a state of chemical equilibrium so that terms in the differential equations involving higher orders of the concentration deviations can be neglected (EIGEN and DeMAEYER, 1974). We thus introduce the following identities into Eq. (5) for the purpose of deriving a linearized rate equation:

$$L = \bar{L}+\delta L, \qquad X = \bar{X}+\delta X, \quad \text{and} \quad S = \bar{S}+\delta S. \tag{6}$$

Then

$$\delta\dot{X} = k_1(\bar{L}\delta S+\bar{S}\delta L) \;\; -k_{-1}\delta X \tag{6a}$$

where $\bar{L}$, $\bar{X}$, and $\bar{S}$ are the equilibrium concentrations and the symbols without the bars are the instantaneous time-dependent values. To effect a solution of Eq. (6a), we require relations between the perturbation variables δL, δX, and δS. Conservation of ligand dictates that

$$\delta L = -\delta X \tag{7}$$

The relationship between δS and δX, however, is not straightforward due to the fact that binding one ligand molecule removes a variable number of potential binding sites from the system (Eq. (4)).

We proceed by noting that the quantity S can be defined in a completely general way in terms of the population of "gaps".

$$S = \sum_{n}^{\infty}(i-n+1)G_i \tag{8}$$

where G_i is the concentration of gaps containing i unoccupied lattice sites and the factor (i-n+1) is the number of potential binding sites for a gap of size i (McGHEE and VON HIPPEL, 1974). We now introduce the following two conservation conditions into Eq. (8).

$$\sum_{o}^{\infty}G_i = X \tag{9}$$

$$\sum_{1}^{\infty}iG_i + nX = S_o \tag{10}$$

Eq. (9) expresses the fact that all gaps necessarily start with a bound ligand (neglecting end effects. See McGHEE and VON HIPPEL, 1974). Thus,

$$S = S_o-(2n-1)X + \sum_{o}^{n-1}(n-i-1)G_i \tag{11}$$

and by differentiating,

$$\delta S = -(2n-1)\delta X + \sum_{o}^{n-1}(n-i-1)\delta G_i \tag{12}$$

$$\delta S = -\delta X \qquad \text{for } n=1 \tag{12a}$$

$$\delta S = -3\delta X + \delta G_o \qquad \text{for } n=2$$

Eqs. (11) and (12) hold under all equilibrium and dynamic conditions and express the essential features of the coupling between the binding and distribution functions for the ligand-lattice interaction. It is seen that only for n=1 does there exist a constant linear relationship between δS and δX (Eq. (12a)). For n>1, however, the number of free sites depends not only upon the amount of bound ligand but upon its localization along the lattice as well. Consider, for example, two adjacent gaps of length 1 (and let n=2). If the central liganded molecule is moved to the left or to the right by one unit, there results a net increase in the number of available binding sites by 1. In fact, this is true of all exchange reactions of the type

$$G_i + G_j \leftrightarrow G_k + G_o \qquad i,j>0 \quad \text{and} \quad k=i+j.$$

The interrelationship between the differential equations for X and G_i, i.e. $\delta\dot{X}$ and $\delta\dot{G}_i$, given by Eq. (12) implies that the relaxation kinetics for any system with n>1 will consist of many components, i.e. many normal modes. This point will be considered further but for the moment, we proceed with the analysis by seeking an alternative expression to Eq. (12). We note that for small deviations from equilibrium, it follows from Eq. (2) that

$$\delta S = f'(\bar{r}) S_o \delta r = f'(\bar{r}) \delta X \quad (13)$$

We now make the assumption, increasingly valid for low r and thus more constant f'(r) (see Fig. 1), that the relationship between δS and δX in Eq. (13) holds throughout the process of relaxation. That is, the equilibrium distribution of bound ligand is always maintained. The relaxation equation is obtained by introducing the identities 6,7, and 13 into Eq. (6a).

$$\delta \dot{X} = -\lambda \delta X \quad (14)$$

where in the reciprocal relaxation time,

$$\lambda = k_1 [S - f'(r) L] + k_{-1} = \tau^{-1} \quad (15)$$

$$\lambda = k_1 [S + L] + k_{-1} \qquad \text{for } n=1 \quad (15a)$$

the bars have been omitted from the equilibrium concentrations for the sake of simplicity, a practice adhered to below as well.

Under these conditions, therefore, there exists a single relaxation time for the system and the quantity $-f'(r)$ (Eq. (4)) appears as a weighting factor for the free ligand concentration (Eq. (15)). For n=1, and only for n=1, $-f'(r) \equiv 1$ (Eq. (4a)) and Eq. (15) reduces to the usual form (Eq. (15a)) for recombination reactions involving independent, equivalent, noninteracting sites. The latter circumstance also arises in the limiting case of r approaching 0, for which

$$\lambda \simeq k_1 (2n-1) [S/(2n-1) + L] + k_{-1} \quad (16)$$

$$\lim_{r \to o} \lambda = k_1 S_o + k_{-1} \quad (16a)$$

where k_1 and S are still defined in basepair units. In physical terms, Eq. (16) represents the non-excluded binding of ligand to non-overlapping sites consisting of three lattice units (for n=2; see also Fig. 1). The reciprocal relaxation times according to both Eqs. (15) and (16) are simulated in Figure 2. It is seen that the effects of excluded binding are to extend the region of ligand concentration over which the relaxation time initially decreases and to reduce the limiting slope at high concentration.

The relaxation amplitude ΔP corresponding to the single step reaction (Eqs. (5) and (14)) is given by

$$\Delta P = \Delta\phi \Delta \bar{X} = \Delta\phi \Gamma \Delta \ln K \quad (17)$$

where $\Delta\phi$ is the specific signal change associated with the binding of 1 M unit of ligand, $\Delta\bar{X}$ is the enforced change in the equilibrium concentration of X, and Γ, the chemical compliance, is given by $\Delta\bar{X}/\Delta \ln K$. (For extensive discussions of relaxation amplitudes see THUSIUS et al., 1973; JOVIN, 1975; EIGEN and DeMAEYER, 1974.) For the temperature-jump,

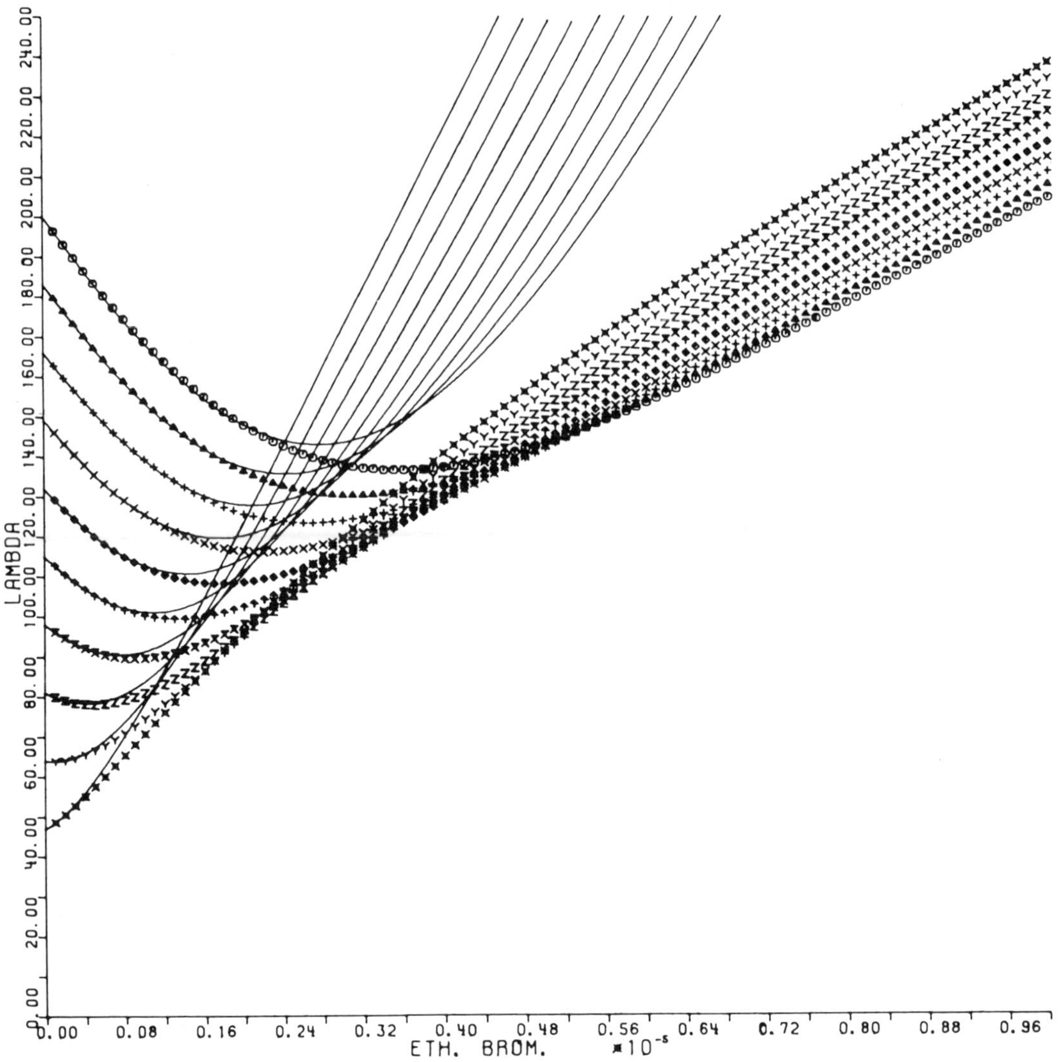

Fig. 2. Simulation of reciprocal relaxation times for excluded and nonexcluded binding as a function of ligand and lattice concentrations. Abscissa: total ligand (ethidium bromide) concentration from 0–10 μM. Ten curves: DNA concentrations from 0.5 to 5 μM (in basepair units) with the lowest concentration showing the lowest intercept (see Eq. (6a)). Curves with symbols: excluded binding with n=2 according to Eq. (15). Smooth lines: excluded binding according to Eq. (16) also evaluated with n=2 (alternatively, Eq. (15a) (n=1) can be used if concentration of sites is given in units of three basepairs and apparent recombination rate constant is set equal to $3k_1$). Kinetic constants used are those for ethidium bromide binding to poly[d(A-T)] given in Table 1

$$\Delta \ln K = (\Delta H^{\circ}/RT^2)\ \Delta T \qquad (18)$$

where $\Delta \ln K$ is the enforced change in the equilibrium constant induced by the shift in temperature ΔT due to the finite reaction enthalpy ΔH°.

The amplitude factor Γ in Eq. (17) is obtained most easily by differentiation of Eq. (1) which yields

$$\Gamma = \Delta\bar{X}/\Delta \ln K = \left[-f'(r)/S + 1/L + 1/X\right]^{-1} \tag{19}$$

It is interesting that $f'(r)$ appears in association with S in Eq. (19) but together with L in Eq. (15).

The last quantity required to fully characterize the relaxation process is the total signal, i.e. a measure of the unperturbed equilibrium state. It follows from the definition of $\Delta\phi$ that

$$\Delta P_o = P - P_o = \Delta\phi \cdot X \tag{20}$$

where P is the actual measured signal and P_o is the algebraic sum of the individual signals attributable to free ligand (L_o) and free lattice (S_o) in the absence of association (JOVIN, 1975). In the case of a single fluorescent species, e.g. the ethidium bromide, P_o is simply the signal derived from the fully unbound ligand. To be consistent, however, ΔP_o should be defined for the temperature achieved after the jump, inasmuch as that is the state under which the relaxation takes place and for which the rate and physical constants are valid. Thus

$$\Delta P_o = a\ \Delta P_o' + \Delta P \tag{21}$$

where a is a factor which includes all changes in the physical parameters (e.g. $\Delta\phi$) due to heating, $\Delta P_o'$ is the total signal observed at the initial temperature and ΔP is the relaxation amplitude (Eq. (17)).

By combining Eqs. (15), (17), (19), and (21) it can be shown that

$$\Delta P/\Delta P_o = k_{-1} \Delta \ln K \cdot \tau \tag{22}$$

This quantity, named the "relative amplitude", is very useful because it provides a unique linear relationship between experimental amplitude factors and the experimental relaxation times valid for all measurements, regardless of concentration, and thus is independent of any calculations performed on the data (JOVIN, 1975). From the slope and k_{-1} determined from the relaxation time one obtains $\Delta \ln K$ and thus ΔH^o (Eq. (18)) directly. The degree of concordance of the data with Eq. (22) can be considered to constitute a qualitative test of the postulated mechanism as well.

In summary, the three canonical equations which characterize the system are

$$\lambda = \tau^{-1} = k_1\left[S - f'(r)\ L + K\right] \tag{23a}$$

$$\Delta P = \Delta\phi \Delta \ln K\left[-f'(r)/S + 1/L + 1/X\right]^{-1} \tag{23b}$$

$$\Delta P_o = \Delta\phi\left[X\right] \tag{23c}$$

plus the interrelationship given by Eq. (22). The four physical, thermodynamic, and kinetic parameters to be determined experimentally are $\Delta\phi$, $\Delta \ln K$, k_1, and k_{-1}.

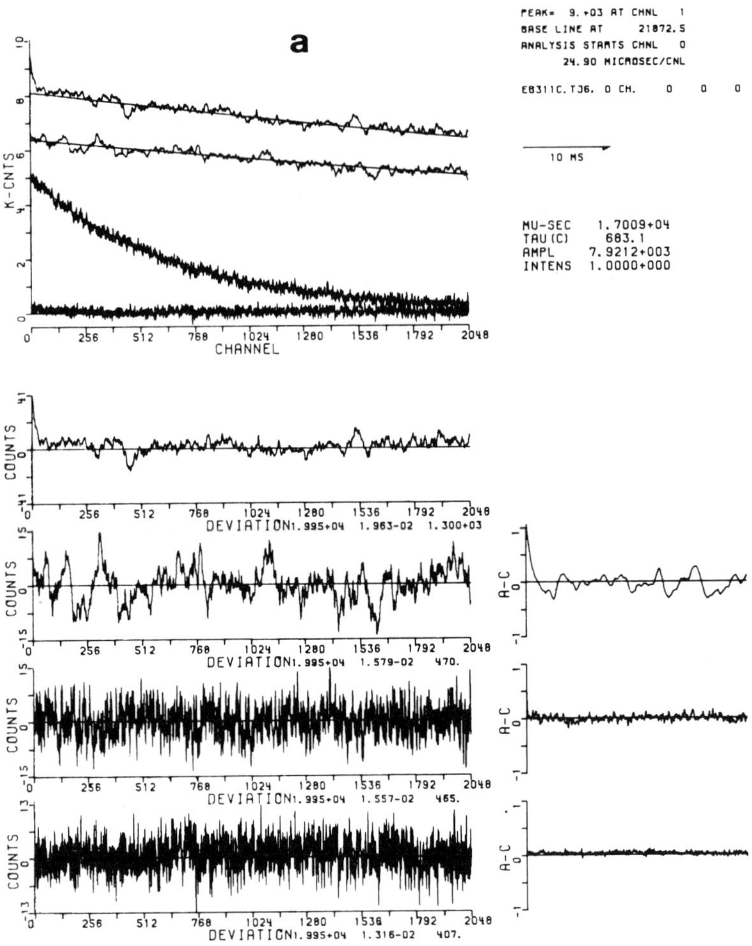

Fig. 3a-c. Examples of experimental relaxation curves and their analyses. Uppermost portion shows four of the five tiers of 2048 measured points each. Arrow: 10 ms in third tier, which in these experiments contained the most information. Theoretical curve for the analysis is drawn as smooth line through data. Below are the four magnified deviation plots, showing difference between measured and theoretical points. On right: autocorrelation plots of these deviations which greatly help in distinguishable between random and systematic error. (a) Example of a good analysis. Note the autocorrelation of third line. (b) Should be rejected, as there is a definite trend in autocorrelation function, indicating that analysis was not successful. Had the fit been forced to a single component, one might suspect the presence of additional exponentials. Analysis was for three components, however, and the program indicated the presence of only one. (c) Data obtained with a longer rise-time filter (0.5 ms instead of 0.05 ms in curves a and b). Curve becomes a great deal smoother, but early convolution effect increases, to the extent that it becomes difficult to use record without explicitly accounting for convolution in analysis

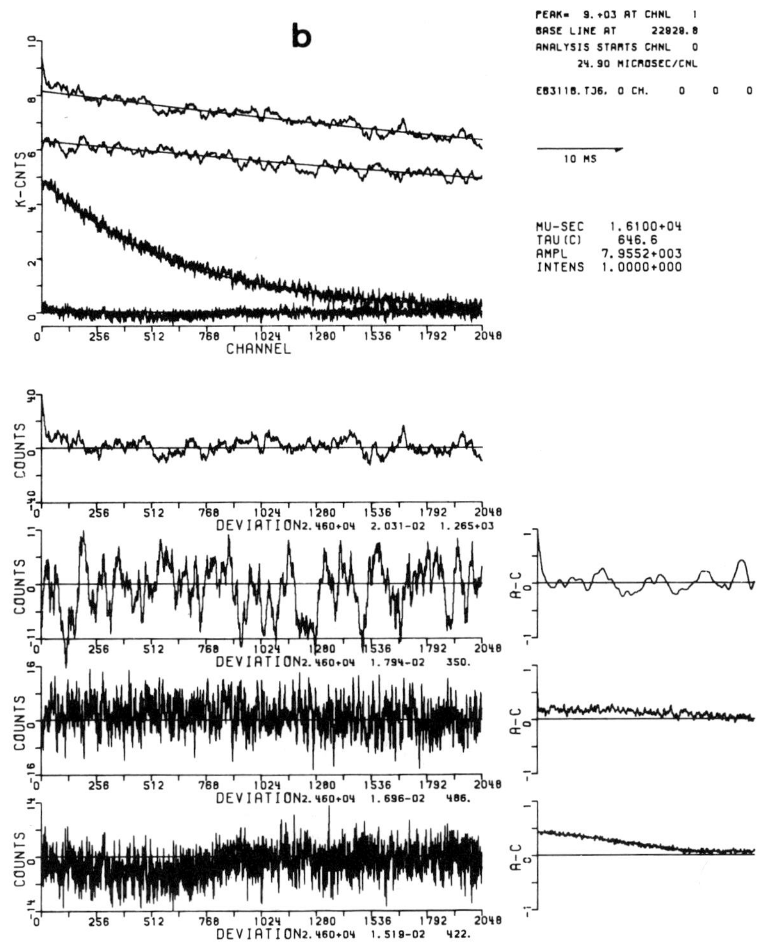

Fig. 3b

The above formulation constitutes an approximate solution for the relaxation kinetic behavior of a system with excluded binding. A complete theoretical and experimental analysis will be presented elsewhere (STRIKER and JOVIN, in preparation). A preliminary derivation and discussion of the coupled differential equations required for the general case, however, is given in the Appendix.

C. Experimental Results and Analysis

The binding of ethidium bromide to poly[d(A-T)] was studied by fluorescence relaxation kinetics with two series of solutions containing 0.4 μM dye and 0 - 11.3 μM polymer at two different temperatures, 21.5 and 22.5°C. As in previous studies with this system (JOVIN, 1975), relaxation curves in the direction of decreased fluorescence (less binding) were observed (Fig. 3).

In general, several T-jumps were made with two separate aliquots of the given solution. For each jump, the total signal was re-

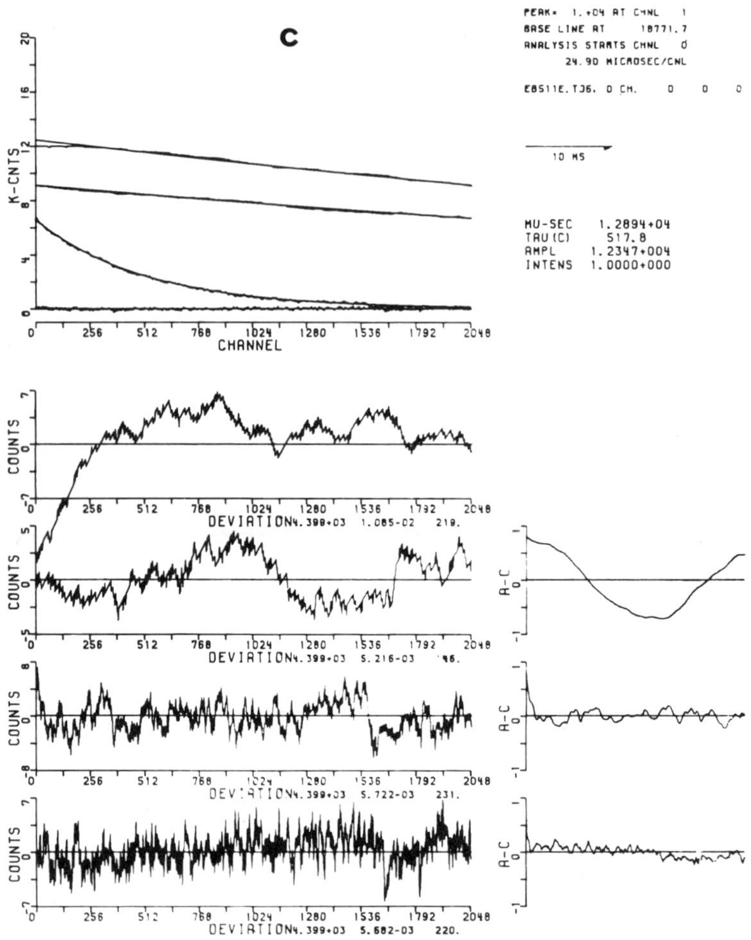

Fig. 3c
Legend see p. 260

corded. Over the extended period of time required to complete the series of measurements, however, instrument drift was sufficiently large so as to make a quantitative comparison between the various members of the concentration set difficult. Thus, enough solution was prepared so that at the end of the experiment, the entire series could be measured (in terms of the total signal) in the T-jump cell within the interval of a few minutes, thereby establishing a calibration. The experimental relaxation amplitudes were then corrected by the multiplicative factor $S_i^f/S_{i,j}$ where S_i^f was the calibration value for solution i and $S_{i,j}$ was the total signal measured for the jth T-jump of solution i (i.e. at the time of the experiment).

All of the data were analyzed using the previously discussed procedure both for three exponentials and for a single component. In a great majority of cases, the three-exponential analysis immediately indicated a single exponential (two of the roots being imaginary or negative). In some of the relaxation curves one or two additional very short times were observed (always of opposite amplitudes in the case of two). As is apparent in Figure 3c, they arise from the convolution effect occuring in most of the relaxations within the first millisecond (or over longer times when the longer risetime filter was employed). In no case were two plausible relaxation times discerned.

In order to minimize the effect of the convolution, the early parts of the curve were deemphasized by reducing the number of modulating functions in the overall analysis. From the analytical point of view, it would be preferable (and mathematically feasible) to introduce a convolution term, but this would require more detailed information as to the nature of the forcing function.

The two resulting series of relaxation times, amplitudes and total signals were kept in mass storage and entered as data into the further analysis programs. The kinetic rate constants were determined using the first of the canonical equations (Eq. (23a)) for the inverse relaxation time, and the other equations (Eqs. (23b,c)) were analyzed for the respective linear constants.

Each of the three canonical equations independently yield by regression analysis a value for the association constant K and some additional value: k_1 (and thus k_{-1}) for the case of the λ, $\Delta\phi\Delta\ln K$ for the amplitudes, and $\Delta\phi$ for the total signal. We note, however, that the analysis for these two variables is in all cases separable. Given a value for K, all of the expressions in brackets (Eqs. (23a-c)) can be evaluated after obtaining r by Newtonian iteration[2] on the mass action equilibrium equation (Eq. (1)) expressed in terms of total concentrations

$$Kf(r)(L_o - rS_o) = r = X/S_o \tag{24}$$

with f(r) given by Eq. (3). The remaining variable in each of the equations is just a multiplicative factor that can be obtained by ordinary linear analysis.

Thus, in place of the process of multivariate nonlinear regression, such as suggested for the solution of Eq. (15a) by VICTOR et al. (1973), it seems advisable to introduce any simple regression method for one variable, e.g. Newtonian or interval reduction, followed at each step by a linear analysis and error determination. Convergence is more certain and more rapid, as one is at least assured of an optimal value of the second parameter for any value chosen for the first. Proper weighting of the dependent variables is essential because inverse values are often involved. It is possible to use the theoretical weighting (e.g. τ^4 or τ^2 in the case of the variable λ, depending upon the assumption of constant absolute or relative error in the measurement of τ), or alternatively, the inverse of the variance of λ derived from individual experiments. We adopted the latter procedure in this study.

---

[2]In the case of n=1, r is determined directly by solving the usual quadratic mass action equation.

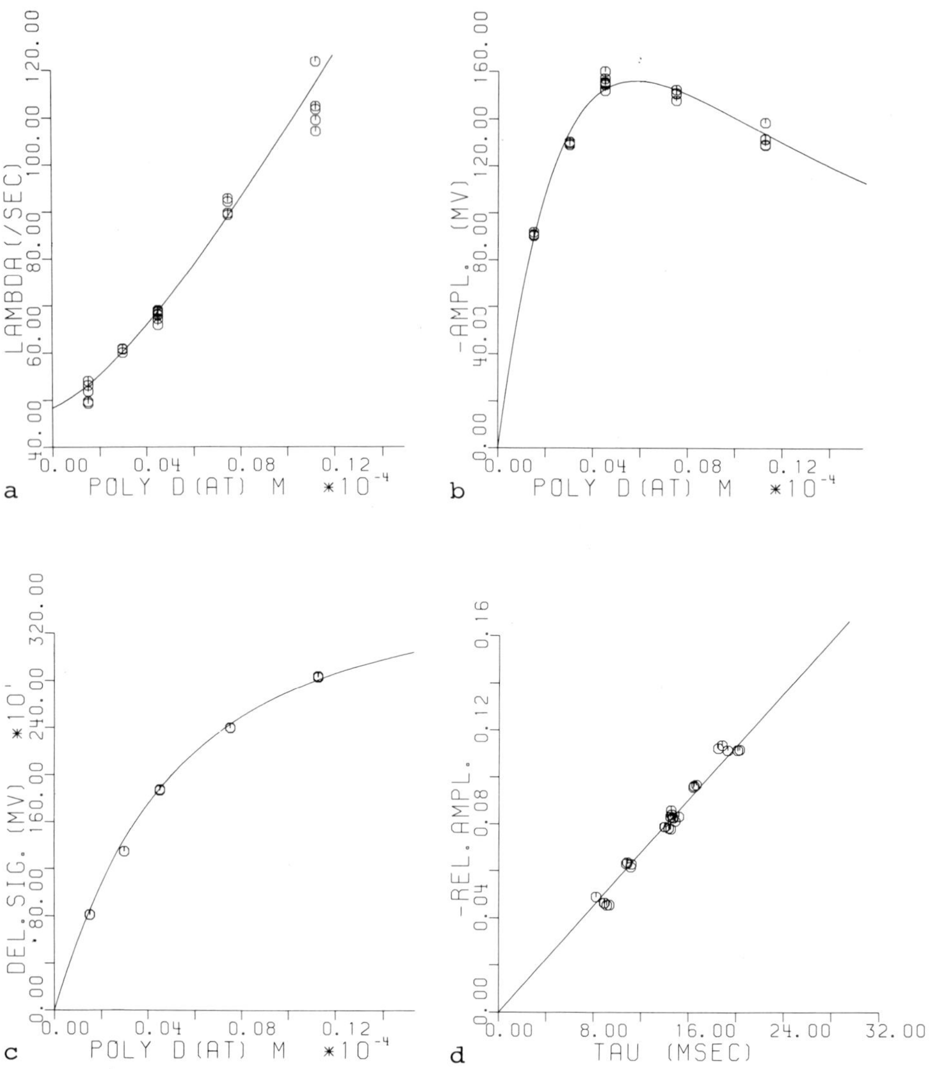

Fig. 4a-d. Relaxation kinetic analysis of ethidium bromide binding to poly [d(A-T)] using fluorescence detection. Experimental conditions are given in the text. (a) reciprocal relaxation time; (b) relaxation amplitude; (c) difference fluorescence signal (Eq. 20); (d) relative amplitude (Eq. 22). Abscissa in (a-c): total DNA concentration in nucleotide units. Total dye concentration was 0.4 μM throughout. Analyses and simulations were made according to Eqs. (23a-c) as described in text

In the course of data reduction, the need for some sort of systematic correction became evident. Although good consistent fits were obtained for both series of experiments with regard to the relaxation times and total signals, the relative amplitude plots were straight lines crossing the x-axis slightly to the right of the origin, and the amplitude plots also showed systematic deviations. It could be shown that all these effects were compatible

Table 1. Kinetic and thermodynamic constants for ethidium bromide binding to poly[d(A-T)] [a]

k_1 [b]	$1.7 \cdot 10^7\ M^{-1}s^{-1}$
k_{-1}	$30\ s^{-1}$
K [b]	$5.7 \cdot 10^5\ M^{-1}$
ΔH^o	-8.6 kcal mol^{-1} [c]
$\Delta\phi$	12 [d]

[a] In 0.1 M KCl, 10 mM Hepes, 0.1 mM DTT, pH 7.5 at 25.3°. See text for definitions of terms. No confidence intervals are given, inasmuch as the problem of their determination is still under study. The use of error estimates from the least linear regression is a misrepresentation (see HAMILTON, 1964).

[b] Concentrations units are in terms of basepairs.

[c] $\Delta \ln K$ was -0.184 and ΔT was 3.8°.

[d] This relatively low value for the fluorescence enhancement upon binding is attributed to the choice of excitation wavelength and broadband detection. However, the selected conditions maximize the absorption of light and the intensity of the detected fluorescence, thus improving the quality of the relaxation data.

with the assumption of a nonlinearity in the relaxation amplitude measurement of the form $\delta \ln \Delta P = -\gamma\ V$ where V is the measured fluorescence signal (0 - 3.5 volts). A value of γ was found (0.08) which led to consistent results for all the data (the source and nature of this presumed nonlinearity are under study). The experimental and simulated curves for one series of measurements is shown in Figure 4.

The final parameters obtained for the system are given in Table 1 and agree reasonably well with previous results (JOVIN, 1975) obtained under conditions where the excluded binding phenomenon would not have been very evident (large excess of polynucleotide). One might wish to reconsider the interpretations of BRESLOFF and CROTHERS (1975), who omitted the influence of excluded binding in their kinetic analysis of data obtained at much higher ligand concentrations (see Eqs. (15), (16) and the associated discussion, and the Appendix).

IV. Relaxation Kinetics of Ethidium Bromide and Poly[d(A-T)] in the Presence of RNA Polymerase

A. Experimental Conditions

Increasing amounts of enzyme were added to a stock solution of standard buffer containing 6.2 μM poly[d(A-T)] and 0.88 μM ethidium bromide. As a consequence, the concentrations of the latter two components diminished slightly (0 - 3%) due to volume dilution,

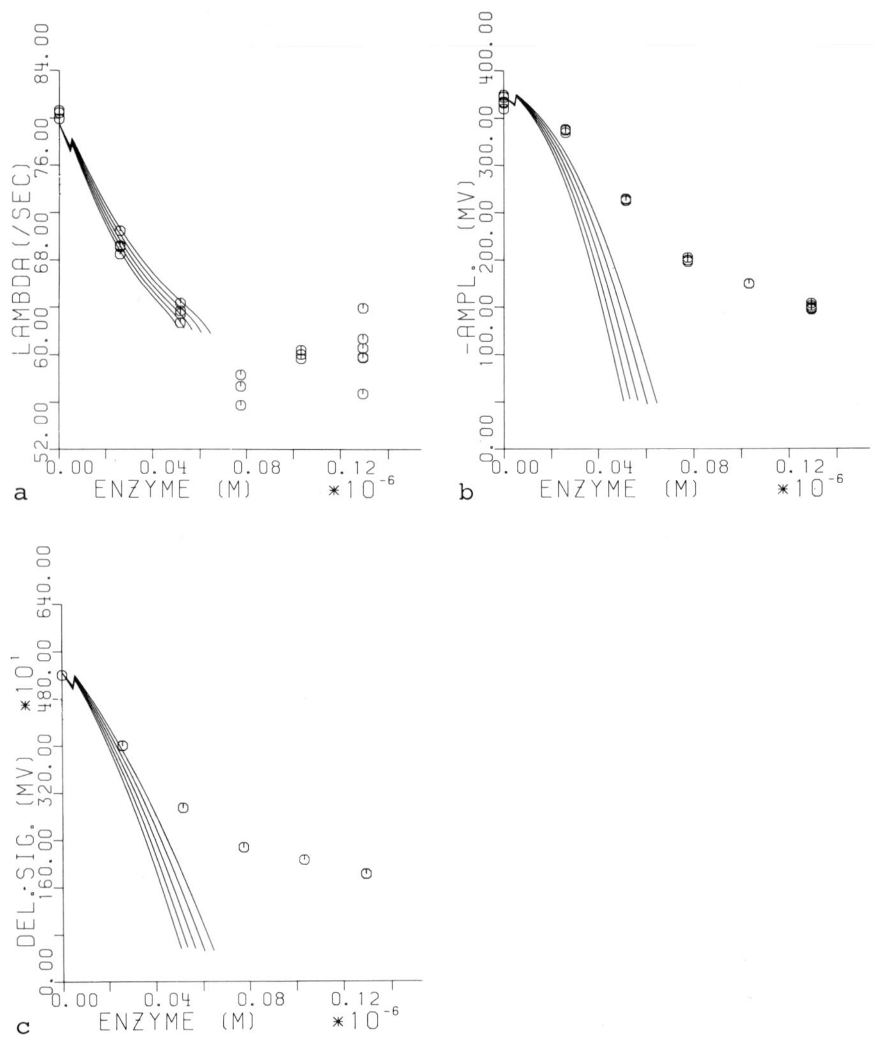

Fig. 5a-c. Relaxation kinetic data for ethidium bromide binding to poly [d(A-T)] in presence of RNA polymerase. (a) reciprocal relaxation time; (b) relaxation amplitude; (c) total signal. Curves were generated according to Model I and correspond in descending order to blocking by enzyme (parameter m) of 43, 46, 49, 52, and 55 basepairs. Discontinuities near origin of simulated curves in this and other figures reflect non-monotonic changes in total ligand and DNA concentrations for the various solutions

an effect which was incorporated into the simulations (see also legend to Fig. 5).

B. Experimental Results, Analysis, and Model Fitting

The relaxation data from experiments performed in the presence of enzyme were analyzed by the same procedure described previ-

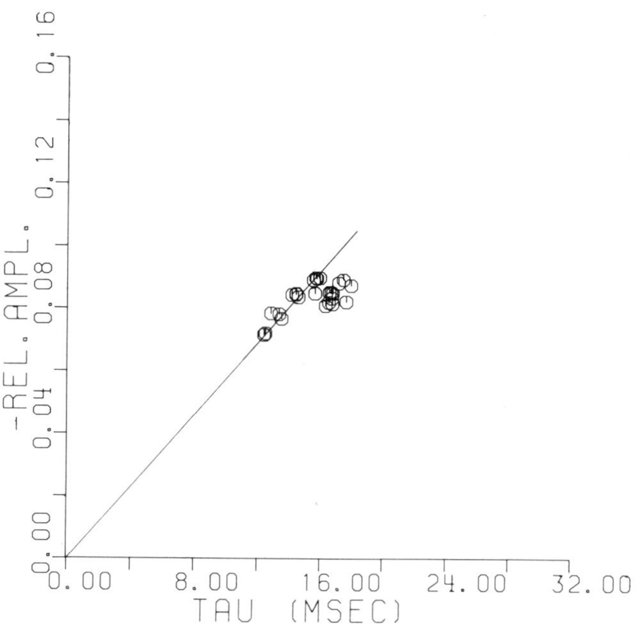

Fig. 6. Relative amplitude plot corresponding to data of Fig. 5

ously (Sect. III). Again, only single exponential decays were observed.

The addition of enzyme to a solution of poly[d(A-T)] and ethidium bromide led to a decrease in all three canonical variables: inverse relaxation time, relaxation amplitude, and total signal (Fig. 5). The first of these quantities (Fig. 5a) showed a minimum at about 80 nM enzyme and then increased slightly. The other variables (Fig. 5b,c) appeared to reach plateau values at the highest enzyme concentrations used indicative of a considerable degree of residual dye binding. The relative amplitudes for enzyme concentrations up to 80 nM lay on a straight line with the expected slope (Fig. 6), although the values derived at higher concentrations fell somewhat below. It was evident from these results that RNA polymerase inhibits the binding of ethidium bromide to the DNA and attempts were made with a number of models to fit the overall behavior.[3]

In all cases, the binding of the enzyme to the polymer was considered to be stoichiometric and not subject to perturbation in the relaxation experiments. Thus, we assume that the DNA lattice is populated by the total given amount of enzyme and that the latter is distributed according to the same excluded binding model discussed above. We denote the number of lattice sites occupied by the enzyme by m and the degree of saturation by ν (these

[3]In separate experiments, no spectroscopic evidence for a direct interaction between enzyme and ethidium bromide was obtained. It was also shown that ethidium bromide is not inhibitory to the enzyme activity at the concentration used in the relaxation experiments (0.8 μM).

parameters are analogous to n and r used above for ethidium bromide binding alone). Again, we introduce the concept of gaps separating bound enzyme molecules. At equilibrium, their fractional distribution is given by (McGHEE and VON HIPPEL, 1974)

$$P_g = \frac{(1-m\nu)^g \nu}{\left[1-(m-1)\nu\right]^{g+1}} \qquad (25)$$

$$P_o = \nu/\left[1-(m-1)\nu\right] \qquad (25a)$$

where P_g is the probability that a gap is g lattice units long.

The bound enzyme in effect partitions the DNA lattice into segments of finite length. As a consequence, it is expected that the binding of ethidium bromide will be altered due to the direct blocking of a certain number of lattice sites by each enzyme molecule but also due to the increased partitioning of the lattice and consequent loss of potential sites at the boundaries. The latter effect manifests itself in the calculation of f(r) (Eq. (3)) which essentially accounts for the statistical distribution of unbound lattice sites. Therefore, we are obliged to introduce a correction which effectively increases the number of "interruptions" of the lattice without otherwise altering the total concentrations. We have approximated this correction by evaluating f(r) and f'(r) (Eqs. (3) and (4)), respectively) not with the true value of r, i.e. X/S_o, but rather with a fictitious quantity given by

$$r' = (X+E')/S_o'+nE') \qquad (26)$$

where X is the concentration of bound ligand as before, S_o' is the total concentration of lattice sites corrected for the fraction (dependent upon the model, see below) blocked by the enzyme, E' is the enzyme concentration reduced by the fraction of zero length gaps (Eqs. (25a)), and n, the length of the ethidium bromide binding site, is set equal to 2. The effect of this correction is to decrease the magnitudes of f(r) and f'(r) and thus the fraction of available sites for binding of the dye.

The first model (designated I) that we considered to account for the experimental results, states that the enzyme simply blocks (directly or by structural transformation) the entire binding site of length m from the dye. Using a value of m=49 basepairs, the first three points in the plot of reciprocal relaxation times (Fig. 5a) are well fit but the DNA lattice rapidly becomes saturated and it is not possible, therefore, to account for the data at higher enzyme levels.[4] Furthermore, this value of m, or any other, is incompatible with the amplitude and total signal data (Fig, 5b,c) since the simulated and measured plots have opposite curvatures.

---

[4]It should be noted that the ratio of polynucleotide to enzyme (basepairs/molecule) was 25 at the highest enzyme concentration used. This number would constitute the upper limit for m, therefore, assuming stoichiometric binding.

Because of the latter finding, a model was sought which would provide for interaction between the effects exerted by adjacent enzyme molecules, although not in the sense of cooperative binding since the latter possibility was not found by simulation to confer the desired properties on the system. We postulate, rather, that the enzyme has a binding site of length m and that a number p of additional adjacent lattice sites (basepairs) are also unavailable for ligand (ethidium bromide) binding. There is no loss of generality in considering these p sites to lie, for example, always to the left of the enzyme and physically, we can conceive that they consist of altered helical structures under the influence of the bound polymerase. A crucial feature is that the regions overlap if two enzyme molecules are close enough together. It can be seen intuitively that the effects on dye binding exerted by individual enzyme molecules will progressively decrease at higher concentrations due to the provision for overlap. The computations are straightforward, as we have the distribution of gaps of any length between enzymes (Eq. (25)) and only gap sizes up to the length n are required explicitly.

The above feature produces changes in the simulated curves, which are in the right direction, but fails to account for the low value of the reciprocal relaxation time at 80 nM enzyme, despite the relatively large amount of free ligand. Therefore, we revise the above postulate and assume that the altered region of p basepairs (hereafter referred to as the β-region, as opposed to the unperturbed, or α-, region for intercalative binding) does bind ethidium bromide but not by a process leading to fluorescence changes. Since only a single relaxation time was observed, this additional mode of binding is assumed to be kinetically and thermodynamically uncoupled to the intercalative binding reaction due to the particular rate constants and reaction enthalpies for the two processes. The effect of "β-binding" is to take up free ligand at the intermediate enzyme concentrations and return it to the system at the higher enzyme levels for which the β-regions become compressed.

This model (designated II) produces better fits of the reciprocal relaxation times (Fig. 7a) using a simple single step mechanism for β-binding and various values for the corresponding equilibrium constant (not shown). We see qualitatively for the first time an upturn in the reciprocal relaxation time at high enzyme concentrations, in accordance with the experimental data. The amplitude and total signal plots, however, are not simulated properly by this model (Fig. 7b,c).

In order to approximate the shapes of all three experimental curves, it was necessary to modify further the postulates regarding the nature of the β-region. We stipulate first that the latter also includes some of the m basepairs constituting the enzyme binding site, so that only j internal lattice units remain in which ligand binding is totally excluded ($j<m$). Finally, we permit residual intercalative binding in a statistical fraction ϕ of the lattice units constituting the β-region. The physical interpretation of this model (designated III) is simply that despite the perturbation of the DNA helix in the vicinity of the enzyme, some of the lattice units (basepairs) are available for

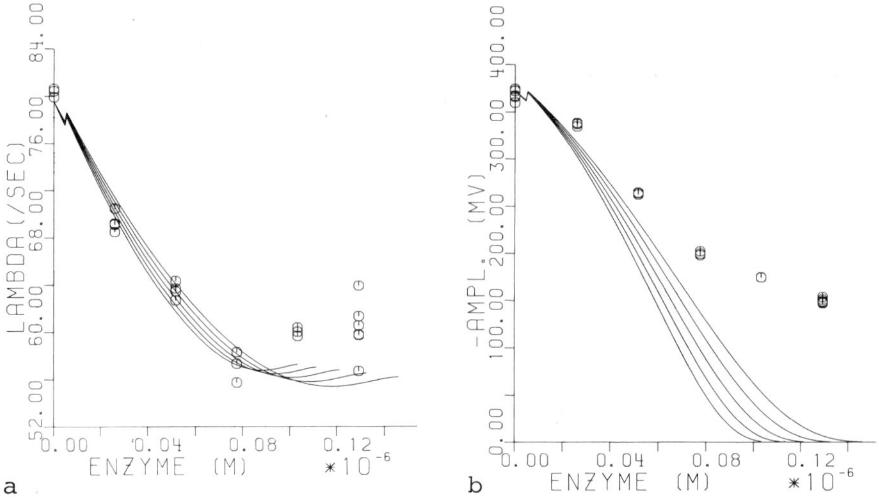

a b

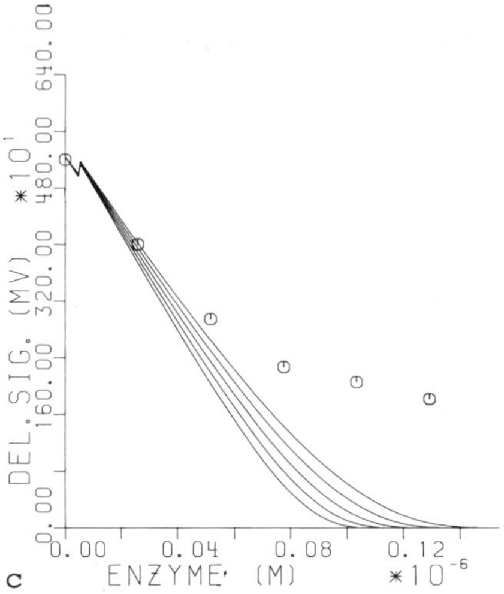

c

Fig. 7a-c. Analysis of relaxation kinetic data of Fig. 5 according to Model II and parameters: p=16 and m=19, 21,23,25,27 (descending order of curves). K_β = $6 \cdot 10^5$ M^{-1}

intercalative binding as in the absence of enzyme. Furthermore, we state that although the distance between the centers of two "touching" enzyme molecules cannot be less than m units, due to stereochemical features, some of the intervening DNA remains accessible to the relatively small dye molecule except for a central kernal of j units. This scheme (Model III) is depicted schematically in Figure 8.

Obviously, model III contains a relatively large number of parameters which cannot be specified uniquely by the limited set of experimental data. Therefore, we regard the results derived from simulation as having more of a qualitative than quantitative significance. It should be noted, however, that elimination of any of the features discussed above materially reduced the quality of the fit. We examined systematically the influence of all the parameters summarized in Figure 8. The effects introduced by

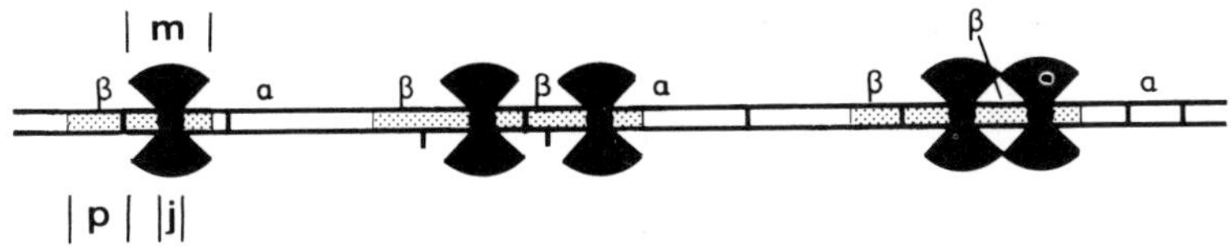

Fig. 8. Schematic representation of interactions between RNA polymerase and ethidium bromide bound to poly[d(A-T)] (Model III). Solid black symbols: enzyme in various states (left to right): alone, two molecules close to each other, and two molecules in closest possible contact. The latter condition is determined by the effective width of the binding sites (m basepairs). The Enzyme alters j basepairs within its binding site so that no dye binding takes place. The remainder of the binding site (m-j basepairs) and a region of p basepairs extending to left is designated as the β-region (shown stippled in the drawing). In this region, intercalactive dye binding (vertical lines between the parallel bars representing the DNA) can still occur at a statistical fraction ϕ of the available sites. In addition, a second mode of non-exclusive binding exists, characterized by an association constant K_β. Regions of DNA unaffected by enzyme are designated as α. Physical constants corresponding to binding in these areas are assumed to be those given in Table 1. Typical values for the various parameters which gave reasonable fits to the data are: $j = 10$, $m = 20$, $p = 27$, $\phi = .4$, $K_\beta = 6 \cdot 10^5\ M^{-1}$

individually altering the values of m, j, and p are shown in Figures 9, 10, and 11, respectively. Reduction of j improves the overall fit of the three curves, but tends to reduce the upturn in the λ curve (see Fig. 7a). The value for K_β in simulations tended to approach that for intercalative binding and thus conceivably could be omitted as a variable, although there is no a priori reason to assume that the two modes of binding would be characterized by the same binding constant. It is also interesting that the best simulations were obtained by postulating that the nonintercalative binding in the β-region is non-exclusive in character, i.e. it affects only a single lattice site.

V. Conclusions

The success of the approach represented in this work depends upon a thorough understanding of the modes of interaction between the DNA and the reporter molecule, in our case ethidium bromide. The excluded site phenomenon clearly complicates the picture theoretically as well as experimentally yet seems susceptible to quantitative treatment. The results presented here, however, bring into question the interpretations of numerous kinetic studies of dye binding to polynucleotide at high ligand concentration, the condition under which a complex relaxation spectrum is predicted. Thus, in the presence of excluded binding, the functional forms of both static and dynamic data are subject to misinterpretation in the sense of invoking numerous binding steps or isomerizations, which in fact do not exist. It is obvious that this consideration applies to other physical phenomena

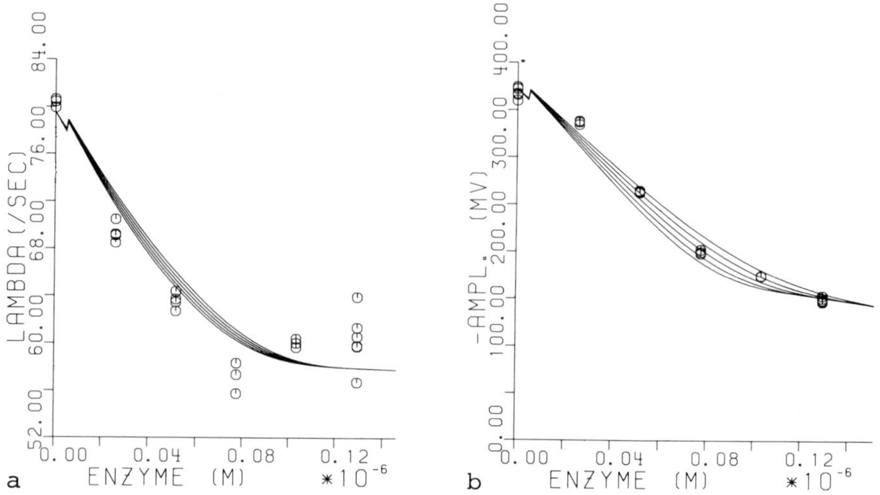

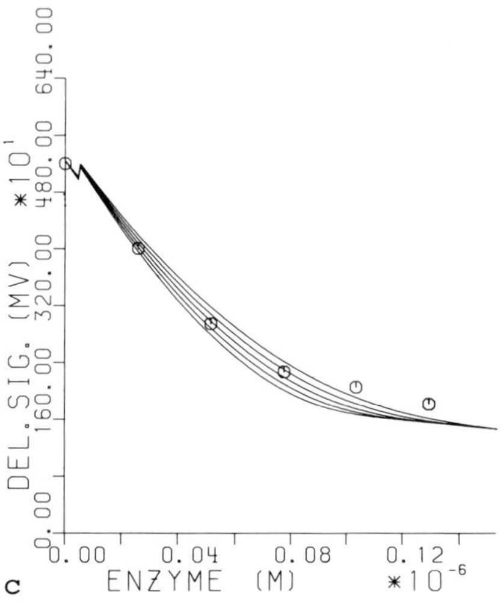

Fig. 9a-c. Data simulation using Model III (Fig. 8) showing effect of m, enzyme binding site. Parameters are: j=7, p=27, ϕ= 0.4, K=6 $\cdot$ 10^5 M^{-1}, and m= 15,17,19,21,23 (descending order of curves)

such as interactions on a two-dimensional lattice, e.g. a membrane. Also instructive is a comparison of our kinetic analysis and that relating to the interaction between the binding and distribution functions in allosteric systems (EIGEN, 1967).

The presence of RNA polymerase drastically alters the binding of ethidium bromide to poly[d(A-T)], indicating that the interaction of enzyme with the DNA duplex can and does affect intercalative processes. While the limited data set reported here is not sufficient to fully characterize the system, a number of conclusions can be made (as usual in kinetic studies, it is easier to exclude models for reaction mechanisms than to prove that any one is correct).

It appears that bound enzyme does not simply exclude a given region of DNA from intercalation by ethidium bromide. To achieve relatively good fits to the data, one must postulate at least

Fig. 10. Data simulation using Model III (Fig. 8) showing the effect of j, the region totally blocked by the enzyme. Paremeters are m=20, with other values as in Figure 9, and j= 6,7,8,9,10 (descending order of the curves). Only the total signal is shown since these small changes in j do not appreciably affect the other variables

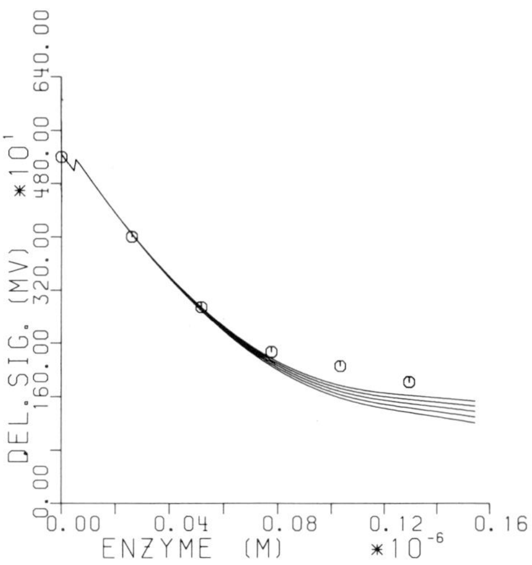

two effects of the enzyme upon dye binding (Fig. 8): (a) a small region of DNA within the binding site is rendered totally unavailable. Reasonable values for the size of this region are <10 basepairs. Physically, this effect might well correspond to the observed extent of denaturation of the DNA induced by the enzyme (ca. 10 basepairs) as discussed in the Introduction. In the case of promoters, this site could be the ubiquitious heptanucleotide sequence left of the initiation site (PRIBNOW, 1975a,b); (b) a larger region (ca. 40 basepairs) is affected in such a way as to reduce the probability of intercalation but allow some other form of binding (e.g. "outside" primarily electrostatic interactions with the phosphate moieties of the DNA backbone). Both modes of binding are observed even with enzyme molecules in closest apposition. Thus, as depicted in Figure 8, we have to stipulate that an individual enzyme molecule blocks >20 basepair units of the DNA lattice from availability for additional enzyme binding, but that ethidium bromide can still interact within this limit, although in an altered fashion.

Thus the enzyme is seen to affect DNA structure over a region extending beyond the confines of its "binding site," although the latter entity cannot be defined unambiguously. As indicated in the Introduction, digestion with pancreatic DNAase of DNA "protected" by bound RNA polymerase yield oligonucleotides about 42 basepairs in length. Additional digestion with micrococcal nuclease or venom phosphodiesterase, however, produces shorter lengths of about 15 basepairs (GIACOMONI et al., 1974) or destabilizes the complex altogether (H. SCHALLER, personal communication). These results support the idea that the apparent length of the binding site reflects the method of measurement to some degree and that the larger probes have less access to regions of the DNA within the external bounds of the DNA-polymerase complex. In any event, the range of acceptable values for m in Figure 8 is certainly compatible with the results obtained with completely different techniques indicated above.

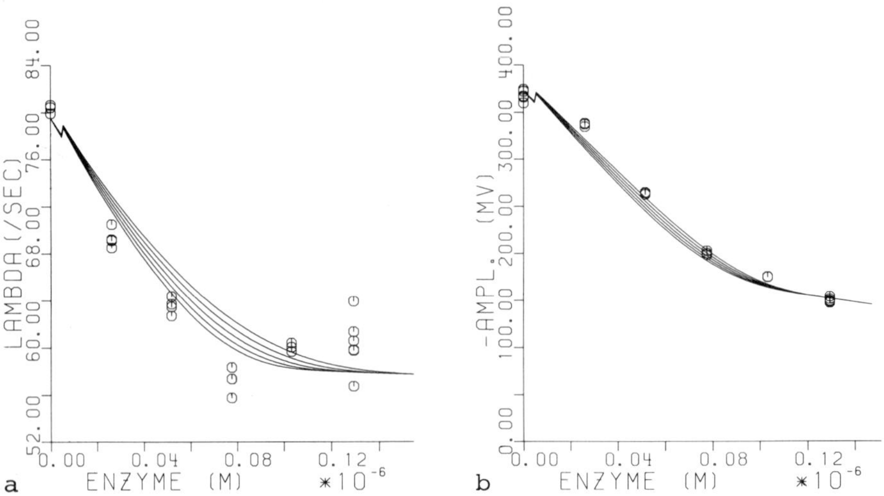

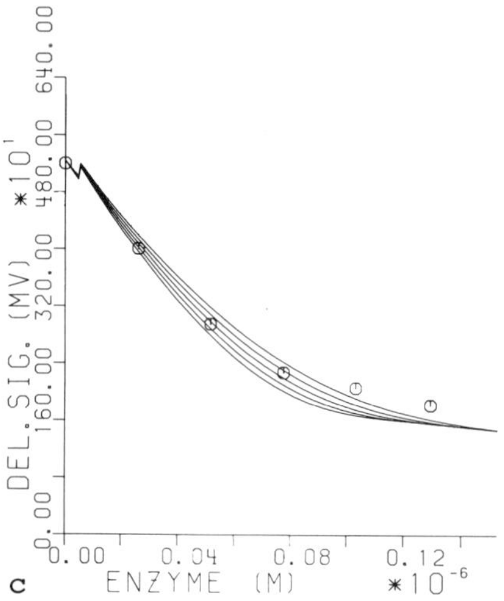

Fig. 11a-c. Data simulation using Model III (Fig. 8) and showing effect of p, the external region influenced by enzyme. Parameters are m=21, with other values as in Figure 9, and p=22,24,26, 28,30 (descending order of curves)

It is important to restate some of the limitations placed upon our interpretations of the data due to the assumptions and experimental conditions. We have asserted, for example, that the enzyme is fully active and quantitatively bound to the poly [d(A-T)]. While compatible with other reports in the literature (cited in the Introduction), the latter point was not verified directly. However, it should be noted that if less enzyme had been bound, the regions of interaction estimated by simulation would be even larger. That is, the enzyme would appear to influence DNA structure considerably beyond the limits of its binding site. We have also assumed that the kinetic parameters for ethidium bromide binding obtained in the absence of enzyme are still applicable. In support of this assumption is the fact that only a single relaxation time was seen under all conditions. If new modes of intercalative binding had arisen as a consequence of enzyme binding, one would have expected in general to have seen additional relaxation processes. Nonetheless, under the condition of near saturation with enzyme, distortions of the polynucleotide structure might affect the rate constants for ethidium bromide binding, thus making the quantitative conclusions of the analysis less reliable. One might expect, for example, that the transition from one region to the other might be gradual and not abrupt as in Figure 8. The experimental conditions need to be explored further, especially by variation of the dye and DNA concentrations. In addition, it would be of interest to perform comparative studies at other ionic strengths, in the presence of Mg^{2+}, and using core enzyme and other DNAs. Finally, it should be obvious that the phenomena reported here suggest an extension of the approach to studies of the enzyme binding kinetics itself, e.g. using stopped-flow instrumentation.

Acknowledgements. We thank Dr. G. RHODES for the supply of RNA polymerase and Drs. F. BARRANTES, R. CLEGG, D. PÖRSCHKE, and G. RHODES for critical reading of the manuscript. We also acknowledge valuable discussions with Drs. M. EIGEN and R. CLEGG on the subject of excluded binding kinetics. The computations were performed with the UNIVAC 1108 of the Gesellschaft für wissenschaftliche Datenverarbeitung Göttingen (GWDG).

VI. Appendix. On the Derivation of General Equations for Relaxation Kinetics of Systems with Excluded Binding

In seeking to avoid the limitations of Eq. (13), we select Eq. (12b) as the point of departure and thus require an expression for δG_o. Since it is no longer possible to assume that G_o instantaneously assumes its equilibrium value during the relaxation, we must consider its time dependence.

$$\dot{G}_o = 2\left[k_1 L \sum_2^{\infty} G_i - k_{-1} G_o\right] \quad \text{(A1)}$$

in which the statistical factor 2 reflects the two ways in which a G_o can be formed by binding within a larger gap or from the fact that two G_os are created by ligand binding to a G_2, as well as the two modes of dissociation (of either of the two molecules comprising a G_o).

McGHEE and VON HIPPEL (1974) showed that we can express

$$\bar{G}_i = (1-\alpha)^i \alpha \bar{X} \qquad \text{with } \alpha = \bar{r}/(1-\bar{r}) \quad \text{(A2)}$$

Thus from Eqs. (9) and (A2)

$$\sum_k^{\infty} G_i = X - \sum_o^{k-1} G_i \quad \text{(A3)}$$

and

$$\sum_k^{\infty} \bar{G}_i = \bar{X} - \sum_o^{k-1} \bar{G}_i = (1-\alpha)^k \bar{X} \quad \text{(A4)}$$

By introducting Eqs. (A3), (A4) and the identities $G_i = \bar{G}_i + \delta G_i$ into the differential Eq. (A1) and cancelling the equilibrium and higher order derivative terms in the usual way, we obtain

$$-\frac{1}{2}\delta\dot{G}_o = k_1\left[(1-\alpha)^2\bar{X}-\bar{L}\right]\delta X + (k_1\bar{L}+k_{-1})\delta G_o + k_1\bar{L}\delta G_1 \quad (A5)$$

At this point, we have various options for dealing with δG_1. Since the excluded binding theory shows that there will always exist more G_o than G_1 (see Eq. (A2)), one could decide to approximate the system with two differential equations, i.e. $\delta\dot{X}$ and $\delta\dot{G}_o$. This can be accomplished by assuming an equilibrium relationship between G_1 and X throughout the relaxation, which in differential form is given (from Eq. (A2)) by

$$\delta G_1 = \beta\delta X \quad \text{with} \quad \beta = 2\alpha(1-\alpha-\alpha^2) \quad (A6)$$

By introducing relationships 7, 12b, and A6 into Eqs. (6a) and (A5), we obtain the two coupled equations

$$\delta\dot{X} = -\left[k_1(3\bar{L}+\bar{S})+k_{-1}\right]\delta X + k_1\bar{L}\delta G_o \quad (A7)$$

$$\delta\dot{G}_o = -2k_1\left[(1-\alpha)^2\bar{X}+(\beta-1)\bar{L}\right]\delta X - 2\left[k_1\bar{L}+k_{-1}\right]\delta G_o \quad (A8)$$

The solution of these equations yields two relaxation times corresponding to two normal modes.

Alternatively, the procedure used to derive Eq. (A4) can be extended by considering the differential equation for G_1, which yields a system of three equations ($\delta\dot{X}$, $\delta\dot{G}_o$, $\delta\dot{G}_1$) containing terms in δG_2 (and thus requiring a simplifying assumption about its relationship to δX in analogy to Eq. (A6)). This process of "bootstrapping" can be continued indefinitely so as to describe the system to any desired degree of accuracy.

It is important to note that our model allows for redistribution of bound ligand only by the combined processes of dissociation and reassociation. If we permit the bound ligand to translate along the lattice, however, an additional kinetic pathway is created, the influence of which depends upon the diffusion constant. In the limit of very rapid diffusion, the assumption of an equilibrium distribution maintained throughout the relaxation process, as expressed in Eq. (13), would be formally justified and the system would exhibit the single relaxation time given by Eq. (15).

References

AKTIPIS, S., KINDELIS, A.: Optical properties of the deoxyribonucleic acid-ethidium bromide complex. Effect of salt. Biochemistry, 12, 1213-1221 (1973)

AKTIPIS, S., PANAYOTATOS, N.: Mechanism of ethidium bromide inhibition of RNA polymerase. Biochem. Biophys. Res. Comm. 68, 465-470 (1976)

ALDEN, C.J., ARNOTT, S.: Visualization of planar drug intercalations in B-DNA. Nucl. Acid Res. 2, 1701-1717 (1975)

ANTHONY, D.D., GOLDTHWAIT, D.A.: Studies with the RNA polymerase III. Enzymatic activity of the monomer form. Biochim. Biophys. Acta 204, 156-167 (1970)

ARNDT-JOVIN, D.J., JOVIN, T.M., BÄHR, W., FRISCHAUF, A.M., MARQUARDT, M.: Covalent attachment of DNA to agarose. Improved synthesis and use in affinity chromatography. Europ. J. Biochem. 54, 411-418 (1975)

ARNOTT, S., CHANDRASEKARAN, R., HUKINS, D.W.L., SMITH, P.J.C., WATTS, L.: Structural details of a double-helix observed for DNAs containing alternating purine and pyrimidine sequences. J. Mol. Biol. 88, 523-533 (1974).

BÄHR, W., STENDER, W., SCHEIT, K.-H., JOVIN, T.M.: Binding of rifampicin to *Escherichia coli* RNA polymerase: thermodynamic and kinetic studies. In: RNA Polymerase. CHAMBERLIN, M., LOSICK, H. (eds.). Cold Spring Harbor, New York: Cold Spring Harbor Lab. 1976, in press

BALDWIN, R.L.: Experimental tests of the theory of deoxyribonucleic acid melting with d(T-A) oligomers. Accounts Chem. Res. 4, 265-272 (1971)

BAUER, W., VINOGRAD, J.: Interaction of closed circular DNA with intercalative dyes II. The free energy of superhelix formation in SV40 DNA. J. Mol. Biol. 47, 419-435 (1970)

BEABEALASHVILLY, R.S., SAVOTCHKINA, L.P.: RNA polymerase-DNA complexes. IV. Influences of the ionic strength on the integrity of the complexes. Biochim. Biophys. Acta 294, 434-441 (1973)

BERG, D., BARRETT, K., CHAMBERLIN, M.: Purification of two forms of *E.coli* RNA polymerase and of sigma component. In: Methods in Enzymology. COLOWICK, S., KAPLAN, N. (eds.). Vol. 21, pp. 506-619. New York: Academic Press 1971

BERG, D., CHAMBERLIN, M.: Physical studies on ribonucleic acid polymerase from *Escherichia coli* B. Biochemistry 9, 5055-5064 (1970)

BITTMAN, R.: Studies of the binding of ethidium bromide to transfer ribonucleic acid: absorption, fluorescence, ultracentrifugation and kinetic investigations. J. Mol. Biol. 46, 251-268 (1969)

BLAKE, A., PEACOCKE, A.R.: The interaction of aminoacridines with nucleic acids. Biopolymers 6, 1225-1253 (1968)

BRAM, S., TOUGARD, P.: Polymorphism of natural DNA. Nature (New Biol.) 239, 128-131 (1972)

BRESLOFF, J.L., CROTHERS, D.M.: DNA-ethidium reaction kinetics: demonstration of direct ligand transfer between DNA binding sites. J. Mol. Biol. 95, 103-123 (1975)

CHAMBERLIN, M.J.: The selectivity of transcription. Ann. Rev. Biochem. 43, 721-775 (1974a)

CHAMBERLIN, M.J.: Bacterial DNA-dependent RNA polymerase. In: The Enzymes. BOYER, P.D. (ed.). Vol. 10, pp. 333-374. New York: Academic Press 1974b

CHAMBERLIN, M.J., LOSICK, H.: RNA Polymerase. Cold Spring Harbor, New York: Cold Spring Harbor Lab. 1976

DAVIES, D.R., BALDWIN, R.L.: X-ray studies of two synthetic DNA copolymers. J. Mol. Biol. 6, 251-255 (1963)

DICKSON, R.C., ABELSON, J., BARNES, W.M., REZNIKOFF, W.S.: Genetic regulation: the lac control region. Science 187, 27-35 (1975)

DOMINGO, E., ESCARMIS, C., WARNER, R.C.: Transcription of *Azotobacter* phage deoxyribonucleic acid. Salt-dependent equilibrium between steps in initiation. J. Biol. Chem. 250, 2872-2877 (1975)

EIGEN, M.: Kinetics of reaction control and information transfer in enzymes and nucleic acids. In: Fast Reactions and Primary Processes in Chemical Kinetics. CLAESSON, S. (ed.). pp. 333-369. New York: Wiley 1967

EIGEN, M., DE MAEYER, L.: Theoretical basis of relaxation spectrometry. In: Techniques of Chemistry, Vol. VI: Investigation of Rates and Mechanisms of Reactions. HAMMES, G.G. (ed.); pp. 63-146. New York: Wiley 1974

FREIFELDER, D.: Electron microscopic study of the ethidium bromide-DNA complex. J. Mol. Biol. 60, 401-403 (1971)

FRISCHAUF, A.M., SCHEIT, K.H.: Affinity labeling of *E.coli* RNA polymerase with substrate and template analogues. Biochem. Biophys. Res. Commun. 53, 1227-1233 (1973)

FUCHS, E., MILLETTE, R.L., ZILLIG, W., WALTER, G.: Influence of salts on RNA synthesis by DNA-dependent RNA-polymerase from *Escherichia coli* . Europ. J. Biochem. 3, 183-193 (1967)

FULLER, W., WARING, M.J.: A molecular model for the interaction of ethidium bromide with deoxyribonucleic acid. Ber. Bunsenges. Physik. Chem. 68, 805-808 (1964)

GIACOMONI, P.U., LE TALAER, J.Y., LE PECQ, J.B.: *Escherichia coli* RNA-polymerase binding sites on DNA are only 14 base pairs long and are located between sequences that are very rich in A + T. Proc. Nat. Acad. Sci. (Wash.) 71, 3091-3095 (1974)

GRINVALD, A., STEINBERG, I.Z.: On the analysis of fluorescence decay kinetics by the method of least-squares. Anal. Biochem. 59, 583-598 (1974)

HAMILTON, W.C.: Statistics in Physical Science. 1-230. New York: Ronald 1964

HEYDEN, B., NÜSSLEIN, C., SCHALLER, H.: Initiation of transcription within an RNA-polymerase binding site. Europ. J. Biochem. 55, 147-155 (1975)

HINKLE, D.C., CHAMBERLIN, M.J.: Studies of the binding of *Escherichia coli* RNA polymerase to DNA. I. The role of sigma subunit in site selection. J. Mol. Biol. 70, 157-185 (1972)

HSIEH, T.-S., WANG, J.C.: Thermodynamic properties of superhelical DNAs. Biochemistry 14, 527-535 (1975)

INMAN, R.B., BALDWIN, R.L.: Helix-random coil transitions in synthetic DNA's of alternating sequence. J. Mol. Biol. 5, 172-184 (1962)

ISHIHAMA, A., HURWITZ, J.: The role of deoxyribonucleic acid in ribonucleic acid synthesis. XVII. Multiple active sites of *Escherichia coli* ribonucleic acid polymerase. J. Biol. Chem. 244, 6680-6689 (1969)

JONES, O.W., BERG, P.: Studies on the binding of RNA polymerase to polynucleotides. J. Mol. Biol. 22, 199-209 (1966)

JOVIN, T.M.: Fluorimetric kinetic techniques: chemical relaxation and stopped-flow. In: Biochemical Fluorescence Concepts. CHEN, R.F., EDELHOCH, H. (eds.); Vol. I, pp. 305-374. New York: Dekker 1975

JOVIN, T.M.: Recognition mechanisms of DNA-specific enzymes. Ann. Rev. Biochem. 45, 889-920 (1976)

KELLER, W.: Determination of the number of superhelical turns in simian virus 40 DNA by gel electrophoresis. Proc. Nat. Acad. Sci. (Wash.) 72, 4876-4880 (1975)

KRAKOW, J.S.: Acrylamide gel electrophoresis as a tool for the study of RNA polymerase and the sigma initiation factor. In: Methods in Enzymology, Vol. 21. GROSSMAN, L., MOLDAVE, K. (eds.); pp. 520-528. New York: Academic Press 1971
KRAKOW, J.S.: Inhibition of *Azotobacter vinelandii* ribonucleic acid polymerase by glutamyl, tyrosyl copolymers. Biochemistry 13, 1101-1105 (1974)
KRAKOW, J.S., VON DER HELM, K.: *Azotobacter* RNA polymerase transitions and the release of sigma. Cold Spring Harbor Symp. Quant. Biol. 35, 73-83 (1970)
KRAKOW, J.S., RHODES, G., JOVIN, T.M.: RNA polymerase: catalytic mechanisms and inhibitors. In: RNA Polymerase. CHAMBERLIN, M., LOSICK, H. (eds.). Cold Spring Harbor, New York: Cold Spring Harbor Lab. 1976; pp. 127-157
LANG, D.: Individual macromolecules: preparation and recent results with DNA. Phil. Trans. Roy. Soc. London B 261, 151-158 (1971)
LePECQ, J.B., PAOLETTI, C.: A fluorescent complex between ethidium bromide and nucleic acids. J. Mol. Biol. 27, 87-106 (1967)
LERMAN, L.S.: Structural considerations in the interaction of DNA and acridines. J. Mol. Biol. 3, 18-30 (1961)
LOEB, J., CAHEN, G.: Extraction a partir des enregistrements de mesures, des paramètres dynamiques d'un système. Automatisme 8, 479-486 (1963)
MAGDE, D., ELSON, E.L., WEBB, W.W.: Fluorescence correlation spectroscopy. II. An experimental realization. Biopolymers 13, 29-61 (1974)
MANGEL, W.F., CHAMBERLIN, M.J.: Studies of ribonucleic acid chain initiation by *Escherichia coli* ribonucleic acid polymerase bound to T7 deoxyribonucleic acid. I. An assay for the rate and extent of ribonucleic acid chain initiation. J. Biol. Chem. 249, 2995-3001 (1974a)
MANGEL, W.F., CHAMBERLIN, M.J.: Studies of ribonucleic acid chain initiation by *Escherichia coli* ribonucleic acid polymerase bound to T7 deoxyribonucleic acid. II. The effect of alterations in ionic strength on chain initiation and on the conformation of binary complexes. J. Biol. Chem. 249, 3002-3006 (1974b)
MANGEL, W.F., CHAMBERLIN, M.J.: Studies of ribonucleic acid chain initiation by *Escherichia coli* ribonucleic acid polymerase bound to T7 deoxyribonucleic acid. III. The effect of temperature on ribonucleic acid chain initiation and on the conformation of binary complexes. J. Biol. Chem. 249, 3007-3013 (1974c)
McGHEE, J.D., VON HIPPEL, P.H.: Theoretical aspects of DNA-protein interactions: co-operative and non-co-operative binding of large ligands to a one-dimensional homogeneous lattice. J. Mol. Biol. 86, 469-489 (1974)
PILET, J., BLICHARSKI, J., BRAHMS, J.: Conformations and structural transitions in polydeoxynucleotides. Biochemistry 14, 1869-1876 (1975)
POHL, F.M., JOVIN, T.M., BÄHR, W., HOLBROOK, J.J.: Ethidium bromide as a cooperative effector of a DNA structure. Proc. Nat. Acad. Sci. (Wash.) 69, 3805-3809 (1972)
PRIBNOW, D.: Bacteriophage T7 early promoters: nucleotide sequences of two RNA polymerase binding sites. J. Mol. Biol. 99, 419-443 (1975a)

PRIBNOW, D.: Nucleotide sequence of an RNA polymerase binding site at an early T7 promoter. Proc. Nat. Acad. Sci. (Wash.) 72, 784-788 (1975b)
PULLEYBLANK, D.E., MORGAN, A.R.: The sense of naturally occurring superhelices and the unwinding angle of intercalated ethidium. J. Mol. Biol. 91, 1-13 (1975)
REINERT, K.E.: DNA stiffening and elongation caused by the binding of ethidium bromide. Biochim. Biophys. Acta 319, 135-139 (1973)
RICHARDSON, J.P.: Some physical properties of RNA polymerase. Proc. Nat. Acad. Sci. (Wash.) 55, 1616-1623 (1966)
RICHARDSON, J.P.: Mechanism of ethidium bromide inhibition of RNA polymerase. J. Mol. Biol. 78, 703-714 (1973)
RIGLER, R., EHRENBERG, M.: Molecular interactions and structure as analyzed by fluorescence relaxation spectroscopy. Quant. Rev. Biophys. 6, 139-199 (1973)
RIGLER, R., RABL, C.R., JOVIN, T.M.: A temperature-jump apparatus for fluorescence measurements. Rev. Sci. Instrum. 45, 580-588 (1974)
SAUCIER, J.M., WANG, J.C.: Angular alteration of the DNA helix by *E.coli* RNA polymerase. Nature (New Biol.) 239, 167-170 (1972)
SCATCHARD, G.: The attractions of proteins for small molecules and ions. Ann. N.Y. Acad. Sci. 51, 660-672 (1949)
SCHELLMAN, J.A.: Cooperative multisite binding to DNA. Israel J. Chem. 12, 219-238 (1974)
SCHELLMAN, J.A.: Macromolecular binding. Biopolymers 14, 999-1018 (1975)
SCHREIBER, J.P., DAUNE, M.P.: Fluorescence of complexes of acridine dye with synthetic polydeoxyribonucleotides: a physical model of frameshift mutation. J. Mol. Biol. 83, 487-501 (1974)
SMITH, D.A., MARTINEZ, A.M., RATLIFF, R.L., WILLIAMS, D.L., HAYES, F.N.: Template-induced dissociation of ribonucleic acid polymerase. Biochemistry 6, 3057-3063 (1967)
SO, A.G., DOWNEY, K.M.: Studies on the mechanism of ribonucleic acid synthesis. II. Stabilization of the deoxyribonucleic acid-ribonucleic acid polymerase complex by the formation of a single phosphodiester bond. Biochemistry 9, 4788-4793 (1970)
STEAD, N.W., JONES, O.W.: Stability of RNA polymerase-DNA complexes. J. Mol. Biol. 26, 131-135 (1967)
STRNISTE, G.F., SMITH, D.A.: Induction of stable linkage between the deoxyribonucleic acid-dependent ribonucleic acid polymerase and $d(A-T)_n \cdot d(A-T)_n$ by ultraviolet light. Biochemistry 13, 485-493 (1974)
STUDIER, F.W.: Sedimentation studies of the size and shape of DNA. J. Mol. Biol. 11, 373-390 (1965)
THUSIUS, D., FOUCAULT, G., GUILLAIN, F.: The analysis of chemical relaxation amplitudes and some applications to reactions involving macromolecules. In: Dynamic Aspects of Conformation Changes in Biological Macromolecules. SADRON, C. (ed.); pp. 271-284. Dordrecht: De Reidel 1973
TICHADOU, J.L., GENEST, D., WAHL, P., AUBEL-SADRON, G.: The use of fluorescence anisotropy decay of poly d(A-T) ethidium bromide complex to estimate the unwinding angle of the double helix. Biophys. Chem. 3, 142-146 (1975)
TRITTON, T.R., MOHR, S.C.: Relaxation kinetics of the binding of ethidium bromide to unfractionated yeast tRNA at low dye/phosphate ratio. Biochem. Biophys. Res. Commun. 45, 1240-1249 (1971)

TRITTON, T.R., MOHR, S.C.: Kinetics of ethidium bromide binding as a probe of transfer ribonucleic acid structure. Biochemistry 12, 905-914 (1973)

TSAI, C.-C., JAIN, S.C., SOBELL, H.M.: X-ray crystallographic visualization of drug-nucleic acid intercalative binding: structure of an ethidium-dinucleoside monophosphate crystalline complex, ethidium: 5-iodouridylyl(3'-5')adenosine. Proc. Nat. Acad. Sci. (Wash.) 72, 628-632 (1975)

VALEUR, B., MOIREZ, J.: Analyse des courbes de decroissance multiexponentielles par la methode des fonctions modulatrices - application a la fluorescence. J. Chim. Phys. 70, 500-506 (1973)

VICTOR, J., HASELKORN, D., PECHT, I.: Direct evaluation of rate constants for an assumed single-step mechanism from chemical relaxation data. Computers Biomed. Res. 6, 121-126 (1973)

WALTER, G., ZILLIG, W., PALM, P., FUCHS, E.: Initiation of DNA-dependent RNA synthesis and the effect of heparin on RNA polymerase. Europ. J. Biochem. 3, 194-201 (1967)

WANG, J.C.: The degree of unwinding of the DNA helix by ethidium I. Titration of twisted PM2 DNA molecules in alkaline cesium chloride density gradients. J. Mol. Biol. 89, 783-801 (1974)

WARING, M.J.: Complex formation with DNA and inhibition of *Escherichia coli* RNA polymerase by ethidium bromide. Biochim. Biophys. Acta 87, 358-361 (1964)

WILLIAMS, R.E., SELIGY, V.L.: The interaction of ethidium bromide with synthetic polydeoxyribonucleic acids. Effect of base composition and sequence on the induced circular dichroism spectra. Can. J. Biochem. 52, 281-287 (1974)

WYMAN, J.: A group of thermodynamic potentials applicable to ligand binding by a polyfunctional macromolecule. Proc. Nat. Acad. Sci. (Wash.) 72, 1464-1468 (1975)

ZUBAY, G., SCHWARTZ, D., BECKWITH, J.: Mechanism of activation of catabolite-sensitive genes: a positive control system. Proc. Nat. Acad. Sci. (Wash.) 66, 104-110 (1970)

Protein Folding and Unfolding

F. M. POHL

I. Introduction

The folding of polypeptide chains into compact globular structures is often the supposition for the expression of their biological function in the cell. The determination of the spatial arrangement of the peptide chain by X-ray diffraction has given the most important information on the static properties of globular proteins. But the elucidation of the pathways of folding, the magnitude of the different thermodynamic forces involed, and the kinetics of such processes are still in a relatively early stage of investigation.

It is not the aim of this short article to provide a review of all the recent results, and for a broader introduction to the problems the reader is referred to other reviews and the work cited there (JOLY, 1965; TANFORD, 1968, 1970; POHL, 1972; ANFINSEN, 1973; ANFINSEN and SCHERAGA, 1975; BALDWIN, 1975). Instead I shall quote some recent developments as illustrations of the present state in this field. The emphasis will be put on kinetic methods, their application to the measurement of steady-state rates, and the importance of transient effects for gaining a more detailed understanding of the different processes involved in the folding and unfolding of proteins. The development of theoretical models for the description of such kinetic results has made considerable progress and some of them will be outlined, although no quantitative description of the experimental results is possible at the moment.

II. Time-Independent Phenomena

The unfolding and refolding of relatively simple proteins, such as ribonuclease, myoglobin, chymotrypsinogen, lysozyme, trypsin, trypsin inhibitor, cytochrome c, Stapholococcus nuclease etc. have served as extensively studied model systems. One reason is that the condition can be found where the unfolding is completely reversible and where association phenomena between peptide chains are negligible. This allows a relatively simple thermodynamic analysis of the experimental data. Figure 1 gives an example of the melting curve of chymotrypsinogen at acid pH (POHL, 1970). The change of the tryptophan absorbance at 293 nm with temperature

Fig. 1. Equilibrium melting curve of chymotrypsinogen A (0.4 mM, pH = 2.0). The change of absorbance per degree is given as function of the temperature. The pronounced asymmetry of this differentiated melting curve is due to the change of the specific heat upon unfolding of the protein

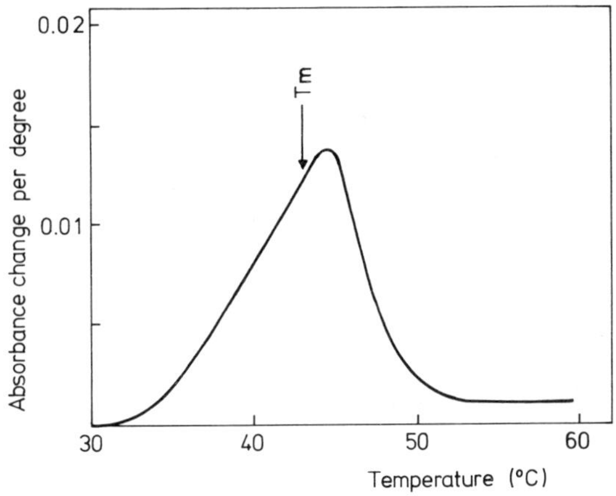

is often a sensitive parameter for measuring such reactions in the spectrophotometer. The melting curve, given in the differentiated form, that is the change of the absorbance per degree, is shown as a function of the temperature. Such a presentation of melting curves is very convenient and helpful, since details of the equilibrium data, like the pronounced asymetry, are more easily recognizable than they are from the integrated melting curves (see Fig. 10).

There is a long-lasting debate on which of the following two extreme types of reaction is a better approximation for such unfolding reactions of globular proteins:

1. The transition takes place between two macroscopic states of the protein, the native state A and the unfolded state B, in a highly cooperative fashion, which means that in the melting region a mixture of fully native molecules and unfolded molecules exists with negligible amounts of intermediate states.

2. There is gradual unfolding of every protein molecule in the transition region with many different intermediate states in such a way that, for example, at the melting temperature T_m, half of every molecule is unfolded, the rest being folded. Such behavior is expected, for example, for the helix-coil transition of very long polypeptides.

Although both hypotheses are extreme pictures of the unfolding of globular proteins, they give rise to rather different kinetic and thermodynamic behavior after a change of external valiables, like temperature ΔT, pressure ΔP, or the concentration of small effector molecules Δn_i. (Probably the strongest support for the value of the two-state hypothesis in the case of globular proteins up to now is the following: the accurate analysis of the unfolding equilibria of chymotrypsinogen A and ribonuclease gives rise to nonlinear van't Hoff plots, which were interpreted by BRANDTS, on the basis of the two-state model, as being due to the change of the specific heat of the protein upon unfolding, a prediction which was subsequently confirmed by direct calorimetric measurements (BRANDTS, 1964; JACKSON and BRANDTS, 1970).

A. Van't Hoff Analysis of Unfolding Reactions

Describing the unfolding reactions as an intramolecular conformational change between two states of the protein

$$A\ (\text{folded}) \longleftrightarrow B\ (\text{unfolded})$$

gives an apparent equilibrium constant K defined by

$$K = (A)/(B) = \Theta/(1-\Theta) \qquad (1)$$

where Θ is the degree of transition. The corresponding change of the free energy is

$$\Delta G = -RT \ln K \qquad (2)$$

and the thermodynamic functions are then derived in the usual way by differentiating the equilibrium data, for example with respect to the temperature. The influence of external variables is generally given by

$$d(\Delta G) = (\delta\Delta G/\delta T)_{P,n_i} + (\delta\Delta G/\delta P)_{T,n_i} + (\delta\Delta G/\delta n_i)_{T,P} \qquad (3)$$

$$= -\Delta S\ dT + \Delta V\ dP + \Sigma\Delta\mu_i\ dn_i$$

The change of the temperature, dT, and pressure, dP, are rather universal external parameters, while the change of the solvent composition dn_i, is a specific one, and the most important in biological systems. The first derivative of the free energy with respect to the temperature gives the change in enthalpy and entropy. With a constant change of the specific heat ΔC_p in the reaction the following expression is derived:

$$\Delta G = \Delta H - T\Delta S = [\Delta H_o + \Delta C_p(T-T_o)] - T[\Delta S_o + \Delta C_p \cdot \ln T/T_o] \qquad (4)$$

where the subscript refers to one standard temperature T_o. Including the pressure dependence with a constant compressibility factor $\beta = (\delta V/\delta P)_T$, a constant thermal expansitivity factor $\alpha = (\delta V/\delta T)_P$, and a constant volume change ΔV_o the following expression for the change in the free energy at constant solvent composition was obtained (HAWLEY, 1971):

$$\Delta G = \Delta H_o + \Delta C_p\ (T-T_o) - T\ (\Delta S_o + \Delta C_p \cdot \ln T/T_o) + \qquad (5)$$
$$+ \Delta\alpha(P-P_o)\ (T-T_o) + \Delta V_o\ (P-P_o) + \Delta\beta(P-P_o)^2/2$$

The van't Hoff analysis of the pressure and temperature dependence of chymotrypsinogen A or ribonuclease A at acid pH leads to similar free energy surfaces for the unfolding reaction (BRANDTS et al., 1970; HAWLEY, 1971). An example is shown in Figure 2, derived from the data obtained by HAWLEY, 1971.

A change of the particle concentration n_i, which may involve ions, protons, substrate molecules denaturants like urea, guanidin HCl etc., offers a very large number of possibilities for influencing the free energy of a protein molecule. A well-studied case is the variation of the pH of the solution. The protein in

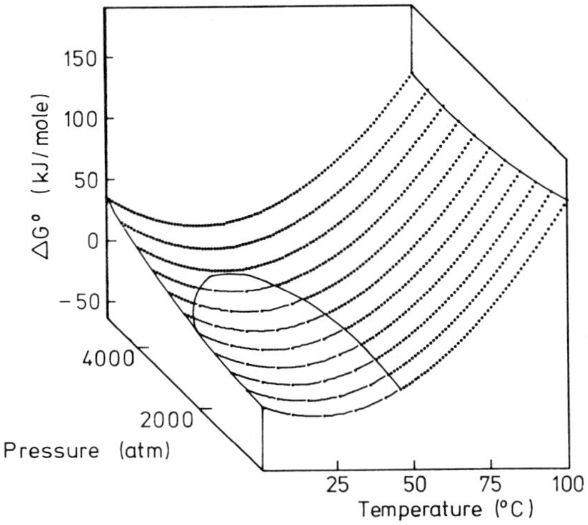

Fig. 2. The free energy change for the unfolding of chymotrypsinogen A at pH = 2.07 as function of the temperature and the pressure as obtained by HAWLEY (1971)

the folded and the unfolded state respectively may involve a different number of binding sites and/or different affinities for the ligand. The general reaction scheme will be given by

$$\begin{array}{ccccccccc} & K_{a,1} & & K_{a,2} & & & & K_{a,n} & \\ A_o & \rightleftarrows & A_1 & \rightleftarrows & A_2 & \rightleftarrows & \cdots\cdots & \rightleftarrows & A_n \\ \downarrow\uparrow & & \downarrow\uparrow & & \downarrow\uparrow & & & & \downarrow\uparrow \\ B_o & \rightleftarrows & B_1 & \rightleftarrows & B_2 & \rightleftarrows & & \rightleftarrows & B_n \\ & K_{b,1} & & K_{b,2} & & & & K_{b,n} & \end{array} \qquad (6)$$

The change of the free energy due to the binding of effector molecules is that given by BRANDTS (1964):

$$\Delta G = -RT \left[\sum_{i=1}^{n} \ln\,(1 + c/K_{a,i}) - \sum_{i=1}^{n} \ln(1 + c/K_{b,i}) \right] \qquad (7)$$

with c as the free concentration or activity of the effector molecules, $K_{a,i}$ the binding constant to the i-th binding site of the folded state, $K_{b,i}$ the corresponding values for the unfolded state, and n the number of binding sites. In some cases, like the pH dependence at acid pH, these parameters are practically independent of temperature and pressure, and can be included in Eq. (5) by a pH-dependent contribution to the entropy change ΔS. Figure 3 shows as an example the free energy surface for chymotrypsinogen A at atmospheric pressure as function of the pH and temperature. An important observation is that at particular temperatures the enthalpy change becomes zero (BRANDTS, 1964). The difficulties encountered in such a van't Hoff analysis are two-fold:

1. The two state mechanism has to be established, since otherwise the whole treatment is only a formal one and no direct physical interpretation of the change in free energy, enthalpy, or specific heat is possible.

2. The derivation of such thermodynamic functions, like the change of the specific heat, involves the second derivative of the ex-

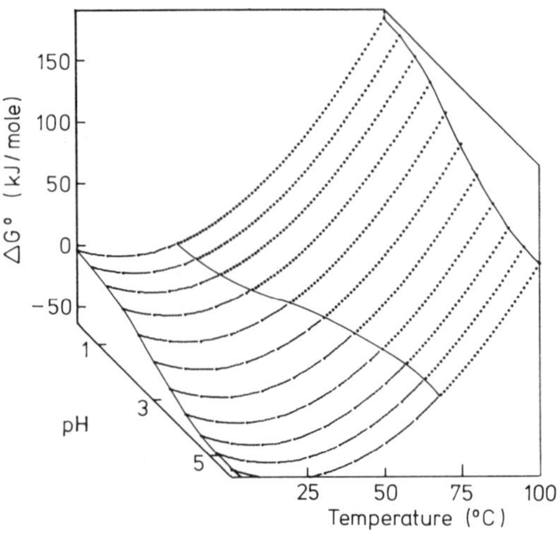

Fig. 3. The free energy change for the unfolding of chymotrypsinogen at atmospheric pressure as function of the temperature and the pH. (...) ΔG>O; (----) ΔG<O

experimental data with respect to the temperature and requires therefore an extremely high accuracy.

Both difficulties can be overcome by direct calorimetric measurements.

B. Recent Calorimetric Results

During the last few years drastic advance in the construction of calorimeters with respect to sensitivity and accuracy has taken place, which allows to test the prediction made by the van't Hoff analysis based on a two-state transition. Figure 4 shows schematically the specific heat of dilute protein solutions as functions of the temperature. The unfolding transition shows up as a peak in the heat uptake (JACKSON and BRANDTS, 1970; TSONG et al., 1970; PRIVALOV and KHECHINASHVILI, 1974).

The peak height or half-width is proportional to the van't Hoff enthalpy change, while the area of the peak corresponds to the calorimetric enthalpy change. The experimental observation by

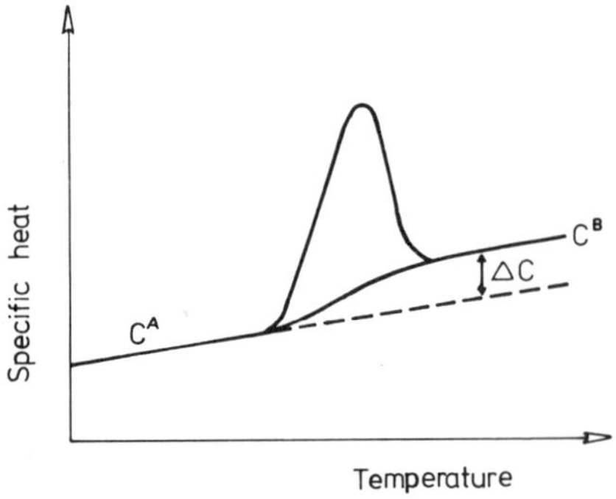

Fig. 4. Schematic presentation of the differential heat uptake of protein solutions in calorimetric measurements; C^A is the specific heat of the native state, C^B the one of the unfolded state and C the change in specific heat upon unfolding. The area under the peak corresponds to the calorimetric enthalpy change, while the width or the height of the peak is proportional to the van't Hoff enthalpy change

Table 1. Comparison of the change of the specific heat in kJ/mol K upon unfolding of globular proteins as derived from calorimetric measurements (PRIVALOV and KHECHINASHVILLI, 1974) and the van't Hoff analysis of optical melting curves. The error in both cases is approximately ± 10%. Values in brackets are obtained from the analysis of the pressure-dependence of unfolding

Protein	ΔC_p (calorimetric)	ΔC_p (van't Hoff)
chymotrypsinogen	16	14 (16)[a,b,c]
α-chymotrypsin	12.5	14[c]
Ribonuclease	5.5	8.5 (7.1)[d,b]
Myoglobin	11.3	10.5[e]
Lysozyme	6.7	8.8[f]

[a]BRANDTS (1964); [b]HAWLEY (1971); [c]POHL (1968a); [d]BRANDTS et al. (1970); [e]ACAMPORA and HERMANS (1967); [f]SEGAWA et al. (1973)

PRIVALOV and KHECHINASHVILI (1974) on the unfolding of five proteins (lysozyme, myoglobin, chymotrypsin, ribonuclease, and cytochrome c) at different pH values shows that the calorimetric ΔH_c is indeed very close to the van't Hoff enthalpy change ΔH:

$$\Delta H_c = 1.05 \pm 0.03 \ \Delta H$$

as is expected in the case where intermediates states are of minor importance. The change of the specific heat upon unfolding ΔC_p, as determined by calorimetry, agree within the error of measurements with those obtained from the van't Hoff analysis (Table 1). A result, not anticipated from previous measurements with other methods, is the linear change of the specific heat of the native proteins, with temperature of about 9 ± 2 mJ/mol K^2 being the same for all five proteins studied so far.

Both observations, the large increase in the specific heat upon unfolding and its increase with temperature can be rationalized when compared with the specific heat of small model compounds in aqueous solution. Figure 5 shows the specific heat of a some small molecules in aqueous solution (COHN and EDSALL, 1943). The pure compounds in the liquid or solid state have all very similar specific heats and also a much smaller temperature dependence. The pecularities observed are mainly due to the interaction with water:

1. The exposure of hydrophobic parts, like methyl groups, increases the specific heat due to the solvation by water by about 60 J/mol K (see the difference between alanine and glycine or between lactamid and glycolamid). This change of the specific heat is practically independent of the polar groups present in the molecule and of the temperature of the solution.

2. The change of the specific heat upon the solvation of polar, but uncharged, groups is close to zero around room temperature, but shows a strong temperature dependence. Thus the contributions

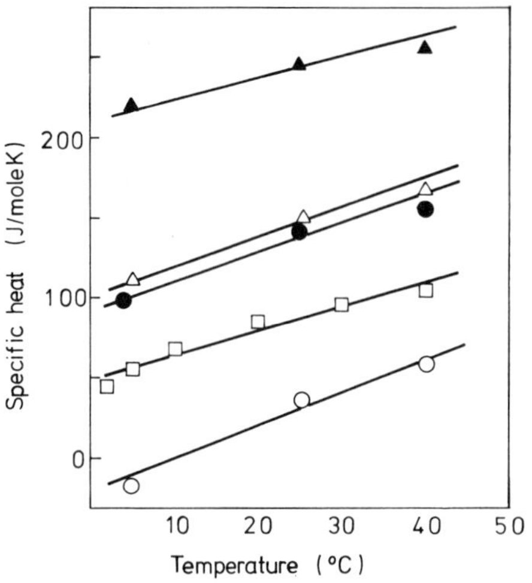

Fig. 5. Specific heat of dilute aqueous solution of small molecules as function of temperature. (Δ) glycolamid, (▲) lactamid, (O) glycine, (●) alanine, (□) urea. The interaction of unpolar groups with water increases the specific heat considerably (e.g. glycine:alanine), while for polar groups the interaction with water at room temperature does not change the specific heat appreciable, but it is accompanied by strong increase with temperature

of the exposure of unpolar or polar groups to water leads to rather different contributions to the specific heat.

The increase in the specific heat upon unfolding appears to be largely due to the exposure of unpolar groups, which in the native state, are mainly in the interior of the molecule. The increase of the specific heat with temperature of the native proteins may be due to the interaction of polar groups on the surface with the surrounding water. Taking into account the exposure of surface groups to water in the native state as calculated from the X-ray structures (LEE and RICHARDS, 1971; SHRAKE and RUPLEY, 1973), nearly a quantitative explanation of the specific heat of globular proteins and its temperature dependence can be obtained from the small molecule data (POHL, unpublished). But it should be mentioned also, that this is not the usual explanation for the increase of the specific heat of a protein solution with temperature; most authors take this increase of C^A as an indication of some kind of conformational fluctuation of the native structure.

Considerable changes of the specific heat appear to be of great importance not only in unfolding reactions of proteins, but also in the association of globular proteins with each other or with coenzymes etc. This has important consequences on studing the temperature dependence of protein reactions:

The enthalpy change involved may be positive, negative or zero, depending in the temperature of measurement. Therefore, if it happens that one is studing a protein reaction at a temperature where $\Delta H = 0$, T-jump studies may not be of great value.

III. "Slow" Temperature-Jump Methods

For measuring the kinetics of reactions in a time-range larger than a few seconds after a temperature jump, the method based on a fast thermostating of a solution in a micro-cell was found to

Fig. 6. A simple arrangement for cooling or heating a solution in a usual 1x1 cm fluorescence cuvette. Arrows indicate the flow of the thermostating liquid through the tube and also the lightpaths for the exiting and emitted light

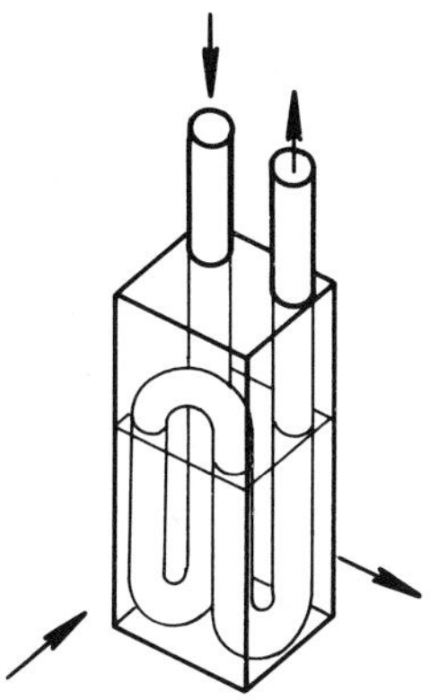

be very satisfactory (POHL, 1968a). Since it uses heat conduction for the transfer of thermal energy it allows the solution to be heated and cooled by an arbitrary amount, which is given by the temperature settings of two water baths. In order to get a reasonable compromise between the heating rate, which is proportional to the square of the dimensions, and the loss of light due to a small aperture of the cell, a careful consideration of geometric and thermal properties of the materials used in the construction of the cells is necessary.

A. Cells for Optical Measurements

The simplest design found up to now is shown schematically in Figure 6, which requires practically no special cells or modifications of the photometer. It was designed mainly for measurements of the fluorescence change where the angle between the exiting and the emitted light is 90 degrees. (If both are at the same wavelength, this cell may also be used for light scattering experiments in a fluorimeter.) A tube of the indicated form, through which the thermostatting liquid flows, is mounted into a teflon stopper and inserted into a usual 1 x 1 cm cuvette. Making the heat-exchange tube from thin-walled glass tubes requires a lot of skill, but with an inert metal tube, this type of heat exchanger can be built very easily. With stopcocks for switching the liquid from one water bath to the other it offers a slow temperature-jump method, the apparatus for which can be built in most laboratories and which should be useful for practical classes.

Two types of cells for absorbance measurements are shown in Figure 7. In Figure 7a the solution comes in contact only with glass. If thin-walled glass tubes with 3 mm diameter are used in the construction a reasonably fast heat exchange (half-time about 1.5 s) can be achieved. Most of the measurements on the unfolding and refolding of globular proteins reported previously and below have been performed with such cells, mounted into a microcell holder of a Zeiss PMQ 3 spectrophotometer. This cell holder allows for adjusting the position of the cell in three dimensions and a small rotation of the cell in three perpenticular directions with respect to the light beam.

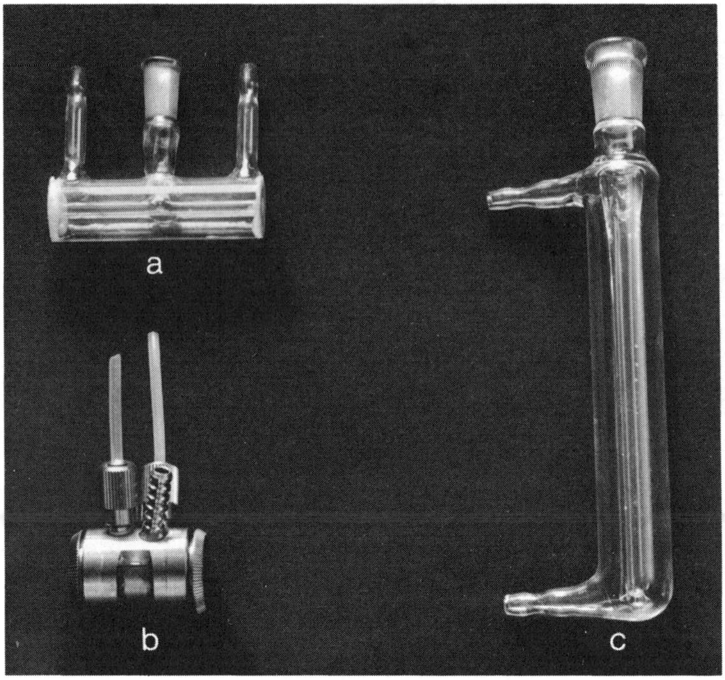

Fig. 7a-c. Photographs of different cells used in "slow-T-jump" experiments. (a) Glass cell for absorbance measurements with 4 cm lightpath and 3 mm diameter of the inner tube. (b) Modification of a commercial flow-through cell for rapid thermostatting the solution in a tube of 1 mm diameter. (c) Cell for following the pH change of a solution with a micro-pH electrode after a temperature jump

A convenient cell for absorbance measurements is shown in Figure 7b, which is a modification of a commercially available flow cell from ZEISS. By cutting a piece from the middle of the cell, a thin-walled tube, for example a platinum tube with 1 mm diameter, can be inserted. The cell body is restored by glueing on a piece of metal, which leaves room around the platinum tube for the thermostatting liquid, and the connections for the inlet and the outlet tubes. With this cell type a rather fast heat exchange rate is obtained (half-time about 0.3 s). The filling of the cell requires only 50 μl of solution and can be done while the cell is mounted in the photometer. This is very convenient, if the solution has to be changed very often. Both cells can be used for temperatures up to 70°C with aqueous solutions. At higher temperatures the formation of bubbles at the walls of the cell – even with degassed solutions – often interferes with the light path and gives rise to artifacts. With completely reversible unfolding reactions the equilibrium data, and the kinetics as function of the temperature, can be obtained within a short time using such a cell.

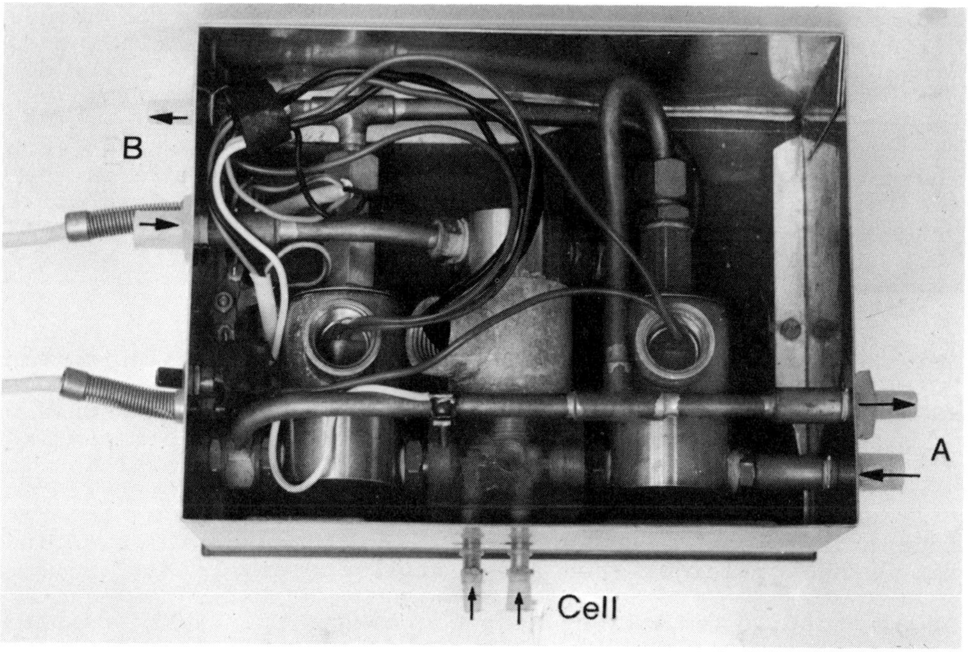

Fig. 8. Switching unit with three magnetic valves for changing the liquid flowing through the cell from water bath A to water bath B

B. Cell for pH-Measurements

The measurement of the change of the hydrogen activity upon the unfolding of proteins is not a very common technique, but has the potential of offering a very sensitive detection in unbuffered solutions near neutral pH. Using a micro-pH electrode (e.g. Schott and Gen., Mainz) together with a thin walled glass vessel, as shown in Figure 7c, the kinetics of the release or uptake of protons during the unfolding or refolding of the protein can be measured after a temperature-jump. For example, the unfolding of chymotrypsin at pH 2 - 3 is accompanied by the uptake of three protons, which follows the same slow kinetics as observed by the change of the absorbance at different wavelengths. Since the time resolution is limited by heat conduction only small amounts of protein solutions are necessary.

C. Switching Unit and Thermostates

A very convenient and fast change of the thermostating liquid around the cell from one waterbath to another is achieved by the use of magnetic valves. Figure 8 shows the arrangement of three magnetic valves. The dead volume in the switching unit should be small and the tubes to the cell rather short for obtaining optimal conditions. The thermostated water baths should provide an high flow rate. This is important for achieving short dead times and fast cooling rates. For accurate measurements the temperature difference between the cell and the water baths should be determined with a thermocouple. The difference can be kept to about 0.01° per degree deviation of the bath temperature from the room temperature by proper thermal isolation of all tubings. This calibration allows to measure the bath temperature with calibrated thermometers and to calculate the actual temperature in the cell, but for most measurements around room temperature this correction can be neglected.

D. Other Methods of Temperature Perturbation

For the measurement of slow T-jump kinetics the solution can also be heated by a pulse of alternating current (SPATZ and CROTHERS, 1969; TSONG et al., 1972a,b). In order to achieve a downward jump of the temperature, which may be important in measuring the rate of renaturation at temperatures below the transition region, one can also use a stop-flow apparatus with the two syringes at different temperature. After mixing the solution, an intermediate temperature is obtained and the kinetics can be followed in the time range of larger than a few milliseconds.

An alternative method has been proposed by F. KAYNE (personal communication) of forcing the solution through a specially designed heat exchanger. In this way it should be possible to achieve heating and cooling rates in the time range of fractions of a second. At present it appears that the steady-state rates of unfolding reactions of globular proteins are in the time range from some fractions of a second to minutes, and therefore the different methods described above should be of value in studying most of these reactions.

Of course, if the perturbation is brought about by the addition of some effector molecules, the stop-flow method offers a complete coverage of this time range.

IV. Kinetics of Unfolding and Refolding

A considerable amount of experimental data have been collected on the kinetics of denaturation of proteins after large perturbations, but only relatively few kinetic measurements have been reported using small perturbations of the unfolding equilibria of globular proteins. Temperature-jump studies are of considerable interest, since a large time range can be covered. Many of the unfolding reactions will be accompanied by some changes of the optical properties, such as an exposure of aromatic side chains observed near 280 nm, or a change of the hypochromism of the peptide bonds near 200 nm.

Kinetic experiments allow to expand the reaction along the time coordinate and give important informations on the reaction mechanism. An "all-or-none" transition ($A \rightleftharpoons B$), for example, will

Fig. 9. Schematic presentation of the kinetics of an "all-or-none" transition after a heating-jump (I→II) and a cooling-jump (II→I)

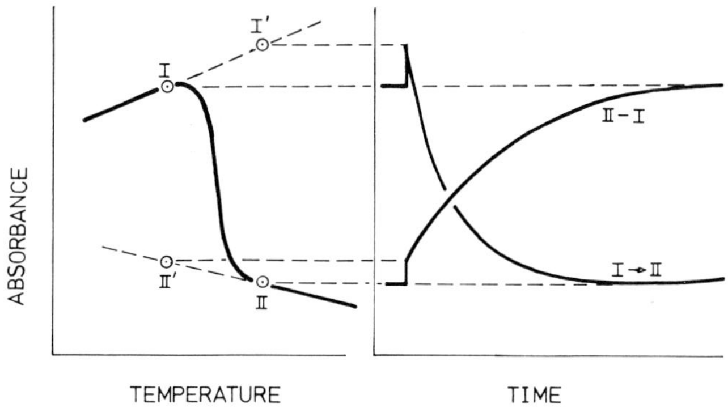

be characterized by a single exponential decay to equilibrium after a step-wise perturbation according to

$$A_t = A_{t_o - t_\infty} \exp(-t/\tau) \qquad (8)$$

A reaction involving a limited number of discret intermediate states of comparable population (A ⟷ B ⟷ C ⟷ D...) may give rise to a spectrum of monomolecular steps with very different time constants. The relaxation to equilibrium will then be the sum of exponential decay curves:

$$A_t = \sum_i A_i \exp(-t/\tau_i) \qquad (9)$$

In this way kinetic experiments allow to exclude quickly some of the possible mechanisms, which is difficult to do on the basis of equilibrium measurement alone.

These formulae also indicate that from relaxation experiments two different parameters are obtained in the same experiment:

1. The amplitudes of the relaxation curves A_i and
2. The relaxation times τ_i.

Figure 9 illustrates in a schematic way the outcome of kinetic experiments, if the unfolding follows strictly an "all-or-none" transition. Since the absorbance of the native and/or the unfolded state is usually a function of the temperature – a trivial one being the change of the solvent density with temperature – this gives rise to fast effects, occurring during the heating or cooling of the solution. (Figure 9 also shows that the amplitudes for a heating-jump are not necessarily the same as for a cooling-jump of the same size.)

Figure 10 gives an example for the difference between the static absorbance change and the relaxation amplitude for the folding of chymotrypsinogen as a function of the temperature. The data are obtained by keeping the temperature of one water bath just below the transition region while the temperature of the other one is changed in every experiment. Curve (a) gives the total change of the absorbance between $t = \infty$ and $t < 0$ (II → I in Fig. 9),

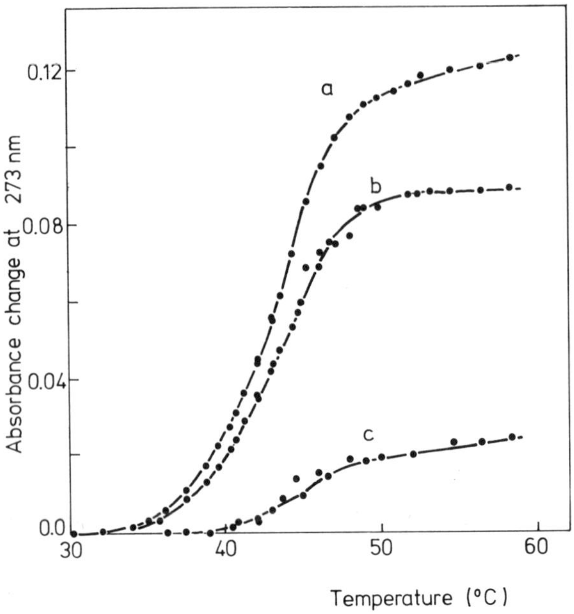

Fig. 10. The absorbance change of chymotrypsinogen at pH 2.0 for T-jumps of different size. (a) The total change of absorbance ($A_{II \to I}$), which can also be obtained by static measurements in the spectrophotometer, at different temperatures. (b) The amplitudes of relaxation curves for jumps from the indicated temperature to 30°C, obtained from the extrapolition of a plot of ln A_t against time back to t = 0 ($A_{II' \to I}$). (c) Difference between the static change and the amplitude after taking into account the density change of the solution at different temperatures. About half of this change can be due to transient effects, occurring faster than the heating rates of the slow T-jump method

curve (b) the change of the absorbance by extrapolating the exponential relaxation curve to t = 0, that is the amplitude (II' → I) for the "cooling-jump" from the indicated temperature to final one. Curve (c) gives the difference in absorbance – after correction for the change in solvent density – which is associated with the temperature dependence of the extinction coefficient and with transient effects. Most of the change in absorbance follows a single exponential in the time range of seconds to minutes. Only in the upper end of the transition curve do transient effects become appreciable.

A. Kinetic Difference Spectra

If the temperature of both water baths is kept constant and the wavelength of observation changed, one obtains from an analysis of the relaxation amplitudes 'kinetic' difference spectra, which can give additional information on the presence of intermediate states in the reaction. Figure 11 gives, again as an example, the difference spectra between the folded and the unfolded state

Fig. 11. "Kinetic" difference spectra for the unfolding of chymotrypsinogen at pH = 2.0 for jumps from 31.4° to 43.0°C and back. (o) Wavelength dependence for the amplitudes of cooling jumps ($A_{II'\to I}$). (Δ) Differences between the amplitudes of heating ($A_{I'\to II}$) and cooling ($A_{II'\to I}$) jumps. (□) Differences between amplitudes ($A_{II'\to I}$) and the static absorbance change ($A_{II\to I}$)

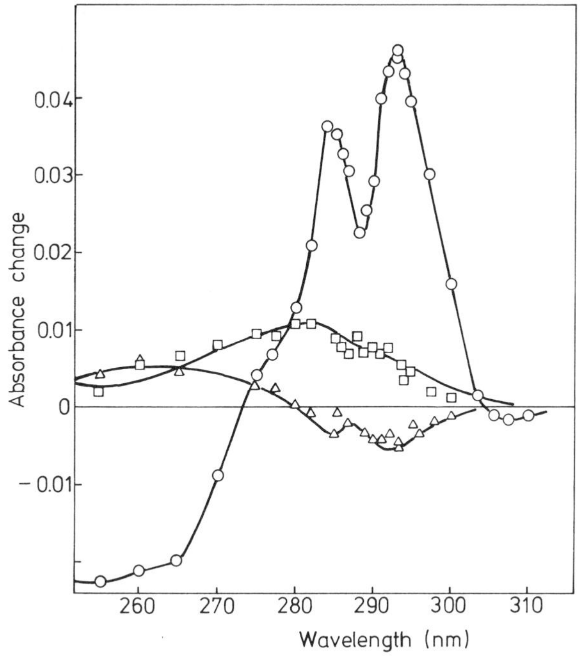

of chymotrypsinogen at acid pH for jumps between 31.4° and 43.0°; that is, from the start of the melting to T_m and back again.

The relaxation time was the same at all these different wavelengths. The difference spectrum (a), obtained from the amplitudes of the relaxation curves for heating-jumps, is typical for the exposure of a number of tryptophan and tyrosine residues to the aqueous solvent upon the unfolding (BELLO, 1970). Spectrum (b) shows the difference spectrum between the amplitudes obtained from heating the solution (I' → II) and the ones for cooling (II' → I). As already discussed above, most of the difference can be accounted for by the temperature dependence of the extinction coefficients in the native and denatured state respectively. Spectrum (c) shows the difference between the relaxation amplitude and the total change of absorbance (II → II').

By comparing spectrum (c) with the one of the protein, it becomes obvious that the major contribution is due to the change of the solvent density. Such an analysis can give rather detailed information on the optical properties of the native and the unfolded state of the protein. Furthermore it allows to select the most favorable wavelength for observing transient effects.

B. Steady State Rates

From the above discussion of the relaxation amplitudes it is obvious that most of the reaction after a perturbation of the equilibrium by a temperature jump follows a single exponential, relatively slow, decay to the final state. At constant solvent com-

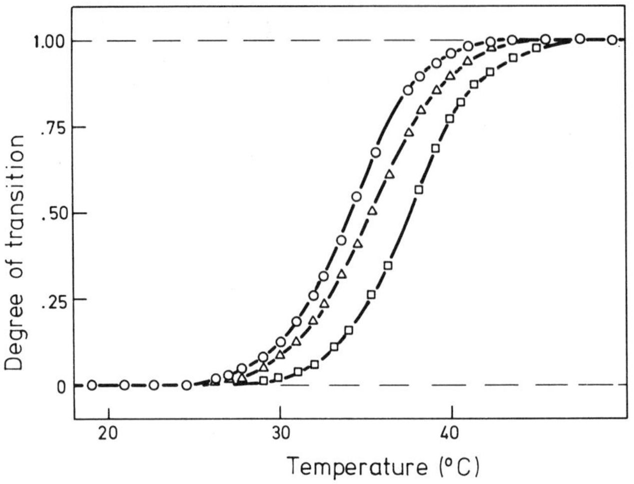

Fig. 12. Melting curves for α-chymotrypsin (o), and covalent modifications by a di-isopropyl-phosphoryl (DIP) group (□) and an anthranyloyl (AN)-group (Δ), respectively (pH = 2.08)

position this relaxation time is independent of the size of the perturbation and the parameter of observation (absorbance, optical rotation, fluorescence, or pH) and is only a function of the final temperature of the jump. This allows to describe the unfolding within a good approximation as an intramolecular conformational change between two macroscopic states of the protein.

Instead of reviewing data obtained on the kinetics of unfolding of different proteins (BRUNORI et al., 1968; POHL, 1968a,b, 1969, 1972; McPHIE, 1972; SEGAWA et al., 1973), an example will be discussed of the influence of a covalent modification of α-chymotrypsin with a di-isopropylphosphoryl (DIP) and by an anthranyloyl (AN)-group (POHL, 1970).

Figure 12 shows that the introduction of groups at the active site of the enzyme leads to a stabilization of the "native" conformation relative to the unfolded one, if all other parameters are kept the same. The over-all or steady-state rate constants for the unfolding $\vec{k}$ and for refolding $\overleftarrow{k}$ are calculated from the two equations

$$K = \Theta/(1-\Theta) = \vec{k}/\overleftarrow{k} \text{ and } \tau^{-1} = \vec{k} + \overleftarrow{k} \qquad (10)$$

where the degree of unfolding Θ can be calculated from the relative amplitudes of the relaxation curves at different temperatures.

The temperature dependence of the relaxation time also reveals the differences between these proteins (Fig. 13). Calculating the overall rate constants shows that both rate constants are smaller in the modified proteins. Figure 14 gives the ratio of the rate constants obtained with the modified proteins relative to the ones of the unmodified enzyme. The major influence is due to a decrease in the rate of unfolding accompanied by a decrease in the activation energy by 20 and 30 kJ/mol, respectively. The difference of the refolding rate is less drastic and shows a nonlinear dependence on the temperature. Thus, the temperature de-played an important role and contributed significantly to the characterization of the binding mechanism (FROESE and SEHON, 1975). The interaction of antibodies with haptens was often found to be characterized by a high rate constant of association (in the range of $10^7 - 10^8$ M^{-1} s^{-1}) and it was argued that such reactions may be diffusion-controlled (DAY et al., 1963). Fast kinetic methods (stopped flow and chemical relaxation) were thus used in many of these studies. In most systems the binding mechanism was found to consist of a single (association) step for which the rate constants and sometimes also the activation parameters were measured. Comparative kinetic studies of homogenous antibodies with different haptens enabled the elucidation of the elementary interactions involved in the binding. In parallel, a search for the kinetic manifestation of hapten-induced conformational changes was carried out but until recently inconclusive results were reported.

The majority of kinetic studies were performed with haptens, which are low molecular weight compounds that represent an antigenic determinant. However, the hapten-antigen analogy should be considered critically, since no induction of immunoglobulin effector functions can be achieved upon hapten binding (METZGER, 1974). This is probably due to the monovalency of haptens and to their interaction with a relatively limited number of contact residues in the site (cf. the review of DAVIES et al., 1975 describing phosphoryl-choline binding to its antibody site). In view of this situation more kinetic studies of antigen-antibody systems are required. The advantage of haptens is that a chemically defined part of the antigen is being used and precipitate formation via cross linking is avoided. The use of a hapten as a model for antibody-antigen interactions is therefore justified, and in spite of its limitations has yielded detailed insight into the binding mechanism.

A recent review has dealt essentially with all the literature on the subject which appeared till 1974 (FROESE and SEHON, 1975). We shall therefore confine ourselves to later results, and to the analysis and discussion of some selected topics that evolve from kinetic studies of antibodies and haptens. First, a general rate-analysis of the association step will be attempted. We shall then proceed to the analysis of the specific rates of dissociation which reflect predominantly the elementary interactions involved in hapten binding, thus reporting about the structure of the antibody combining site. Finally, kinetically observed conformational equilibria in antibody-hapten systems will be reported and discussed.

II. The Kinetics of the Association Step

The aim of this section is to try to characterize in detail the energy profile of the association step in antibody-hapten reactions. An attempt will be made to rationalize the measured rate constants in terms of this profile, and to correlate the activation parameters with molecular properties. We shall also deal with the question of whether the rate of the forward reaction is controlled by the diffusion of the reactants.

Hapten-immunoglobulin association may be described by the following scheme:

$$A + B \underset{k_{off}}{\overset{k_{on}}{\rightleftharpoons}} C \qquad (1)$$

where A is the immunoglobulin (antibody), B is the hapten (or antigen), C is the complex formed between them, and k_{on} and k_{off} are observed the rate constant for the association and dissociation, respectively. Table 1 summarizes most of the data on k_{on} and k_{off} documented by FROESE and SEHON (1975) together with more recent results, all listed in the order of decreasing k_{on}. Included in the Table are also data on activation parameters where available.

From Table 1 it is evident that while k_{on} changes only over three orders of magnitude ($5 \cdot 10^5$ – $5 \cdot 10^8$), k_{off} is much more diverse and may change from $2.7 \cdot 10^{-5}$ to $6 \cdot 10^3$ (more than 8 orders of magnitude). This seems to be due to the fact that for noncovalent binding $\Delta G^{\ddagger}_{on}$ is never very large, as no high energy bonds are broken upon association; $\Delta G^{\ddagger}_{off}$, on the other hand, may attain a wide range of values, depending on the number and magnitude of the elementary interactions that have to be disrupted in the process of dissociation. The latter fact forms the basis of the relation between k_{off} and the overall free energy of binding which underlies the kinetic analysis of elementary interactions in Section III. The factors that determine k_{on} and are responsible for its high value and limited variability will be discussed below.

Scheme (1) is experimentally found to represent the overall binding of haptens to immunoglobulins in all investigated systems, either as the only step in the reaction (HASELKORN et al., 1974; FROESE and SEHON, 1975) or when coupled to a protein isomerization step (LANCET and PECHT, 1976; see Sect. IV). Still, a rigorous kinetic analysis of the association requires the use of a more detailed mechanism, as previously proposed by HASELKORN et al. (1974). This mechanism actually applies to any association reaction in solution (EIGEN, 1974) and has the following form:

$$A + B \underset{k_{-1}}{\overset{k_1}{\rightleftharpoons}} AB \underset{k_{-2}}{\overset{k_2}{\rightleftharpoons}} C \qquad (2)$$

where A is the binding site of the immunoglobulin (antibody), B is the hapten or antigen, AB is an encounter complex in which A and B form a part of the solvation sphere of each other with no elementary interactions yet formed, and C is the final noncovalent complex. The first step of this scheme is the *encounter step* while the second is the *actual binding step*. Some ligand-protein interactions, such as hydrophobic ones, may be formed in the mere process of A becoming part of the solvation sphere of B, in which case one should take AB to be an "outer sphere" encounter complex. This, however, does not change the basic validity of the following discussion.

For systematic analysis in terms of scheme (2), one has to characterize its two steps separately. Kinetic measurements are informative only with respect to the overall behavior scheme (1) and more information is therefore needed for at least one of the two steps of the more detailed scheme. Such information is indeed available for the encounter step, which may be characterized given the geometry and diffusion properties of the reactants, even without detailed knowledge of chemical attributes. This will enable us to set limits on the values of its rate and activation parameters. Then, using the relations between the overall and detailed kinetic parameters (as derived below), and the overall experimental kinetic data, an attempt will be made to deduce the kinetics of the actual binding step, and to correlate it with chemical properties.

In the treatment that follows we shall deal with the case of noncharged reactants, i.e. use the general equations for the limit where the charge equals zero. However, the same analysis may be carried out for charge molecules. The rate constants of the encounter step are given by (cf. EIGEN, 1974):

$$k_1 = \frac{4\pi N}{1000} (D_A + D_B) R_{AB} \qquad (3)$$

$$k_{-1} = 3(D_A + D_B) \frac{1}{R_{AB}^2} \qquad (4)$$

where N is Avogadro number, D_A and D_B are the diffusion coefficients of the antibody and ligand respectively, and R_{AB} is the encounter distance.

The equíbrium constant for this step is therefore:

$$K_1 = \frac{k_1}{k_{-1}} = \frac{4\pi R_{AB}^3}{3} \cdot \frac{N}{1000} \qquad (5)$$

Legend to Table 1 (see pp. 310 and 311).

Abbreviations: ADHB 4-(3-aminophenyl)-2,6-diphenyl-N-(4-hydroxyphenyl) Pyridinium; TNP 2,4,6-Trinitrophenyl; DNP 2,4-Dinitrophenyl; NP 4-Nitrophenyl; DHNDS 4,5-dihydroxy-3-(p-nitrophenylazo)-2,7 Naphthalene disulphonate; DNP-n, m-NDS 2-(2,4 dinitrophenylazo)-1-naphthol-n,m-disulphonate; AC Amino caproate; N-R 1-naphthol-4(4-(4'-azobenzeneazo)phenyl arsonate); DMP-R' p-(p-dimethylaminophenylazo)-benzene arsonate; DT Diphtheria Toxin; DNS 4-Dimethylaminoaphthalene-1 Sulphonyl-NH-; IM_m-NPFI (Isomaltose)$_n$-1(m-nitrophenyl)flavazole; Hb Hemoglobin; Lactose-dye N-(α-N-Acetyl,ε-N-DNP-L-Lysyl)-p-aminophenyl-β-lactoside.

[a]From fast relaxation time data (see Sect. IV). [b] Calculated from k_{on} and overall K. [c]Fast association (see text, Sect. II). [d]Slow association (see text, Sect. II). [e]These are average values, since considerable temperature dependence of $\Delta H^{\ddagger}_{on}$ was found. [f]Calculated from k_{on} and $\Delta H^{\ddagger}_{on}$ by Eqs. (8), and (9).

Table 1. Kinetic parameters for antibody-hapten association

	Immunoglobulin or antibody	Hapten or antigen	k_{on} (M^{-1} s^{-1})	k_{off} (s^{-1})
1	Anti ADHB	ADHB	$6.0 \cdot 10^8$	6000
2	MOPC 315	$DNP-NH-CH_3$	$5.2 \cdot 10^8$	540
3	MOPC 315	$DNP-NH-nC_4H_{10}$	$3.4 \cdot 10^8$	110
4	MOPC 315	α-N-DNP-Glycine	$1.9 \cdot 10^8$	1300
5	Anti NP	DHNDS-NP	$1.8 \cdot 10^8$	760
6	MOPC 460[a]	ε-N-DNP-Lysine	$1.3 \cdot 10^8$	580
7	MOPC 315	ε-N-DNP-Lysine	$1.3 \cdot 10^8$	53
8	Anti TNP (T1+2P)	TNP-AC	$1.0 \cdot 10^8$	1.6
9	Anti DNP	DNP-AC	$9.7 \cdot 10^7$	1.1[b]
10	Anti TNP (T1+2P)	DNP-AC	$8.7 \cdot 10^7$	6.7
11	Anti DNP	DNP-Lysine	$8.4 \cdot 10^7$	1.1[b]
12	Anti DNP	DNP-3,6-NDS	$8.0 \cdot 10^7$	1.4[b]
13	Anti DNP (D64P)	DNP-AC	$7.4 \cdot 10^7$	2.4
14	Anti Fluorescein	Fluorescein	$6.0 \cdot 10^7$	$9 \cdot 10^{-5}$
15	HOPC 8	Phosphorylcholine	$4.1 \cdot 10^7$	110
16	MOPC 315	$DNP-(Lysine)_9$	$3.3 \cdot 10^7$	32
17	Anti DNP (D64P)	TNP-AC	$3.1 \cdot 10^7$	27
18	Anti P-Azobenzene Arsonate	N-R'	$2.0 \cdot 10^7$	50
19	Anti DNP	DNP-2,5-NDS	$1.4 \cdot 10^7$	410
20	Anti Digitoxin	Digitoxin	$1.4 \cdot 10^7$	$2.4 \cdot 10^{-4}$
21	Anti Digitoxin	Ouabain	$1.3 \cdot 10^7$	$1.4 \cdot 10^{-2}$
22	Anti DNP	ε-N-DNP-Lysine	$1.1 \cdot 10^7$	0.5[b]
23	Anti P-Azobenzene Arsonate	DMP-R'	$1.1 \cdot 10^7$	$1.4 \cdot 10^3$
24	Anti Digitoxin	Digoxin	$1.1 \cdot 10^7$	$1.2 \cdot 10^{-3}$
25	Anti DNP	DNP-2,5-NDS	$9.5 \cdot 10^6$	76
26	Anti DNP	DNP-insulin	$4.7 \cdot 10^6$	0.4[b]
27	Anti poly Alanine	$Ala-NH-(CH_2)_2NH-DNS$	$4.0 \cdot 10^6$	2
28	Anti Lactose IgM[c]	Lactose-dye	$3.9 \cdot 10^6$	29
29	Anti Lactose IgM[d]	Lactose-dye	$1.1 \cdot 10^6$	5
30	Anti ovalbumin	Ovalbumin	$1.0 \cdot 10^6$	
31	Anti Cytochrome c	Cytochrome c	$1.0 \cdot 10^6$	
32	Anti Dextran	IM_6-NPFI	$6.1 \cdot 10^5$	1.7
33	Anti-Hb-s(site specific)	Hb-s	$5.8 \cdot 10^5$	$2.7 \cdot 10^{-5}$
34	Anti Dextran	IM_4-NPFI	$4.4 \cdot 10^5$	3.9
35	Anti Staph. nuclease	Nuclease	$4.1 \quad 10^5$	$4.9 \cdot 10^{-4}$

k_{on}/k_{off} (M^{-1})	$\Delta H_{on}^{\ddagger}$ Kcal/mol	$-T\Delta S_{on}^{\ddagger}$ Kcal/mol[f]	T°C	Reference
$1.0 \cdot 10^5$			25	HAUSTEIN (1971)
$9.6 \cdot 10^5$	6.4[e]	-0.86	21	HASELKORN (1975)
$3.1 \cdot 10^6$	6.6[e]	-0.81	21	HASELKORN (1975)
$1.5 \cdot 10^5$			25	PECHT et al. (1972)
$2.4 \cdot 10^5$			25	FROESE and SEHON (1965)
$2.2 \cdot 10^5$			25	LANCET and PECHT (1976)
$2.4 \cdot 10^6$			25	PECHT et al. (1972)
$6.2 \cdot 10^7$	5.9	0.61	25	BARISAS et al. (1975)
$8.8 \cdot 10^7$	4.4	2.13	25	DAY et al. (1963)
$1.8 \cdot 10^7$	4.7	1.9	25	BARISAS et al. (1975)
$7.6 \cdot 10^7$	4.1	2.52	25	Day et al. (1963)
$5.7 \cdot 10^7$			25	DAY et al. (1963)
$8.5 \cdot 10^6$	4.3	2.4	25	BARISAS et al. (1975)
$6.6 \cdot 10^{11}$	7.1	-0.28	18.5	LEVISON et al. (1975)
$3.3 \cdot 10^5$			25	HARTMANN et al. (unpubl.)
$1.0 \cdot 10^6$			21	HASELKORN (1975)
$1.1 \cdot 10^6$	2.5	4.7	25	BARISAS et al. (1975)
$4.0 \cdot 10^5$			25	FROESE et al. (1962)
$3.4 \cdot 10^4$			23	FROESE (1968)
$5.8 \cdot 10^{10}$			22	SMITH and SKUBITZ (1975)
$9.3 \cdot 10^8$			22	SMITH and SKUBITZ (1975)
$2.2 \cdot 10^7$			25	KELLY et al. (1971)
$7.9 \cdot 10^3$			25	FERBER (1965)
$9.2 \cdot 10^9$			22	SMITH and SKUBITZ (1975)
$1.2 \cdot 10^5$			25	KELLY et al. (1971)
$1.2 \cdot 10^7$			25	LEE and FROESE (1973)
$2 \cdot 10^6$			25	LICHT (unpubl.)
$1.3 \cdot 10^5$	6.1	2.3	25	BLATT et al. (1976)
$2.2 \cdot 10^5$	5.1	4.1	25	BLATT et al. (1976)
	12.0	-2.7	25	LEVISON et al. (1968)
	7	2.25	20	NOBLE et al. (1969)
$3.6 \cdot 10^5$			25	LANCET et al. (unpubl.)
$2.1 \cdot 10^{10}$			2	NOBLE et al. (1972)
$1.1 \cdot 10^5$			25	LANCET et al. (unpubl.)
$8.3 \cdot 10^8$			25	SACHS et al. 1972

(the concentration units are mol/l for k_1, K_1). K_1 is equal to the molar volume (in l) of a species with a molecular radius R_{AB}. With R_{AB} in the range of 3 - 15 Å, K_1 values range between 0.1 and 10 M^{-1}. At the concentrations of A and B normally employed in studies of ligand-protein systems (where essentially always the overall association constants $K > 10^3\ M^{-1}$), AB concentration is negligible compared to those of all other species and the stationary state approximation may be assumed for it, yielding the following relations between the rate constants of schemes (1) and (2).

$$k_{on} = \frac{k_1\ k_2}{k_{-1} + k_2} \tag{6}$$

$$k_{off} = \frac{k_{-1}\ k_{-2}}{k_{-1} + k_2} \tag{7}$$

Under these conditions the overall reaction will behave as a single elementary step (EIGEN and DE MAEYER, 1974, p. 71), in line with the above-mentioned experimental results (cf. HASELKORN et al., 1974).

Two extreme cases may be considered (see EIGEN, 1974):

1. $k_2 \ll k_{-1}$, yielding $k_{on} = k_2k_1/k_{-1} = K_1k_2$ and $K_{off} = k_{-2}$. This is a case of fast diffusional preequlibrium, where (since $K_1 \sim 1$) k_2 determines the rate of association. The latter will therefore depend on the exact free energy path by which the elementary interactions involved in C are formed, and may vary significantly from one system to another.

2. $k_2 \gg k_{-1}$, where we obtain $k_{on} = k_1$ and $k_{off} = k_{-1}k_{-2}/k_2 = k_{-1}/K_2$. The rate of association is equal in this case to the rate of encounter complex formation, and is independent of the chemical nature of the system. Such reactions are said to be diffusion-controlled.

It is noteworthy that k_{on} reflects the actual binding step in case 1 and the encounter step in case 2, while k_{off} is always an expression of the actual binding step. This is so since k_{off} is never larger than k_{-2}, i.e. it is limited by the rate of disruption of the elementary interactions.

Equations 3 and 4 enable the calculation of k_1 and k_{-1}. Taking a typical ligand of molecular weight 400, the approximate values of $R_{AB} = 7.5$ Å and $D_B = 5 \cdot 10^{-5}$ cm^2/s may be used; $D_A = 5 \cdot 10^{-7}$ cm^2/s, as found for immunoglobulins. These yield a value of about $3 \cdot 10^9$ for both k_1 and k_{-1} ($M^{-1}\ s^{-1}$ and s^{-1} respectively) and $K_1 = 1\ M^{-1}$.

When the reactant A is a macromolecule, k_1 depends on Ω, the solid angle of approach of B to A. With 4π being the whole sphere, it was previosly proposed to take $\Omega = 2\pi$ namely half sphere (ALBERTY and HAMMES, 1958). This is correct only if one assumes that the site has not depth. Yet, since the binding site on globular macromolecules is invariably found to be a depression, it seems more plausible that ligand approach is limited by a cone with an apex angle $0^\circ < \alpha < 180^\circ$. This "approach cone" defines a solid angle given by:

$$\Omega = 2\pi \; (1 - \cos \frac{\alpha}{2}) \tag{8}$$

The higher limit for k_1, obtained with $\Omega = 2\pi (\alpha = 180^\circ)$, is thus $1.5 \cdot 10^9$ M^{-1} s^{-1}. A reasonable lower limit is obtained with $\alpha = 45^\circ$, where $\Omega = 0.038 \cdot 4\pi$ (only 3.8% of the sphere is available for ligand approach) and $k_1 \cong 1 \cdot 10^8$ M^{-1} s^{-1}. An orientation factor, presumably not much lower than 0.3 (as the rotational relaxation time of low molecular weight ligands in water is comparable to the lifetime of the encounter complex $1/k_{-1} = 3 \cdot 10^{-10}$ s) may decrease k_1 to $3 \cdot 10^7$ M^{-1} s^{-1}.

The value of k_{-1}, on the other hand, is not expected to change with the solid angle of approach. This is so since $1/k_{-1}$ is related to the time needed for B to attain a random-walk distance from A equal to R_{AB} on the line that connects their centers (cf. BENSON, 1960, p. 495). A constant k_{-1} is also in line with the notion that K_1 will decrease in proportion to k_1 with decreasing α, as less volume is available for the residence of B near A. Hence, k_{-1} has a more or less fixed value of $3 \cdot 10^9$ s^{-1}.

The activation parameters of the encounter step may be derived assuming that it is a regular chemical reaction for which the Eyring equation (GLASSTONE et al., 1941) applies:

$$k = \frac{k_b \; T}{h} \; e^{-\Delta G^\ddagger/RT} \tag{9}$$

where k_b is Boltzmann's constant, h is Planck's constant, T the absolute temperature and R the gas constant, and where

$$\Delta G^\ddagger = \Delta H^\ddagger - T\Delta S^\ddagger \tag{10}$$

with $\Delta G^\ddagger$, $\Delta H^\ddagger$ and $\Delta S^\ddagger$ being respectively the free energy, enthalpy and entropy of activation. The encounter complex is considered an intermediate with a standard free energy roughly equal to that of the reactants ($K_1 \sim 1$, see above). It is separated from the reactants by a diffusional free energy barrier which is due to the unfavorable process of pushing solvent molecules apart when entering or leaving the solvent cage of the encounter complex (cf. NORTH, 1964, p. 5). It is also separated from the final complex by a barrier which is the highest free energy state that the ligand-site system attains when forming the elementary interactions.

$\Delta G_1^\ddagger$ and $\Delta G_{-1}^\ddagger$ may thus be calculated from k_1, k_{-1} using Eq. (9). $\Delta H_{\pm 1}^\ddagger$ are obtained from the temperature dependence of $k_{\pm 1}$. Both rate constants depend on temperature in the same way, since the main source of dependence is due to $D_A + D_B$, with R_{AB} chaning very little over the normally studied temperature range. Diffusion coefficients in water were shown by LONGSWORTH (1954) to depend on temperature with an apparent activation energy $E_D = 4.7 \pm 0.3$ kcal/mol irrespective of molecular weight. Hence $H_{\pm 1}^\ddagger = E_D - RT = 4.1 \pm 0.3$ kcal/mol (at 25°C), and therefore $\Delta H_1 = 0$.

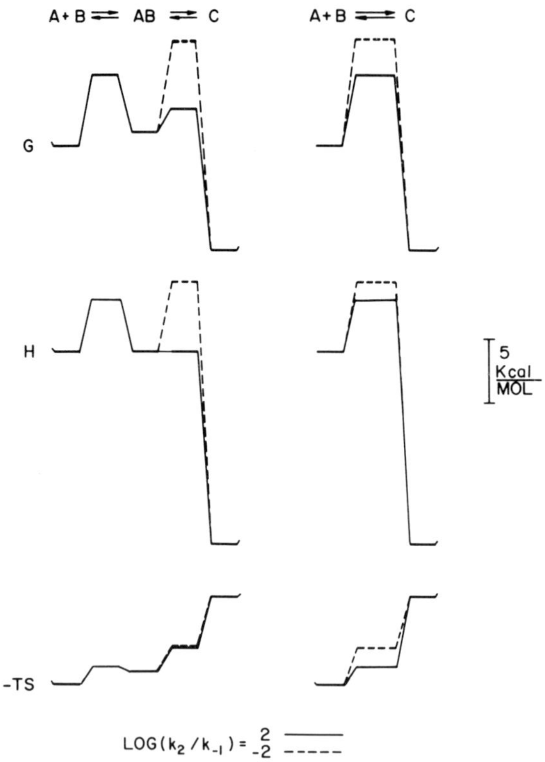

Fig. 2. Energy profiles for the cases of k_2/k_{-1} = 100 and k_2/k_{-1} = 0.01, according to Figure 1. The numerical values are summarized in Table 2

to be due to a change of $\Delta H_2^{\ddagger}$ at a constant $\Delta S_2^{\ddagger}$, (a boundary condition $\Delta H_2^{\ddagger} = 0$ at $k_2/k_{-1} = 10^2$ is chosen, thus determining also the value of $-T\Delta S_2^{\ddagger}$). Since k_{-1} is constant, Figure 1 actually shows how the effected change of k_2 is reflected in the change of $\Delta G_{on}^{\ddagger}$ (proportional to $-\log(k_{on})$ by Eq. (9)), $\Delta H_{on}^{\ddagger}$ and $-T\Delta S_{on}^{\ddagger}$. $\Delta G_{on}^{\ddagger}$ is calculated by Eqs. (6) and (9), $\Delta H_{on}^{\ddagger}$ by Eq. (11) and $-T\Delta S_{on}^{\ddagger}$ is then obtained using Eq. (10). The overall equilibrium parameters are chosen as those characteristic of dinitrophenyl binding to homogeneous immunoglobulins (JOHNSTON et al., 1974). Full energy profiles for two k_2/k_{-1} values, namely $k_2/k_{-1} = 10^2$ (diffusion control) and $k_2/k_{-1} = 10^{-2}$ (fast preequilibrium) are depicted in Figure 2 and Table 2.

The following points emerge from this simulation: the passage from the case of diffusion control to that of diffusional preequilibrium is characterized by a gradual change of the overall activation parameters. In the range of $k_2 \ll k_{-1}$ the overall activation parameters $\Delta X_{on}^{\ddagger}$ are constant, independent of the change of $\Delta H_2^{\ddagger}$, and have the value $\Delta X_{on}^{\ddagger} = \Delta X_1^{\ddagger}$. In the range of $k_2 \gg k_{-1}$, $\Delta G_{on}^{\ddagger}$ and $\Delta H_{on}^{\ddagger}$ change in the same way as $\Delta H_2^{\ddagger}$, while $\Delta S_{on}^{\ddagger}$ is constant (since $\Delta S_2^{\ddagger}$ does not change). The value of the overall activation parameters in this range is given by $\Delta X_{on}^{\ddagger} = \Delta X_1 + \Delta X_2^{\ddagger}$. Thus, in the two limiting cases the activation parameter is equal to the height of the barrier corresponding to the rate-limiting step, as measured from the state of the reactants. It is not necessarily equal to the highest barrier in the parameter profile. This is most pronounced in the middle range ($k_2 \sim k_{-1}$), where clearly $\Delta H^{\ddagger}_{on}$ is lower than ΔH_1.

The change of $\Delta S^{\ddagger}_{on}$ in the middle range is worth noting. It is effected by the change of $\Delta H^{\ddagger}_2$. $\Delta S^{\ddagger}_{on}$ is equal to $\Delta S^{\ddagger}_1$ as long as $k_2 \gg k_{-1}$. When the second step becomes rate-limiting ΔS_{on} increases, as the entropy of the activated complex for the overall reaction is now that corresponding to actual binding. This change occurs even though the entropy profile is kept constant throughout.

The conclusions of this simulation together with the limits set above for the parameters of the encounter step will now be used to analyze some experimental results. Only those cases of Table 1 for which the activation parameters are known will be considered.

1. Reactions No. 2 and 3 (Table 1): the binding of DNP derivatives to the homogeneous immunoglobulin MOPC 315 (HASELKORN, 1974). These have very large values of k_{on} and may be taken to be diffusion-controlled. However, $\Delta H^{\ddagger}_{on}$ of 6.5 kcal/mol leaves no doubt that this is not the case, since if $k_2 \gg k_{-1}$, ΔH_{on} must be strictly equal to $\Delta H^{\ddagger}_1$ (4.1 ± 0.3 kcal/mol). Still, the high k_{on} seems to imply that the reaction is "not far from diffusion control", namely $k_2 \sim k_{-1}$, which is on the borderline between the extreme cases dealt with above. In this case, ΔH_{on} of 6.5 kcal/mol requires a high $\Delta H^{\ddagger}_2$, about 9 kcal/mol (by Eq. (11)). The high k_2 ($\cong k_{-1} = 3 \cdot 10^9$ s^{-1}) must then be due to a large negative $-T\Delta S^{\ddagger}_2$ = -4.5 kcal/mol which is indeed relfected in the experimental negative $-T\Delta S^{\ddagger}_{on}$. We may conclude that the reaction of simple DNP derivatives with the homogeneous myeloma protein MOPC 315 is "almost diffusion-controlled". It has a high enthalpy barrier which is contributed by step 2: the breaking of ligand-water or protein-water hydrogen bonds, or a conformational barrier may be involved. It also has a negative entropic contribution to the free energy of activation, i.e. positive $\Delta S^{\ddagger}$, probably due to water release (hydrophobic interactions, KAUZMANN, 1959).

2. Reactions 8, 10, 13 and 17 of Table 1 (BARISAS et al., 1975): anti DNP and TNP with the homologous and heterologous haptens. If we write this series in the order of decreasing k_{on}, all other parameters also behave monotonically. Cases 8 and 17, which are the reactions of TNP-AC with anti TNP and anti DNP, have $\Delta H^{\ddagger}_{on}$ of 5.9 and 2.5 kcal/mol respectively (very different from $\Delta H^{\ddagger}_1$) and are obviously not diffusion-controlled. It is reasonable to state, therefore, that the reactions of the slightly different hapten DNP-AC with the same antibodies are not diffusion-controlled either. Let us also assume, as a first approximation, that the parameters of step 1 are equal for all four reactions (a reasonable assumption for very similar ligands and heterogeneous antibodies in which the gross geometrical properties of the sites are averaged out). This series may then be considered as a case of gradual changes in the parameters of step 2 (at constant step 1 parameters), similar to the above-presented simulation (Fig. 2). Here only the region of $k_2/k_{-1} \leq 1$ is spanned and therefore the parameters of the second step influence the overall behavior throughout the range. Reaction 8 may be similar to reactions 2,3

treated above ("almost diffusion-controlled"), but as we proceed through the series, $\Delta G_2^{\ddagger}$ becomes higher and the case of diffusional preequilibrium is approached. It is interesting that $\Delta H^{\ddagger}_{on}$ decreases but $-T\Delta S^{\ddagger}_{on}$ rises so as to overcompensate for it, yielding an increasing $\Delta G^{\ddagger}_{on}$. Reaction 17 may already be an example of a fast diffusional preequilibrium with $\Delta H^{\ddagger}_{on} = \Delta H^{\ddagger}_2$. In view of the above simulation it is not surprising that $\Delta H^{\ddagger}_{on}$ is only 2.5 kcal/mol as $\Delta H^{\ddagger}_2$ may assume any value (even negative) provided that $\Delta G^{\ddagger}_2 > 0$. For Reaction 17, $\Delta S^{\ddagger}_2$ is high enough to yield $k_2 \ll k_{-1}$, as reflected in the high $-T\Delta S^{\ddagger}_{on} = 4.7$ kcal/mol.

3. Cases 9 and 11: the stopped-flow study of DAY et al. (1963) on anti DNP antibodies. The $\Delta H^{\ddagger}_{on}$ and k_{on} values both agree with the possibility of being diffusion-controlled. Still, the values are very close to those of reactions 10 and 13 in group 2 obtained for similar antibodies and ligands, where there are indications that they are not diffusion-controlled (see above).

4. Cases 28, 29: the unpublished results of BLATT et al. for IgM anti lactose antibodies which display restricted heterogeneity and yield two distinct relaxation times. The inverse values of both times increase linearly with the sum of free ligand and free protein concentrations, apparently representing two independent associations - dissociation equilibria. The values of k_{on} are 1 - 2 orders of magnitude lower than those of the first three groups, and in this respect they are similar to reactions 27, 32 and 34 (anti polyalanine and anti dextran). If cases 28 and 29 are taken to represent this group, it may be said that the reason for the low forward rate constants is the fairly high $\Delta H^{\ddagger}_{on}$, together with a positive $-T\Delta S^{\ddagger}_{on}$.

All five cases are obviously non-diffusion-controlled, and since the observed positive $-T\Delta S^{\ddagger}_{on}$ are characteristic for association (JENCKS, 1975, p. 286) low k_{on} should be attributed to the high $\Delta H^{\ddagger}_2$ and to the lack of a negative $-T\Delta S^{\ddagger}_2$ contribution. Both are understandable since the haptens in question are hydrophilic: they form hydrogen bonds with water which have to be broken before binding and there is no compensating positive entropy change due to hydrophobic interactions. It is noteworthy that reaction 28 resembles very much the case of $k_2/k_{-1} = 10^{-2}$ in Table 2 and Figure 2.

5. Reaction 14: anti fluorescein antibodies (LEVISON et al., 1975). $\Delta H^{\ddagger}_{on}$ is high, excluding the possibility of diffusion control. The relatively high k_{on} is the result of the negative value of $-T\Delta S^{\ddagger}_{on}$. All this is in line with the chemical nature of fluorescein, being able to form hydrogen bonds with water and still having hydrophobic properties, thus being an intermediate between the haptens of groups 1 and 4.

6. Reactions 26, 30, 31, 33 and 35 are examples for the interaction of an antibody with an antigen (i.e. a macromolecule). This type of reaction requires special analysis since the parameters of the encounter step are different from those calculated above: for the interaction of spherical reactants of equal radius (r), k_1 is independent of r since $R_{AB} = 2r$ while D_A and D_B are inversely proportional to r (GLASSTONE, 1946, p. 261). The value of k_1 =

$3 \cdot 10^9$ s^{-1} M^{-1} is therefore applicable also to the interaction of two macromolecules. On the other hand, the limitation of the solid angle of approach applies here to both reactants, which may decrease the minimal k_1 by as much as an order of magnitude. Still, if we judge from reactions 30 and 31 for which activation parameters are available, the high $\Delta H^{\ddagger}_{on}$ excludes the possibility of diffusion control. The activation parameters of reaction 31 are similar to those described in Section 4, and may imply a similar type of interaction, while the high $\Delta H^{\ddagger}_{on}$ and negative $-T\Delta S^{\ddagger}_{on}$ of reaction 30 may indicate that more hydrogen bonds are broken in the reaction path and that the activated complex involves more hydrophobic interactions.

Concluding Remarks

The two-step model for hapten-immunoglobulin association (scheme 2) makes it possible to extract more information from the measured overall kinetic and activation parameters. This is mainly true for the forward (association) reaction which, according to the case, may reflect either the encounter step or the formation of the elementary interactions in the actual binding step.

For the encounter step it is found that k_{-1}, $\Delta H^{\ddagger}_{-1}$ and $S^{\ddagger}_{-1}$ may be considered to be constant for all hapten-immunoglobulin reactions, whereas k_1 and $\Delta S^{\ddagger}_1$ range between certain limits due to differences in the solid angle of approach and in the orientation factor. For the case of diffusion control, all the parameters of the overall forward reaction are expected to be identical to those of the forward reaction of the encounter step. Knowledge of the latter enables the identification of diffusion-controlled reactions. Thus, necessary experimental conditions for diffusion control are: (1) $\Delta H^{\ddagger}_{on} = 4.1 \pm 0.3$ kcal/mol; (2) k_{on} larger than about $3 \cdot 10^7$ M^{-1} s^{-1}. There may still be reactions in which these two conditions hold, yet $k_2 \gg k_{-1}$ does not. This will happen when $\Delta H^{\ddagger}_2$ is in the range of 4 kcal/mol and $-T\Delta S^{\ddagger}_2$ is not very large (≤ 2.5 kcal/mol). In such cases it may be impossible to decide whether a particular single reaction is diffusion-controlled. However, a study of a series of different haptens with one homogeneous immunoglobulin (cf. HASELKORN et al., 1974) may be helpful in this respect: a value of k_{on} independent of the chemical nature of the hapten implies diffusion control. For reactions proven to be diffusion-controlled by this method, it is then possible to obtain Ω, the solid angle of approach, from the magnitude of k_{on} (or $\Delta S^{\ddagger}_{on}$), assuming that no orientation restrictions exist.

Of the reactions listed in Table 1 only cases 1 - 17 may be diffusion-controlled according to their k_{on}. Still the value of $\Delta H^{\ddagger}_{on}$ together with variation of k_{on} for different haptens seems to indicate that for none of these the condition $k_2 \gg k_{-1}$ strictly applies, and that some of them may be at best "almost diffusion-controlled" ($k_2 \sim k_{-1}$). Essentially all haptens in this group are simple aromatic compounds and their high rate of association to the antibody may be due to the hydrophobic nature of the interactions involves, and to the fact that no strong interactions with water have to be disrupted.

The transition from diffusion control to diffusional preequilibrium is effected by a decrease of k_2. This may be the result of either higher $\Delta H_2^{\ddagger}$ or higher $-T\Delta S_2^{\ddagger}$, or both. One possible reason for an increase of $\Delta H_2^{\ddagger}$ is hydration of the reactants which has to be overcome before binding can occur. Other effects that may lower k_2 are: (1) nonspecific electrostatic interactions that have to be broken before binding (2) a conformational equilibrium between different forms of the hapten. This may be the case in oligo-peptide or oligo-saccharides which do not have a preferred conformation in solution. In such cases $-T\Delta S_2^{\ddagger}$ will be higher, yet no distinct conformational step will necessarily be resolved; (3) conformational changes in the protein. These will be dealt with in Section IV.

III. Kinetic Expression of the Elementary Interactions

In the preceding paragraphs, where the specific rate of hapten binding was discussed, it became evident that at a constant k_{off} the overall association constant of hapten binding can be augmented by increasing this rate only up to the limit set by the rate of diffusion. Thus, for a single-step equilibrium, the major kinetic expression of variation in the affinity is usually the specific rate of dissociation. This has indeed been recognized already in the very first kinetic measurements of antibody-hapten reactions. Table 1 gives some general confirmation of this situation out of the variety of antibody specificities that were investigated till now. Illustrative are reactions 1 - 10 where the k_{on} changes only six-fold and k_{off} varies over more than three orders of magnitude.

Table 3. Specific rates of binding and dissociation of isomaltose oligosaccharides to fractionated antidextran and of oligoalanines to fractionated anti-polyalanine antibodies

Hapten	Antibody fraction	k_{on} M^{-1} s^{-1}	k_{off} s^{-1}		k_{on}/k_{off} M^{-1}
IM_4-NPF	IM_3 eluate	$(4.4 \pm 0.3) \cdot 10^5$	3.9	0.6	$1.1 \cdot 10^6$
IM_6-NPF	Im_3 eluate	$(6.1 \pm 0.3) \cdot 10^5$	1.7	0.4	$3.6 \cdot 10^6$
D(Ala)-R	$D(Ala)_3$ eluate	$(5.7 \pm 0.6) \cdot 10^6$	2.0		$2.9 \cdot 10^6$
D(Ala)-R	$D(Ala)_4$ eluate	$(3.0 \pm 0.3) \cdot 10^6$	1.5		$2.0 \cdot 10^6$
$D(Ala)_2$-R	$D(Ala)_3$ eluate	$(4.3 \pm 0.4) \cdot 10^6$	1.8		$2.4 \cdot 10^6$
$D(Ala)_2$-R	$D(Ala)_4$ eluate	$(1.7 \pm 0.2) \cdot 10^6$	3.8		$4.5 \cdot 10^5$

IM isomaltose; NPF1 1(m-Nitrophenyl) flavazole (LANCET et al., to be published). (See HARISDANGKUL and KABAT, 1972). R= -HN-$(CH_2)_2$NH-dimethylamino naphthalene sulphonyl. (LICHT, A., PECHT, I., to be published).

The heterogeneity of the normally induced antibodies did not allow the systematic kinetic analysis of the affinity of a given combining site towards a homologous series of different ligands. In this context, it is of interest to consider the data obtained with normally induced anti-dextran and anti-polyalanine antibodies with their respective haptens oligo-isomaltose or oligoalanine (derivatized with a chromophore so as to enable the spectroscopic monitoring of the interactions with the antibody). In Table 3 the specific rates of binding and dissociation of the respective haptens to different fractions of the normal heterogeneous population prepared by elution with haptens of increasing length. No significant correlation between "size" of the antibody site (which changes according to the fraction) and the rates could be achieved. This is probably so because the relatively small differences existing for individual sites are not resolved in the fractions which are still heterogenous.

When the differences in affinity are more pronounced this is clearly expressed in the rates of dissociation even for a heterogeneous population. This is illustrated in the recent study of the rates of interactions between three cardiac glycosides and their specific antibodies (SMITH and SKUBITZ, 1975) carried out by rapid separation of free and antibody-bound radioactive hapten using the approach originally conceived by BERSON and YALOW (1959). This method, which is suitable for kinetic measurement of very high affinity antibodies, produced results which show the essential invariance of the specific rates of binding for the three types of antibodies (Table 4), all in the range of 10^7 M^{-1} s^{-1}.

Table 4. Specific rates of interactions betwen haptens and normally induced antibodies

Ab-System	Hapten	k_{on} (M^{-1} s^{-1}) x 10^7	k_{off} (s^{-1})
Anti digoxin	Digoxin	0.93	$1.9 \cdot 10^{-4}$
Anti digoxin	Ouabain	1.5	$1.7 \cdot 10^{-2}$
Anti digoxin	Digitoxin	1.4	$7.2 \cdot 10^{-4}$
Anti ouabain	Ouabain	0.87	$1.5 \cdot 10^{-3}$
Anti ouabain	Digoxin	1.3	$6.4 \cdot 10^{-3}$
Anti ouabain	Digitoxin	1.4	$3.8 \cdot 10^{-3}$
Anti digitoxin	Digitoxin	1.4	$2.4 \cdot 10^{-4}$
Anti digitoxin	Ouabain	1.3	$1.4 \cdot 10^{-2}$
Anti digitoxin	Digoxin	1.1	$1.2 \cdot 10^{-3}$
Anti fluorescein	Fluorescein	1 to 7	$5.2\text{-}700 \cdot 10^{-4}$

From SMITH and SKUBITZ, 1975 the antiglycosides antibodies were studied at 22°C in phosphate-buffered saline (0.15 M NaCl, 0.01 M sodium phosphate buffer pH 7.4). Those on antifluorescein at 18°C same buffer yet pH 7.0. LEVISON et al., 1975.

The pronounced affinity of these antibodies for their respected haptens is brought about by the rather slow rates of dissociation (10^{-2} - 10^{-4} s^{-1}) and their variation accounts for the affinity differences. Interesting is the examination of cross reactivity between these antibodies and the different glycosides which carry a significant basic similarity in their structure: again a variation over two orders of magnitude in k_{off} is found, which fully accounts for the affinity difference. No correlation between systematic structural changes in one hapten and its expression in the values of k_{off} was attempted in this case. Though these antibodies were of restricted heterogeneity, they are heterogeneous to a degree which would make such a correlation difficult and limited.

For a homogeneous antibody-combining site, this analysis of relation between systematic variation in structure of the hapten and the specific rates of binding and dissociation is expected to produce a meaningful correlation with structural features of the site. Protein 315 excreted by plasmacytoma cells is one of the first homogeneous immunoglobulins for which binding specificity has been determined (EISEN et al., 1968). This specificity is for 2,4 di- or 2,4,6 trinitrophenyl derivatives and an extensive correlation between the rates of interactions of several series of derivatives of 1-amino 2,4 dinitrobenzen has been carried out (PECHT et al., 1972; HASELKORN et al., 1974; PECHT, 1975).

Variation in the structure of the aromatic nitrophenyl nucleus itself is limited by the difficulty of resolving effects caused by shifting the electronic density in the ring or in ring substituents from steric changes. One example is the difference observed between 2,4-dinitroluene and 2,4,6-trinitrotoluene (k_{on} = $1.4 \cdot 10^8$ and $2.2 \cdot 10^8$ M^{-1} s^{-1} and k_{off} = 1050 and 284 s^{-1} respectively). The markedly slower rate of dissociation of the trinitrotoluene cannot be interpreted by involving interactions with the third nitro group since rather weak binding is observed for the 2,6-dinitrotoluene. Thus this rate difference most probably reflects the better charge-transfer interaction which the trinitro derivatives is capable of forming.

The largest body of derivatives examined were those with different side chains attached at the 1-amino group of 2,4 dinitroaniline. Although the main binding energy for the hapten is associated with the polynitro aromatic ring and contributions from the attached side chains are significantly smaller, the detailed mapping of the site was achieved. Moreover, having the ring anchored at one subsite gave a reliable topographic reference point to relate the other subsites.

Hydrophobic subsites were scanned by a series of alkyl derivatives of increasing length (Table 5). The branched chains and those derivatives carrying the bulkier aromatic or aliphatic rings as well as the bis-substituted amine served to indicate the width of the site at increasing distance from the DNP subsite (Fig. 3). The presence of contact residue(s) in the site which carry net electrostatic charge was indicated already by the very low equilibrium constant of 2,4 dinitrophenolate. Again for examining this

Table 5. Rate and equilibrium constants for the interaction between protein 315 and a homologous series of DNP-alkylamines and DNP-derivatives containing branched, cyclic or aromatic side chains

	Hapten		$k_{on} \cdot 10^{-8}$ M^{-1} Sec^{-1}	k_{off} Sec^{-1}	$K=(k_{on}/k_{off}) \cdot 10^{-6}$ M^{-1}	$K_{st} \cdot 10^{-6}$ M^{-1}
1)	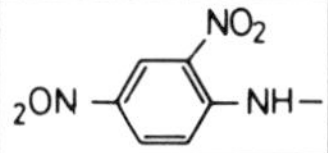	$-CH_3$	3.5 ± 0.24	373 ± 30	0.94	
2)	-"-	$-CH_2-CH_3$	2.8 ± 0.18	143 ± 24	1.95	2.04
3)	-"-	$-CH_2-CH_2-CH_3$	2.6 ± 0.18	94 ± 11	2.70	
4)	-"-	$-CH_2-CH_2-CH_2-CH_3$	2.5 ± 0.3	79 ± 10	3.20	3.40
5)	-"-	$-CH_2-CH_2-CH_2-CH_2-CH_2-CH_3$	1.2 ± 0.12	123 ± 18	1.00	1.10
6)	-"-	$-CH<(CH_3)(CH_3)$	3.7 ± 0.22	108 ± 8.6	3.40	
7)	-"-	$-CH<(CH_2-CH_2)(CH_2-CH_2)>CH_2$	2.55±0.25	116 ± 10	2.20	
8)	-"-	$-CH_2-CH<(CH_3)(CH_3)$	2.0 ± 0.14	83 ± 11	2.40	2.70
9)	-"-	$-C(CH_3)<(CH_3)(CH_3)$	3.6 ± 0.3	58 ± 6	6.10	
10)	-"-	$-CH_2-$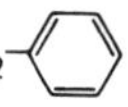	1.8 ± 0.15	47 ± 9	3.80	

elementary interaction a homologous series of derivatives was prepared and measured. This series carried straight alkyl chains with charged negative (carboxylate) or positive (ammonium) groups at their ends at increasing distance from the DNP ring (Table 6). While a carboxylate group positioned 2.5 to 3 Å away from the ring causes the value of k_{off} to drop more than thirty-fold, a positive ammonium group at similar distance increases it. The dependence of rate and association constants on the distance of a negative group from the DNP center is shown in Figure 4.

Further refinement in our probing of this site was achieved by studying the rates of interactions with the series of single DNP carrying oligopeptides. Here the length, stereochemistry and electrostatic charges were changed and led to a self-consistent picture of the spatial arrangement of the different subsites (HASELKORN et al., 1974). A series of oligolysines carrying a single DNP on the ε-amino group of their carboxy-terminal residue

Table 6. Rate and equilibrium constants for the interaction between protein 315 and a homologous series of DNP derivatives containing electrostatically charged side chains

	Hapten	$k_{on} \cdot 10^{-8}$ M^{-1} Sec^{-1}	k_{off} Sec^{-1}	$K=(k_{on}/k_{off}) \cdot 10^{-5}$ M^{-1}	$K_{st} \cdot 10^{-5}$ M^{-1}
1)	$_2ON$-(2-NO_2-phenyl)-NH-CH_2 ‖ CH_3	2.8 ± 0.18	143 ± 24	19.5	20.4
2)	-"- ‖ COO^-	1.7 ± 0.18	1340 ± 100	1.27	2.0
3)	$_2ON$-(2-NO_2-phenyl)-NH-CH(CH_3) ‖ COO^-	2.4 ± 0.2	664 ± 80	3.60	2.67
4)	$_2ON$-(2-NO_2-phenyl)-NH-CH_2-CH_2 ‖ CH_3	2.6 ± 0.18	94 ± 11	27.0	
5)	-"- ‖ OH	2.5 ± 0.3	370 ± 50	6.75	6.6
6)	-"- ‖ COO^-	1.8 ± 0.3	285 ± 50	6.40	
7)	-"- ‖ NH_3^+	0.65 ± 0.1	735 ± 65	0.89	1.3
8)	$_2ON$-(2-NO_2-phenyl)-NH-CH_2-CH_2-CH_2 ‖ CH_3	2.5 ± 0.3	79 ± 10	32.0	34.0
9)	-"- ‖ COO^-	3.8 ± 0.2	45 ± 9	84.0	
10)	-"- ‖ NH_3^+	0.83 ± 0.08	850 ± 100	0.98	
11)	$_2ON$-(2-NO_2-phenyl)-NH-CH_2-CH_2-CH_2-CH_2 ‖ COO^-	1.2 ± 0.3	55 ± 5	22.0	19.0
12)	$_2ON$-(2-NO_2-phenyl)-NH-CH_2-CH_2-CH_2-CH_2-CH_2 ‖ CH_3	1.2 ± 0.12	123 ± 18	10.0	11.0
13)	-"- ‖ COO^-	1.2 ± 0.12	60 ± 10	20.0	
14)	$_2ON$-(2-NO_2-phenyl)-NH-CH_2-CH_2-CH_2-CH_2-CH_2-CH_2 ‖ NH_3^+	0.9 ± 0.02	120 ± 10	7.5	

(Table 7) served to indicate that beyond the second lysyl residue specific interactions with the binding site were small as k_{off} did not change. The decrease in the overall association constant could be attributed to a decrease of k_{on} probably due to non-specific electrostatic attraction which increases the free energy barrier in step 2 of scheme 2 (cf. Sect. II).

The kinetic mapping resulted in the construction of a hypothetic model for the binding site illustrated in the schematic drawing of Figure 5a. The limits of the site as judged from the dimensions of the ligands bound to it are 11 - 12 Å deep and 6 Å wide at the

Fig. 3a-c. The dependence of k_{on}, k_{off} and K on the length of normal aliphatic side chains measured from the center of the DNP-ring (a) and (b). The dependence of k_{off} on the distance and dimensions of a hydrophobic nucleus from the center of the DNP ring (see Table 6). Two types of models were analyzed, the compact model and the extended one (c)

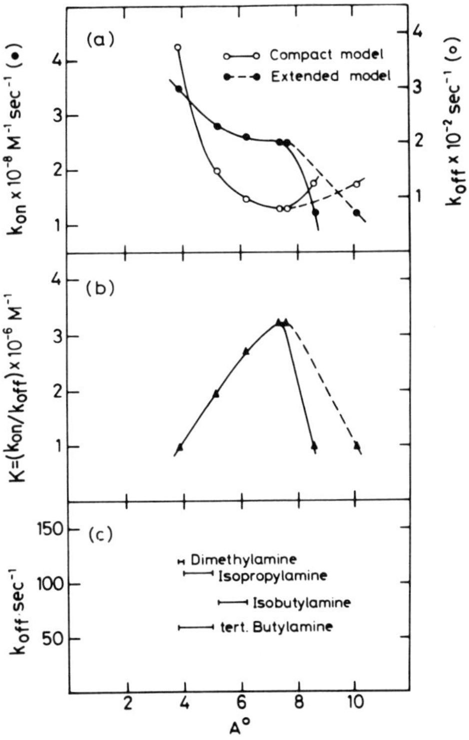

Table 7. Specific rates of interactions between protein 315 and ε-N-(2,4 dinitrophenyl)lysyl derivatives

Hapten	k_{on} 10^{-7} M^{-1} s^{-1}	k_{off} s^{-1}	k_{on}/k_{off} M^{-1} 10^{-6}
ε-N-DNP-lysine	13.0 ± 1.5	53 ± 5	2.4
α-N-acetyl;ε-N-DNP-lysine	12.0 ± 1	51 ± 7	2.4
$(lys)_1$ε-N-DNP-lysine	8.7 ± 0.8	35 ± 4	2.5
$(lys)_2$ε-N-DNP-lysine	8.1 ± 0.6	27 ± 4	3.0
$(lys)_5$ε-N-DNP-lysine	4.8 ± 0.8	29 ± 6	1.7
$(Lys)_2(Ala)_3$ε-N-DNP-lysine	8.8 ± 0.8	27 ± 4	3.2
$(lys)_8$ε-N-DNP-lysine	3.3 ± 0.4	32 ± 4.5	1.0
$(Lys)_5(Ala)_3$ε-N-DNP-lysine	6.0 ± 0.5	25 ± 5	2.4

All measurements were carried out in 0.05 M sodium phosphate buffered 0.1 M NaCl pH 7.4, at 21°C.

subsite close to the DNP, becoming somewhat wider (9 Å) further away (7 - 8 Å) from the DNP subsite. The latter region comprises a distinct hydrophobic subsite as expressed by the perferential interactions with the various hydrocarbon residues (e.g. the pro-

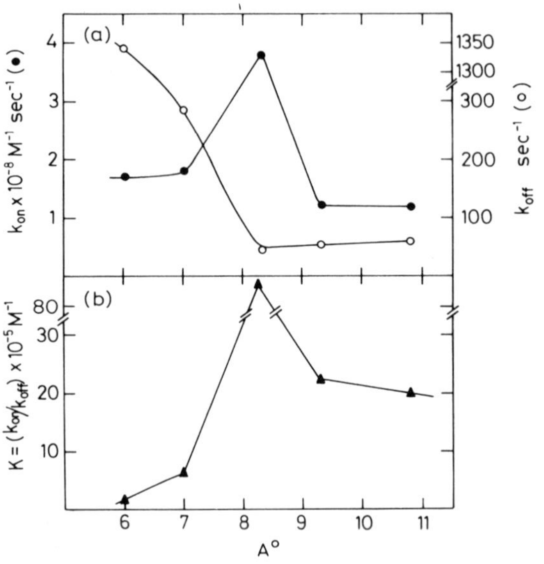

Fig. 4a and b. The dependence of k_{on}, k_{off}, and K on the distance of the carboxyl group of the aliphate side chain from the center of the DNP ring (a) and (b) (see Table 7). Only the extended model of the molecule was analyzed

nounced long residence time (k_{off}^{-1}) observed for the benzyl and isobutyl derivatives – Table 6). This is also corroborated by the residence times of the charged derivatives. Those carrying the charge at close distance to the DNP are strongly unfavored irrespective of the sign of their charge (Table 6) whereas clear discrimination is observed for those haptens having their charge further away from the ring: the positive ones are repelled even more strongly and the negative reach the longest residence times, which defines a positive subsite. Both these effects fade away when the side chain reaches a length of 7 Å, putting the limit to the range of specific electrostatic interactions.

No direct structural data from crystallography are available for protein 315. Still, by combining the crystallographic three-dimensional structure of a homologous immunoglobulin (McPC 603 Fab') together with the known amino acid sequence of protein 315 it was possible to construct a plausible molecular model of the variable part (PADLAN et al., 1977). This model of the combining site shows a marked agreement with several structural features indicated by the kinetic mapping (PADLAN et al., 1977). The site in this model (Fig. 5b) is a pronounced cavity dominating the middle of the hypervariable surface. Compared with the site of McPC 603 it contains substantially more aromatic residues and the dimensions fit those predicted by the kinetic data (Fig. 5a). More specific points are: (1) the possibility of hydrogen bond formation between the 2 and 4 nitro groups and Asn 36 (L) and Asn 36 (H) which explains the preference of the di over the mononitro and the steric specificity; (2) An extensive hydrophobic region where both the DNP ring itself as well as the branched or ring side chains carried on it reside; (3) The polar or charged side chains may interact with a number of residues which are at the more distant position; Lys 52 (H) may be the residue forming the positive subsite discussed above.

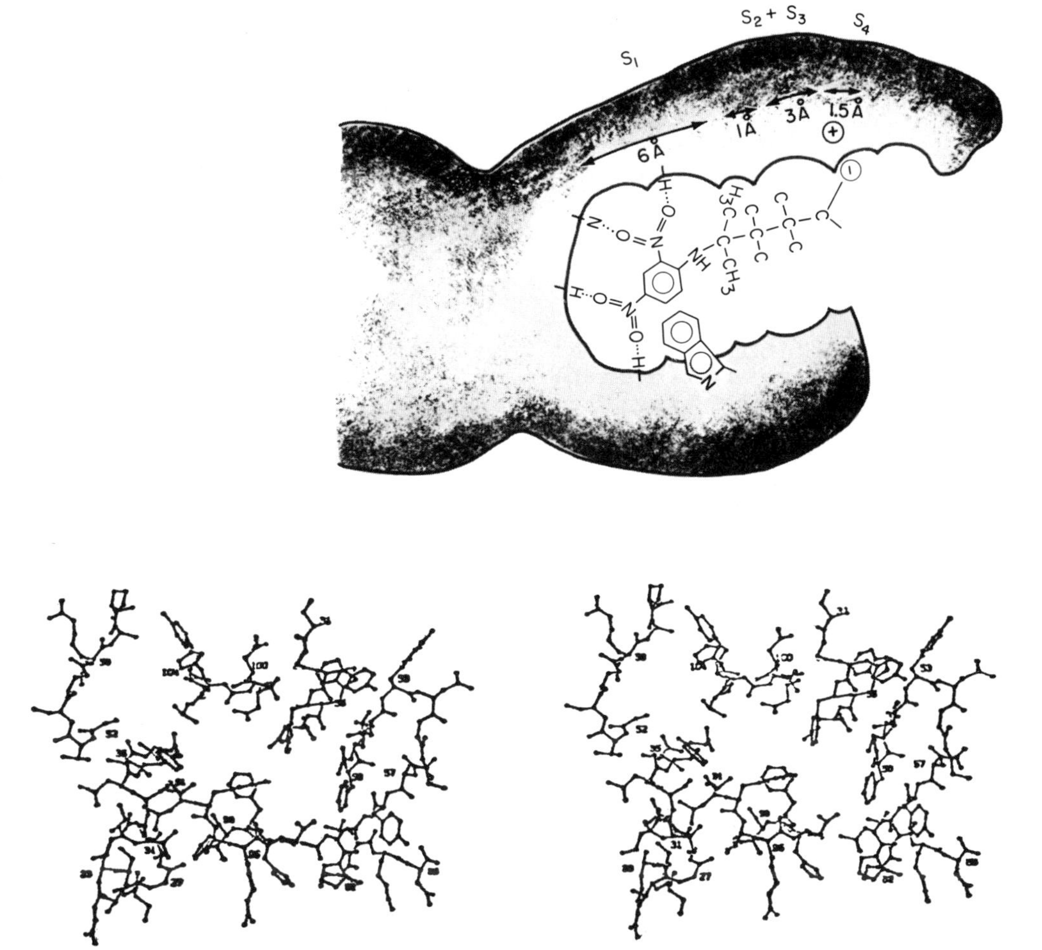

Fig. 5. (a) A hypothetic model view of the binding site of MOPC 315 indicating the different subsites of interaction. (b) Stereodrawing of the hypervariable residues projecting into the 315 hapten-binding cavity. A drawing of 2,4 dinitrophenyl amino-ethyl acetate is included in the figure to show the hapten-protein interactions and the possibility of affinity-labeling Tyr 34 (L)

IV. Conformational Transitions Induced by Hapten Binding

The binding of hapten (or antigen) to the immunoglobulin may be accompanied by changes in the spatial arrangement of residues in the binding site. These changes may be restricted to the contact residues ("restricted conformational changes") or involve a larger part of the protein ("long range conformational changes"). In the latter case, other loci on the immunoglobulin may be affected, i.e. allosteric phenomena will arise (MONOD et al., 1965). An allosteric mechanism for the induction of physiological activities of antibodies by antigen binding (cf. METZGER, 1974) thus requires evidence for the involvement of long range conformational changes. The model proposed by HUBER et al. (1976) based on X-ray crystallographic studies implies that these changes include the formation of non-covalent interactions between the immunoglobulin domains, together with changes in the conformation of the hinge peptides and in the position of the Fab regions relative to the Fc. Spectroscopic and other static methods indeed indicated the occurrence of conformational changes, yet their nature and significance has not been resolved (see METZGER, 1974; SCHLESSINGER et al., 1975 and references therein).

Kinetic methods may, in some cases, be more revealing than static ones for the study of conformational transitions. Thus, they may enable a distinction between restricted and long range conformational changes, as the latter would be expected to have higher $\Delta G^{\ddagger}$ and will therefore be slower, while both types of reaction may have similar overall ΔG. Also, as no homotropic site interaction is observed in hapten-immunoglobulin reactions, hapten saturation curves are noninformative, and kinetic methods remain as the only direct chemical way to study hapten-induced conformational changes.

Kinetic studies may yield both direct and indirect information; the conformational transition may either constitute monomolecular elementary step in the reaction mechanism, or be incorporated in an observed single association step. Mechanism 1 of Table 8 describes the general case of an association followed by a conformational isomerization. When T_1 concentration is comparable to those of the other species, two distinct elementary steps will be observed, and direct evidence for a conformational isomerization will arise. For the most common case, where the association is much faster than the isomerization, two relaxation times will be observed if the fast association step is accompanied by a spectral change. On the other hand, when the spectral change is due to the slower isomerization, only one relaxation time will be seen, yet it will have a typical concentration dependence, leveling off to a constant value with increasing reactant concentrations.

Mechanism 1 of Table 8 is completely different from scheme 2 in Section II. The first step of the former, just like any other association, includes both the encounter and the actual binding step. The monomolecular step of scheme 2 is not a conformational isomerization but expresses the formation of the elementary interactions (see Sect. II). Only in the special case where T_1 concentration is negligibly small will all three processes (the encounter, the formation of non-covalent interactions and the conformational transition) be included in an observed single association step. Still, whenever an analysis of this case in terms of separate, unresolvable steps is carried out, mechanism 1 of Table 8 has to be used. For the case of negligible T_1, assuming a fast associative preequilibrium, i.e. a relatively slow conformational change, the overall observed k_{on} will be equal to k_1K_T and the overall observed k_{off} will equal k_{-1} (in analogy to the formuli obtained for fast diffusional preequilibrium in Sect. II).

Table 8. Formuli are referenced by the column number and the row letter. Symbols denote species as well as their equilibrium concentrations: H: hapten, R_o and R_1: free and bound protein in the better binding state, T_o and T_1: free and bound protein in the worse binding state. P_t: total protein concentration.
The overall association constant is defined as $K = (R_1 + T_1)/H \cdot (R_o + T_o)$. Individual equilibrium constants are defined as $K_i = k_i/k_{-i}$ and the direction of k_i is defined to be from reactants to products.
The associations are assumed to be much faster than the isomerizations. For mechanism 3 the association to the T state is assumed to be much faster than that to the R state (see text).
g_{ij} is the element of the matrix g on the i-th row and j-th column. g_i is the principal partial matrix of order i derived from g, and $|g_i|$ is its determinant.
$\Delta \ln K_i = \Delta H_i \Delta T/RT^2$ where ΔT is the temperature difference in the jump, ΔH_i is the molar enthalpy change of the i-th step, R is the gas constant and T is the absolute temperature. Δf_i is the normalized molar fluorescence change of the i-th step (see LANCET and PECHT, 1976).

		1	2	3
A	MECHANISM	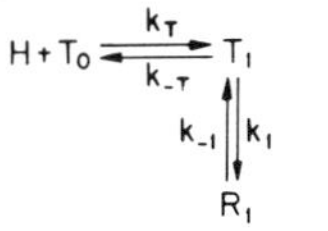	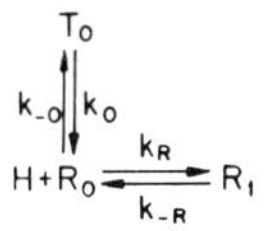	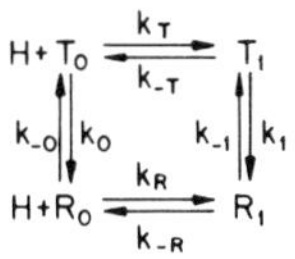
B	OVERALL ASSOCIATION CONSTANT	$K = K_T(1+K_1)$	$K = K_R/(1+K_o^{-1})$	$K = K_T(1+K_1)/(1+K_o)$
C	FAST TIME	$\frac{1}{\tau_F} = k_T(T_o+H)+k_{-T}$	$\frac{1}{\tau_F} = k_R(R_o+H)+k_{-R}$	$\frac{1}{\tau_{F1}} = k_T(T_o+H)+k_{-T}$ $\frac{1}{\tau_{F2}} = k_R\left(H+R_o\frac{1+K_TH}{1+K_T(H+T_o)}\right)+k_{-R}$
D	SLOW TIME	$\frac{1}{\tau_S} = k_1\frac{K_T(T_o+H)}{1+K_T(T_o+H)}+k_{-1}$	$\frac{1}{\tau_S} = k_{-o}\frac{K_RR_o+1}{K_R(R_o+H)+1}+k_o$	$\frac{1}{\tau_S} = \frac{k_o+k_1K_THX}{1+K_THX}+\frac{k_{-o}+k_{-1}K_RHY}{1+K_RHY}$ $X = \frac{T_o+R_o+H+K_R^{-1}}{K_TK_R^{-1}T_o+R_o+H+K_R^{-1}}$ $Y = \frac{T_o+R_o+H+K_T^{-1}}{T_o+K_RK_T^{-1}R_o+H+K_T^{-1}}$
E	$\lim_{H\to 0}\frac{1}{\tau_S}$	$k_1/(1+K_T^{-1}T_o^{-1})+k_{-1}$	k_o+k_{-o}	k_o+k_{-o}
F	$\lim_{H\to\infty}\frac{1}{\tau_S}$	k_1+k_{-1}	k_o	k_1+k_{-1}
G	g MATRIX	$\begin{bmatrix}\frac{1}{T_o}+\frac{1}{T_1}+\frac{1}{H} & -\frac{1}{T_1}\\ -\frac{1}{T_1} & \frac{1}{T_1}+\frac{1}{R_1}\end{bmatrix}$	$\begin{bmatrix}\frac{1}{R_o}+\frac{1}{R_1}+\frac{1}{H} & -\frac{1}{R_o}\\ -\frac{1}{R_o} & \frac{1}{T_o}+\frac{1}{R_o}\end{bmatrix}$	$\begin{bmatrix}\frac{1}{T_o}+\frac{1}{T_1}+\frac{1}{H} & \frac{1}{H} & \frac{1}{T_o}\\ \frac{1}{H} & \frac{1}{R_o}+\frac{1}{R_1}+\frac{1}{H} & -\frac{1}{R_o}\\ \frac{1}{T_o} & -\frac{1}{R_o} & \frac{1}{T_o}+\frac{1}{R_o}\end{bmatrix}$
H	FAST AMPLITUDE	$A_F = \frac{1}{g_{11}P_t}\Delta f_T\Delta\ln K_T$	$A_F = \frac{1}{g_{11}P_t}\Delta f_R\Delta\ln K_R$	$A_{F1} = \frac{1}{g_{11}P_t}\Delta f_T\Delta\ln K_T$ $Q_{12} = -\frac{g_{12}}{g_{11}}$ $A_{F2} = \frac{g_{11}}{\lvert g_2\rvert P_t}(Q_{12}\Delta f_T+\Delta f_R)(Q_{12}\Delta\ln K_T+\Delta\ln K_R)$
I	SLOW AMPLITUDE	$A_S = \frac{g_{11}}{\lvert g_2\rvert P_t}(Q_{12}\Delta f_T+\Delta f_1)\cdot$ $\cdot(Q_{12}\Delta\ln K_T+\Delta\ln K_1)$ $Q_{12} = -\frac{g_{12}}{g_{11}}$	$A_S = \frac{g_{11}}{\lvert g_2\rvert P_t}(Q_{12}\Delta f_R+\Delta f_o)\cdot$ $\cdot(Q_{12}\Delta\ln K_R+\Delta\ln K_o)$ $Q_{12} = -\frac{g_{12}}{g_{11}}$	$A_S = \frac{\lvert g_2\rvert}{\lvert g_3\rvert P_t}(Q_{13}\Delta f_T+Q_{23}\Delta f_R+\Delta f_o)\cdot$ $\cdot(Q_{13}\Delta\ln K_T+Q_{23}\Delta\ln K_R+\Delta\ln K_o)$ $Q_{13} = \frac{1}{\lvert g_2\rvert}(g_{12}g_{23}-g_{13}g_{22})$ $Q_{23} = \frac{1}{\lvert g_2\rvert}(g_{12}g_{13}-g_{11}g_{23})$

As T_1 is assumed to be very small, K_T may not be much larger than 10^3 M^{-1}, and as the specific rate for conformational isomerizations is practically always found to be lower than 1000 s^{-1} (HAMMES, 1968), k_{on} will be 10^6 M^{-1} s^{-1} or lower. Also, for a series of ligands with one immunoglobulin, k_{on} is expected to change while k_{off} will remain constant. Such behavior, together with a low k_{on}, may therefore serve as indirect kinetic indication for a hapten-induced conformational change.

The application of fast kinetic methods to hapten-immunoglobulin equilibria (Table 1) has practically always resulted in the observation of a simple, single association step mechanism (FROESE and SEHON, 1975). Still, there are several reports of more than one elementary step in the binding mechanism, which include:

1. Relaxation studies of normally induced anti-DNP antibodies with DNP ligands (FROESE, 1968). A low-amplitude slow relaxation time (20 - 50 ms) was observed in addition to the faster relaxation time of the association step. The heterogeneity of the antibodies and the difficulty in establishing the concentration dependence makes it impossible to draw definite conclusions from these results.

2. In the preliminary stopped flow experiments of the binding of bifunctional DNP derivatives to anti-DNP antibodies (BARISAS et al., 1973) a slow concentration-independent step with $t^{1/2} = 5$ s was found. The authors attribute this slow process to a conformational transition predecing the intramolecular ring closure in a dimer of two antibody molecules bridged by two bifunctional haptens. The brief abstract reporting this interesting observation does not allow the discussion of the proposed and alternative kinetic interpretations.

3. The relaxation study of bovine antibodies of the IgM class against a polynuclear aromatic solvatochrome ADHB (HAUSTEIN, 1971; cf. Table 1). Apart from the fast relaxation time which was attributed to the association step, a slow (22 ms) concentration independent time was also observed. In addition to the straightforward interpretation of a conformation transition which is coupled to hapten binding, the authors consider the possibility of a "non-hapten-dependent" conformational transition monitored by a spectral change of the environment-sensitive hapten. The latter alternative seems to us somewhat irrelevant since if the hapten has different absorption when bound to the different conformations, it will most probably also bind differently to them. This is so since the spectral changes must be due to different spatial relations between the hapten and the functional groups in the binding site, and the same relations are also responsible for the binding energy. It may be helpful to check the wavelength dependence of the amplitudes: if it is identical for both relaxation times then it almost certainly is a case of a conformational change coupled to hapten binding.

The above results were all obtained with preparations of normal heterogeneous antibodies which often display kinetic heterogeneity (BARISAS et al., 1975; BLATT et al., unpubl.; cf. cases 28, 29 of Table 1 and their discussion in Sect. II). This will result in a multiexponential relaxation spectrum even when only an association step is involved and will make the resolution of a slow monomolecular relaxation time more difficult. In view of this we chose to look for kinetic manifestations of conformational equilibria in systems of homogeneous immunoglobulins with known hapten specificity. The kinetic behavior of two of these systems indeed brought evidence for such equilibria and one of them (MOPC 460) was extensively characterized and analyzed. The system of the myeloma homogeneous immunoglobulin XRPC-24 with its oligogalactose haptens (Gal_3 and Gal_4) yielded a relaxation spectrum which could be analyzed in terms of two time constants, one of which is in the range of 20 - 80 ms and other 150 - 400 ms. The full kinetic analysis of this system is now being carried out.

The homogeneous myeloma immunoglobulin MOPC 460 which binds nitroaromatic compounds was found to display two distinct relaxation times, the faster (τ_F) in the range of 0.25 - 1.0 ms and the slower (τ_S) in the range of 10 - 18 ms (LANCET and PECHT, 1976). This protein has been characterized before (JAFFE et al., 1971; ROSENSTEIN et al., 1972; JOHNSTON et al., 1974). It is of special interest that ROSENSTEIN et al. (1972) found selective influence of certain treatments or modifications of the protein on the binding of one hapten without affecting that of the other. Also HSIA and RUSSEL-LITTLE (1973) reported pronounced differences in the immobilization of different spin-labeled haptens as judged from ESR spectra. Both these reports may indicate that the different haptens bind at somewhat different site. However, as the latter authors note, this may also be due to the stabilization of different conformations by different haptens.

The well-resolved relaxation spectrum of protein MOPC-460 with its haptens, as seen via the quenching of the protein tryptophan fluorescence (Fig. 6a,b), enabled us to confirm the hypothesis that two conformational forms of protein 460 exist, and that the equilibrium between them is shifted by the binding of haptens. This was borne out primarily by the fact that while τ_f behaves as a normal bimolecular relaxation time ($1/\tau_f$ increasing with the sum of free reactant concentration), $1/\tau_s$ shows typical monomolecular behavior, decreasing to a constant value with increasing reactant concentration. The concentration dependence of both relaxation times for the hapten DNP-lysine is depicted in Figures 7 and 8. Other haptens [2,4-dinitronapth-1-ol and 3 methyl 1,4-naphthoquinon (menadione)] also yield the same relaxation spectrum, although with the latter hapten amplitudes are small due to low values of ΔH and the spectral change (JOHNSTON et al., 1972). Of particular interest is the fluorescent DNP-analog 4-α-N-alanine-7-nitro-benz-2-oxa-1,3-diazole (NBDA), for which the same relaxation spectrum was also observed in fluorescence and absorption changes of the hapten. The wavelength dependence of the amplitudes of both relaxation times (when followed by hapten absorption) was identical to the static difference spectrum. This strongly suggests that the slow step is coupled to hapten binding and is seen through the fast readjustment of the fractional saturation by the hapten which follows every increment of the slow conformational transition.

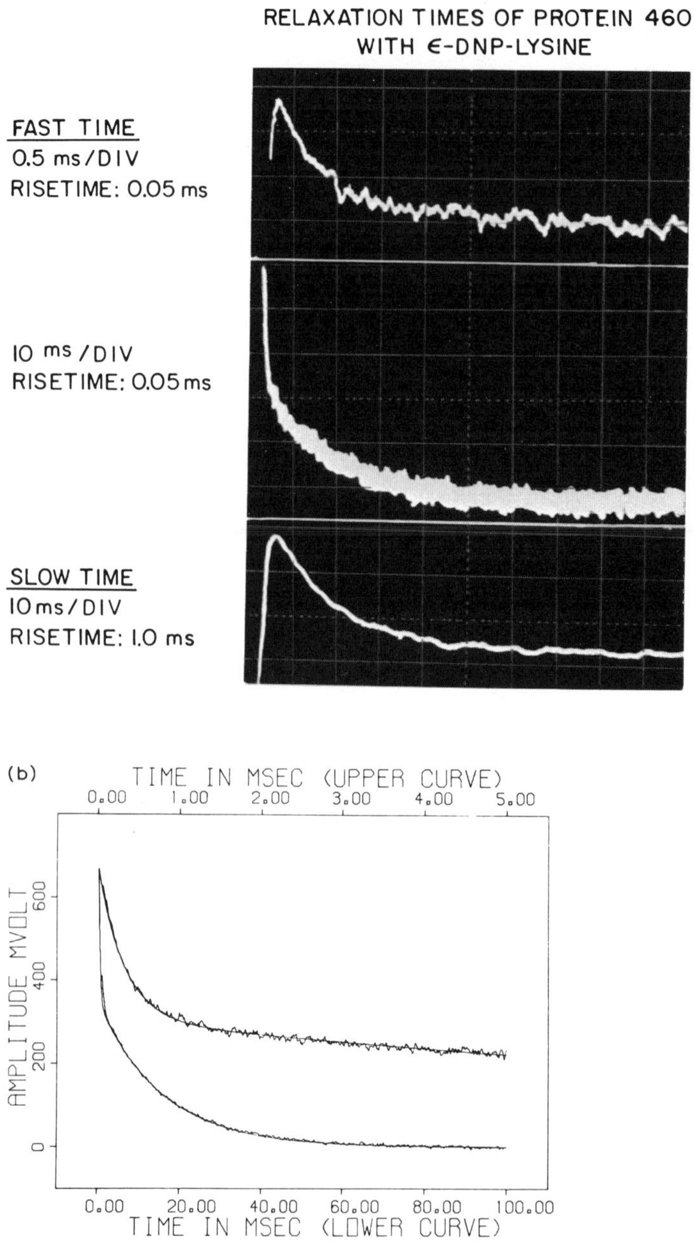

Fig. 6. (a) Experimental relaxation traces for protein 460 with DNP-lysine. The vertical scale is 100 mv/division. The middle picture depicts the full relaxation trace. (b) A computer-drawn relaxation trace in the system of protein 460 and DNP-Lys. The noisy curve represents the data and the smooth curve is calculated using the fit parameters. Two different sweep rates are shown so that both relaxation times are displayed on an appropriate time scale. Concentrations used: DNP-Lys = $1.93 \cdot 10^{-5}$ M, Protein 460 = $1.65 \cdot 10^{-6}$ M sites. (The deviation in the initial part of the lower curve is due to the rise time filter)

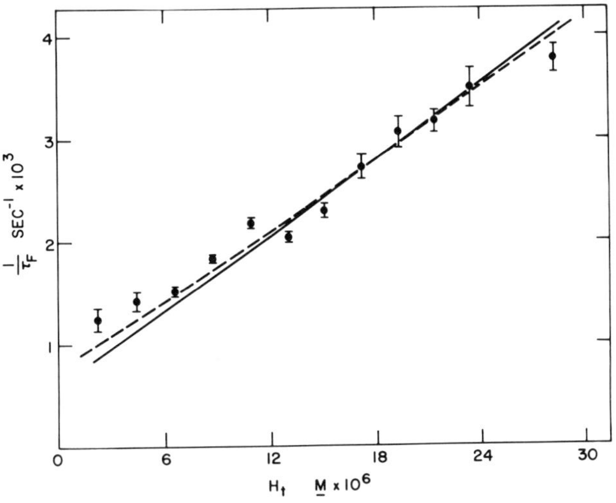

Fig. 7. The dependence of the inverse fast relaxation time on total DNP-Lysine concentration. Broken and full lines are calculated using the best fit parameters for mechanisms 2 and 3 respectively. The concentration of protein 460 was 1.5 - 1.7 • 10^{-6} M sites. The bars denote approximate standard deviation in the group of jumps represented by one point

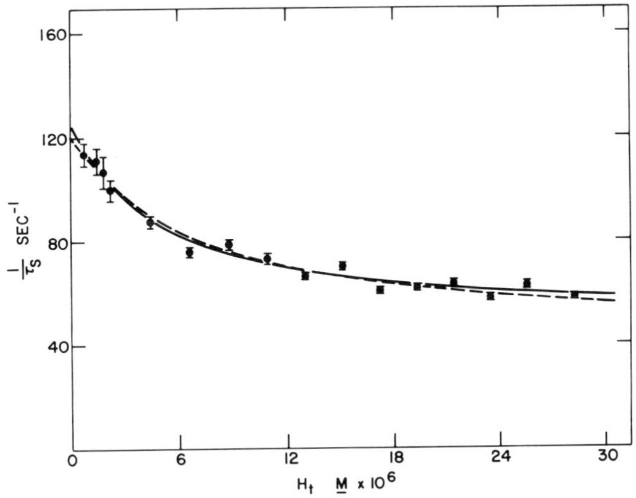

Fig. 8. The dependence of the inverse slow relaxation time on total DNP-Lys concentration. All other details are the same as in Figure 7

Row A of Table 8 depicts the simplest three mechanisms that may be proposed for a system of ligand-protein association linked to a conformational transition. The formulas for the fast and slow relaxation times, derived according to the general formalism of CASTELLAN (1963) using the g matrices (row G), are given in rows C and D. These yield rows E and F from which it may be seen that mechanism 1 does not fit the observed decrease of $1/\tau_S$ with increasing concentration. The decision between the other two mechanisms depends mainly on the analysis of the relaxation amplitudes. This was done using the formulas of Table 8 rows H and I, which were derived according to the formalism of JOVIN (1975), using the above-mentioned g matrices. Figure 9 shows the optimal fit of mechanisms 2 and 3 to the experimental fast and slow amplitude data. Mechanism 3 is found to best fit the data and is

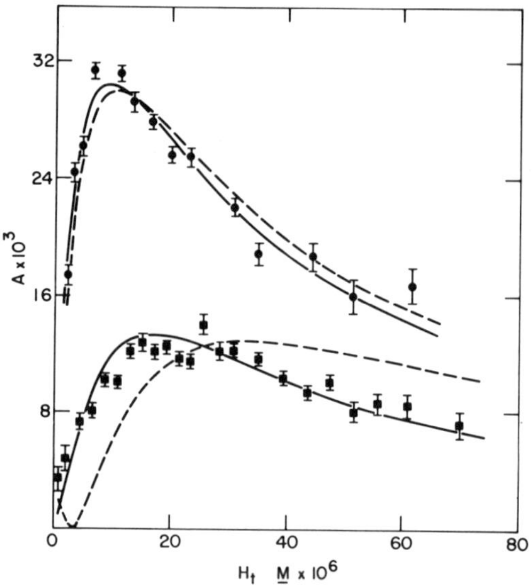

Fig. 9. The dependence of fast relaxation amplitude (upper curve) and slow relaxation amplitude (lower curve) on total DNP concentrations. All other details are the same as in Figure 7

established as the operating mechanism for this system. The resultant parameters are summarized in Table 9. For the details of the experimental procedures and theoretical derivations, the reader is referred to the original paper (LANCET and PECHT, 1976).

It should be emphasized that mechanism 3 of Table 8 is the most general representation of ligand-induced conformational transitions in proteins and that it forms the basic mechanistic element of the allosteric model (MONOD et al., 1965). Thus the above kinetic analysis is significant in the general context of ligand-protein interactions.

The monomolecular rate constants obtained for the protein isomerization steps in this system are in the range of $10 - 100\ s^{-1}$. This corresponds to $\Delta G^{\ddagger} = 14 - 16$ kcal/mol. Preliminary temperature-dependence experiments show that the activation enthalpies for the conformational transitions are in the range of 10 kcal/mol yielding $-T\Delta S^{\ddagger}$ of about 5 kcal/mol ($\Delta S^{\ddagger} = -17$ e.u.). The high free energy and enthalpy of activation may imply that the observed slow process represents a long-range conformational transition and that it is not restricted to a few contact residues. The negative entropy of activation means that the transition state is more ordered than the two thermodynamically stable conformations.

A similar relaxation spectrum is obtained for reduced and alkylated monomer IgA as well as for the nonreduced oligomer and the Fab fragment. This may mean that the conformational change does not extend beyond the Fab region. However, as the evaluation of the concentration dependence for the latter two cases is underway, quantitative differences are not yet ruled out. Also, as protein 460 is an IgA, it interacts with complement via the alternative pathway at a site presumably located on the Fab (GÖTZE

Table 9. Experimental values of parameters for the reaction between MOPC-460 and DNP-Lysine

Mechanism	i	K_i	k_i	k_{-i}	Δf_i	ΔH_i	ΔG_i [f]	ΔS_i [g]
2	R	$1.73 \cdot 10^5$ [b]	$1.20 \cdot 10^8$ [a]	690 [a]	-4.8 [c†]	-15.9 [c†]	-7.12	-29.5
	O	0.56 [b]	43 [b]	77 [b]	-	-	-	-
3	T	$1.90 \cdot 10^4$ [e]	$1.3 \cdot 10^8$ *	6800 *	-0.47 [d]	-14.1 [d]	-5.81	-27.8
	R	$2.18 \cdot 10^5$ [e]	$1.27 \cdot 10^8$ [a]	580 [a]	-0.50 [d §]	-15.0 [d §]	-7.25	-26.0
	O	0.30 [e]	29 [e]	96 [e]	0.02 [d]	0.64 [d]	0.71	-0.23
	1	3.40 [e]	44 [e]	13 [e]	-0.01 [d]	-0.26 [d]	-0.72	1.55

Units: k_T and k_R - s^{-1} M^{-1}, all other k_i, k_{-i} - s^{-1}; K_T and K_R - M^{-1}, K_O, K_1 and Δf_i - dimensionless; ΔH_i and ΔG_i - kcal/mol, ΔS_i - cal/mol•degree. Error ± 20%.
Source of values: [a] - τ_F, [b] - τ_S, [c] - A_F, [d] - A_S, [e] - τ_S and A_S, [f] - by $\Delta G_i = -RT\ln K_i$, [g] - by $\Delta S_i = (\Delta H_i - \Delta G_i)/T$. * According to the assumption $k_T = k_R$, see text. † Only the product $\Delta H_R \Delta f_R$ is determinable and the listed values are obtained with the assumption $\Delta f_R = \Delta f$ (overall). § The fast amplitude fit yields a somewhat higher value for $\Delta H_R \cdot \Delta f_R$ (10.7 as compared with 7.5 kcal/mol here). This, however, corresponds to less than 20% difference in each of the two parameters, within the estimated error. For definition of symbols see Table 8.

and MÜLLER-EBERHARD, 1971). Thus it is possible, in this case, that physiologically significant changes will be restricted to the Fab.

It is noteworthy that in two of the three systems studied by others (see above), as well as in protein 460, the slow reaction has a time constant of about 20 ms. Since the study of BARISAS et al. (1973) suggests that changes involving the relative position of the two Fab towards each other are much slower, one may speculate that the 20 ms reflect the change in the relative position of the V and C domains of the Fab (HUBER et al., 1976). A kinetic study of an Fv fragment (INBAR et al., 1972) from protein 460 may confirm this hypothesis.

The inability to observe a slow conformational step in hapten-immunoglobulin systems may be due to the limited analogy between haptens and antigens (see Sect. I): some haptens may bind in a way that does not lead to the conformational changes caused by the corresponding whole antigens. Kinetic investigations in more immunoglobulin-hapten system, and the use of larger haptens or even of whole antigenic determinant may help to resolve this question.

References

ALBERTY, R.A., HAMMES, G.G.: Application of the theory of diffusion-controlled reactions to enzyme kinetics. J. Phys. Chem. 62, 154-159 (1958)

BARISAS, B.G., SINGER, S.J., STURTEVANT, J.M.: Thermodynamics of the binding of 2,4 dinitrophenyl and 2,4,6-trinitrophenyl haptens to the homologous and heterologous rabbit antibodies. Biochemistry 11, 2741-2744 (1972)

BARISAS, B.G., SINGER, S.J., STURTEVANT, J.M.: Kinetic evidence for a conformational change by an anti-hapten antibody upon binding multivalent hapten. Abstr. 9th Intern. Congr. Biochemistry, Stockholm, 312 (1973)

BARISAS, B.G., SINGER, S.J., STURTEVANT, J.M.: Kinetics of binding of 2,4 dinitrophenyl and 2,4,6-trinitrophenyl haptens to homologous and heterologous rabbit antibodies. Immunochemistry 12, 411-421 (1975)

BERSON, S.A., YALOW, R.S.: Quantitative aspects of the reaction between insulin and insulin-binding antibody. J. Clin. Invest. 38, 1996-2016 (1959)

BLATT, Y., KARUSH, F., PECHT, I.: unpublished (1976)

CASTELLAN, G.W.: Calculation of the spectrum of chemical relaxation times for a general reaction mechanism. Berich. Bunsenges. 67, 898-908 (1963)

DANDLIKER, W.B., LEVISON, S.A.: Investigation of antigen-antibody kinetics. Immunochemistry 5, 171-183 (1968)

DAVIES, D.R., PADLAN, E.A., SEGAL, D.M.: Immunoglobulin structures at high resolution. Contemp. Topics Mol. Immunol. 4, 127-155 (1975)

DAY, L.A., STURTEVANT, J.M., SINGER, S.J.: The kinetics of the reactions between antibodies to the 2,4 dinitrophenyl group and specific haptens. Ann. N.Y. Acad. Sci. 103, 611-625 (1963)

EDELMAN, G.M., GALL, W.E.: The antibody problem. Ann. Rev. Biochem. 38, 415-466 (1969)

EIGEN, M.: Diffusion control in biochemical reactions. In: Quantum statistical mechanics in the nature sciences. MINTZ, S.L., WIEDERMAYER, S.M. (eds.). New York: Plenum Press 1974, p.37-61

EIGEN, M., DE MAEYER, L.: Theoretical basis of relaxation spectrometry. In: Techniques of Chemistry. WEISSBERGER, A., HAMMES, G.G. (eds.); Vol. VI, Part 2, p. 63-146. New York: Wiley 1973

EISEN, H.N., SIMMS, E.S., POTTER, M.: Mouse myeloma proteins with anti-hapten antibody activity: The protein produced by plasma cell tumor MOPC 315. Biochemistry 7, 4126-4134 (1968)

FROESE, A.: Kinetic and equilibrium studies on 2,4 dinitrophenyl hapten-antibody systems. Immunochemistry 5, 253-264 (1968)

FROESE, A., SEHON, A.H.: Kinetic and equilibrium studies of the reaction between anti-p-nitrophenyl antibodies and a homologous hapten. Immunochemistry 2, 135-143 (1965)

FROESE, A., SEHON, A.H.: Kinetics of antibody-hapten reactions. Contemp. Topics Mol. Immunol. 4, 23-54 (1975)

FROESE, A., SEHON, A.H., EIGEN, M.: Kinetic studies on protein-dye and antibody hapten interactions with the temperature-jump method. Can. J. Chem. 40, 1786-1797 (1962)

GLASSTONE, S.: Textbook of Physical Chemistry. Princeton: Van Nostrand 1946

GÖTZE, O., MÜLLER EBERHARD, J.: The C3-activator system: an alternate pathway to complement activation. J. Exp. Med. 134, 90S-108S (1971)
HAMMES, G.G., SCHIMMEL, P.R.: Rapid reactions and transient states. In: The Enzymes. BOYER, P.D. (ed.); Vol. II, 3rd ed., pp. 112. New York: Academic Press 1970
HARISDANGKUL, V., KABAT, E.A.: Interaction of 1-(m-nitrophenyl)-flavazoles of isomaltose oligosaccharides with purified anti-dextran. J. Immunol. 108, 1232-1243 (1972)
HARTMANN, U., PECHT, I.: unpublished data
HASELKORN, D.: Ph.D Thesis Feinberg Graduate School, The Weizmann Inst. Sci., Rehovot, Israel (1975)
HASELKORN, D., FRIEDMAN, S., GIVOL, D., PECHT, I.: Kinetic mapping of the antibody combining site by chemical relaxation spectrometry. Biochemistry 13, 2210-2222 (1974)
HAUSTEIN, D.: Dissertation, Univ. Freiburg, W. Germany (1971)
HSIA, J.C., RUSSEL LITTLE, J.: Structural properties of the ligand binding sites of murine myeloma proteins. FEBS Lett. 31, 80-84 (1973)
HUBER, R., DEISENHOFER, J., COLMAN, P., MATSUSHIMA, M., PALM, W.: X-ray diffraction analysis of immunoglobulin structure. 27th Mosbach Coll. Berlin-Heidelberg-New York: Springer 1976
INBAR, D., HOCHMAN, J., GIVOL, D.: Localization of antibody combining sites within the variable portions of heavy and light chains. Proc. Nat. Acad. Sci. (Wash.) 69, 2659-2664 (1972)
JAFFE, B.M., SIMMS, E.S., EISEN, H.N.: Specificity and structure of the myeloma protein produced by mouse plasmacytoma MOPC 460. Biochemistry 10, 1693-1699 (1971)
JENCKS, W.P.: Binding Energy, Specificity, and Enzymic Catalysis: The circe effect. Advan. Enzymol. 43, 219-410 (1975)
JOHNSTON, M.F.M., BARISAS, B.G., STURTEVANT, J.M.: Thermodynamics of hapten binding to MOPC 315 and MOPC 460 mouse myeloma proteins. Biochemistry 34, 390-396 (1974)
JOVIN, T.: Fluorimetric kinetic techniques: Chemical relaxation and stopped-flow. In: Biochemical Fluorescence / Concepts. CHEN, R.F., EDELHOCH, H.: (eds.); Vol. I, pp. 305-374. New York: Dekker 1975
KABAT, E.A.: The nature of the antigenic determinant. J. Immunol. 97, 1-11 (1966)
KABAT, E.A., WU, T.T.: Attempts to locate complementarity determining residues in the variable positions of light and heavy chains. Ann. N.Y. Acad. Sci. 190, 382-390 (1971)
KARUSH, F.: The affinity of antibody: Range, variability and the role of multivalence. In: Comprehensive Immunology (in press, 1976).
KAUZMANN, W.: Some factors in the interpretation of protein denaturation. Advan. Protein Chem. 14, 1-63 (1959)
KELLY, K.A., SEHON, A.H., FROESE, A.: Kinetic studies on antibody-hapten reactions. Immunochemistry 8, 613-625 (1971)
LANCET, D., KABAT, E.A., PECHT, I.: unpublished data.
LANCET, D., PECHT, I.: Kinetic evidence for hapten induced conformational transition in immunoglobulin MOPC 460. Proc. Nat. Acad. Sci. (Wash.) 73, 3548-3553 (1976)
LEE, F.H., FROESE, A.: Kinetics of the reaction between DNP-Insulin and homologous antibody. Immunol. Comm. 2, 565-571 (1973)

LEVISON, S.A., HICKS, A.N., PORTMANN, A.J., DANDLIKER, W.B.: Fluorescence polarization and intensity kinetic studies of anti-fluorescein antibody obtained at different stages of immune response. Biochemistry 14, 3778-3786 (1975)
LEVISON, S.A., JANCSI, A.N., DANDLIKER, W.B.: Temperature effects on the kinetics of the primary antigen-antibody combination. Biochem. Biophys. Res. Comm. 33, 942-948 (1968)
LONGSWORTH, L.E.: Temperature dependence of diffusion in aqueous solutions. J. Phys. Chem. 58, 770-773 (1954)
METZGER, H.: Effect of antigen binding on the properties of antibody. Advan. Immunol. 18, 167-207 (1974)
MONOD, J., WYMAN, J., CHANGEUX, J.P.: On the nature of allosteric transitions: A plausible model. J. Mol. Biol. 12, 88-118 (1965)
NOBLE, R.W., REICHLIN, M., GIBSON, Q.H.: The reactions of antibodies with hemoprotein antigens. J. Biol. Chem. 244, 2403-2411 (1969)
NOBLE, R.W., REICHLIN, M., SCHREIBER, R.D.: Studies on antibodies directed towards antigenic sites on globular proteins. Biochemistry 11, 3326-3332 (1972)
NORTH, A.M.: The Collision Theory of Chemical Reactions in Liquids. New York: Methuen 1964
PADLAN, E.A., DAVIS, D.R., PECHT, I., GIVOL, D., WRIGHT, C.E.: Model building studies of antigen binding sites: The hapten binding site of MOPC-315. Cold Spring Harbor Symp. Quant. Biol. Vol. XLI (1977)
PECHT, I.: Antibody combining sites as a model for molecular recognition. In: Protein-Ligand Interactions. SUND, H., BLAUER, G. (eds.); pp. 356-371. Berlin: de Gruyter 1975
PECHT, I., GIVOL, D., SELA, M.: Dynamics of antiboy-hapten interaction. Studies on a myeloma protein with anti DNP specificity. J. Mol. Biol. 68, 241-247 (1972)
PECHT, I., HASELKORN, D., FRIEDMAN, S.: Kinetic mapping of antibody binding sites by chemical relaxation spectrometry. FEBS Lett. 24, 331-334 (1972)
POLJAK, R.: X-ray diffraction studies of immunoglobulins. Advan. Immunol. 21, 1-33 (1975)
ROSENSTEIN, W.R., MUSSON, R.A., ARMSTRONG, M.Y.K., KONIGSBERG, W.A., RICHARDS, F.F.: Contact regions for DNP and menadione haptens in an immunoglobulin binding more than one antigen. Proc. Nat. Acad. Sci. (Wash.) 69, 877-881 (1972)
SACHS, D.H., SCHECHTER, A.N., EASTLAKE, A., ANFINSEN, C.B.: Inactivation of staphylococcal nuclease by the binding of antibodies to a distinct antigenic site. Biochemistry 11, 4268-4273 (1972)
SCHLESSINGER, J., STEINBERG, I.Z., GIVOL, D., HOCHMAN, J., PECHT, I.: Antigen-induced conformational changes in antibodies and their Fab fragments studied by circular polarization of fluorescence. Proc. Nat. Acad. Sci. (Wash.) 72, 2775-2779 (1975)
SMITH, T.W., SKUBITZ, K.M.: Kinetics of interactions between antibodies and haptens. Biochemistry 14, 1496-1502 (1975)

Glutamate Dehydrogenase Self-Assembly. An Application of the Light Scattering Temperature-Jump Technique to the Study of Protein Aggregation

D. THUSIUS

I. Introduction

Glutamate dehydrogenase catalyzes the reversible oxidative deamination of L-glutamate to α-ketoglutarate and ammonia with pyridine nucleotides as coenzyme

$$HOOC-CH_2-CH_2-\underset{\displaystyle NH_2}{\underset{|}{CH}}-COOH + NAD(P)^+ + H_2O \rightleftharpoons$$

$$HOOC-CH_2-CH_2-\underset{\displaystyle O}{\underset{\|}{C}}-COOH + NAD(P)H + NH_3 + H^+$$

The above reaction plays a key role in linking nitrogen metabolism to the tricarboxylic acid cycle (Fig. 1).

The most thoroughly studied glutamate dehydrogenase is the beef liver enzyme. Here it is known that the nucleotides ADP and GTP, which participate in processes coupled to GDH catalysis in vivo, activate and inhibit GDH activity, respectively. This observation suggests feedback control, and numerous possibilities for such regulation are evident in Figure 1.

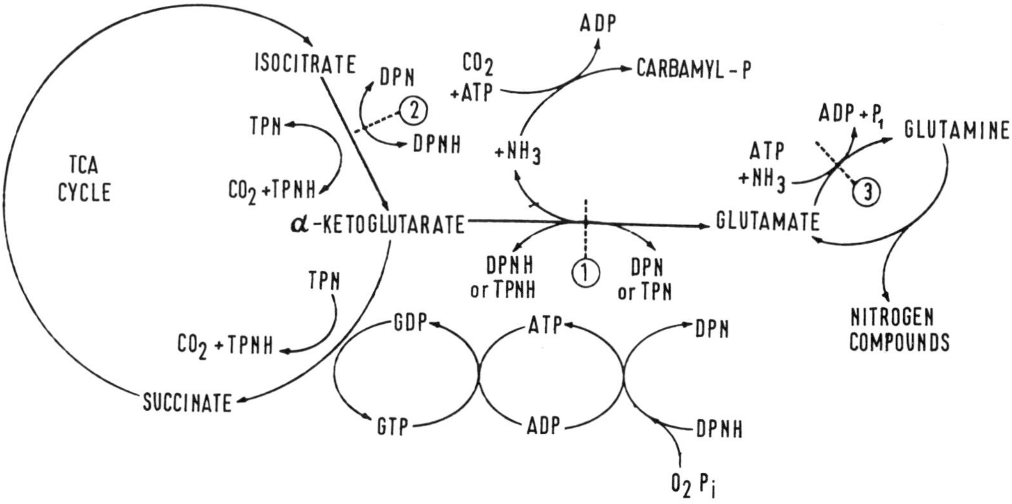

Fig. 1. Role of glutamate dehydrogenase (1) in cellular metabolism. Circles: points of feedback control. (From STADTMAN, 1966)

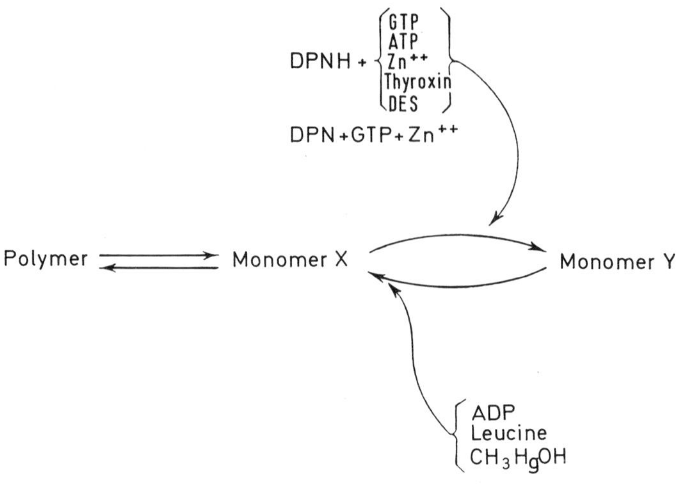

Fig. 2. Ligand induced association-dissociation of glutamate dehydrogenase. (See STADTMAN, 1966; after TOMKINS et al., 1965)

The ability of GDH to undergo a reversible, concentration dependent association-dissociation was first recognized by OLSEN and ANFINSEN (1952) more than 20 years ago. Subsequent studies (FRIEDEN, 1963; TOMKINS et al., 1965) demonstrated that the state of aggregation is strongly influenced by metabolites which also modify enzymic activity. For example, in the presence of co-enzyme, GTP promotes dissociation of GDH aggregates to monomer units, while addition of ADP restores the associated state.

The reciprocal relation between the degree of protein aggregation and enzymic activity suggests that GDH self-assembly may be involved in a control mechanism of biological significance. An intriguing possibility is the proposal of FRIEDEN (1963) and TOMKINS et al. (1965) given in Figure 2. Here the GDH monomer is assumed to exist in a state Y, which has no glutamate dehydrogenase activity, and an active form X, which can undergo polymerization. Reagents promoting polymerization are assumed to bind only form X, while those leading to dissociation bind only form Y. Although recent results (reviewed by FISHER, 1973) show that the relation between activity and degree of polymerization is more subtle than realized in the early work on GDH self-assembly, it can be said that the dramatic effect of metabolites on the state of aggregation, the occurrence of high GDH concentrations in liver mitochondria and the absence of self-assembly in most other oligomeric enzymes argue for a biological role. Even in the event that GDH self-asembly is shown to be an evolutionary accident, this system has and continues to serve as a beautiful model for the investigation of reversible protein-protein interactions.

In the present work the spontaneous association-dissociation is studied with a light-scattering temperature-jump technique. The kinetic results not only corroborate earlier conclusions based on structural and thermodynamic measurements, but resolve certain mechanistic features of self-assembly which could not be elucidated with equilibrium methods. Since our experimental technique

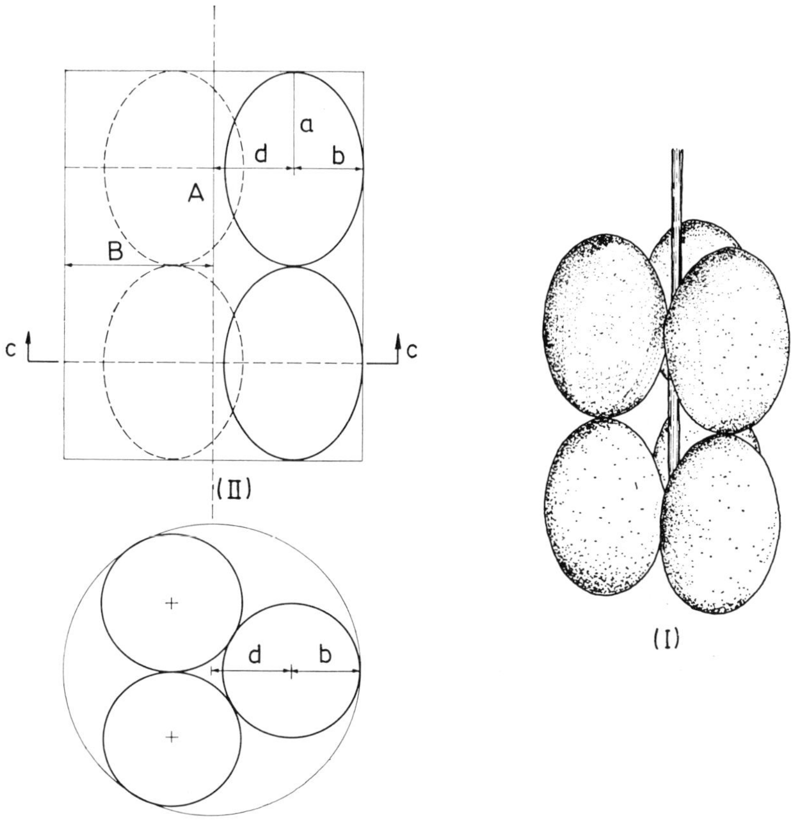

Fig. 3. Macroscopic model of glutamate dehydrogenase oligomer. The calculated major and minor axes of the ellipsoids are 2a=66.5Å and 2b=43Å. Polymerization proceeds in the direction of the oligomer three-fold axis of symmetry. (From REISLER and EISENBERG, 1970)

is relatively recent, a significant part of this Chapter is devoted to theoretical and practical aspects of light-scattering detection in chemical relaxation.

II. Structural Features[1]

In its simplest active form, GDH exists as an oligomer of six apparently identical polypeptide chains. Dissociation into unfolded subunits occurs in 5 M guanidine hydrochloride and at extreme pH, indicating that the oligomer quaternary structure is

[1]For details on the material summarized in this section, the reader is referred to three recent reviews on glutamate dehydrogenase (EISENBERG, 1971; SUND et al., 1975; EISENBERG et al., 1976).

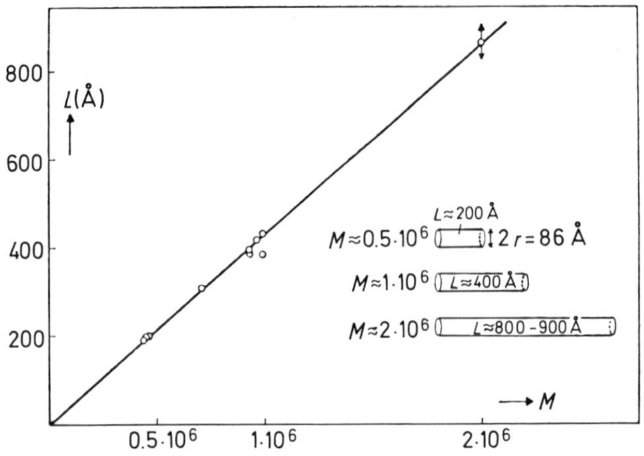

Fig. 4. Average length of glutamate dehydrogenase particles as function of average molecular weight. (From SUND et al., 1969)

stabilized only by non-covalent bonds[2]. Complete loss of enzymic activity accompanies dissociation. Following a number of conflicting reports, it is now generally accepted that the oligomer molecular weight is $3.36 \cdot 10^5$ daltons.

A physical model for the oligomer has been proposed on the basis of X-ray small angle results, electron microscopy, light scattering and hydrodynamic measurements (Fig. 3). It is believed that the active enzyme unit is built of two layers, each composed of three elongated subunits approximated by prolate elipsoids of revolution. The overall length and diameter of the model are 133 and 86 Å, respectively. Although trimers have been observed with electron microscopy, units smaller than the hexamer have not been detected in solution under conditions in which enzymatic activity is conserved.

Above ∿0.3 mg/ml, the oligomer forms polymers of high molecular weight. The first indication of aggregation was an increase in sedimentation coefficient with protein concentration. It was not until recently, however, that a number of converging lines of evidence established that oligomer units undergo reversible end-to-end association to form elongated cylinders. As an example of the physical techniques involved in these studies, Figure 4 presents X-ray small angle results which strongly imply that the length per unit mass of the associated protein is constant. In agreement with this conclusion, rod-like structures have been detected directly with electron microscopy (Fig. 5).

---

[2]The smallest unit participating in reversible association-dissociation has been referred to in the literature as "monomer", "hexamer", "oligomer" and "unimer". Here we shall use the terms "oligomer" and "monomer".

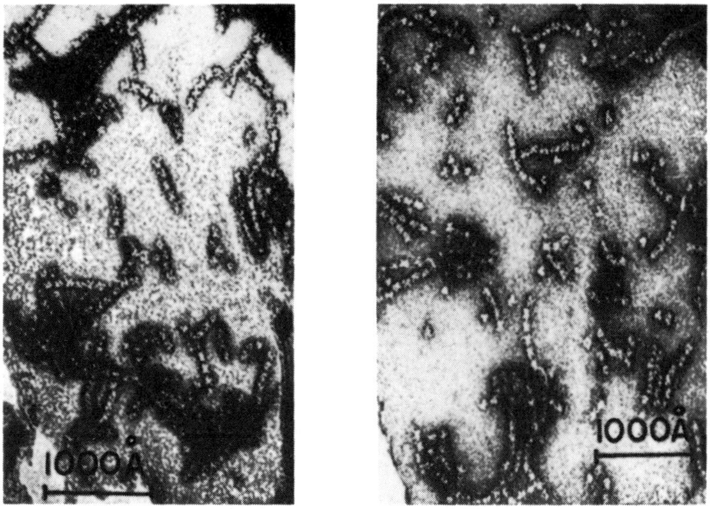

Fig. 5. Polymer chains of glutamate dehydrogenase (JOSEPHS, 1971). Chains consisting of up to 10 or more oligomer units can be observed

Table 1

$nP_1 \rightleftarrows P_n$	(1)	$2P_1 \rightleftarrows P_2$	
$2P_1 \rightleftarrows P_2$	(2)	$P_1+P_2 \rightleftarrows P_3$	
$2P_2 \rightleftarrows P_4$		$\vdots \quad\quad \vdots$	(3)
$2P_4 \rightleftarrows P_8$		$P_1+P_n \rightleftarrows P_{1+n}$	

III. Thermodynamics of Self-Assembly

A. Models

From studies of molecular weight as a function of protein concentration, several schemes have been suggested for the mode of GDH self-assembly, some of which are presented in Table 1.

Eq. (1) is the limiting case of a closed association-dissociation without detectable intermediates. Although early evidence suggested a closed monomer-tetramer equilibrium (COLMAN and FRIEDEN, 1966), a later comparison of experimental light scattering data with calculated curves argued against this model (SUND and BURCHARD, 1968). A scheme in which only dimers and tetramers occur as intermediates (Eq. (2)) has also been considered (SUND and WEBER, 1966), but this type of association could not rationalize molecular weights obtained over a wide protein concentration range (SUND and BURCHARD, 1968).

The most recent working model for GDH self-assembly is Eq. (3), in which rod growth is represented as a sequential addition of monomers (DESSEN and PANTALONI, 1969; REISLER and EISENBERG, 1971; SUND et al., 1972; CHUN et al., 1972; MALENCIK and ANDERSON, 1972). The simplest situation occurs when the equilibrium constants are the same for each step. This in turn would result in an open polymerization; i.e. chain length would have no upper limit, or

$$P_1 + P_i \rightleftarrows P_{1+i}$$

$$K = \bar{P}_{1+i}/\bar{P}_1\bar{P}_i \qquad \text{(Scheme I)}$$

$$i = 1,2\ldots\ldots\infty$$

The assumption of identical equilibrium constants is plausible in view of the end-to-end nature of the aggregation and the large distance between association sites (≥130 Å). The weight average molecular weight for the above scheme can be expressed as a simple function of the association equilibrium constant, K, and the total protein weight concentration, C_T (ADAMS and WILLIAMS, 1964)

$$\bar{M}_W = \Sigma C_i M_i \,/\, \Sigma C_i = \Sigma C_i M_i / C_T \tag{1}$$

$$\bar{M}_W = M_1(1+4KM^{-1}C_T)^{1/2} \tag{2}$$

where C_i and M_i are respectively the weight concentration and molecular weight of species P_i.

A fit of GDH light scattering data to Eq. (2) is shown in Figure 6 (MARKAU et al., 1971). The dotted line represents theoretical molecular weights calculated from the extrapolated value of M_1 and a trial-and-error estimate of K. Up to about 2 mg/ml, the observed molecular weights correspond closely to those predicted for scheme I. Equilibrium sedimentation results for dilute GDH solutions are also in reasonable agreement with this model (REISLER et al., 1970). Under the conditions of the kinetic experiments described later in this chapter, the association equilibrium constant has a value of $3.3 \cdot 10^5\ M^{-1}$ (THUSIUS et al., 1975).

B. Nonideality

The deviations from theoretical molecular weights evident in Figure 6 have been attributed to thermodynamic nonideality (MARKAU et al., 1971). The simplest type of nonideality in solutions of macromolecules arises from "excluded volumes" type (TANFORD, 1961). Large molar volumes result in a loss of solute translational entropy relative to an infinitely dilute solution. Therefore, even in the absence of long-range solute-solute interactions, molecular weights of large particles determined with thermodynamic methods (light scattering, osmotic pressure, equilibrium sedimentation) may show significant deviations from true values.

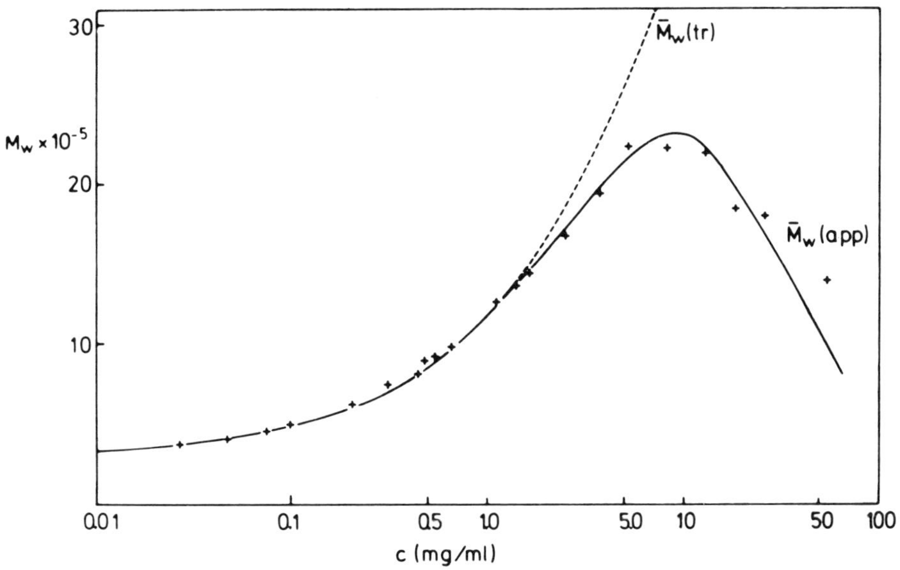

Fig. 6. Dependence of the weight average molecular weight of glutamate dehydrogenase on protein concentration (MARKAU et al., 1971). Measurements were in 0.066 M phosphate buffer, pH 7.6 at 20°C. Dotted line: calculated from Eq. (2) with M_1 = 310,000 and $K = 1.1 \cdot 10^6$ M^{-1}. Solid line: calculated from Eq. (5) with $B = 8 \cdot 10^{-9}$ M g^{-2}

It is known that the relation between the true and experimental molecular weight of a monodisperse macromolecule often conforms to the equation

$$1/M^{app} = 1/M + 2BC \qquad (3)$$

where the constant B (second virial coefficient) is a measure of solution nonideality. If it is assumed that only geometric exclusion contributes to B, the virial coefficients for simple molecular structures may be calculated with statistical mechanics. The result for a cylinder having a large length-to-diameter ratio is (TANFORD, 1961)

$$B = \pi N_A d l^2 / 4M^2 \qquad (4)$$

where N_A is Avogadros' number, l is length and d is diameter. Since the parameter d and the ratio l/M do not change with molecular weight, it may be concluded that the virial coefficients of polymers formed via linear aggregation of an elongated monomer will be nearly identical.

Several authors (EISENBERG and TOMKINS, 1968; MARKAU et al., 1971; CHUN et al., 1972) have adopted this point of view and rationalized GDH molecular weights at high concentrations in terms of a single virial coefficient and single equilibrium constant. Extension of Eq. (3) to scheme I then gives:

$$1/\bar{M}_W^{app} = 1/\bar{M}_W + 2BC_T \qquad (5)$$

where $\bar{M}_W$ is the weight average molecular weight defined in Eq. (2). The solid line of Figure 6 represents molecular weights calculated with Eq. (5). Reasonable agreement with the experimental data was achieved by assuming the M_1 and K values determined in dilute solutions and a virial coefficient close to that calculated from Eq. (4). It is also noteworthy that to within 10% the same virial coefficient is found in 0.03 M phosphate, 0.5 M phosphate and in 5 M ethylene glycol, even though the degree of polymerization in these media varies considerably (GAUPER et al., 1974). This observation corroborates the assumption that nonideality in the GDH system originates predominantly in excluded volume effects; contributions from long-range electrostatic forces between protein molecules or from protein-solvent interactions would vary with ionic strength and solvent polarity.

Alternatively, downward curvature of molecular weights for a sequential addition of monomers can be explained by assuming that the association equilibrium constants decrease with polymer chain length. Although MARKAU et al. (1971) have pointed out that this hypothesis cannot explain the *decrease* in GDH molecular weights above 10 mg/ml, we would emphasize that the rather good fit of molecular weights to Eq. (5) does not exclude the possibility that K is only an apparent value, which varies at intermediate protein concentration. The square-root dependence of $\bar{M}_W$ on the association constant in Eq. (2), together with the increased importance of the BC_T term in Eq. (5), combine to make the experimental molecular weights rather insensitive to the value of K. At 10 mg/ml, for example, a factor of 2 variation of the equilibrium constant modifies $\bar{M}_W^{app}$ by only 10%. Thus, although nonideality is no doubt important in GDH association-dissociation, and is the only simple explanation of the molecular weight behavior at high protein concentrations, the evidence for a unique equilibrium constant over a wide concentration range is not compelling. REISLER and EISENBERG (1971) have also cautioned against a two-parameter fit, but from a somewhat different point of view.

REISLER and EISENBERG (1970) have side-stepped the above ambiguity by taking advantage of a fortunate circumstance which occurs in GDH solutions saturated with toluene. For a given protein concentration, toluene enhances the degree of polymerization by a factor of 10 - 20, resulting in the formation of large aggregates in dilute solution. Thus, over a considerable range of molecular weights, it is possible to neglect the virial coefficient contribution in Eq. (5) and identify the observed molecular weights with true values. Under these conditions it could be shown that light scattering measurements are in excellent agreement with a sequential addition of monomers defined by a single equilibrium constant (REISLER and EISENBERG, 1970, 1972). However, the extent to which toluene may modify the spontaneous association-dissociation mechanism is still an open question.

C. Nature of Interactions Between Oligomers

Reaction enthalpies and entropies can provide valuable clues concerning the forces involved in protein-protein bond formation.

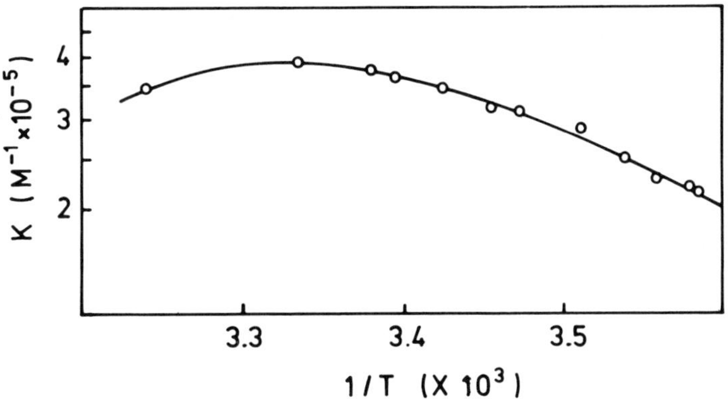

Fig. 7. Van't Hoff plot of beef liver glutamate dehydrogenase association-dissociation equilibrium constant (THUSIUS et al., 1975). Conditions: 0.2 M phosphate, pH 7.5

Thermodynamic functions for GDH self-assembly have been determined by studying the temperature dependence of the association equilibrium constant at protein concentrations where nonideality is negligible (THUSIUS et al., 1975; REISLER and EISENBERG, 1971; GAUPER et al., 1974). Figures 7 and 8 demonstrate that the association enthalpy and entropy vary strongly with temperature.

At low temperatures the driving force for self-assembly is a large, positive entropy change which more than compensates for an unfavorable reaction enthalpy. Both ΔH^0 and ΔS^0 become less positive with increasing temperature : near 28°C, ΔH^0 changes sign, while $T\Delta S^0$ slowly approaches zero. Consequently, for a given protein concentration the degree of association increases with temperature up to this region, after which further heating promotes dissociation.

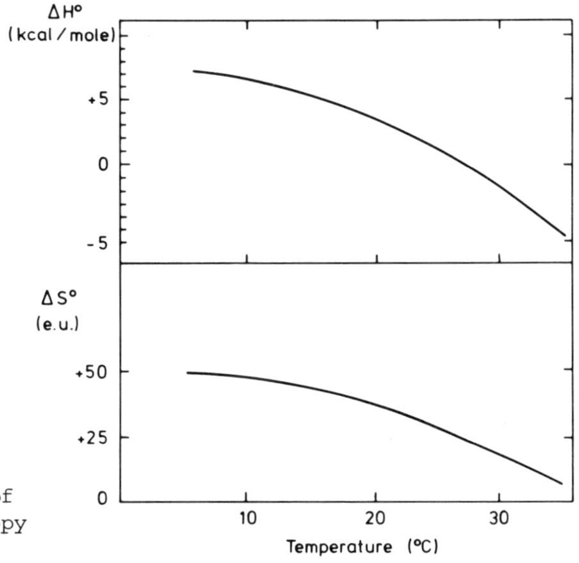

Fig. 8. Temperature dependence of self-assembly enthalpy and entropy changes (THUSIUS et al., 1975)

The low temperature behavior strongly implies hydrophobic interactions. According to the current view of hydrophobic forces (NEMETHY and SCHERAGA, 1962), bond formation results from the breakdown of an "iceberg" water structure associated with nonpolar residues. In this interpretation positive enthalpies represent primarily the potential energy needed to disrupt the hydrogen bonds of the ice-like structure, positive entropies originate in the increased freedom of water molecules when transfered to bulk solvent, and large, negative heat capacities arise from the weakening of ordered water structure with temperature. Additional evidence pointing to hydrophobic bonding comes from the effect of organic solvents on the association-dissociation equilibrium constant (GAUPER et al., 1974). At low ionic strength and elevated temperatures, nonhydrophobic interactions become important (REISLER and EISENBERG, 1971; GAUPER et al., 1974).

From the entropy data of Figure 8 it was estimated (THUSIUS et al., 1975) that below 10°C approximately 50 water molecules are released to solvent in the course of hydrophobic bond formation between two GDH particles. It was then concluded that the implied nonpolar region responsible for aggregation represents less than 10% of the total protein cross section (i.e. 4b in Fig. 3).

This interpretation now receives independent support from measurements of reaction volumes. If hydrophobic bonds are in fact formed, GDH self-assembly should show a positive reaction volume, ΔV^0, originating in the transfer of "ice-like" water to bulk solvent (NEMETHY and SCHERAGA, 1962). In addition, ΔV^0 should increase with decreasing temperature. In agreement with these predictions, recently reported volumes of reaction for GDH association-dissociation range from + 10 ml/mol at 40°C to +33 ml/mol at 10°C (HEREMANS, 1974). If we assume 0.5 - 1 ml/mol for the volume change accompanying the transfer of a single water molecule from a hydrophobic environment (NEMETHY and SCHERAGA, 1962), the value of the volume change at 10°C implies the participation of 33 - 66 water molecules. The excellent agreement with the earlier estimate suggests that losses in translational and rotational freedom when bringing together two GDH particles are negligible, and corroborates the assumption (THUSIUS et al., 1975) that the observed entropy change at low temperatures is dominated by the contribution from changes in water structure. The conclusion that GDH self-assembly derives from the existence of a very small hydrophobic region on the protein "end surface" conveniently explains the specificity of polymerization (absence of copolymers with other globular proteins; strictly end-to-end chain growth), and is consistent with the geometry of the macroscopic model of the oligomer (Fig. 3).

D. Polymer Distribution

The above thermodynamic description of GDH solutions allows one to calculate the polymer distribution at equilibrium. Assuming

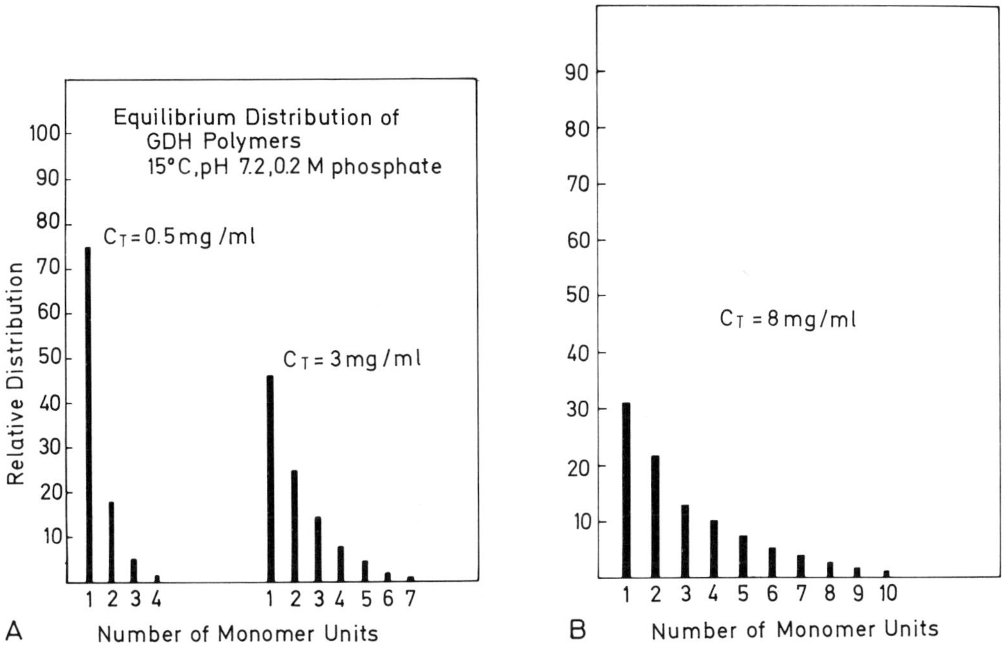

Fig. 9A and B. Equilibrium distribution of glutamate dehydrogenase polymers. Vertical axis: percentage of all particles present as n-mer

the open association-dissociation of scheme I, we may derive the following expression for the equilibrium concentration of species P_i.

$$\bar{P}_i = \bar{P}_1^i K^{i-1} \qquad (6)$$

where the oligomer concentration $\bar{P}_1$ is given in terms of K and C_T (THUSIUS, 1975c).

$$\bar{P}_1 = K^{-2}\left[1+2M_1^{-1}KC_T - (1+4KM_1^{-1}C_T)^{1/2}\right]/2C_TM_1^{-1} \qquad (7)$$

Figure 9 presents the results of calculations at 15°C for three protein weight concentrations. At 0.5 mg/ml the enzyme is highly dissociated, with 95% of the molecules in the form of monomers and dimers. At larger C_T values the relative concentration of monomers decreases, and higher molecular weight species become significantly populated. In all cases, however, the relative populations decrease monotonically with molecular weight — a feature characteristic of an indefinite polymerization defined by a unique equilibrium constant (see also DESSEN, 1973).

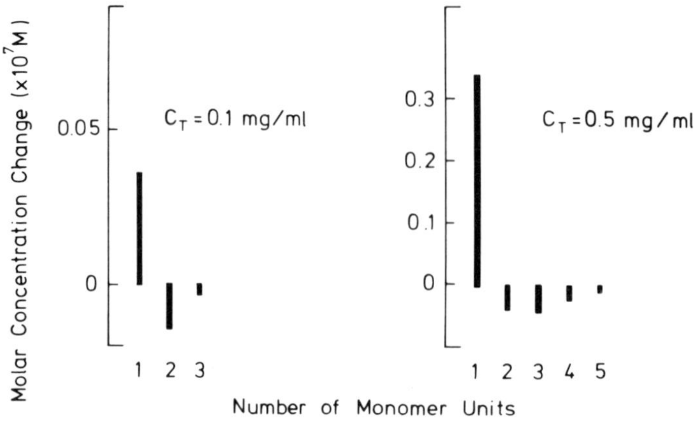

Fig. 10

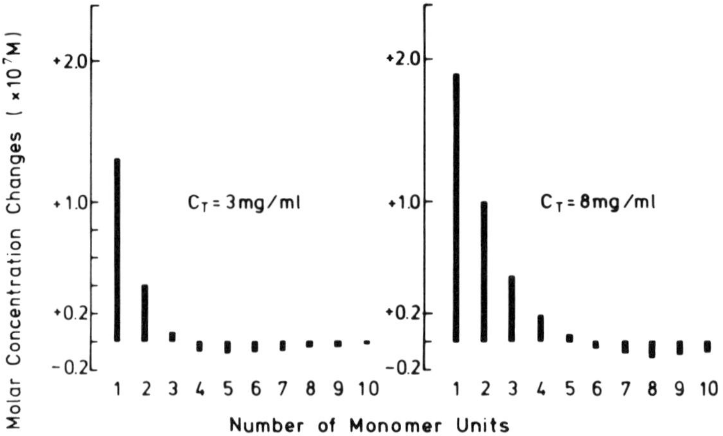

Fig. 11

Fig. 10 and 11. Equilibrium distribution of glutamate dehydrogenase polymers following a $4^{\circ}C$ temperature perturbation. The calculations assumed the equilibrium constant and enthalpy change determined in 0.2 M phosphate buffer, pH 7.2 at $15^{\circ}C$ (Figs. 7 and 8)

Even at a protein concentration eight-fold higher than the dissociation equilibrium constant (Fig. 9B), more than 50% of the molecules exist as monomer or dimer.[3] Nevertheless, the second-order dependence of scattered light intensity on molecular weight (Eq. (12)) shows that relatively small amounts of high molecular weight species can be detected. For example, although the heptamer concentration at 3 mg/ml is only $\sim$2% of the monomer concentration, these particles make equal contributions to the total scattered radiation.

[3]The dissociation equilibrium constant in units of mg/ml is the product $K^{-1}M_1$.

From the experimental equilibrium constants and association enthalpies (Figs. 7 and 8) one may calculate the change in molar equilibrium concentrations induced by a small temperature perturbation at a given initial temperature and total protein concentration. This information is pertinent to the later discussion on the dynamics of self-assembly as studied by the temperature-jump relaxation technique. Figures 10 and 11 give the overall concentration changes for "jumps" between 11 and 15°C. In extremely dilute solutions ($<$0.1 mg/ml) self-assembly conforms to a monomer-dimer equilibrium; a given variation of dimer concentration is accompanied by an approximately two-fold larger variation of monomer concentration. At higher C_T values, coupling between steps prevent a simple physical interpretation. It is evident that whether the equilibrium concentration of a polymer increases or decreases depends on total protein concentration. On the other hand, the sign of the monomer concentration change remains the same.

IV. Scattered Light Detection in Chemical Relaxation Experiments

The use of scattered light as a probe for following rapid reactions is relatively recent. The sensitivity of the method to molecular weight changes allows the selective detection of macromolecular aggregation phenomena without interference from ligand binding or small conformational changes. Chemical relaxation instruments for fluorescence measurements can of course also be used for 90° light scattering studies. The 90° detection mode is considerably more sensitive than the zero-angle configuration (turbidity mode). Interest in this approach to the study of rapid reactions will increase in the near future due to the recent appearance of high performance stopped-flow (RIESNER and BUENEMANN, 1973) and temperature-jump instruments (RIGLER et al., 1974) for 90° observation, and the wide potential applications to self-associating systems.

Scattered light detection is, however, characterized by certain complications which do not arise in other optical methods. Below these factors are considered in relation to kinetic measurements near equilibrium.

A. Angular Dependence

If the molecular dimensions of light-scattering particles approach the value of the incident light wavelength, the intensity of scattered radiation shows an angular dependence due to destructive interference. In static experiments, observations are made at several angles and the results extrapolated to zero-angle. The molecular weight of a monodisperse solute molecule in an ideal solution is related to the scattered light intensity at zero-angle by the well-known relation

$$I_{\theta=o} = \gamma MC \qquad (8)$$

where M is molecular weight, C is weight concentration and γ is an optical constant,

$$\gamma = 2\pi^2 n^2 \ (dn/dc)^2 (1 + \cos^2\theta) I_\lambda^o / r^2 \lambda^4 N_A$$

n = solution index of refraction
λ = incident light wavelength
N_A = Avogadro's number
r = distance from observation cell to photomultiplier
I_λ^o = incident light intensity
θ = observation angle

Although rapid kinetic measurements at several angles could in principle be realized with multichannel detection and data acquisition, no relaxation instrument with this capability has yet been reported. With presently available equipment, rate investigations are limited to 90° observation. We must therefore ask, to what extent must the observation angle be taken into account when interpreting relaxation decay curves?

Of course the problem of angular dependence may be avoided entirely by choosing an incident wavelength much larger than the molecular dimensions of the particles being studied. However, long wavelengths can result in a severe degradation of signal-to-noise ratios, since γ in Eq. (8) is proportional to λ^{-4}. The high signals needed in chemical relaxation experiments will often require wavelengths where angular dependence is significant.

Eq. (8) may be generalized to any observation angle by introducing the particle-scattering factor, $\rho(\theta)$ (TANFORD, 1961).

$$I = \gamma M \rho(\theta) C \tag{9}$$

When the particle dimensions are no larger than about 30% of the incident radiation wavelength, $\rho(\theta)$ is assumed to have the form

$$\rho(\theta) = (1 - \frac{16\pi^2}{3\lambda^2} R^2 \sin^2\theta/2) \tag{10}$$

where R is the scatterer radius of gyration. We see that the conditions $\theta \rightarrow 0$ or $\lambda >> R$ is sufficient to reduce the general expression to the limiting form of Eq. (8).

Regarding the detection of concentration changes in a kinetic experiment, it proves useful to define the "specific signal" of an arbitrary species, P_i, as (THUSIUS, 1972; THUSIUS et al., 1973; THUSIUS, 1975a)

$$\Phi_i = \partial I / \partial \bar{P}_i \tag{11}$$

If we set $\bar{C}_i = M_i \bar{P}_i$, Eq. (9) becomes

$$I = \gamma M_i^2 \rho_i(\theta) \bar{P}_i \tag{12}$$

Application of Eq. (11) gives the specific signal of species P_i for light-scattering detection when thermodynamic nonideality can be ignored.

$$\Phi_i = \gamma M_i^2 \rho_i(\theta) \tag{13}$$

It is known from the theory of chemical relaxation that near equilibrium, the time-dependent deviations of reactant concentrations from their final values is an exponential sum. It is not difficult to show that the signal change in scattered light measurements is given by

$$\delta I = \sum_{i=1}^{R} \delta I_i^o \, e^{-t/\tau_i} \tag{14}$$

where R is the number of elementary steps in the reaction mechanism, τ_i are relaxation times and the δI_i^o are scattered light amplitudes (THUSIUS, 1975c).

$$\delta I_i^o = \sum_{j=1}^{N} \Phi_j \beta_{ji}\,, \quad N = \text{number of chemical species} \tag{15}$$

Here Φ_j is the specific signal given in Eq. (13), and β_{ji} is a function of specific rate constants and overall concentration changes. We therefore conclude that, whereas the observation angle must be considered in the interpretation of relaxation amplitudes, relaxation times are insensitive to this parameter. Thus, the extraction of specific rate constants from the concentration dependencies of relaxation times proceeds in the usual way. On the other hand, angular dependence does enter into the treatment of a large number of *closely spaced* relaxation times, as will be seen in the later discussion on GDH self-assembly kinetics.

When the particle scattering factor is not far from unity, the specific signal Φ_i is squared, proportional to the molecular weight M_i^2. Even if not directly accessible, molecular weights are usually implied in an assumed stoichiometry. In contrast, specific signals in absorbance and fluorescence detection are proportional to molar extinction coefficients and quantum yields; these latter parameters cannot usually be calculated a priori.

When angular dependence is significant, a quantitative amplitude treatment relies on estimates of the radii of gyration of reactants, which in turn assumes that structural information is available. For rod-like molecules, R_i in the expression $\rho_i(\theta)$ may be calculated from the radius r_i and length l_i (TANFORD, 1961):

$$R_i^2 = \frac{r_i^2}{2} + \frac{l_i^2}{12} \tag{16}$$

B. The Role of Virial Coefficients

Even in dilute solutions the experimental molecular weight of a macromolecule determined by light scattering may show deviations from the true value, due to thermodynamic nonideality. To the extent that Eq. (3) is obeyed, we may express the observed intensity of scattered radiation in terms of the second virial coefficient

$$I = \gamma M\rho(\theta)C/(1+2BMC) \quad (17)$$

It follows from the above relation and Eq. (11) that the specific signals of a self-associating system are:

$$\Phi_i = \gamma M_i^2\rho_i(\theta)/(1+2B_iM_iC_i)^2 \quad (18)$$

where the usual assumption of a unique γ has been made (EISENBERG and TOMKINS, 1968). This is a general expression which takes into account both angular dependence and nonideality. It is clear from Eq. (18) that the Φ_i associated with light scattering may vary with concentration, and thus complicate the analysis of kinetic data obtained over large extents of reaction. On the other hand, the small perturbation condition of chemical relaxation assures that the Φ_i remain constant during the return to equilibrium for a given set of initial concentrations.

Although Eq. (18) is useful for the interpretation or simulation of scattered light relaxation amplitudes, it gives no information on how nonideality will modify relaxation times. This question is probably most conveniently treated in terms of activity coefficients.

A non-negligible BC term in Eq. (3) indicates that the activity coefficient of the solute molecule is varying with concentration. From absolute rate theory it is known that specific rate constants are functions of reactant and transition state activity coefficients. When activity coefficients vary with reactant concentrations, rate constants also change, making kinetic analyses awkward. In the case of small electrolytes, nonideality usually originates in long-range charge interactions between molecules. It is therefore standard procedure in solution kinetics to add a large excess of an inert salt; the ionic environment of reaction partners is then nearly constant, resulting in activity coefficients which are independent of reactant concentrations. Of course rate constants determined in these conditions are functions of the ionic medium.

The same approach is applicable to kinetic studies of macromolecular polyelectrolytes. At least in principle, charge contributions to virial coefficients (and therefore to activity coefficients) can be rendered negligible with a sufficiently large excess of supporting electrolyte. On the other hand, we have seen that solutions of macromolecules are inherently nonideal due to geometric exclusion. This is a purely statistical phenomenon which persists in the absence of long-range interactions, and therefore cannot be suppressed by the addition of an inert salt.

It is often assumed that the activity coefficients of species participating in a self-assembly process may be expressed as a simple function of total solute concentration and a unique virial coefficient (ADAMS and WILLIAMS, 1964):

$$\ln y_i = M_iBC_T \quad (19)$$

In many cases Eq. (19) has been used without justification to obtain solutions to otherwise intractable equilibrium expressions. More recently, SCHWARZ has rigorously derived the above relation for the case of thin rods and excluded volume nonideality (G. SCHWARZ, personal communication). From the definition of the thermodynamic equilibrium constant for the association of two molecules P_i and P_j having molecular weights M_i and M_j, we see from Eq. (19) that variations of the y with C_T will not modify the ratio of equilibrium concentrations:

$$K = \frac{\bar{a}_{i+j}}{\bar{a}_i \bar{a}_j} = \frac{\bar{P}_{i+j}}{\bar{P}_i \bar{P}_j} \quad \frac{\bar{y}_{i+j}}{\bar{y}_i \bar{y}_j}$$

$$= \frac{\bar{P}_{i+j}}{\bar{P}_i \bar{P}_j}$$

where a denotes activity. This is a unique situation not found in reactions between small molecules, where activity coefficient variations usually result in apparent, concentration-dependent equilibrium constants.

In regard to reaction rates, we note that an invariant K requires concentration-independent rate constants, except in the very special case where the forward and reverse rate constants are identical functions of concentration. Intuitively, then, it appears that when Eq. (19) is applicable, relaxation times are insensitive to solution nonideality.

It is perhaps more satisfying to develop a formal demonstration of this conclusion by introducing Eq. (19) into rate equations near equilibrium. For a simple dimerization,

$$2\ P_1 \rightleftarrows P_2$$

it can be shown that the reciprocal relaxation time is in general

$$1/\tau = 4k_a^o(\bar{y}_{P_1}^2/y^{\ddagger}) \quad (1 + \frac{\delta \ln \bar{y}_{P_1}}{\delta \ln \bar{P}_1})\ \bar{P}_1 + k_d^o\ (\frac{\bar{y}_{P_2}}{y^{\ddagger}})\ (1 + \frac{\delta \ln \bar{y}_{P_2}}{\delta \ln \bar{P}_2}) \quad (20)$$

Here k_a^o and k_d^o are true (concentration-independent) rate constants, $y^{\ddagger}$ is the activity coefficient of the activated complex, and the ratios appearing in the bracketed terms represent deviations of the y_i from their equilibrium values for a small variation in $\bar{P}_i$. However, it follows from Eq. (19) that for a physical perturbation at constant weight concentration, $\delta \ln \bar{y}_{P_1} = \delta \ln \bar{y}_{P_2} = 0$, or

$$1/\tau = 4\ k_a^o\ (\bar{y}_{P_1^2}\ /\ y^{\ddagger})\ \bar{P}_1 + k_d^o\ (\bar{y}_{P_2}\ /\ y^{\ddagger})$$

Furthermore, although the activity coefficients are functions of total solute concentration in regions of nonideal behaviour, the reasonable assumption that the molecular weight of the activated complex is close to that of the dimer gives

$$\frac{\bar{y}_{P_1}^2}{y^{\ddagger}} = \frac{\bar{y}_{P_2}}{y^{\ddagger}} = 1, \text{ or}$$

$$1/\tau = 4k_a^o\bar{P}_1 + k_d^o \qquad (21)$$

which is the dimerization relaxation time for an ideal solution. Therefore, at least in the limiting conditions under consideration here (excluded volume nonideality for thin rods), one can assume the usual expressions for relaxation times. Like angular dependence, nonideality modifies the definition of the amplitudes in the exponential sum of Eq. (14), but not the meaning of the time constants. In this respect light scattering relaxation methods allow one to probe the stoichiometry and underlying mechanism of a self-associating system without being hindered by ambiguities inherent in thermodynamic molecular weight measurements (light scattering, osmotic pressure, sedimentation equilibrium and viscosity).

It should be emphasized that the assumption of geometric exclusion in the above derivation is subject to experimental tests. For example, virial coefficients should be insensitive to ionic strength and solvent polarity. Furthermore, experimental values can be compared with calculated virial coefficients. Finally, excluded volume effects can be made dominant over other types of nonideality by choosing high ionic strength and working near the isoelectric point of the aggregating macromolecule.

Apparently the only other discussion on the influence of thermodynamic nonideality on light-scattering detection in chemical relaxation is the work of TAI and KEGELES (1971). In contrast to the above conclusion, these authors suggest that relaxation times for macromolecular interactions should be corrected for activity coefficient effects. Stopped-flow results on hemocyanin dimerization were consequently analyzed with the expression (KEGELES and TAI, 1973)

$$1/\tau = (4k_a^o\bar{P}_1 + k_d^o)\left(1 + \frac{d\ln y}{d\ln C}\right) \qquad (22)$$

where y, the macroscopic activity coefficient of the system (TAI and KEGELES, 1971), was determined from ultracentrifuge data.

Examination of the assumption used in deriving Eq. (22) shows that excluded volume nonideality is implied. The basic difference between the earlier treatment and the one presented here appears to be the view (TAI and KEGELES, 1971) that activity coefficient effects intervene in the same way for diffusion controlled and non-diffusion controlled reactions. The driving force of chemical diffusion is a negative gradient of chemical potential which is corrected for nonideality by the factor $(1 + d\ln y/d\ln C)$ appearing in the above equation (TANFORD, 1961). However, the rate of a nondiffusion controlled reaction is determined by both the chemical potential gradient and the microscopic exchange rate (PRIGOGINE, 1955). By treating the problem in terms of absolute rate theory and linearizing near equilibrium (Eq. (20)), both

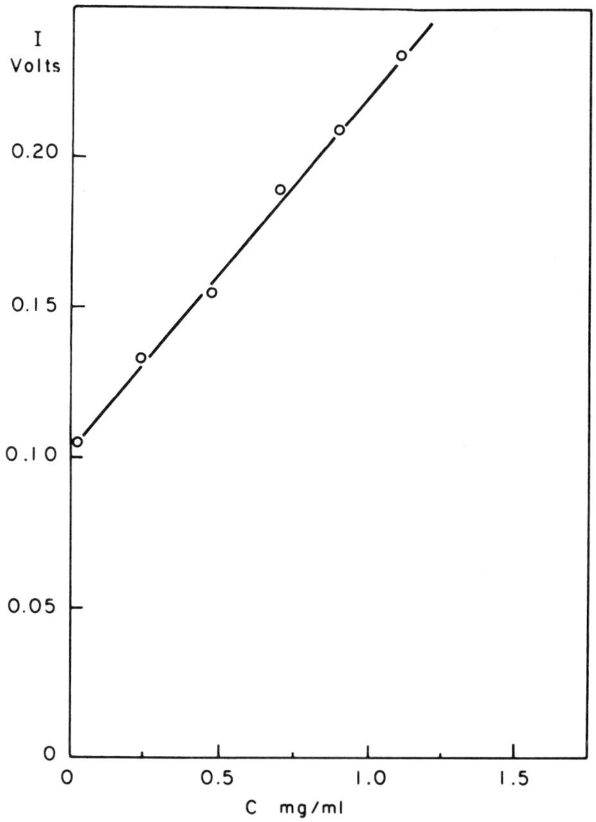

Fig. 12. Static temperature-jump light scattering measurements at 90°. The signals refer to solutions of trypsin (molecular weight 24,000) in $4 \cdot 10^{-3}$ M HCl. Photomultiplier; IP28, photomultiplier voltage: 350 volts, incident light wavelength: 368 nm, lamp: 100 w, HBO 100 W/2, monochromotor: Bausch and Lomb, "A" fixed slits. The nonzero signal at infinite dilution is the background scattering of solvent. It will be noted that the precision in determining protein concentration is limited by the error introduced when taking the difference between total scattered radiation and the background signal, which can be appreciable even at low noise levels

contributions are taken into account. At least in the context of Eq. (19), it appears that relaxation time corrections are negligible.

V. Kinetics of Self-Assembly

Because GDH self-assembly is reversible and rapid (equilibrium reached in less than 1 sec), relaxation methods offer a powerful experimental approach to the elucidation of the underlying reaction mechanism.

It should be emphasized that light scattering kinetic experiments can often be realized through modest revisions of existent relaxation instruments. Our light-scattering temperature-jump measurements on GDH association-dissociation only required a simple modification of an early Messanlagen single beam machine; an additional window (5 mm diameter) was added to the original temperature-jump cell, and the photomultiplier housing together with focusing lense was mounted at right angles to the incident beam. With the purpose of collecting the maximum amount of scattered radiation, no slit or bandpass filter was placed between the photomultiplier and sample cell. This straightforward arrangement was found satisfactory for both fluorescence (THUSIUS et al., 1973; THUSIUS, 1975a) and light scattering detection (THUSIUS et al., 1975; THUSIUS, 1975b). Although a high-sensitivity photomultiplier (such as the EMI 9558 Q) is essential for weak signal detection, less demanding applications can be realized with the original RCA IP28. The intensity of the 100 w mercury lamp was stable to ± 1% over a 24-h period. Figure 12 presents static light scattering signals obtained on a solution of a non-associating protein of medium molecular weight. In the GDH study we chose an incident wavelength of 368 nm. The inverse fourth power dependence on γ and the occurence of a high intensity peak at this wavelength resulted in a sufficient signal-to-noise ratio for detection of self-assembly below 0.5 mg/ml.

A. The Sequential Model

At least in dilute solutions, the thermodynamic behavior of GDH self-assembly can be rationalized by assuming a sequential addition of oligomers to growing polymer chains with identical equilibrium constants for each step (DESSEN and PANTALONI, 1969; REISLER and EISENBERG, 1971; SUND et al., 1972; CHUN et al., 1972; MALENCIK and ANDERSON, 1972). As this model is based on equilibrium experiments only, scheme I represents the stoichiometry of self-assembly, but not necessarily the microscopic reaction mechanism. Kinetic experiments are therefore needed to further probe the detailed mode of aggregation.

The sequential model implies a unique role for the oligomer, P_1. By making the reasonable assumption that the association and dissociation rate constants are the same for each step, we may extend scheme I to the following kinetic model

$$P_1 + P_i \underset{k_d}{\overset{k_a}{\rightleftharpoons}} P_{1+i} \qquad i = 1,2\ldots\infty$$

The above mechanism will in general give rise to an infinite number of time constants, $1/\tau_i$. In the event that a large number of amplitudes are significant, it is not possible to resolve the observed exponential sum into its individual components.

A practical approach to the analysis of a large number of overlapping relaxation times has been developed by SCHWARZ (1968), who has shown that the time-axis intercept of the initial signal change, τ^*, can be expressed explicitly in terms of elementary thermodynamic and kinetic parameters. In the case of light scattering detection, application of the SCHWARZ formalism to the sequential mechanism yields

$$1/\tau^* = k_d \; (1 + 4KM_1{}^{-1}C_T)^{1/2} M_1\bar{P}_1/C_T \qquad (23)$$

where the specific signals of Eq. (13) have been assumed with $\rho(\theta) = 1$. Clearly $1/\tau^*$ becomes smaller as total protein concentration is increased (Fig. 13).

Fig. 13. Inverse mean relaxation time for an infinite sequential addition of monomers. The line was calculated from equation 23 with $K = 3.0 \cdot 10^5\ M^{-1}$ and $k_d = 5\ s^{-1}$. The dissociation rate constant was estimated by extrapolating glutamate dehydrogenase association-dissociation rates to infinite dilution (FISHER and BARD, 1969; KEMPFLE and WINKLER, 1973; THUSIUS et al., 1975)

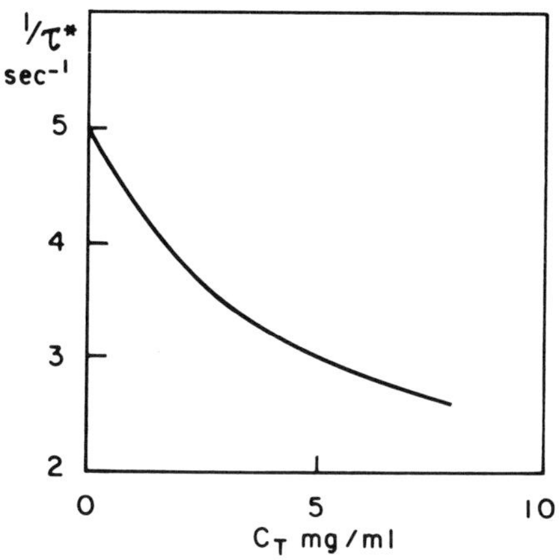

Although Eq. (23) offers a means of testing the sequential mechanism with a kinetic technique, experimental precision must be very high for an accurate estimate of initial slopes. In addition, whenever feasible it is of interest to test the time course of the entire signal change. For these reasons a complete analysis was undertaken in which individual relaxation times and light scattering amplitudes were calculated with a digital computer (THUSIUS, 1975c).

The numerical simulation of the relaxation spectrum of the sequential model was facilitated by using a series of subroutines written by ILGENFRITZ (this volume). The number of species as-

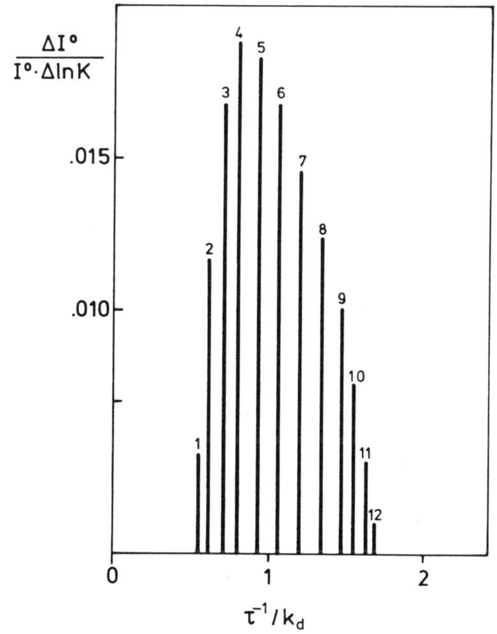

Fig. 14. Simulated relaxation spectrum of the sequential model assuming (GDH) = 0.1 mg/ml. The light scattering amplitudes are normalized with respect to the static signal, I^O, and the forced shift in the values of the thermodynamic function, $\Delta \ln K$

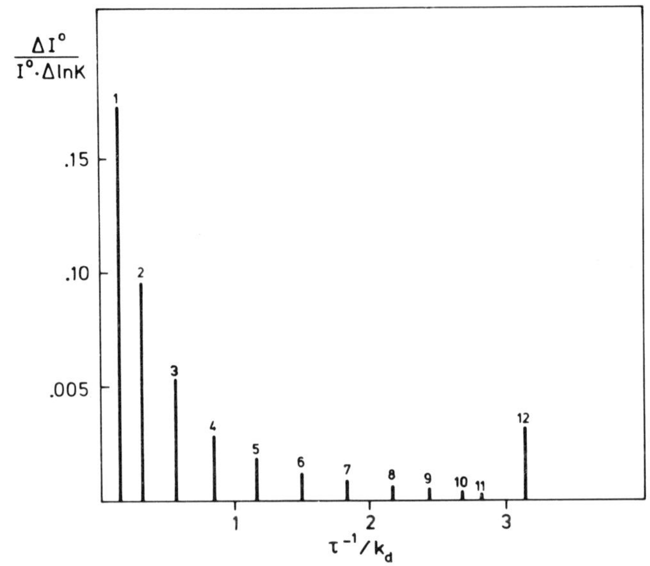

Fig. 15. Simulated relaxation spectrum of the sequential model for (GDH) = 2 mg/ml

sumed in the calculation was increased until certain criteria for convergence were satisfied (THUSIUS, 1975c). Figures 14 and 15 present results obtained for two different protein concentrations. In dilute solutions, where the monomer-dimer process predominates, the distribution of inverse relaxation times is narrow and centered at k_d. Increasing concentration splits the spectrum into a rapid time constant with a small amplitude, and a slow, broad distribution of time constants in which the smallest $1/\tau_i$ are associated with the largest amplitudes. By summing over

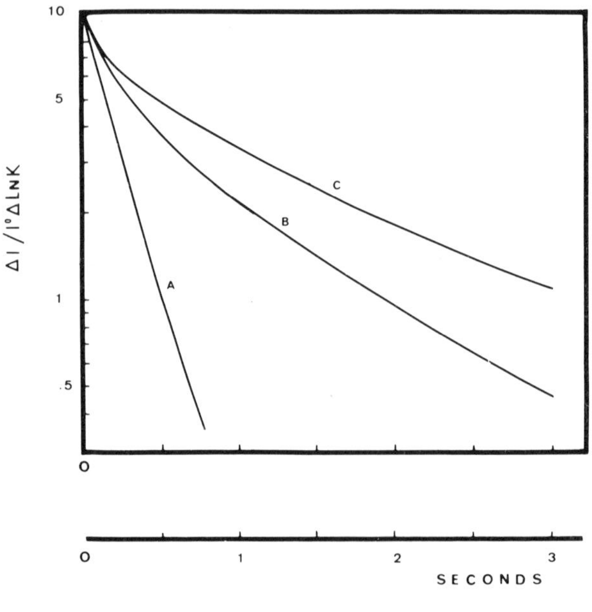

Fig. 16. Simulated light scattering progress curves for glutamate dehydrogenase self-assembly proceeding via sequential addition of oligomers. The signal changes have been normalized to give an overall amplitude of 10. A 0.1 mg/ml; B 2 mg/ml; C 4 mg/ml

all exponential terms we may generate theoretical light-scattering progress curves which can be directly compared with experimental oscillograms (Fig. 16). Examination of these curves leads to two experimental tests of the sequential model: (1) Above 1 mg/ml first-order plots should become non-linear. (2) The overall reaction half-life should increase with increased total protein concentration.

Regarding the plausibility of sequential rod growth at the microscopic level, we note that indirect evidence for two oligomer forms differing in polymerization capacity (Fig. 2) at least suggests that oligomer could be an essential participant at all stages of GDH assembly. For example, a reasonable detailed mechanism for linear polymerization incorporating a conformational change of the oligomer is the following scheme

$$P_1 \rightleftharpoons P_1^* \xrightleftharpoons{P_i} P_{1+i} \qquad i = 1,2........\infty$$

In the limit of a weakly populated intermediate P_1^*, the macroscopic stoichiometry of the above model corresponds to scheme I. If the isomerization is not rate-limiting, the kinetic behavior near equilibrium is identical to that predicted in Figures 14 - 16.

B. The Random Association Model

In contrast with the predicted behavior for a sequential addition of oligomers, GDH light-scattering decay curves obtained with stopped-flow (FISHER and BARD, 1969; THUSIUS et al., 1975) or temperature-jump methods (KEMPFLE and WINKLER, 1973; THUSIUS et al., 1975) are strictly exponential over a wide temperature and concentration range. Furthermore, the same studies have demonstrated that the association-dissociation half-life decreases with increasing C_T (Figs. 17 and 18). These observations exclude the sequential model, and argue strongly against a unique role for oligomer in GDH self-assembly.

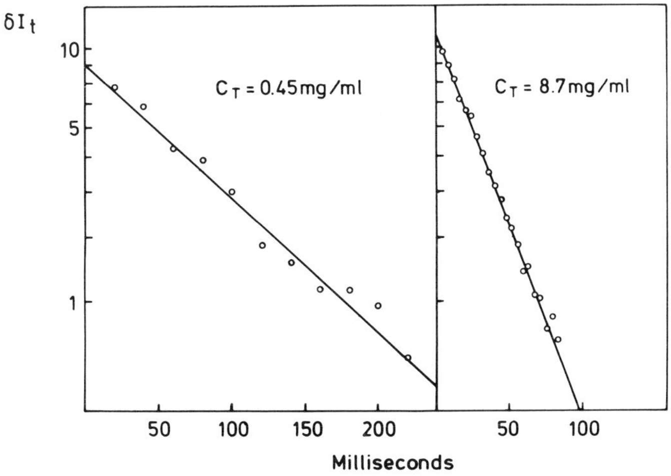

Fig. 17. First-order plots of light scattering signals observed in temperature-jump experiments. The data were acquired directly in digital form and later analyzed with a non-linear regression program. Conditions: 0.2 M phosphate buffer, pH 7.5 at 15°C

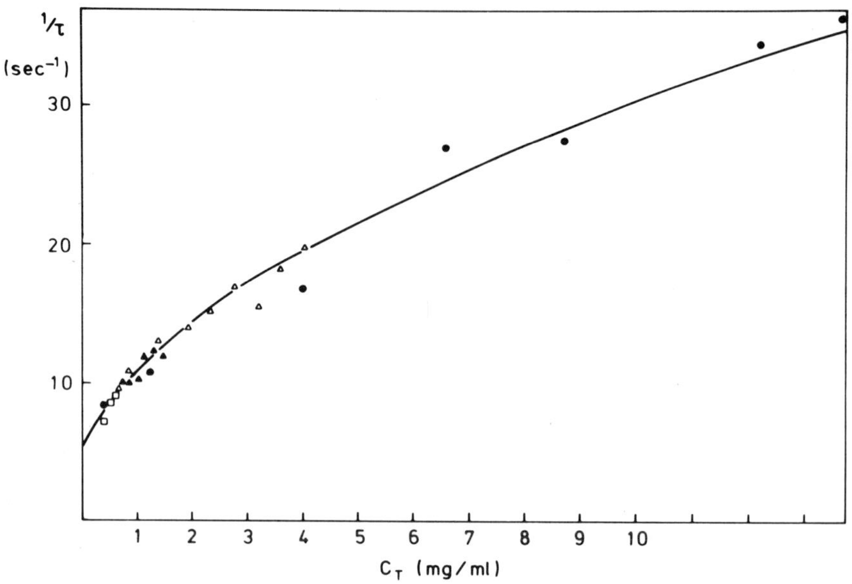

Fig. 18. Concentration dependence of reciprocal relaxation times for the reversible self-assembly of glutamate dehydrogenase

An obvious alternative mechanism is the following set of bimolecular reactions occurring between any two protein species

$$P_i + P_j \underset{k_d}{\overset{k_a}{\rightleftharpoons}} P_{i+j} \qquad \text{(Scheme II)}$$

$$K = \bar{P}_{i+j}/\bar{P}_i\bar{P}_j$$
$$i = 1,2,\ldots.\infty$$

Identical rate constants for all steps would result in a "random association" process. This is the extreme case of linear aggregation proceeding via polymers possessing identical and independent association sites.

Although the mechanistic implications of Schemes I and II are quite different, the two models are equivalent in the thermodynamic sense. This can perhaps be best appreciated by considering microscopic equilibrium constant expressions. Obviously Scheme II involves a large number of association-dissociation steps in addition to the reactions of Scheme I (Table 2). However, these additional equilibria are thermodynamically redundant. As shown in Table 2, the equilibrium constants of the processes in question can be expressed as products of the restricted set of constants which define Scheme I. For a given total concentration and a given set of independent constants K_1, K_2 ..... K_N, the polymer equilibrium distribution is identical for the two mechanisms. Thus, it is not possible to distinguish between sequential and random association modes of self-assembly with thermodynamic methods. Fortunately a distinction can be made with kinetic techniques.

Table 2. Microscopic equilibria of scheme II

$2\ P_1 \rightleftharpoons P_2$	$K_1 = \bar{P}_2/\bar{P}_1^2$
$P_1 + P_2 \rightleftharpoons P_3$	$K_2 = \bar{P}_3/\bar{P}_1\bar{P}_2$
$P_1 + P_3 \rightleftharpoons P_4$	$K_3 = \bar{P}_4/\bar{P}_1\bar{P}_3$
$2\ P_2 \rightleftharpoons P_4$	$K'_3 = K_2K_3/K_1$
$P_1 + P_4 \rightleftharpoons P_5$	$K_4 = \bar{P}_5/\bar{P}_1\bar{P}_4$
$P_2 + P_3 \rightleftharpoons P_5$	$K'_4 = K_3K_4/K_1$
$P_1 + P_5 \rightleftharpoons P_6$	$K_5 = \bar{P}_6/\bar{P}_1\bar{P}_5$
$2\ P_3 \rightleftharpoons P_6$	$K'_5 = K_3K_4K_5/K_1K_2$
$P_2 + P_4 \rightleftharpoons P_6$	$K''_5 = K_4K_5/K_1$
.	.
.	.
.	.

C. Treatment of Kinetic Results

We have seen that the reversible rate of GDH association-dissociation obeys simple first-order kinetics. Exploitation of the full information content of the kinetic results involves an analysis of the relaxation time, relaxation amplitude and the static signal I^o. The latter is recorded just before applying the physical perturbation.

$$I(t) = \delta I^o\ e^{-t/\tau} + I^o, (\delta I^o << I^o) \tag{24}$$

Static signals measured in temperature-jump experiments on GDH self-assembly in the range 0.5 to 1.5 mg/ml are presented in Figure 19A. The monotonic increase of I^o with C_T is characteristic of an open aggregation; unlike a closed equilibrium, the observed molecular weight of the system does not saturate. From

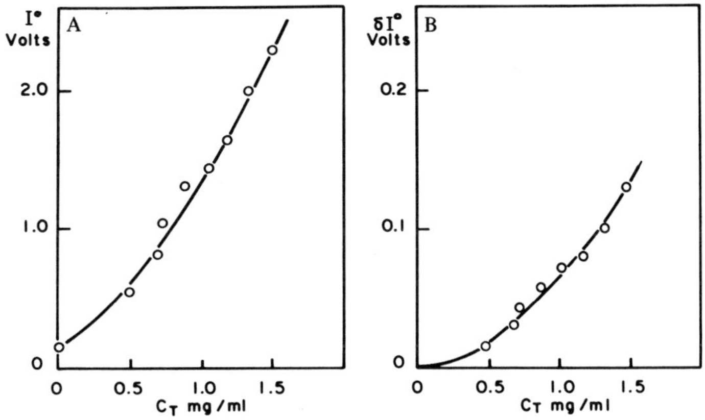

Fig. 19A and B. Static signals (A) and relaxation amplitudes (B) determined in dilute glutamate dehydrogenase solutions

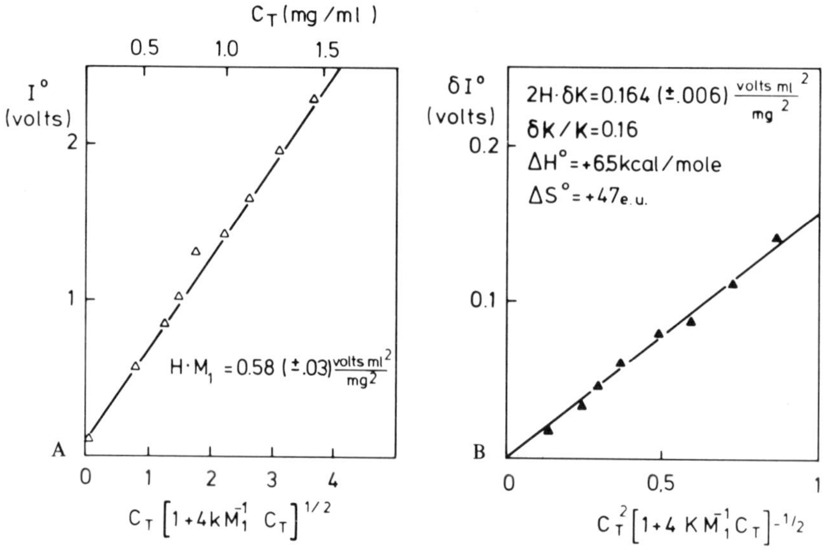

Fig. 20A and B. Linear representations of static signals (A) and relaxation amplitudes (B). The instrument constant H is equivalent to γ in the text. A value of $3 \cdot 10^5$ M^{-1} was assumed for K (see Fig. 7)

earlier light scattering work (EISENBERG and TOMKINS, 1968) it can be shown that angular dependence and nonideality can be ignored to a good first approximation at the incident wavelength and in the concentration range of these experiments. As Scheme II is thermodynamically equivalent to Scheme I, static signals for both the sequential and random association models should follow the relation [from Eqs. (2) and (9)]

$$I^{0} = \gamma M_1 C_T \sqrt{1 + 4\ K M_1^{-1} C_T} \tag{25}$$

The data of Figure 20A (THUSIUS et al., 1975) demonstrate that the above equation is in fact satisfied.

Although static signals give limited information on reaction mechanisms, they serve to verify an assumed stoichiometry. In addition, since the intensity of scattered radiation is sensitive to factors which do not alter reaction rates (the presence of dust, lamp fluctuations, small amounts of denatured protein), a satisfactory fit of static signals indicates that the kinetic experiment has been carried out under closely controlled conditions. We also recall that the concentration dependence of I^{0} yields the instrument constant, γ, which is used in the analysis of light scattering relaxation amplitudes.

The temperature-jump amplitude of a true or apparent one-step transformation is a function of the reaction enthalpy, ΔH^{0}. The amplitude concentration dependence can yield precise estimates of enthalpies over a narrow temperature range, and is therefore a valuable means of determining temperature-dependent enthalpies. On the other hand, the overall signal change is a function only of thermodynamic parameters. Thus, the relaxation amplitudes for both Scheme I and Scheme II are given by the same expression,

$$\delta I^{o} = \gamma \sum_{i} M_i^2 \ \delta \bar{P}_i \qquad (26)$$

where the total molar concentration changes, $\delta \bar{P}_i$, are given in Figures 10 and 11 for some representative C_T values. As in the case of initial signals, I^o, the second-order dependence on molecular weights and the unlimited nature of the association-dissociation produce a rapid increase in δI^o with total protein concentration (Fig. 19 B). It can be shown (THUSIUS et al., 1975) that the sum of Eq. (26) reduces to the following function of K and C_T.

$$\delta I^{o} = 2 \ \gamma \ M_1^{-1} \ \left[C_T^2 \ / \ (1 + 4 \ KM_1^{-1} \ C_T)^{1/2} \right] \ \delta K \qquad (27)$$

The good fit of the amplitude data to Eq. (27) (Fig. 20 B) and acceptable agreement with the enthalpy change evaluated from equilibrium experiments at the same temperature (Fig. 8) provide additional evidence for an open aggregation defined by a unique equilibrium constant in dilute protein solutions.

It would also be of interest to analyze the amplitude concentration dependence at higher protein concentrations, where the particle scattering factor $\rho(\theta)$ is no longer equal to unity, and the virial coefficient contribution to light scattering intensity is non-negligible. In principle this could be done by substitution of the general specific signal expressions [Eq. (18)] into Eq. (26).

Relaxation times are more sensitive than static signals or amplitudes to microscopic events involved in self-assembly. In regard to Scheme II, several authors have pointed out that an infinite polymerization taking place by random association-dissociation can be formally represented as a simple dimerization (PÖRSCHKE and EGGERS, 1972; RASSING, 1972; DESSEN, 1973).

$$2A \rightleftharpoons B$$

where A denotes free association sites and B denotes bonds between monomer units. The two-state nature of the mechanism results in a highly degenerate relaxation spectrum defined by a single observable relaxation phenomenon, the time constant being

$$1/\tau = k_a \left[(1 + 4KM_1^{-1}C_T)^{1/2} - 1 \right] K^{-1} + k_d \qquad (28)$$

In contrast to the sequential mechanism, the half-life of a random association process decreases monotonically with C_T. Conveniently, the square of the inverse relaxation time in Eq. (28) is a linear function of total protein concentration (RASSING, 1972; KEMPFLE and WINKLER, 1973; THUSIUS et al., 1975).

$$1/\tau^2 = 4k_a k_d M_1^{-1} C_T + k_d^2 \qquad (29)$$

which allows a graphical analysis without introducing an estimate of the equilibrium constant.

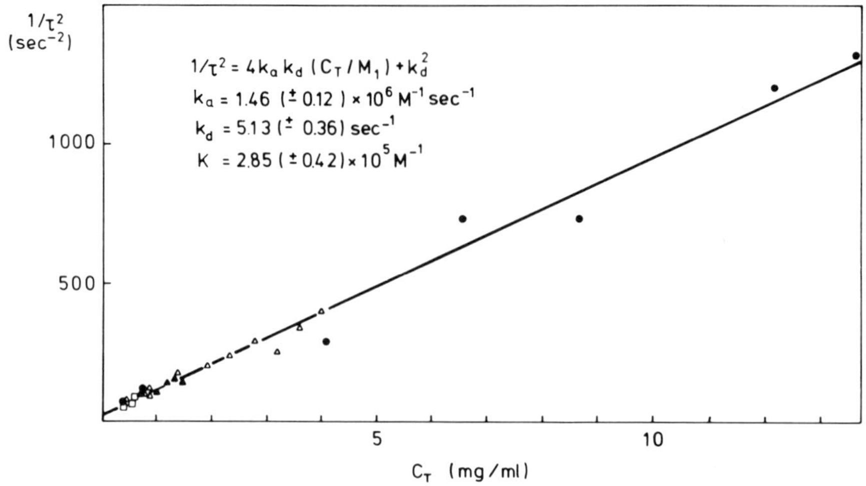

Fig. 21. Inverse relaxation times as a function of total glutamate dehydrogenase concentration

The strictly first-order nature of the observed kinetic decays together with the excellent fit of the relaxation times to Eq. (29) (Fig. 21) strongly favor a random association-dissociation mechanism for GDH self-assembly (THUSIUS et al., 1975).

Earlier in this Chapter it was argued that in the event of excluded volume nonideality, rate constants for the formation of thin rods will be concentration-independent. Also cited was experimental evidence indicating that deviations from ideal behavior in GDH solutions does in fact originate in excluded volumes. It is therefore significant that the parameter $1/\tau^2$ remains a linear function of C_T at enzyme concentrations where large deviations from solution ideality are known to occur (Fig. 6). We take this to mean that the specific rate constants k_a and k_d, and therefore the equilibrium constant K, are to a very good approximation independent of polymer chain length both in dilute and concentrated protein solutions. Thus, the assumption of a unique equilibrium constant over a wide concentration range (MARKAU et al., 1971; CHUN et al., 1972) receives strong independent support from the kinetic results. Since equilibrium molecular weight determinations were less conclusive regarding this point and do not permit a distinction between a sequential addition of monomers or a random association process, the present work provides new and more stringent tests of the self-assembly mechanism.

D. Summary of Kinetic Results

Kinetic and thermodynamic results for GDH association-dissociation at 15°C are summarized in Table 3.

Table 3. Kinetic and thermodynamic parameters for glutamate dehydrogenase association-dissociation (0.2 M phosphate, pH 7.2, 15°C)

	Thermodynamic parameters		
	M^{-1}	kcal/mol	cal/deg mol
	K	ΔH^{o}	ΔS^{o}
Static light Scattering	$3.1 \cdot 10^{5}$	+5.2	+ 44
Relaxation Amplitudes	-	+6.5	+ 47
Relaxation times	$2.9 \cdot 10^{5}$	-	-

Kinetic parameters			
k_a	$1.6 \cdot 10^{6}\ M^{-1}\ s^{-1}$	k_d	$5\ s^{-1}$
$\Delta H_a^{\neq}$	12 kcal/mol	$\Delta H_d^{\neq}$	7 kcal/mol
$\Delta S_a^{\neq}$	+ 13 e. u.	$\Delta S_d^{\neq}$	- 30 e. u.

The good agreement between parameters determined with different techniques demonstrate that the results are internally consistent.

The association rate constant k_a is 2 - 3 orders of magnitude smaller than the value usually cited for a diffusion-controlled reaction. If spherical symmetry is assumed, the calculated encounter-controlled rate constant for dimerization of a protein the size of GDH is about $5 \cdot 10^{8}\ M^{-1}\ s^{-1}$ (THUSIUS et al., 1975). This is, however, only an order of magnitude approximation. In particular, it is assumed that each collision has the correct symmetry for productive bond formation. If steric effects are in fact rate controlling, one would expect an unfavorable activation association entropy. But the positive $\Delta S_a^{\ddagger}$ and substantial $\Delta H_a^{\ddagger}$ argue against this interpretation.

In view of the evidence pointing to the importance of ordered water structure during the formation of GDH protein-protein bonds, it is natural to consider the extent to which the rate determining step may involve the breakdown of "ice-like" water. Even if the residence time of a single water molecule is considerably shorter than the half-life of protein association-dissociation, the disruption of an entire sheath of ordered solvent molecules could conceivably be rate determining. The activation parameters are at least consistent with a cooperative expulsion of water, analogous to the melting of ice. The activation entropy $\Delta S_a^{\ddagger}$ is far from the overall value of + 50 e.u. while $\Delta H_a^{\ddagger}$ is larger than the entire potential energy input of self-association. This suggests that formation of the transition state requires rupture of a small but critical number of strong hydrogen bonds, after which the entire solvent structure separating the protein contact sites rapidly collapses (THUSIUS et al., 1975). Of course, until more experimental results on the microscopic mechanism become available, this interpretation can only be considered tentative.

Acknowledgment. I thank P. DESSEN for his comments on the manuscript, G. SCHWARZ for communicating unpublished results, T. JOVIN, H. BUC and J. YON for their support.

References

ADAMS, E.T., WILLIAMS, J.W.: Sedimentation equilibrium in reacting systems. II. Extensions of the theory to several types of association phenomena. J. Am. Chem. Soc. 86, 3454-3461 (1964)

CHUN, P.W., KIM, S.J., WILLIAMS, J.D., COPE, W.T., TANG, L.H., ADAMS, E.T.: Graphical approach to non-ideal cases of self-associating protein systems. Biopolymers 11, 197-214 (1972)

COLMAN, R.F., FRIEDEN, C.: On the role of amino groups in the structure and function of glutamate dehydrogenase. II. Effect of acetylation on molecular properties. J. Biol. Chem. 241, 3661-3670 (1966)

DESSEN, P.: Polymérisation linéaire et fixation de ligands. Biochimie 55, 405-411 (1973)

DESSEN, P., PANTALONI, D.: Glutamate deshydrogenase. Structure quaternaire et propriétés rotatoires. Europ. J. Biochem. 8, 292-302 (1969)

EISENBERG, H.: Glutamate dehydrogenase: Anatomy of a regulatory enzyme. Accounts Chem. Res. 4, 379-385 (1971)

EISENBERG, H., JOSEPHS, R., REISLER, E.: Bovine liver glutamate dehydrogenase. Advan. Protein Chem. 30, 101-181 (1976)

EISENBERG, H., TOMKINS, G.: Molecular weight of the subunits, oligomeric and associated forms of bovine liver glutamate dehydrogenase. J. Mol. Biol. 31, 37-49 (1968)

FISHER, H.F.: Glutamate dehydrogenase-ligand complexes and their relationship to the mechanism of reaction. Advan. Enzymol. 39, 369-417 (1973)

FISHER, H.F., BARD, J.R.: The kinetic measurement of the α- β dissociation of glutamate dehydrogenase. Biochim. Biophys. Acta 188, 168-170 (1969)

FRIEDEN, C.: Glutamate dehydrogenase. V. The relation of enzyme structure to the catalytic function. J. Biol. Chem. 238, 3286-3299 (1963)

GAUPER, F.P., MARKAU, K., SUND, H.: The influence of salts and organic solvents on the thermodynamic parameters of the association-dissociation of beef-liver glutamate dehydrogenase. Europ. J. Biochem. 49, 555-563 (1974)

HEREMANS, K.: High pressure light scattering studies on glutamate dehydrogenase. Proc. Intern. Conf. High Pressure. Kyoto, Japan, 1974

JOSEPHS, R.: Electron microscope studies on glutamic dehydrogenase subunit structure of individual molecules and linear associates. J. Mol. Biol. 55, 147-153 (1971)

KEGELES, G., TAI, M.: Rate constants for the hexamer-dodecamer reaction of lobster hemocyanin. Biophys. Chem. 1, 46-50 (1973)
KEMPFLE, M., WINKLER, H.: Relaxations Kinetische Untersuchungen des Assoziationgleichgewichts der Glutamat-Dehydrogenase aus Rinderleber. Z. Physiol. Chem. 354, 816-818 (1973)
KRAUSE, J., MARKAU, K., MINSSEN, M., SUND, H.: Quaternary Structure and Enzymic Properties of Beef Liver Glutamate Dehydrogenase. In: Pyridine Nucleotide-Dependent Dehydrogenases. SUND, H. (ed.), pp. 279-291. Berlin-Heidelberg-New York: Springer 1970
MALENCIK, D.A., ANDERSON, S.R.: Fluorescence polarization studies on the self-association of beef liver glutamate dehydrogenase. Biochemistry 11, 3022-3027 (1972)
MARKAU, K., SCHNEIDER, J., SUND, H.: The mechanism of the association-dissociation equilibrium of beef-liver glutamate dehydrogenase. Europ. J. Biochem. 24, 393-400 (1971)
NEMETHY, G., SCHERAGA, H.A.: The structure of water and hydrogen bonding in proteins. III. The thermodynamic properties of hydrogen bonding in proteins. J. Chem. Phys. 66, 1773-1789 (1962)
OLSEN, J.A., ANFINSEN, C.B.: The crystallization and characterization of L-glutamic acid dehydrogenase. J. Biol. Chem. 197, 67-79 (1952)
PILZ, I., SUND, H.: Studies of glutamate dehydrogenase. The X-ray small-angle investigation of the beef liver glutamate dehydrogenase oligomer. Europ. J. Biochem. 20, 561-568 (1971)
PÖRSCHKE, D., EGGERS, F.: Thermodynamics and kinetics of base-stacking interactions. Europ. J. Biochem. 26, 490-498 (1972)
PRIGOGINE, I.: Thermodynamics of Irreversible Processes, 3rd ed., p. 58. New York: Wiley 1955
RASSING, J.: Hydrogen bond kinetics and relaxation spectrometry. Advan. Mol. Relaxation Processes 4, 55-86 (1972)
REISLER, E., EISENBERG, H.: Studies on the viscosity of solutions of bovine liver glutamate dehydrogenase and on related hydrodynamic models; effect of toluene on enzyme association. Biopolymers 9, 877-889 (1970)
REISLER, E., EISENBERG, H.: Bovine liver glutamate dehydrogenase association and dependence of association on temperature. Biochemistry 10, 2659-2663 (1971)
REISLER, E., EISENBERG, H.: Solubility of toluene in bovine liver glutamate dehydrogenase solutions and enhancement of enzyme action. Biochim. Biophys. Acta 258, 351-357 (1972)
REISLER, E., POUYET, J., EISENBERG, H.: Molecular weights, association, and frictional resistance of bovine liver glutamate dehydrogenase at low concentration. Equilibrium and velocity sedimentation, light-scattering studies, and settling experiments with macroscopic models of the enzyme oligomer. Biochemistry 9, 3095-3102 (1970)
RIESNER, D., BUENEMANN, H.: A stopped-flow apparatus with light scattering detection and its application to biochemical reactions. Proc. Nat. Acad. Sci. (Wash.) 70, 890-893 (1973)
RIGLER, R., RABLE, C., JOVIN, T.: A temperature-jump apparatus for fluorescence measurements. Rev. Sci. Instrum. 45, 580-588 (1974)
SCHWARZ, G.: Kinetic analysis by chemical relaxation methods. Rev. Mod. Phys. 40, 206-218 (1968)
STADTMAN, E.R.: Allosteric regulation of enzyme activity. Advan. Enzymol. 28, 41-154 (1966)

SUND, H., BURCHARD, W.: Sedimentation coefficient and molecular weight of beef liver glutamate dehydrogenase at the microgram and milligram level. Europ. J. Biochem. 6, 202-206 (1968)
SUND, H., MARKAU, K., KOBERSTEIN, R.: Glutamate dehydrogenase. In: Subunits in Biological Systems. TIMASHEFF, S.N., FASMAN, G.D. (eds.), pp. 225-287. New York: Marcel Dekker 1975
SUND, H., MARKAU, K., MINSSEN, M., SCHNEIDER, J.: Physico-chemical studies of glutamate dehydrogenase in solution. In: Structure and Function of Oxidation Reduction Enzymes. AKESON, A., EHRENBERG, A. (eds.), pp. 681-690. New York: Pergamon Press 1972
SUND, H., PILZ, I., HERBST, M.: Studies of glutamate dehydrogenase. 5. The X-ray small-angle investigation of beef liver glutamate dehydrogenase. Europ. J. Biochem. 7, 518-525 (1969)
SUND, H., WEBER, K.: The quaternary structure of proteins. Angew. Chem. Intern. ed. 5, 231-241 (1966)
TAI, M., KEGELES, G.: Subunit interactions of lobster hemocyanin. II. Temperature-jump kinetic studies. Arch. Biochem. Biophys. 142, 258-267 (1971)
TANFORD, C.: Physical Chemistry of Macromolecules. New York: Springer 1961.
THUSIUS, D.: Relaxation amplitudes for two coupled equilibria. J. Am. Chem. Soc. 94, 356-364 (1972)
THUSIUS, D.: The analysis of chemical relaxation amplitudes. Proc. NATO Advan. Study Inst. Chem. Biol. Applications Chem. Relaxation. Boston: Reidel 1975a
THUSIUS, D.: Kinetics of glutamate dehydrogenase self-assembly. Proc. NATO Advan. Study Inst. Chem. Biol. Applications Chem. Relaxation. Boston: Reidel 1975b
THUSIUS, D.: Mechanism of bovine liver glutamate dehydrogenase self-assembly. II. Simulation of relaxation spectra for an open linear polymerization proceeding via a sequential addition of monomer units. J. Mol. Biol. 94, 367-383 (1975c)
THUSIUS, D., DESSEN, P., JALLON, M.: Mechanism of bovine liver glutamate dehydrogenase self-assembly. I. Kinetic evidence for a random association of polymer units. J. Mol. Biol. 92, 413-432 (1975)
THUSIUS, D., FOUCAULT, G., GUILLAIN, F.: The analysis of relaxation amplitudes and some applications to reactions involving macromolecules. In: Dynamic Aspects of Conformational Changes in Macromolecules. SADRON, C. (ed.), pp. 271-284. Boston: De Reidel 1973
TOMKINS, G.M., YIELDING, K.L., CURRAN, J.F., SUMMERS, M.R., BITENSKY, M.W.: The dependence of the substrate specificity on the conformation of crystalline glutamate dehydrogenase. J. Biol. Chem. 240, 3793-3798 (1965)

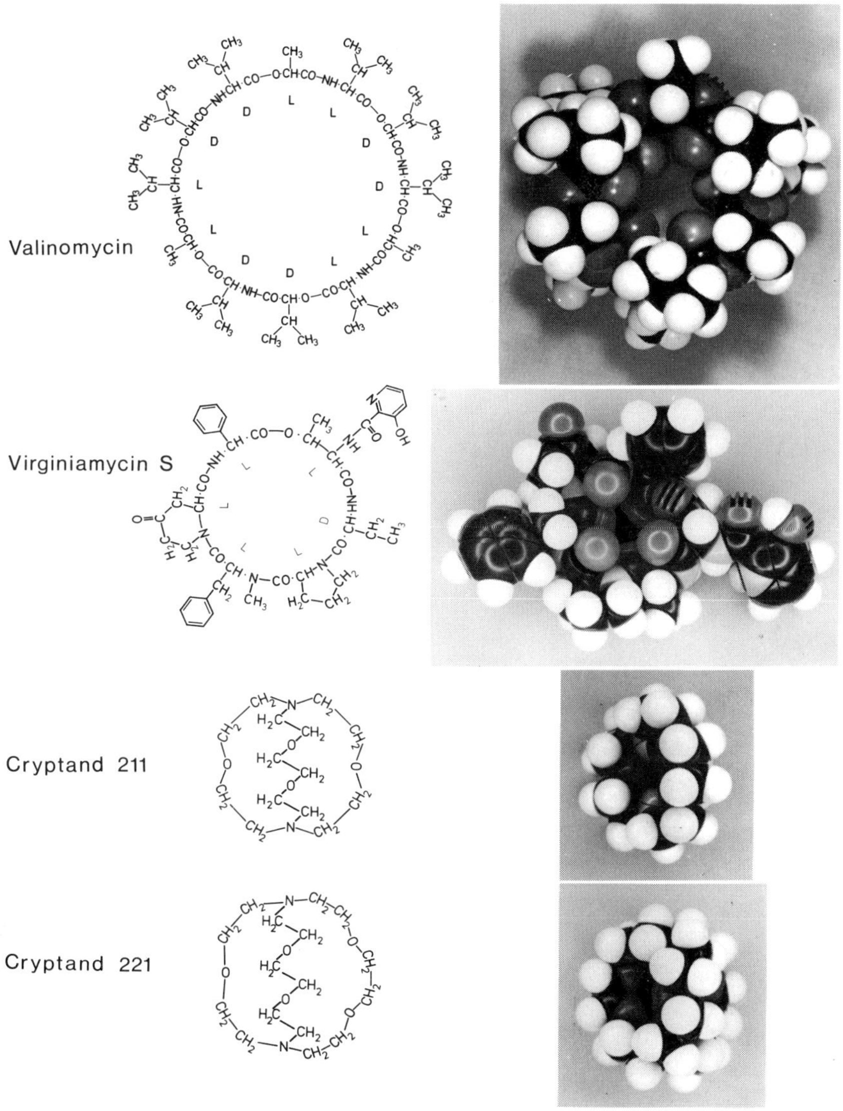

Fig. 2. Chemical formulas and space-filling models of valinomycin, virginiamycin S, cryptand-211 and cryptand-221

In order to elucidate the mechanism for the cation specificity and carrier action of these antibiotics in terms of elementary steps, relaxation methods have been employed on lipid bilayer membranes in homogeneous solution.

II. Kinetic Studies on Lipid Bilayer Membranes

For the investigation of the dynamics of valinomycin-mediated cation transport across lipid bilayer membranes, an electric

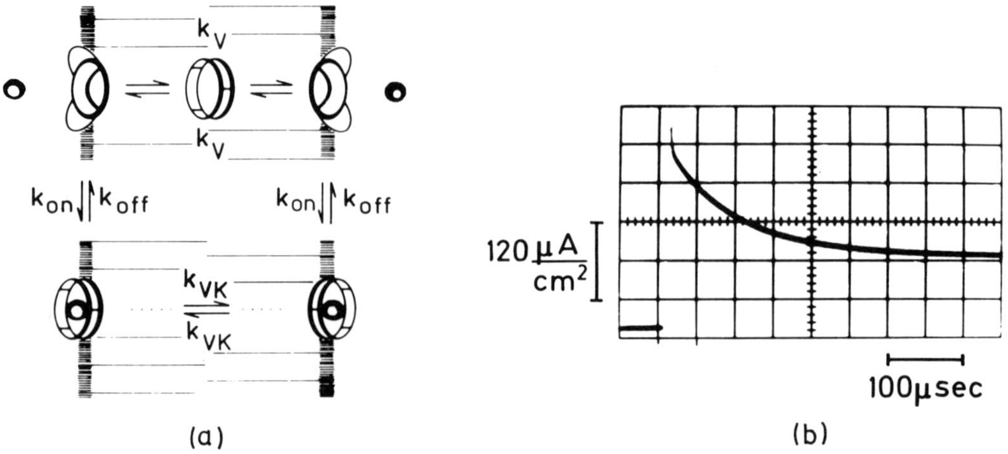

Fig. 3. (a) Transport of a cation through a lipid bilayer membrane mediated by the carrier valinomycin. (b) Relaxation of the electrical current for a phosphatidylserine membrane in the presence of valinomycin and 1 M KCl; 12°C. (After BAMBERG et al., 1974)

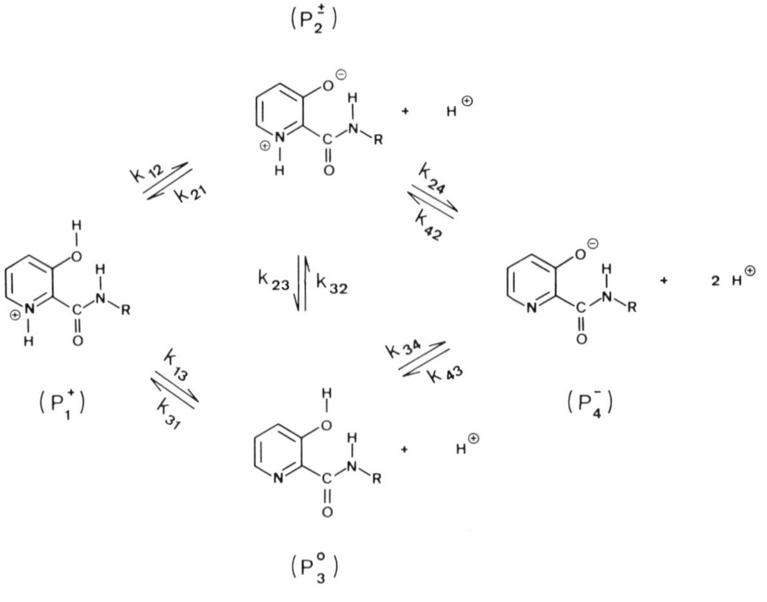

Fig. 4. Protolytic states of virginiamycin S (R peptide residue); $K'_{12} = k_{12}/k_{21}$, $K'_{13} = k_{13}/k_{31}$, $K'_{24} = k_{24}/k_{42}$, $K'_{34} = k_{34}/k_{43}$

Table 1. Thermodynamic and kinetic parameters of K^+ transport through a phosphatidylinositol bilayer membrane mediated by valinomycin. (After STARK et al., 1971)

K' (M^{-1})	k_{on} $(M^{-1}\ s^{-1})$	k_{off} (s^{-1})	k_V (s^{-1})	k_{VK} (s^{-1})
1	$5 \cdot 10^4$	$5 \cdot 10^4$	$2 \cdot 10^4$	$2 \cdot 10^4$

field jump technique with current detection has been developed (STARK et al., 1971). A voltage pulse, with a rise time of 1 μs and an amplitude of about 30 mV, is applied to the cell containing the membrane, the resulting current being followed on an oscilloscope. For a typical oscillogram of such a relaxation experiment see Figure 3b. From the time and amplitude of this relaxation process and stationary conductance data the individual rate constants can be calculated (STARK et al., 1971; Table 1).

The rate constants, k_{on} and k_{off}, characterize the K^+ loading and unloading of the membrane-bound valinomycin whereas k_{VK} and k_V characterize the diffusion of the valinomycin-K^+ complex and the uncomplexed antibiotic across the membrane (Fig. 3a).

The rate constant for cation complex formation at the membrane interface is three orders of magnitude smaller than the value observed in methanol employing chemical relaxation spectrometry (GRELL and FUNCK, 1973). The difference between these kinetic parameters cannot only be due to the dependence of k_{on} on the surface potential of the membrane in the case of the heterogeneous reaction. In order to clarify this apparent discrepancy further experimental investigations are required (cf. Sect. III.B.3).

In addition, it is interesting to note that the apparent stability constant, K', of the membrane-bound valinomycin-K^+ complex which has been confirmed by CD (circular dichroism) titrations of valinomycin bound to 1,2-dimyristoyl-L-lecithin vesicles (GRELL et al., 1975) is nearly 10^5 times smaller than that determined in methanol (Table 4). The thermodynamic and kinetic parameters shown in Table 1 are optimal for a fast and efficient cation transfer across the membrane. If the valinomycin-cation complex formed a stronger complex at the membrane surface than is experimentally observed, then the unloading rate, k_{off}, would be rate-limiting for the cation transfer, thus leading to less favorable conditions.

III. Elementary Steps Involved in Carrier-Mediated Cation Transport

A. Elementary Steps Relevant to Proton Transport

1. Thermodynamic Parameters

Since the antibiotic virginiamycin S acts as a proton carrier in lipid bilayer membranes, proton transfer must be considered as an important elementary step in the overall transport process. The four protolytic states of the 3-hydroxypicolinyl residue of virginiamycin S are shown in Figure 4. The 3-hydroxypicolinyl residue exists in two different monoprotonated states, $P_2^{\pm}$ and P_3^{o}, in aqueous solution or methanol/water mixtures between pH 5 and 3. The two states differ in the location of the proton which can be bound either to the phenolic oxygen or to the ring nitrogen.

Stationary conductance measurements on 1,2-dioleoyl-L-lecithin bilayer membranes in the presence of the antibiotic (GRELL et al., 1977) indicate that the protolytic state P_4^- acts as a proton acceptor at the membrane interphase. Since the permeability of the membrane-bound antibiotic is larger in state P_3^o than in state $P_2^{\pm}$, the proton is preferentially carried across the membrane by the state P_3^o. After the unloading process the carrier action is completed by back diffusion of the antibiotic in the protolytic state P_4^-. Since the state P_3^o predominantly carries the proton across the membrane, the transport process must be coupled to a preequilibrium of the membrane-bound antibiotic between states $P_2^{\pm}$ and P_3^o where proton transfer occurs. This proton transfer could occur via the states P_1^+ or P_4^- by H^+ or OH^- catalysis respectively, via intermolecular proton transfer with participation of the buffer molecules or by intramolecular proton transfer. It is therefore important to determine the rate constants of the various reaction paths (cf. Sect. III.A.2) With respect to the reaction at the membrane surface it could be that the proton transfer occurs in a region of the membrane interface where the local concentrations of H^+, OH^- and buffer molecules are small compared to their concentrations in the bulk solution. This implies that intramolecular proton transfer dominates. The thermodynamic parameters of the protolytic reactions of virginiamycin S must be determined before kinetic studies can be undertaken.

The two apparent macroscopic dissociation constants, K_I' and K_{II}' are determined by spectrophotometric titrations in the wavelength region 250 to 450 nm where the absorption spectrum is related only to the 3-hydroxypicolinyl chromophore. An example of a spectrophotometric pH titration of membrane-bound virginiamycin S is shown in Figure 5a together with a plot of extinction coefficient versus pH (Fig. 5b). In order to calculate the four microscopic dissociation constants K_{12}', K_{13}', K_{24}' and K_{34}' (the indices of these equilibrium constants refer to the protolytic states as numbered in Fig. 4), the equilibrium constant K_{23}' must be known (cf. Eqs. (1) and (2)):

$$K_I' = K_{12}' + K_{13}' \qquad K_{II}'^{-1} = K_{24}'^{-1} + K_{34}'^{-1} \tag{1}$$

$$K_{23}' = \frac{k_{32}}{k_{23}} = \frac{K_{12}'}{K_{13}'} = \frac{K_{34}'}{K_{24}'} \tag{2}$$

The equilibrium constant K_{23}' is evaluated from the areas of the absorption bands of the states $P_2^{\pm}$ and P_3^o which are obtained by computer simulation of the absorption spectra in the form of Gaussian curves.

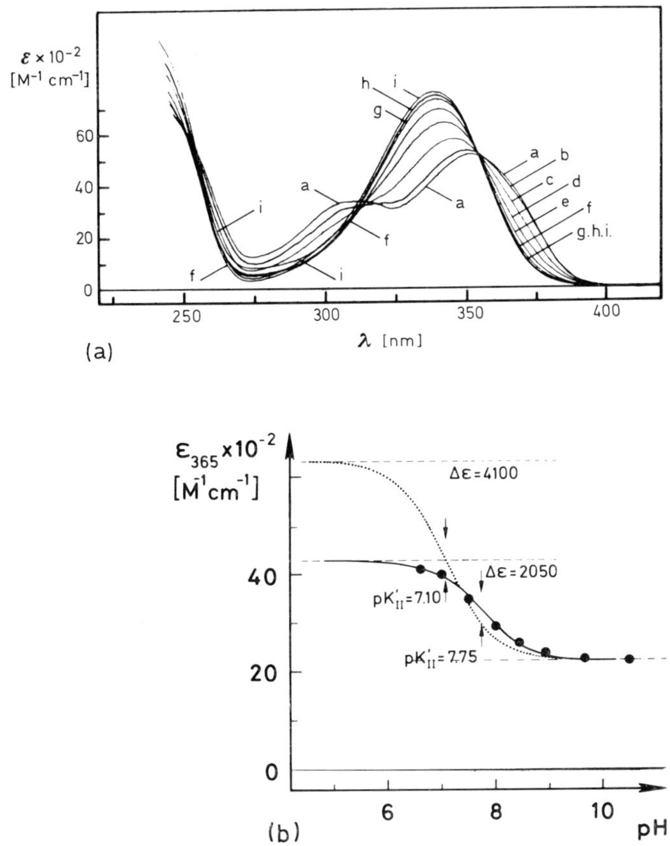

Fig. 5. (a) Spectrophotometric pH-titration of virginiamycin S in the presence of a lecithin dispersion ($2.24 \cdot 10^{-2}$ M 1,2-dimyristoyl-L-lecithin), $2 \cdot 10^{-2}$ M tetramethylammonium chloride, $5 \cdot 10^{-3}$ M N-ethylmorpholine with 0.35 M tetramethylammonium hydroxyde ranging from pH 6.62 (spectrum a) to pH 10.50 (spectrum i); 27°C. (b) Plot of extinction coefficient at 365 nm versus pH in the absence of lipid (·····) and in the presence of $2.24 \cdot 10^{-2}$ M 1,2-dimyristoyl-L-lecithin (●——●). The theoretical titration curves have been calculated for the indicated pK'-values; (experimental conditions correspond to those specified in Fig. 5a)

In Table 2 the thermodynamic parameters of virginiamycin S as determined in various solvent systems and when bound to a 1,2-dimyristoyl-L-lecithin membrane are given. From the results shown in Table 2 it can be concluded that the thermodynamic parameters of the protolytic reactions of virginiamycin S in 33.8% water in methanol (v/v) are remarkably different from those obtained in methanol or water alone, they are however nearly identical to those of the membrane-bound antibiotic. Thus, it is to be expected that the membrane-bound antibiotic is preferentially located at the membrane surface. The kinetic studies have therefore been carried out in 33.8% water in methanol (v/v).

Table 2. Thermodynamic parameters related to protolytic reactions of virginiamycin S in different solvents and of membrane-bound compound 25 - 27°C)

Solvent	pK'_{I}	pK'_{12}	pK'_{13}	K'_{23}	pK'_{II}	pK'_{24}	pK'_{34}
Methanol	0.33	1.43	0.36	0.09	8.60	7.50	8.56
33.8% water in methanol (v/v)	0.84	1.20	1.09	0.79	7.80	7.44	7.54
Water	1.27	1.37	1.98	4.07	7.08	6.98	6.37
Membrane-bound state	--	--	--	0.77	7.80	7.44	7.55

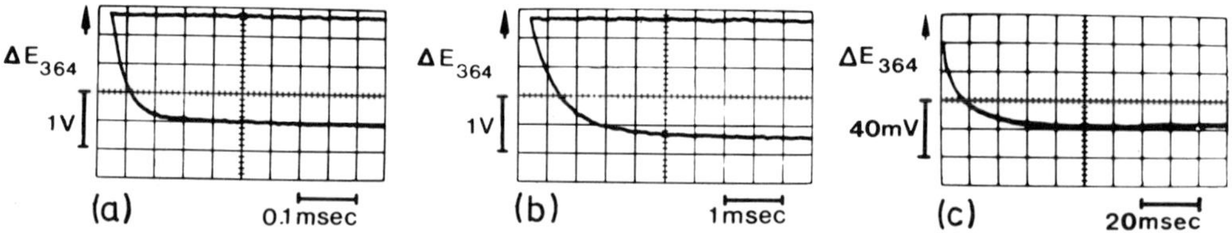

Fig. 6a-c. Kinetics for proton transfer and conformational rearrangement of virginiamycin S. Oscillograms of temperature jump relaxation experiments ($\delta T \sim 5^{\circ}C$; total signal $\sim$ 25 V) with: (a) $6.25 \cdot 10^{-5}$ M virginiamycin S in 33.8% water in methanol (v/v) in the presence of 0.2 M tetraethylammonium chloride, $5 \cdot 10^{-2}$ M acetic acid/tetraethylammonium hydroxyde pH 6.42; 23°C; (b) same experimental conditions but in the absence of the acetate buffer; 23°C; (c) $8 \cdot 10^{-5}$ M virginiamycin S in 40% water in methanol (v/v) in the presence of 0.2 M tetraethylammonium chloride, $5 \cdot 10^{-2}$ M acetic acid/tetraethylammonium hydroxyde pH 5.45; 23°C

2. *Kinetics of Proton Transfer*

The protolytic reactions of the 3-hydroxypicolinyl residue of virginiamycin S can be followed by absorption spectroscopy (Fig. 5). Thus absorption detection can also be used for the study of the kinetics of proton transfer of virginiamycin S employing temperature jump relaxation spectrometry. The temperature jump method has been described by EIGEN and DE MAEYER (1973).

Three relaxation processes can be detected. The relaxation time of the fastest process ($\tau_1 \leq 3$ µs) cannot be resolved and is attributed to the diffusion controlled recombination of the populated protolytic states with H^+ and OH^-. The longer of the two resolvable relaxation times ($\tau_3 \sim 10$ ms, Fig. 6c) is due to a pH-independent, conformational rearrangement of the peptide part of the antibiotic. This relaxation time cannot be attributed to a slow proton transfer reaction; it is independent of the concentration of virginiamycin S and pH, does not depend on the presence of buffer and cannot be observed with the model compound 3-hydroxypicolinamide. The sum of the forward and backward rate constants of the conformational rearrangement is $1.5 \pm 1 \cdot 10^2$ s^{-1}.

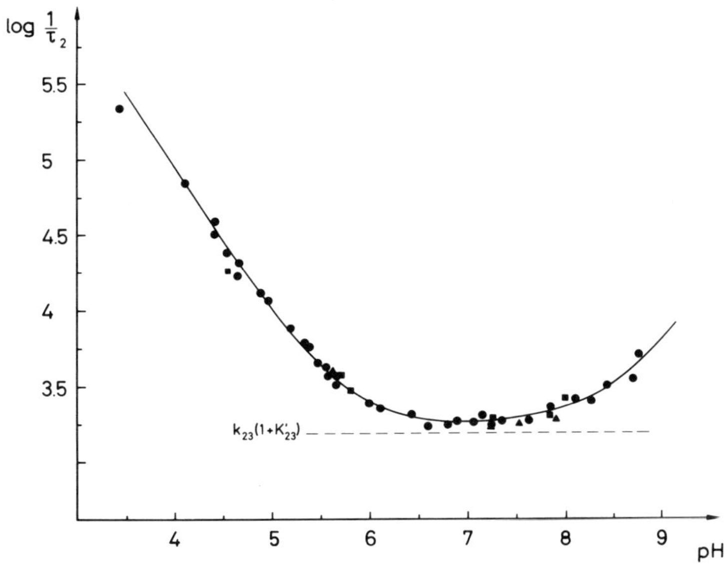

Fig. 7. Kinetics of proton transfer of virginiamycin S: Plot of reciprocal relaxation time τ_2 versus pH

The second relaxation time ($\tau_2 \sim 500$ μs at a pH of approximately 7; Fig. 6b) will be considered in more detail since it can be attributed to the overall proton transfer reaction. Within the pH range 3.5 to 9 the relaxation time τ_2 is independent of the virginiamycin S concentration but is sensitive to pH as shown by Figure 7. However τ_2 is smaller in the presence of buffer (cf. Fig. 6a) indicating that the buffer molecules participate in the proton transfer reactions.

In the absence of buffer, several parallel reactions contribute to the overall proton transfer process. In addition to the intramolecular proton transfer, which is characterized by the rate constants k_{23} and k_{32}, there are intermolecular proton transfer reactions in which H^+ or OH^- participate:

$$H^+ + P_2^{\pm} \underset{k_{12}}{\overset{k_{21}}{\rightleftharpoons}} P_1^{+} \underset{k_{31}}{\overset{k_{13}}{\rightleftharpoons}} P_3^{0} + H^+ \tag{3}$$

$$HO^- + P_1^{+} \underset{k_{21}^*}{\overset{k_{12}^*}{\rightleftharpoons}} P_2^{\pm} \underset{k_{42}}{\overset{k_{24}}{\rightleftharpoons}} P_4^{-} + H^+ \tag{4}$$

$$OH^- + P_1^{+} \underset{k_{31}^*}{\overset{k_{13}^*}{\rightleftharpoons}} P_3^{0} \underset{k_{43}}{\overset{k_{34}}{\rightleftharpoons}} P_4^{-} + H^+ \tag{5}$$

$$HO^- + P_3^{0} \underset{k_{43}^*}{\overset{k_{34}^*}{\rightleftharpoons}} P_4^{-} \underset{k_{24}^*}{\overset{k_{42}^*}{\rightleftharpoons}} P_2^{\pm} + OH^- \tag{6}$$

Furthermore, intermolecular proton transfer reactions can be expected to take place as a consequence of collisions between antibiotic molecules which are present in different protolytic states:

$$P_2^{\pm} + P_1^{+} \underset{k_{-A}}{\overset{k_A}{\rightleftharpoons}} P_3^{o} + P_1^{+} \quad (7)$$

$$P_2^{\pm} + P_4^{-} \underset{k_{-D}}{\overset{k_D}{\rightleftharpoons}} P_3^{o} + P_4^{-} \quad (8)$$

The contribution of the latter processes to the mathematical expression for the relaxation time τ_2 should lead to a dependence of τ_2 on the antibiotic concentration. However, within the experimental error of the present investigations τ_2 is observed to be independent of the virginiamycin S concentration between pH 3.5 and 9. Thus the contributions of reactions (7) and (8) can be neglected and only upper limits for the rate constants k_A, k_{-A}, k_D and k_{-D} can be given (Table 3).

Table 3. Kinetic parameters of proton transfer reactions of virginiamycin S (25°C)

Solvent	K'_{23}	k_{23} (s^{-1})	k_{32} (s^{-1})	k_H ($M^{-1}s^{-1}$)	k_{-H} ($M^{-1}s^{-1}$)	k_{OH} ($M^{-1}s^{-1}$)	k_{-OH} ($M^{-1}s^{-1}$)	k_A; k_{-A} ($M^{-1}s^{-1}$)	k_D; k_{-D} ($M^{-1}s^{-1}$)
Water	4.07	$9.8 \cdot 10^2$	$4 \cdot 10^3$						
33.8% water in methanol (v/v)	0.79	$8.5 \cdot 10^2$	$6.7 \cdot 10^2$	$4.4 \cdot 10^8$	$3.5 \cdot 10^8$	$2.8 \cdot 10^8$	$2.2 \cdot 10^8$	$<5 \cdot 10^8$	$<3 \cdot 10^6$

Under steady-state conditions with respect to the protolytic states P_1^+ and P_4^-, the following expression for the second relaxation time τ_2, which can be attributed to the remaining proton transfer reactions, is obtained according to SCHUSTER et al. (1977) (cf. Chap. 3 of this volume) and SCHUSTER et al. (1976):

$$\tau_2^{-1} = k_{23} + k_{32} + \frac{(k_{13}+k_{13}^* c_{OH^-})(k_{21} c_{H^+}+k_{21}^*) + (k_{31}^*+k_{31} c_{H^+})(k_{12}+k_{12}^* c_{OH^-})}{k_{13}+k_{13}^* c_{OH^-}+k_{12}+k_{12}^* c_{OH^-}} + \frac{(k_{43} c_{H^+}+k_{43}^*)(k_{24}+k_{24}^* c_{OH^-})+(k_{34}+k_{34}^* c_{OH^-})(k_{43} c_{H^+}+k_{42}^*)}{k_{43} c_{H^+}+k_{43}^*+k_{42} c_{H^+}+k_{42}^*} \quad (9)$$

A similar, but less general treatment has also been derived by AHRENS (1973). For an evaluation of the present data it is necessary to examine whether some of the terms in Eq. (9) can be neglected under special experimental conditions. Except for k_{23} and k_{32}, the magnitudes of the individual terms can be estimated from the known values of general protolytic recombination rate constants and by using the microscopic dissociation constants (Table 2) to calculate the rate constants of the reverse processes. It is taken into consideration that the recombination rate constants depend on electrostatic contributions due to the charge of the recombining species. For pH ranging from 3 to 6 (pH<7) $k_{13}^* c_{OH^-}$, $k_{12}^* c_{OH^-}$, $k_{24}^* c_{OH^-}$, $k_{34}^* c_{OH^-}$ and the rate constants k_{21}^*, k_{31}^*, k_{43}^* and k_{42}^* can be neglected in Eq. (9) leading to an approximation — Eq. (10)

$$\text{pH}<7 \qquad \tau_2^{-1} = k_{23} + k_{32} + \frac{k_{13}k_{21} + k_{12}k_{31}}{k_{12} + k_{13}} c_{H^+} + \frac{k_{24}k_{43} + k_{34}k_{42}}{k_{42} + k_{43}} \quad (10)$$

$$C_1 = \frac{k_{24}k_{43} + k_{34}k_{42}}{k_{42} + k_{43}}$$

The constant term, C_1, in Eq. (10) is estimated to be about $1.0 \cdot 10^3\ s^{-1}$ and cannot be neglected.

In the alkaline pH region ranging from 8 to 10 (pH>7) $k_{13}^* c_{OH^-}$, $k_{12}^* c_{OH^-}$, $k_{43} c_{H^+}$, $k_{42} c_{H^+}$ and the rate constants k_{21}^*, k_{31}^*, k_{24}, k_{34} in Eq. (9) are small compared to the other terms and can be neglected. Thus an approximation, Eq. (11), is obtained:

$$\text{pH}>7 \qquad \tau_2^{-1} = k_{23} + k_{32} + \frac{k_{13}k_{21} + k_{12}k_{31}}{k_{12} + k_{13}} c_{H^+} + \frac{k_{24}^*k_{43}^* + k_{34}^*k_{42}^*}{k_{42}^* + k_{43}^*} c_{OH^-} \quad (11)$$

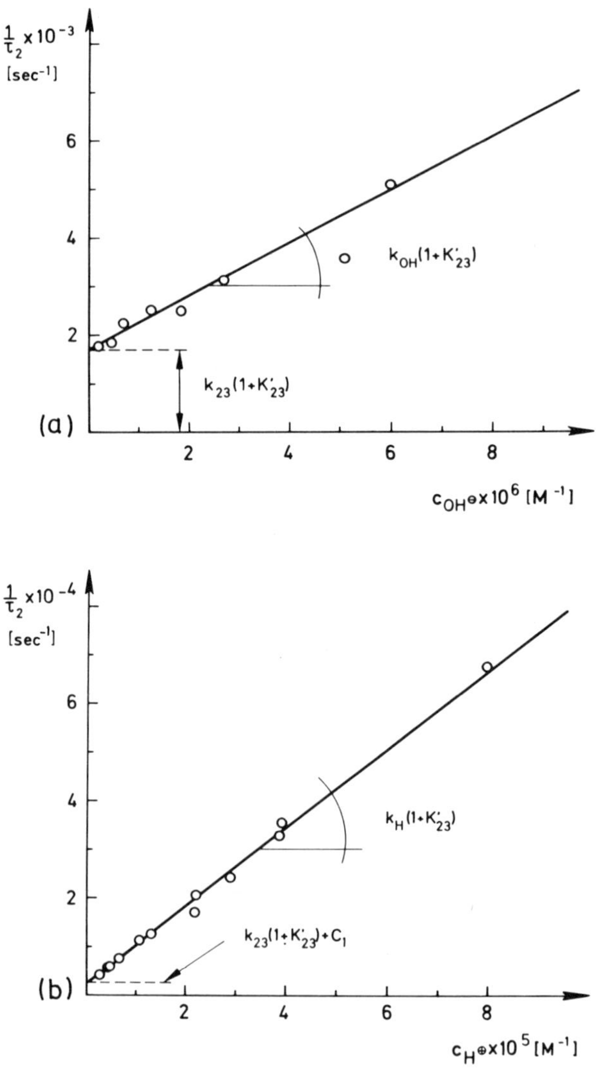

Fig. 8a and b. Evaluation of kinetic parameters for proton transfer of virginiamycin S: Plot of reciprocal relaxation times versus concentration of hydroxyl ions (a) and protons (b)

The third term in Eq. (11) decreases with increasing pH and is about 10^2 s^{-1} at pH 8. It is therefore small compared to the other terms and can be neglected so that the following approximation results where the reciprocal relaxation time τ_2 is linearly dependent on c_{OH^-}:

pH > 7

$$\tau_2^{-1} = (1 + K'_{23})(k_{23} + k_{OH} c_{OH^-}) \qquad k_{OH} = \frac{k^*_{24}\ k^*_{43}}{k^*_{42} + k^*_{43}} \qquad k_{-OH} = \frac{k^*_{34}\ k^*_{42}}{k^*_{42} + k^*_{43}} \tag{12}$$

According to Eq. (10) it follows for the acidic pH range:

pH<7

$$\tau_2^{-1} = (1 + K'_{23})(k_{23} + k_H c_{H^+}) + C_1 \qquad k_H = \frac{k_{13}\ k_{21}}{k_{12}+k_{13}} \qquad k_{-H} = \frac{k_{12}\ k_{31}}{k_{12}+k_{13}} \tag{13}$$

Eq. (12) and (13) allow an approximative evaluation of the present experimental data under the limiting conditions mentioned above. The corresponding dependence of the reciprocal relaxation time τ_2 as a function of c_{H^+} or c_{OH^-} in the acidic or alkaline pH ranges respectively is shown in Figure 8. Since K'_{23} is known, the intercept of the plot in Figure 8a yields k_{23}. Thus C_1 can be calculated from the intercept of the plot in Figure 8b and is found to be $1.1 \cdot 10^3\ s^{-1}$ which is consistent with the estimated value. The overall rate constants, k_H and k_{OH} are calculated from the slopes in the relevant figure. The kinetic parameters which have been obtained in the absence of buffer are summarized in Table 3.

According to these kinetic data, it is evident that in the acidic and alkaline pH regions intermolecular proton transfer occurs with dominant participation of H^+ or OH^- respectively. On the other hand, at about pH 7 the overall proton transfer is mainly due to the intramolecular process. The rate constants of the intramolecular proton transfer k_{23} and k_{32} are comparable to those of the estimated protolytic dissociation rate constants k_{24} and k_{34}. With respect to the intramolecular process this indicates that eg. the proton bound to the phenolic oxygen must first dissociate before it can be transfered to the ring nitrogen, implying that no special intramolecular proton transfer mechanism, such as participation of the neighboring carbonyl group, exists. It has already been mentioned that τ_2 does not depend on the virginiamycin S concentration as is observed for the model compound 3-hydroxypicolinamide. It is therefore concluded that the corresponding rate constants k_A, k_{-A} and k_D, k_{-D} (cf. Eq. (7), (8)) of the antibiotic must be considerably smaller than those of the model compound (cf. experimental data of SCHUSTER et al. 1977; AHRENS, 1973). This is probably due to the fact that steric hindrance leads to a decrease in the rate constant for the direct proton transfer between the 3-hydroxypicolinyl residues of two antibiotic molecules. The unexpectedly low value of k_H can also be explained on the basis of steric hindrance.

B. Elementary Steps Relevant to Alkali Ion Transport

1. Conformational Properties and Localization of Alkali Ion Specific Antibiotics in Membranes

The elementary steps which are relevant to the action of an alkali ion carrier in a membrane are the selective binding of a cation at the membrane interface and the diffusion of the carrier-cation complex through the hydrophobic interior region of the membrane. In addition, we have to consider the back diffusion of the unloaded carrier through the surface and the hydrophobic interior region of the membrane. Within the surface region the carrier is in contact with polar water molecules and must exist in a conformation suitable for cation coordination. All these elementary steps are related to a remarkable conformational adaptation of the carrier which depends on the size of the cation and the polarity of the surrounding medium. The diffusion steps of the unloaded carrier and its cation complex can only be studied under conditions where the carrier is bound to a membrane. However the principal properties of the other elementary steps can also be investigated in homogeneous solution. The ability of valinomycin and similar molecules to undergo conformational rearrangements will first be discussed.

Alkali ion carriers such as valinomycin and enniatin B can exist in a variety of conformational states, the predominant population being dependent on the polarity of the solvent. This has been investigated by employing various spectroscopic techniques (SHEMYAKIN et al., 1969; IVANOV et al., 1969; GRELL et al., 1972a; GRELL et al., 1972b). It is to be expected that the conformational properties of valinomycin may also be important for its carrier action in membranes so that a comparable carrier population can be maintained in the various parts of a phospholipid membrane which differ in polarity.

According to the authors mentioned above, valinomycin adopts a totally hydrogen-bonded conformation with a bracelet-like structure (cf. conformation A in Fig. 23) in nonpolar solvents such as n-heptane, this conformation resembles that of the ligand in its K^+ complex in solution and in the crystalline state. This conformation is characterized by the formation of six intramolecular hydrogen bonds between amide C = O and NH groups. Two groups of three intramolecular hydrogen bonds are located in two parallel planes of the molecule. In solvents of intermediate polarity it has been suggested that a conformational state with three intramolecular hydrogen bonds in one of the planes exists (cf. conformation B in Fig. 23). On the other hand an open conformation predominates in polar solvents where intramolecular hydrogen bonds in both planes of the valinomycin molecule are broken (cf. conformation C in Fig. 23).

An illustration of such a conformational behavior is shown in Figures 9 and 10 employing circular dichroism. On the addition of ethanol to a solution of valinomycin in n-heptane where the bracelet-like conformation predominates the Cotton effect at 220 nm decreases. A constant level for the molecular ellipticity, $[\theta]$, is reached at a mole fraction (γ) of ethanol of 0.6 (cf. broken line in Fig. 9b). This reflects a conformational transition from conformation V_A to at least one other conformational state V_B, which may correspond to the half-opened states where only three intramolecular hydrogen bonds remain (GRELL et al., 1972b, 1975). The beginning of an additional conformational transition can be deduced from Figure 9b for $\gamma_{ethanol}$ larger than 0.6. The continuation of the second transition can be followed in the solvent systems ethanol/methanol and methanol/water (cf. Fig. 10). The final state V_C of the second transition is characterized by an intense negative Cotton effect at 215 nm (Fig. 10a) due to the open conformation of valinomycin. The conformational behavior

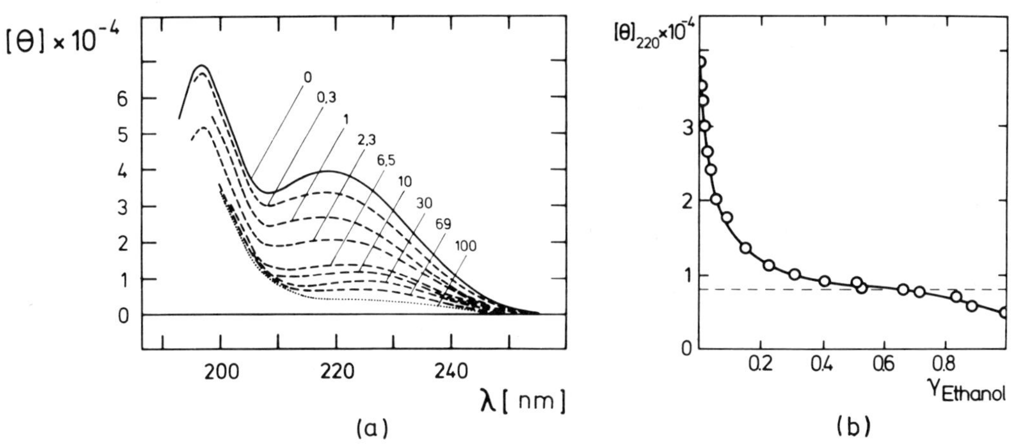

Fig. 9a and b. Conformational transitions of valinomycin: (a) Circular dichroism of valinomycin in the solvent system n-heptane/ethanol; 25°C (composition is given as % ethanol in n-heptane (v/v)). (b) Plot of the molecular ellipticity at 220 nm versus the mole fraction of ethanol. Thin broken line: molecular ellipticity of the conformer V_B

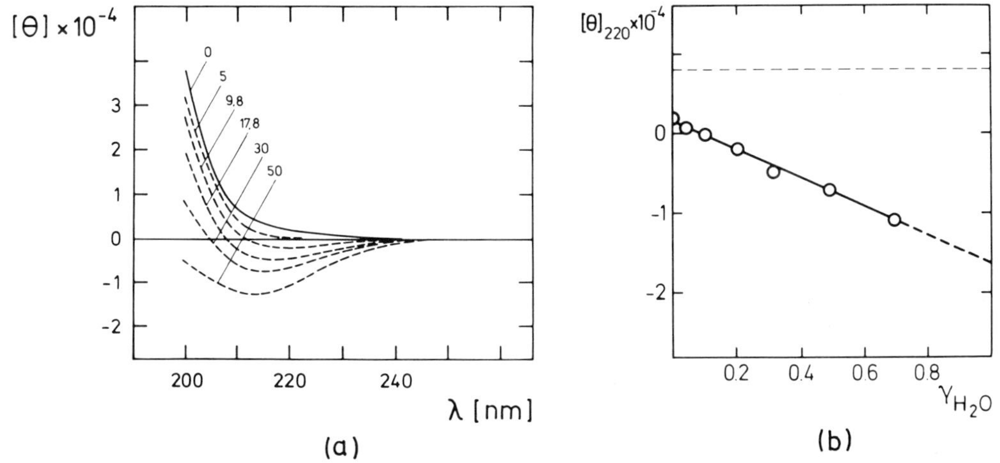

Fig. 10a and b. Conformational transitions of valinomycin: (a) Circular dichroism of valinomycin in the solvent system methanol/water; 25°C (composition is given as % water in methanol (v/v)). (b) Plot of the molecular ellipticity at 220 nm versus the mole fraction of water. Thin broken line: molecular ellipticity of the conformer V_B

of valinomycin in these solvent systems can be described by the following conformational equilibria

$$V_A \cdots \underset{}{\overset{K_{AB}}{\rightleftharpoons}} \cdots V_B \cdots \underset{}{\overset{K_{BC}}{\rightleftharpoons}} \cdots V_C \qquad (14)$$

where $K_{AB} = cv_A/cv_B$ and $K_{BC} = cv_B/cv_C$ are the corresponding equilibrium constants which depend on solvent composition. An evaluation of these equilibrium constants is given in Part III.B.3.

These equilibria must however be considered as a very crude approximation of the real situation which is more closely approached in ultrasonic absorption studies in the frequency range of 0.5 to 100 MHz (GRELL et al., 1972a,b, 1975). This method not only provides information about the time constants of the individual conformational transitions but also indicates the minimum number of populated conformational states. According to the above studies, the ultrasonic absorption properties of valinomycin demonstrate a complexity of dynamic behavior that cannot simply be described by three different conformations. Even in solvents of pronounced nonpolar character such as n-hexane, fast configurational rearrangements with relaxation times of $2 \cdot 10^{-9}$ s and $2 \cdot 10^{-8}$ s occur. Furthermore, these kinetic studies indicate that at least three different conformers exist in this solvent. In more polar media additional fast relaxation processes due to further conformational changes have been observed indicating that in solvents such as methanol a considerable number of conformational states have to be considered with rate constants in the range $8 \cdot 10^{8}$ s^{-1} to about $5 \cdot 10^{6}$ s^{-1}.

Optical methods such as circular dichroism have also been employed to investigate membrane-bound antibiotics (GRELL et al., 1972b, 1975). When bound to a 1,2-dimyristoyl-L-lecithin bilayer membrane, about 60% of the valinomycin molecules are in contact with water and thus located at the membrane surface. The conformations of valinomycin in this state are of the more open type where only a few intramolecular hydrogen bonds are formed. The remaining percentage of the valinomycin molecules is not in contact with water molecules and thus located within the hydrophobic interior of the membrane (Fig. 3a). In the latter state the bracelet-like conformer, typical for nonpolar solvents, is adopted which is characterized by the formation of six intramolecular hydrogen bonds. These results clearly indicate that the remarkable conformational adaption of valinomycin to the polarity of the medium found in homogeneous solution can also be observed when the antibiotic is bound to a membrane; this leads to an approximately equal partition of the uncomplexed antibiotic between the nonpolar interior of the membrane and the polar interface.

2. Thermodynamic Aspects of Alkali Ion Specificity and Structure of Cation Complexes

It has already been mentioned that carrier molecules such as valinomycin and enniatin B can also form alkali ion complexes in homogeneous solution. Before the complex formation properties of these compounds were discovered, no other ligands were known which could form such strong and specific complexes with alkali ions. Recently synthetic molecules such as macrocyclic polyethers (PEDERSEN, 1967) and cryptands (LEHN and SAUVAGE, 1971; DIETRICH et al., 1973) which can also selectively coordinate alkali ions have been investigated. Only the former compounds act as cation carriers in membranes but the cryptands are considered as inter-

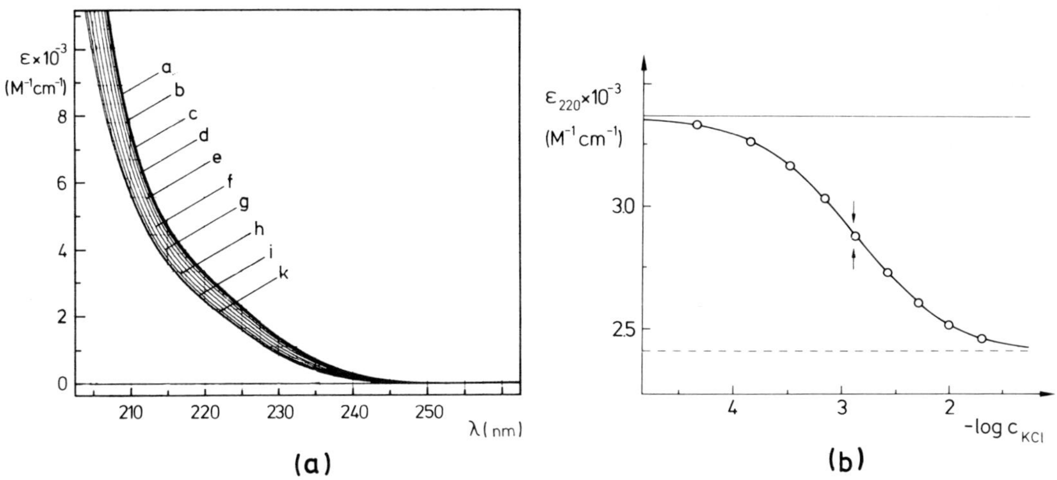

Fig. 11. (a) Spectrophotometric titration of $4.75 \cdot 10^{-5}$ M valinomycin in 20% water in methanol (v/v) (spectrum a) with KCl (spectra b - k; conc. of KCl in spectrum k: 0.02 M); 25°C. (b) Plot of extinction coefficient at 220 nm versus -log c_{KCl}. The theoretical curve has been calculated for an apparent stability constant (K') of 870 M^{-1}. Thin solid line: extinction coefficient of the uncomplexed valinomycin; Thin broken line: of its K^+-complex

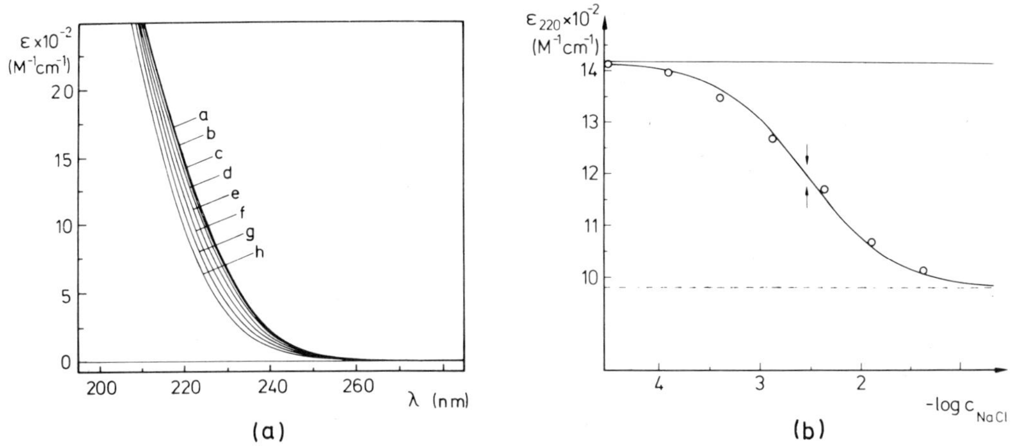

Fig. 12. (a) Spectrophotometric titration of $1.65 \cdot 10^{-4}$ M cryptand-211 in water (spectrum a) with NaCl (spectra b - h; conc. of NaCl in spectrum h: 0.04 M); 25°C. (b) Plot of extinction coefficient at 220 nm versus -log c_{NaCl}. The theoretical curve has been calculated for an apparent stability constant (K') of 350 M^{-1}. Thin solid line: extinction coefficient of the uncomplexed cryptand; Thin broken line: of its Na^+-complex

esting model compounds for the elucidation of the mechanism of alkali ion specificity.

Ultraviolet (UV) absorption measurements have revealed that the spectra of valinomycin, enniatin B (cf. GRELL et al., 1972a,b)

Table 4. Apparent stability constants (K') of alkali ion complexes

Ligand	Solvent (Temp)	Cation	K' (M^{-1})	References
Valinomycin	Ethanol (25°C)	K^+	$1.5 \cdot 10^6$	this article
	Methanol (25°C)	Na^+	4.7	GRELL et al. (1972b)
		K^+	$8.0 \cdot 10^4$	
		Rb^+	$1.8 \cdot 10^5$	
		Cs^+	$2.6 \cdot 10^4$	
	10% water in methanol (25°C)	K^+	$4.7 \cdot 10^3$	GRELL et al. (1975)
	20%	K^+	$8.7 \cdot 10^2$	
	30%	K^+	$9.3 \cdot 10^1$	
	50%	K^+	$1.05 \cdot 10^1$	
	Membrane-bound (25°C)	Na^+	$<1 \cdot 10^{-2}$	GRELL and FUNCK (1973)
		K^+	1.2	
		Rb^+	1.4	
		Cs^+	$3 \cdot 10^{-1}$	
Enniatin B	Methanol (25°C)	Li^+	$1.9 \cdot 10^1$	GRELL et al. (1972b)
		Na^+	$2.6 \cdot 10^2$	
		K^+	$8.3 \cdot 10^2$	
		Rb^+	$5.5 \cdot 10^2$	
		Cs^+	$2.2 \cdot 10^2$	
	10% water in methanol (25°C)	K^+	$2.8 \cdot 10^2$	GRELL et al. (1975)
	20%	K^+	$1.2 \cdot 10^2$	
	30%	K^+	$4.4 \cdot 10^1$	
	50%	K^+	$1.45 \cdot 10^1$	
	Membrane-bound (25°C)	Na^+	$<2 \cdot 10^{-2}$	GRELL and FUNCK (1973)
		K^+	8	
		Rb^+	2.5	
		Cs^+	$1 \cdot 10^{-1}$	
Nonactin	Methanol (30°C)	Na^+	$2.7 \cdot 10^2$	ZUEST et al. (1973)
		K^+	$5.0 \cdot 10^3$	
		Rb^+	$4.2 \cdot 10^3$	
		Cs	$9.3 \cdot 10^2$	
Cryptand-222	5% water in methanol	Li^+	$2.5 \cdot 10^1$	LEHN and SAUVAGE (1971); DIETRICH et al. (1973)
		Na^+	$1.6 \cdot 10^7$	
		K^+	$5.6 \cdot 10^9$	
		Rb^+	$2.5 \cdot 10^8$	
		Cs^+	$3.6 \cdot 10^3$	
Cryptand-211	Water (25°C)	Na^+	$3.5 \cdot 10^2$	this article
		Ca^{++}	$4.5 \cdot 10^2$	

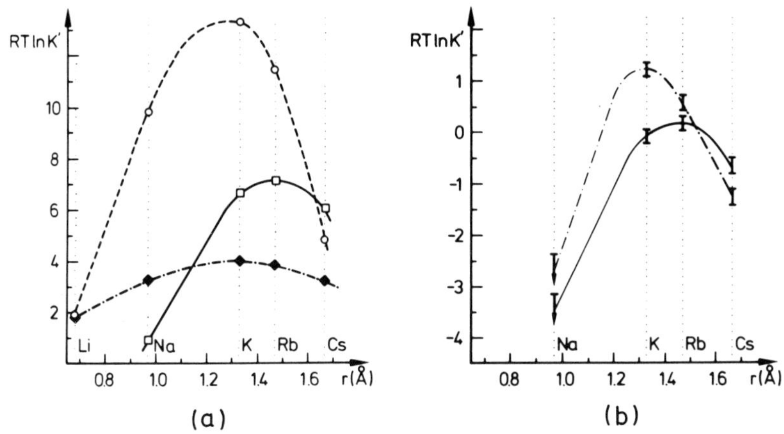

Fig. 13a and b. Alkali ion specificities of valinomycin (———) in methanol (a) and when bound to 1,2-dimyristoyl-L-lecithin (b), enniatin B (-·-·-) in methanol (a) and bound to 1,2-dimyristoyl-L-lecithin (b) and cryptand-222 (----) in 5% water in methanol (a); 25°C. Plot of RT ln K' versus ionic radius (cf. data in Table 4)

and the cryptands are sensitive to the presence of alkali ions and differ slightly with the size of the cation. An advantage of this method is that the stability constants of cation complexes can be determined by spectrophotometric titrations. Titrations of valinomycin are shown in Figures 11a and 12a. The evaluation of the corresponding stability constants is depicted in Figures 11b and 12b. In Table 4 some representative stability constants, K', for antibiotics in homogeneous solution and in the membrane-bound state are given. The latter values were determined by CD titrations of the antibiotics bound to 1,2-dimyristoyl-L-lecithin vesicles (GRELL and FUNCK, 1973; GRELL et al., 1975).

In Figure 13a the free enthalpies of complex formation for the various 1:1 alkali ion complexes are plotted as a function of the cation radius. It is evident that the ligands exhibit a remarkable alkali ion specificity which is similar to those of biological membrane systems shown in Figure 1. Maximum affinities are not observed with the smallest or the largest alkali ions but with cations of intermediate metal ion radius. This indicates that in complex formation the difference between the free energies of ligand binding in the absence of a solvent and cation solvation must pass through a maximum for the selected cation.

The K^+ complex of valinomycin is more stable in methanol than that of enniatin B but the opposite behavior is found for the membrane-bound antibiotics (Fig. 13b). That this is not due to a special effect of the membrane is seen by virtue of the fact that in water containing small amounts of methanol, enniatin B also forms the more stable complex (GRELL et al., 1975) (Fig. 14). However, it is evident that the extrapolated values of the stability constants in pure water are very small compared to those in methanol (Fig. 14). In addition it is important to note that

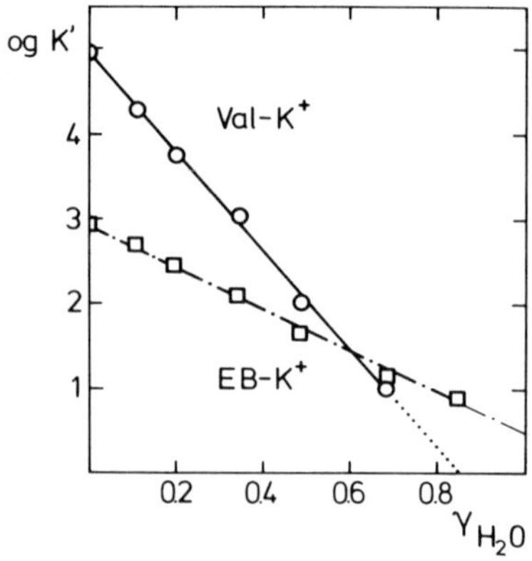

Fig. 14. Plot of the apparent stability constants (K') of the valinomycin-K^+ complex (——) and of the enniatin B -K^+ complex (-·--·-) in the solvent system methanol/water versus the mole fraction of water; 25°C

since both valinomycin and enniatin B are uncharged cyclodepsipeptides, one would expect that plots of log K' for their respective K^+ complexes against solvent composition would have similar slopes. The deviation from this expected behavior (Fig. 14) is probably mainly due to the fact that the two uncomplexed antibiotics show different changes in free energy when transferred from methanol to water. This may be related to the conformational transition observed for valinomycin in the same medium (Fig. 10). This consideration will be important in evaluating the kinetic data in the same solvent system (cf. Sect. III.B.3).

It is to be expected that the alkali ion specificities of valinomycin and similar compounds are related to the structural properties of their cation complexes. According to the crystal structure of the valinomycin-K^+ complex (PINKERTON et al., 1969) and spectroscopic studies of the complex in solution (IVANOV et al., 1969; SHEMYAKIN et al., 1969; GRELL et al., 1973) all six ester carbonyl groups are coordinated to the cation which is located in the central cavity of the ligand as illustrated in Figure 23 (state VI). Thus all the solvent molecules from the inner coordination sphere of the cation are replaced. The conformation of the ligand in the complex is stabilized by the formation of six intramolecular hydrogen bonds. The properties and structure of the valinomycin-Na^+ complex differ substantially from those of the K^+ complex and depend on the polarity of the medium (GRELL et al., 1973; OVCHINNIKOV et al., 1974). Since not all of the ester carbonyl groups are coordinated to Na^+ and no coordination of amide carbonyl groups can be detected, it is assumed that the coordinated Na^+ ion remains partially solvated (Fig. 15). In polar solvents such as methanol the opened structures of the Na^+ complex (states III and IV in Fig. 15). are preferentially populated whereas in less polar solvents (e.g. 1-butanol) a closed form (such as state V in Fig. 15) exists. It is evident that this conformational adaptation of valinomycin to cations of variable size contributes to its outstanding selectivity in comparison with the lower selectivity of other ion-specific antibiotics.

In order to rationalize the alkali ion selectivities, the following basic principles must be considered (also discussed by

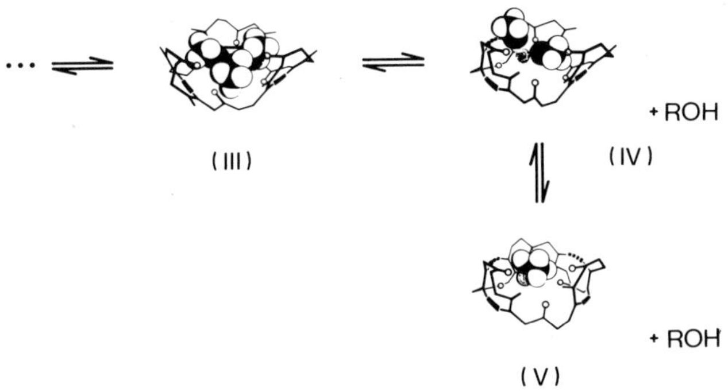

Fig. 15. Schematic diagram of the various structures of the valinomycin-Na^+ complex. Only the backbone of the ligand is shown with the oxygen atoms of the ester carbonyl groups (side chains omitted). Structure III represents the half-opened ligand conformation with three intramolecular hydrogen bonds. Structures III and IV are observed in polar solvents, whereas structure V represents the closed form of the Na^+-complex which predominates in nonpolar solvents

EIGEN and WINKLER, 1970). The electrophilic groups of the carrier molecule (eg. coordinating carbonyl groups) are located inside the molecule, leading to the formation of an inner cation binding site. All solvent molecules in the inner coordination sphere of the cation have to be replaced by the coordinating groups of the carrier. The carrier must form a fixed cavity which depends on the size of the selected cation but the carrier should also exhibit sufficient conformational flexibility to allow a stepwise substitution of the solvent molecules within the coordination sphere of the cation. The dynamic aspects of this process are discussed in the following section.

3. The Kinetics and Mechanism of Complex Formation with Alkali Ions and Its Relevance to Cation Specificity

Details of the molecular mechanism of complex formation and alkali ion selectivity can be derived from kinetic experiments employing relaxation methods such as temperature jump and ultrasonic absorption techniques. For the temperature jump studies complex formation reactions have been followed either directly, by recording the absorption changes of synthetic compounds or antibiotics in the ultraviolet region of the spectrum (FUNCK et al., 1972; GRELL et al., 1972b; CHOCK, 1972; BURGERMEISTER et al., 1974), or by employing an indirect method, such as the metal ion indicator murexide (EIGEN and WINKLER, 1970; WINKLER, 1972; BURGERMEISTER et al., 1974).

Figure 16 shows the complex formation rate constants for the substitution of water molecules in the inner coordination sphere of alkali and alkaline earth ions (cf. DIEBLER et al., 1969) and the overall rate constants for complex formation, k_{on}, of cation carriers and macrocyclic model compounds determined in water and

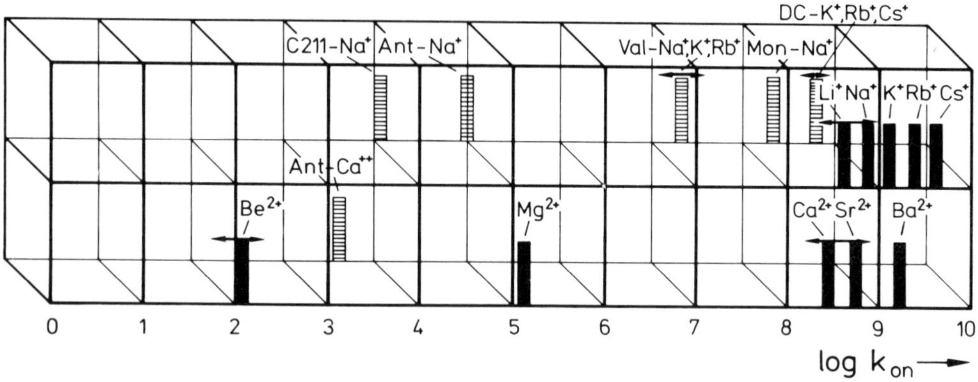

Fig. 16. Characteristic rate constants (after DIEBLER et al., 1969) for the substitution of water molecules from the inner cation solvation sphere (at the front) and overall complex formation constants of carriers with cations (at the back). The data are given in Table 4

methanol by relaxation methods (data and references are given in Table 5). Since the carriers are not water soluble most of the investigations on them have to be carried out in nonaqueous solvents such as methanol. Furthermore, the cation complexes are more stable in methanol than in water, thus allowing a more precise determination of stability and rate constants. The substitution behavior of the cations in methanol can be considered to represent that in water too, since the solvation of cations in water and methanol is quite similar. For the macrotetrolide monactin and dibenzo-30-crown-10 high rate constants for complex formation are observed which are similar to those of the rate constants for simple complex formation reactions (Fig. 16) with the corresponding cations. These rate constants are close to those of diffusion-controlled reactions; this is to be expected if there is a fast stepwise substitution of the inner sphere water molecules of the cation by the coordinating groups of the ligand (cf. EIGEN and WINKLER, 1970). On the other hand, the formation of complexes between cations and cryptands or antamanide is characterized by substantially lower formation rate constants; this may indicate that a different mechanism is involved. The complex formation rate constants of valinomycin lie between these two extreme situations.

A schematic illustration of a subsequent, stepwise ligand-exchange mechanism for the formation of a complex between a multidentate ligand (L) such as a cation carrier and a solvated cation is shown in Figure 17 (Mechanism A). This mechanism requires that the ligand exhibits a conformational flexibility sufficient to allow subsequent, stepwise substitution of all the solvate molecules in the inner, cation coordination sphere by the ligand's own coordinating groups. Since the rate constants of complex formation of the macrotetrolides and dibenzo-30-crown-10 appear to be close to those of diffusion controlled reactions, this implies that the conformational rearrangements occurring during these

Table 5. Kinetic parameters (25°C). Overall rate constants for forward (k_{on}) and reverse reaction (k_{off}); the apparent stability constants (K') were determined separately and are given for the same medium

Ligand	Solvent (Medium)	Cation	K' (M^{-1})	k_{on} ($M^{-1}s^{-1}$)	k_{off} (s^{-1})	References (Method)
Valinomycin	Ethanol	K^+	$1.5 \cdot 10^6$	$3.5 \cdot 10^8$	$2.3 \cdot 10^2$	this article (^{13}C-NMR)
	Methanol	Na^+	4.7	$1.3 \cdot 10^7$	$1.8 \cdot 10^6$	FUNCK et al. (1972) (US,TJ)
	Methanol (0.1 M TBAP)	K^+	$3 \cdot 10^4$	$3.5 \cdot 10^7$	$1.2 \cdot 10^3$	
		Rb^+	$6.5 \cdot 10^4$	$5.5 \cdot 10^7$	$7.5 \cdot 10^2$	
		Cs^+	$8 \cdot 10^3$	$2 \cdot 10^7$	$2.2 \cdot 10^3$	
	10% water in methanol (0.1 M LiCl)	K^+	$4.5 \cdot 10^3$	$1.3 \cdot 10^7$	$2.9 \cdot 10^3$	this article (TJ)
	20% water in methanol (0.1 M LiCl)	K^+	$6 \cdot 10^2$	$5.5 \cdot 10^6$	$1.3 \cdot 10^4$	
		Rb^+	$1.3 \cdot 10^3$	$7.5 \cdot 10^6$	$6 \cdot 10^3$	
		Cs^+	$1.5 \cdot 10^2$	$3 \cdot 10^6$	$2 \cdot 10^4$	
	30% water in methanol (0.15 M LiCl)	K^+	$6.5 \cdot 10^1$	$1.3 \cdot 10^6$	$2.8 \cdot 10^4$	
	Water (extrapolated values)	K^+	$1.5 \cdot 10^{-1}$	$5 \cdot 10^4$	$5 \cdot 10^5$	this article
Monactin	Methanol (TBAP)	Na^+	$5 \cdot 10^2$	$3 \cdot 10^8$	$6 \cdot 10^5$	WINKLER et al. (1972) (US,TJ)
Dinactin	Methanol (TBAP)	Na^+	$1 \cdot 10^3$	$6 \cdot 10^7$	$5 \cdot 10^4$	
Trinactin	Methanol (TBAP)	Na^+	$2 \cdot 10^3$	$7 \cdot 10^7$	$4 \cdot 10^4$	
Dibenzo-30-crown-10	Methanol (0.15 M LiCl)	Na^+	$1.3 \cdot 10^2$	$>1.6 \cdot 10^7$	$>1.3 \cdot 10^5$	CHOCK (1972) (TJ)
		K^+	$3.7 \cdot 10^4$	$6 \cdot 10^8$	$1.6 \cdot 10^4$	
		Rb^+	$4.4 \cdot 10^4$	$8 \cdot 10^8$	$1.8 \cdot 10^4$	
		Cs^+	$1.7 \cdot 10^4$	$8 \cdot 10^8$	$4.7 \cdot 10^4$	
Antamanide	Methanol (TBAP)	Na^+	$5 \cdot 10^2$	$1.1 \cdot 10^5$	$2.1 \cdot 10^2$	BURGERMEISTER et al. (1974) (TJ)
		Ca^{++}	$2.5 \cdot 10^1$	$4.9 \cdot 10^3$	$1.9 \cdot 10^2$	
Cryptand-211	Water	Na^+	$3.5 \cdot 10^2$	$1.2 \cdot 10^4$	$1.5 \cdot 10^2$	this article (TJ, ^{13}C-NMR)

Abbreviations: NMR nuclear magnetic resonance; TBAP tetra-n-butylammonium perchlorate; TJ temperature jump relaxation spectrometry; US ultrasonic absorption

Mechanism A

$$Me^+(H_2O)_n + L \rightleftharpoons Me^+(H_2O)_{n-1}\cdots L \rightleftharpoons Me^+(H_2O)_{n-2}\text{---}L \rightleftharpoons Me^+(H_2O)_{n-3}\text{—}L \rightleftharpoons Me^+\text{—}L$$

$$+H_2O \qquad +H_2O \qquad +H_2O \qquad +(n-3)\,H_2O$$

Mechanism B

$$Me^+(H_2O)_n + L \rightleftharpoons Me^+ + L \rightleftharpoons Me^+\text{—}L$$

$$+\,n\,H_2O$$

Mechanism AB

$$Me^+(H_2O)_n + L \rightleftharpoons Me^+(H_2O)_{n-1}\cdots L \rightleftharpoons Me^+(H_2O)_{n-3}\cdots L \rightleftharpoons Me^+(H_2O)_{n-3}\text{—}L \rightleftharpoons Me^+\text{—}L$$

$$+H_2O \qquad +2H_2O \qquad +(n-3)\,H_2O$$

Fig. 17. Mechanisms of complex formation of multidentate ligands with cations

ligand-exchange processes must also be very fast. The stepwise substitution process is of advantage from a dynamic point of view since the activation energies involved are much smaller than those in a situation where the cation is completely desolvated in a single step which preceeds the actual complex formation reaction (Mechanism B in Fig. 17). In the latter mechanism extremely low complex formation rate constants should be expected. A more realistic situation could be that shown in mechanism AB (Fig. 17) where features of mechanisms A and B are combined (cf. GRELL et al., 1975). In mechanism AB, the first solvate substitution from the inner coordination sphere of the cation is followed by a step where two or more of the remaining solvate molecules of the initial cation complex must be removed before additional coordinations and further stepwise substitution processes can occur. This initial partial desolvation process, where more than one solvate molecule must dissociate prior to additional coordinations, must have a rather high activation energy compared to the subsequent stepwise substitutions as shown in mechanism A. Thus, this partial desolvation process must represent the rate limiting step of mechanism AB. The complex formation rates for mechanism AB are therefore also expected to be lower than those for mechanism A but not as low as those for mechanism B. Mechanism AB could be valid for nonflexible compounds with a sterically fixed cavity, such as the cryptands and antamanide which will be discussed at the end of this section.

Detailed kinetic studies related to complex formation with alkali ions have been carried out on valinomycin in methanol employing temperature jump relaxation spectrometry and ultrasonic absorption techniques (FUNCK et al., 1972; GRELL et al., 1972b, 1975). The results of the kinetic experiments have been evaluated on the basis of the following two-step model of complex formation:

$$V_o + \; \overset{+}{Me} \; \cdots \underset{k_{21}}{\overset{k_{12}}{\rightleftharpoons}} \cdots \underset{\text{I}}{\left[V'\text{-}\overset{+}{Me} \right]} \cdots \underset{k_{32}}{\overset{k_{23}}{\rightleftharpoons}} \cdots \underset{\text{II}}{V\text{-}Me^{+}} \qquad (15)$$

In the first step the solvated cation forms an intermediate complex (I) with one or more of the conformers present in methanol. At this stage of the analysis it is assumed that the various conformers do not show any difference in their ability to form the intermediate I. In the second step of the model the formation of the final cation complex (II) is completed by a first order rearrangement. This model predicts two relaxation times τ_1 and τ_2 for complex formation; only τ_2 could be resolved employing temperature jump relaxation spectrometry. Under conditions where there is a fast rearrangement of the uncomplexed antibiotic and a fast equilibration of the reaction step leading to the unstable intermediate $|k_{21} \gg k_{23}, k_{12}(c_{V_o} + c_{Me^+})|$, a simple equation is obtained for τ_2^{-1} as a function of the concentrations of the uncomplexed antibiotic (c_{V_o}) and metal ion (c_{Me^+}):

$$\tau_2^{-1} = k_{on}\,(c_{V_o} + c_{Me^+}) + k_{off} \qquad \text{where} \qquad (16)$$

$$k_{on} = \frac{k_{12}\;k_{23}}{k_{21}} \qquad k_{off} = k_{32} \qquad (17)$$

The overall formation and dissociation rate constants (k_{on} and k_{off}) can be evaluated from the linear concentration dependence of the reciprocal relaxation time τ_2 for the second reaction step. This linear dependence of τ_2^{-1} can only be expected when concentrations of the free reactants are small. It is important to note that the overall formation rate constant is the product of the rate constant for the relatively slow interconversion of the intermediate I into the final complex II multiplied by the stability constant of the intermediate. Thus neither k_{23} nor k_{12}/k_{21} can be individually determined from k_{on} unless additional data are available. The values for the overall rate constants are given in Table 5, together with the apparent stability constants which have, like the temperature jump experiments, also been determined in methanol containing 0.1 M tetra-n-butylammonium perchlorate.

The overall rate constants for complex formation of valinomycin with alkali ions correspond to those of the slowest interconversions between the different conformers of the antibiotic itself (GRELL et al., 1972a, 1975). It has therefore been concluded that the rate-limiting step of complex formation must be a conformational rearrangement of this multidentate ligand which occurs during the sequestration of the solvated cation.

This interpretation has been further supported by results obtained from ultrasonic absorption measurements (FUNCK et al., 1972; GRELL et al., 1972b) which will now be briefly summarized. The kinetics of complex formation of valinomycin with Na^+ and NH_4^+ and for the formation of the intermediate I in K^+ complexation were followed. Two relaxation times for complex formation were found for the binding of Na^+ and the faster relaxation time (τ_1), which characterizes the formation of the intermediate K^+-complex, could also be followed. These results are consistent with the kinetic model (Eq. (15)) and allow the evaluation of all four rate constants. Formation rate constants k_{12} of $7 \cdot 10^7$ M^{-1} s^{-1} for Na^+ and $4 \cdot 10^8$ M^{-1} s^{-1} for K^+ have been obtained for the fast equilibration between the uncomplexed antibiotic and the intermediate complex I. These values are similar to the overall formation rate constants of the macrotetrolides and dibenzo-30-crown-10 (Table 5); they are typical of a subsequent stepwise initial substitution which is followed by the formation of intermediate I. The corresponding dissociation rate constants, k_{21}, are $2 \cdot 10^7$ s^{-1} for Na^+ and $1 \cdot 10^8$ s^{-1} for K^+. The preequilibrium constant $K_{12} = k_{12}/k_{21}$, is 3.5 M^{-1} for the initial Na^+ complex and 4.0 M^{-1} for the corresponding K^+ complex, this clearly indicates that the outstanding K^+/Na^+ specificity of valinomycin has not been reached at the stage where the intermediate I is formed. The rate constant k_{23} of the rate limiting conformational rearrangement preceding the final, rapid, stepwise solvate substitution process is found to be $4 \cdot 10^6$ s^{-1} for the Na^+ and $1 \cdot 10^7$ s^{-1} for the K^+ complex. The difference between these two values may be due to a more constrained conformational transition during the complexation of the small Na^+ ion. This is also consistent with the observation that the structures of the Na^+ and K^+ complex are different, too (cf. Sect. III.B.2). The corresponding dissociation rate constants, k_{32}, are $2 \cdot 10^6$ s^{-1} for the Na^+ and $1.3 \cdot 10^3$ s^{-1} for the K^+ complex. The difference between these two values reflects the main contribution to the cation specificity of valinomycin and thus is also related to the conformational porperties of this antibiotic.

In the present article an extension of the kinetic studies of the complex formation of valinomycin with alkali ions in a solvent less polar than methanol has been investigated by employing ^{13}C-nuclear magnetic resonance (NMR) spectroscopy. In addition, complex formation in solvent systems more polar than methanol was investigated by employing temperature jump relaxation spectrometry. These studies should show whether the cations discriminate between the various conformers of the uncomplexed antibiotic, the predominant population of which depends on the solvent system used. In Section III.B.2 it has already been shown that the larger decrease of the stability constant of the valinomycin-K^+ complex upon its transfer from methanol to water compared to that of enniatin B (Fig. 14) must, among other effects, be related to the conformational transitions of the uncomplexed antibiotic occurring in the same solvent system (Fig. 10).

In addition, on the basis of general stereochemical considerations, it is to be expected that the uncomplexed bracelet-like conformer of valinomycin could not form a cation complex with such high rate constants as are found for the half-open conformer which exists in solvents such as methanol. If the reaction occurred via the former conformer, most of the solvate molecules would have to dissociate from the cation before the cation could slip into the cavity. On the other hand, a completely opened structure, which exists in aqueous media could be unfavorable

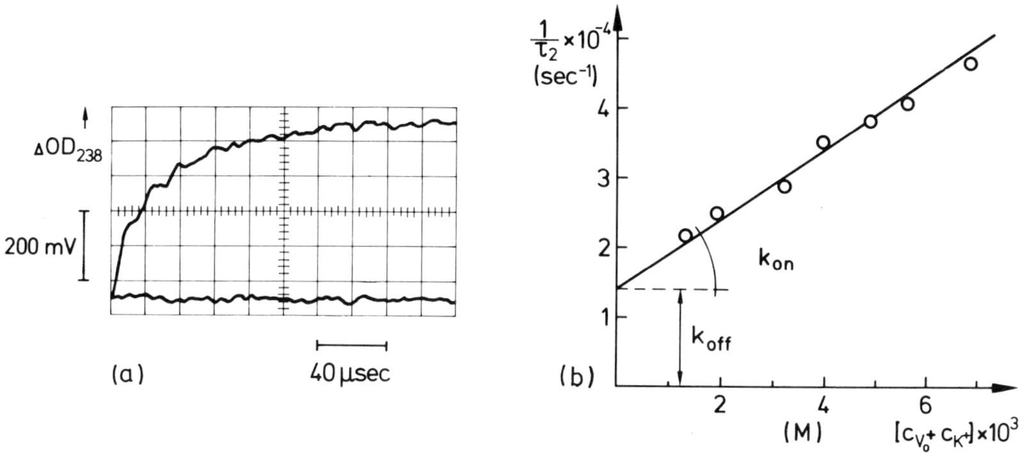

Fig. 18a and b. Kinetics of complex formation between valinomycin and K^+ in 20% water in methanol (v/v) containing 0.1 M LiCl. (a) Oscillogram of temperature jump relaxation experiment ($\Delta T \sim 4^{\circ}C$; total signal $\sim$ 30 V) in the presence of $1.33 \cdot 10^{-3}$ M valinomycin and $1.33 \cdot 10^{-3}$ M KCl; 24°C. (b) Plot of the reciprocal relaxation time τ_2 versus the sum of the free reactant concentrations

for complex formation since the half-open conformers should exhibit a higher affinity for the cation due to the partially preformed cavity. Therefore, it would be understandable why complex formation could proceed faster via particular conformation states of the uncomplexed antibiotic.

In order to clarify this situation kinetic experiments have been carried out in mixtures of water and methanol by temperature jump spectrometry. Figure 18a shows an oscillogram of a typical relaxation experiment of valinomycin. An unresolvable change in optical density, due to fast conformational rearrangements and complex formation, is followed by a slower process with a relaxation time, τ_2, of about 50 μs.

The second process was evaluated according to the kinetic model proposed for methanol (Eq. (16)) under the limiting conditions which were specified earlier. The overall rate constants k_{on} and k_{off} were calculated from the linear dependence of the reciprocal relaxation time τ_2 on the sum of the free concentrations of valinomycin and K^+ (Fig. 18b). The kinetic parameters plotted versus the mole fraction of water, γ_{H_2O}, in the solvent system methanol/water are shown in Figure 19a. The overall formation rate constant k_{on} decreases sharply with increasing water content of the medium. No measurements could be carried out when the water content was very high due to the limited solubility of valinomycin, this means that data in water had always to be derived by extrapolation as shown in the relevant figures.

The stability constant K_{12} of the intermediate complex I in methanol/water should be substantially reduced with increasing water

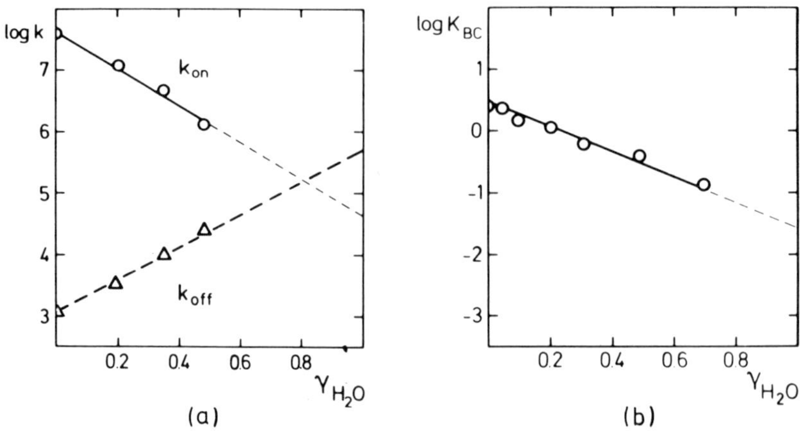

Fig. 19a and b. Kinetics of complex formation between valinomycin and K^+ in the solvent system methanol/water; 25°C. (a) Plot of overall formation (k_{on}) and dissociation (k_{off}) rate constants versus the mole fraction of water. (b) Plot of the conformational equilibrium constant K_{BC} versus the mole fraction of water in the same solvent system

content, similar to the behavior shown by the apparent stability constant, K', in the same system (Fig. 14). The overall formation rate constant, k_{on}, is the product of K_{12} and the rate constant, k_{23}, of the rate-limiting rearrangement of the ligand during solvate substitution; since this step should be independent of solvent composition we can predict that k_{on} will decrease with increasing water content of the medium, this is consistent with the behavior shown in Figure 19a. It is evident that k_{off} must increase with the decreasing stability of the final complex (Fig. 19a). In addition, it is important to note that the decrease of k_{on} and K' as a function of the medium composition can also be correlated with the similar behavior of the equilibrium constant K_{BC} (Fig. 19b) which characterizes the population of the conformers V_B and V_C in the same solvent system; the values for K_{BC} have been calculated form the spectra of the circular dichroism (Figs. 9 and 10).

If we neglect the change in free energy of the uncomplexed antibiotic upon its transfer from methanol to water, the main contribution to the decrease of K' and K_{12} and thus also of k_{on} must be attributed to the change in free energy of the K^+ ion. According to ISMAILOV (1963), the solvation energy of K^+ is 2 kcal/mol higher in water than in methanol. One would therefore expect that the stability constant, K', of the valinomycin-K^+ complex in water should be about two orders of magnitude smaller than that in methanol. This estimation is consistent with results obtained for the enniatin B-K^+ complex (Fig. 14). Furthermore, since k_{on} depends on K_{12}, the estimated decrease of k_{on} upon the same transfer in solvent should not be larger than two orders of magnitude. However, the experimentally observed decreases of K' (Fig. 14) and k_{on} (Fig. 19a) exceed these predicted decreases by several orders of magnitude. It is therefore necessary to consider whether the change in free energy of the cation complex could account for the additional decrease.

Since the structure of the valinomycin-K^+ complex is independent of the solvent composition in the system methanol/water and the bound K^+ ion and the polar groups of the coordinated valinomycin are protected against strong solvent interactions, the difference in free energy of the complex upon its transfer from methanol to water must be smaller than that of the free K^+ ion. Thus, the difference between the predicted and experimentally determined values for the decrease in K' and k_{on} can only be attributed to a small extent to the medium-induced change in free energy of the complex.

Therefore, it can be concluded that the remaining contribution to the free energy, namely that of the uncomplexed antibiotic, upon the transfer from methanol to water must contribute substantially to the decrease of K' and k_{on} as a function of the composition of this solvent system. This implies that the proposed reaction must be modified; thus a new model is suggested where only the conformational state V_B which is predominant in methanol (cf. Sect. III.B.1) and may have a half-opened structure (cf. conformation B in Fig. 23), coordinates the cation. The conformer V_C, which is unsuitable for complex formation, is preferentially populated in water and could have an opened conformation such as that of conformer C in Figure 23.

$$\begin{array}{ccccccc} V_B & + & Me^{+} & \underset{k_{21}}{\overset{k_{12}}{\rightleftharpoons}} & [V'\text{-}Me^{+}] & \underset{k_{32}}{\overset{k_{23}}{\rightleftharpoons}} & V\text{-}Me^{+} \\ K_{BC} \updownarrow & & & & \text{I} & & \text{II} \\ V_C & & & & & & \end{array} \qquad (18)$$

Based on this new model the mathematical expression for k_{on} must be extended by an equilibrium constant for an additional preceding equilibrium which characterizes the populations of the conformers involved.

These considerations are further supported by the results from thermodynamic and kinetic studies of complex formation between valinomycin and K^+ in ethanol (Table 5) which is considerably less polar than methanol. Although the difference between the solvation energies of K^+ in ethanol and methanol is comparable (2.5 kcal/mol according to ISMAILOV, 1963) to that for the transfer of K^+ from methanol to water, the values of K' and k_{on} decrease much less upon transfer of K^+ from ethanol to methanol than on its transfer from methanol to water. The stability constant K' is about twenty times higher in ethanol than in methanol and correlates well with the difference in solvation energy of the uncomplexed K^+ in these two solvents. The rate constant k_{on} for complex formation between valinomycin and K^+ in ethanol has been calculated from the known value of K' and the dissociation rate constant as determined by ^{13}C-NMR. The formation rate constant in ethanol is only about ten times higher than that in methanol; this is in good agreement with the behavior of k_{on} as a function of K_{BC} since the latter equilibrium constant only changes from 2.5 to 7 upon the transfer from methanol to ethanol. These results indicate that the experimental data are consistent with the above model.

In order to test the effect of solvent-dependent, conformational equilibria on the thermodynamic and kinetic parameters of complex formation of valinomycin with a cation, the following hypothetical model has been investigated:

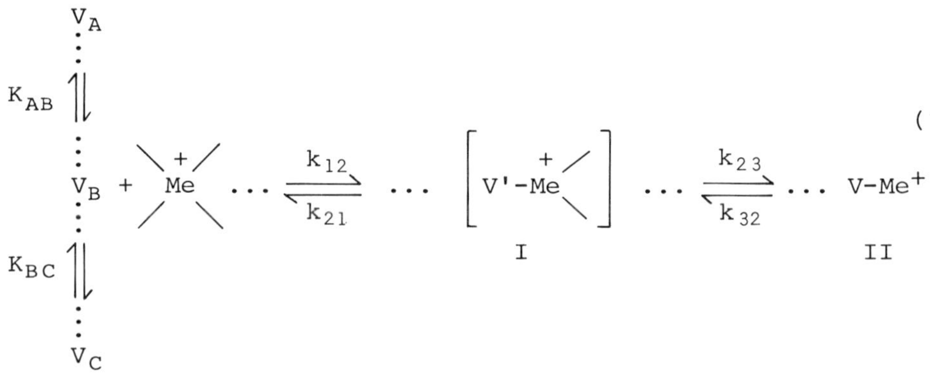

(19)

In this model it is assumed that the uncomplexed valinomycin exists in three main conformations (V_A, V_B, V_C). Only the recombination of conformer V_B with the solvated cation can be expected to lead to the formation of the intermediate I. Conformation V_A predominates in medium I while V_B is preferentially populated in medium II. V_A could correspond to the bracelet-like conformation of valinomycin as observed in nonpolar solvents and V_C could be one of the opened structures of the antibiotic which are found in polar solvent systems. V_B could be attributed to a half-opened conformer of valinomycin. An analysis of the populations of V_A, V_B and V_C as assumed from the medium composition (plot of relative concentrations versus mole fraction of component II in the medium I/II; $c_{V_o} = c_{V_A} + c_{V_B} + c_{V_C}$) is shown in Figure 20a. The dependence of the corresponding conformational equilibrium constants on the medium composition can be seen in Figure 20b. The apparent stability constant of the complex K' ($K' = c_{V-K^+}/c_{V_o}c_{K^+}$) is then related to the equilibrium constant K* of the complex ($K^* = c_{V-K^+}/c_{V_B}c_{K^+}$), which is independent of the population of the conformers by the following equation:

$$K' = K^* \left(1 + K_{AB} + \frac{1}{K_{BC}}\right)^{-1} \quad \text{where} \quad K_{AB} = \frac{c_{V_A}}{c_{V_B}}, \; K_{BC} = \frac{c_{V_B}}{c_{V_C}} \qquad (20)$$

K* is related to the equilibrium constants of the two reaction steps involved in complex formation:

$$K^* = K_{12} \; K_{23} \quad \text{where} \quad K_{12} = \frac{k_{12}}{k_{21}}, \; K_{23} = \frac{k_{23}}{k_{32}} \qquad (21)$$

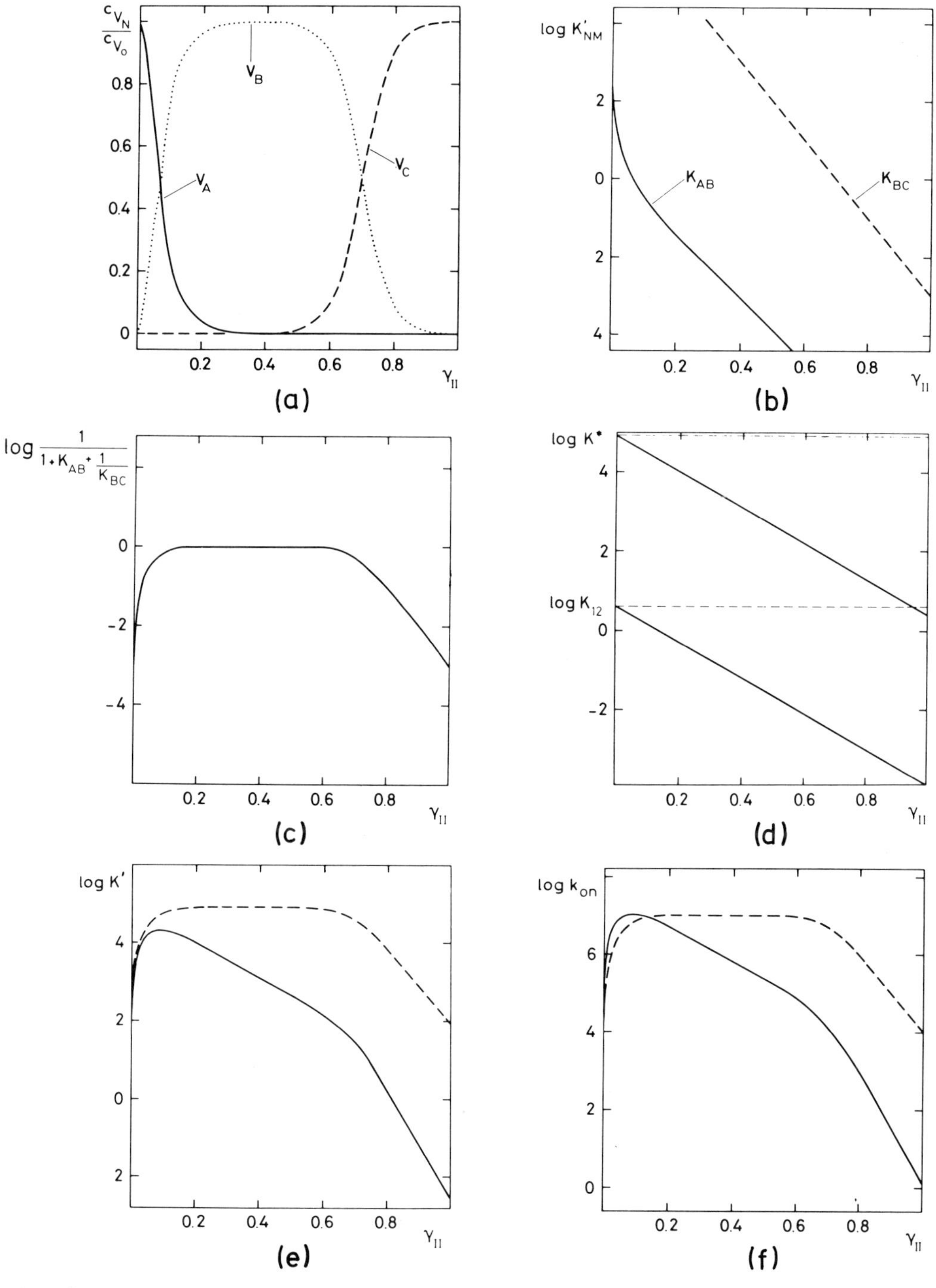

Fig. 20a-f. Effect of solvent-dependent, conformational equilibria on the thermodynamic and kinetic parameters of complex formation of valinomycin with a cation: hypothetical model (details in the text)

The dependence of the second factor in Eq. (20) on γ_{II}, which describes the contribution of the medium-dependent, chemical equilibria is depicted in Figure 20c. The value of this factor reaches a plateau between γ_{II} values ranging from about 0.1 to 0.7 but decreases with higher and lower values of γ_{II}. Furthermore, if it is assumed that K' is $8 \cdot 10^4$ M^{-1} in medium I and log K* decreases linearly as a function of γ_{II} (Fig. 20d), one can calculate the corresponding dependence for K', as shown in Figure 20e (solid line). The broken line in Figure 20e results when K* remains constant in this solvent system. If an additional assumption is introduced, namely that the entire dependence of K* on γ_{II} is related to K_{12} (cf. Fig. 20d, K_{12} is assumed to be 4 M^{-1} in medium I) thus implying that K_{23} is independent of the medium compositon, then the dependence of the overall rate constant, k_{on}, on solvent composition can be predicted by the following equation:

$$\tau_2^{-1} = K_{12} \; k_{23} \; \left(\frac{1}{1 + K_{AB} + \frac{1}{K_{BC}}}\right) \; (c_{V_o} + c_{K^+}) + k_{32} \quad \text{where} \tag{22}$$

$$k_{on} = K_{12} \; k_{23} \; \left(1 + K_{AB} + \frac{1}{K_{BC}}\right)^{-1}, \quad k_{off} = k_{32} \tag{23}$$

The conditions which specify the validity of this expression have been given earlier (Eq. (16)). In addition, it is assumed that the relaxation times related to the conformational transitions are fast in comparison with τ_2 which is consistent with results obtained from ultrasonic absorption techniques (cf. Sect. III.B.1). The calculated dependence of log k_{on} on γ_{II} (k_{23} is independent of solvent composition and assumed to be $4 \cdot 10^7$ M^{-1} s^{-1}) is shown in Figure 20f, firstly for the case where K_{12} is dependent on the composition of the solvent (solid line) and secondly for a constant value of K_{12} (broken line). It is evident that the assumption that K_{12} is independent of solvent composition cannot be valid since the free energy of solvation of the cation must depend on the medium composition. As shown in Figure 20f (solid line) k_{on} first increases with low values of γ_{II}, then decreases due to the dependence of K_{12} on the composition of the medium and finally shows a much more pronounced decrease due to the contribution of the transition between the conformations V_B and V_C. This dependence reflects the expected but, however, hypothetical behavior of k_{on} for the complexation of K^+ by valinomycin over the whole polarity range. The dependence of log k_{on} on γ_{II} between γ_{II} values of about 0.5 and 1 resembles that observed in ethanol/methanol and methanol/water.

Since this model is consistent with the experimental data, it will be used for the calculation of the various equilibrium constants in the solvent system methanol/water. The values of K_{AB} are much smaller than 1 in the system methanol/water and the values for K_{BC} are the same as in Figure 19b. The value for the rate constant k_{23} of the rate-limiting, conformational transition in methanol is taken as $1.5 \cdot 10^7$ M^{-1} s^{-1} which is very close to the experimentally determined value. K_{12} can now be calculated from the experimentally determined values of k_{on} since k_{23} and

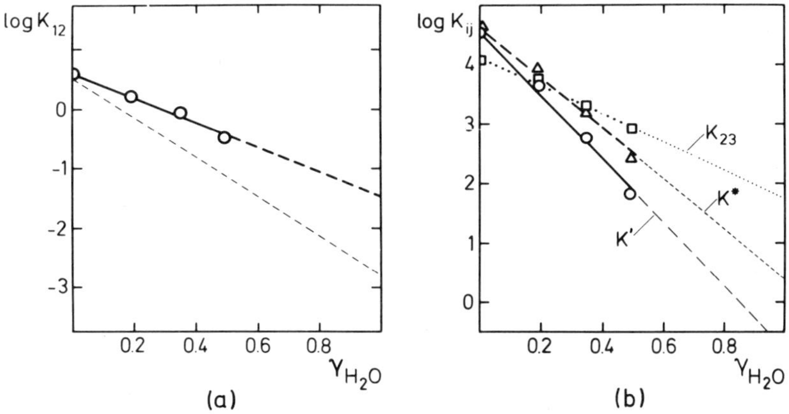

Fig. 21a and b. Complex formation between valinomycin and K^+ in the solvent system methanol/water: Plot of thermodynamic parameters (K_{12} (a); K', K* and K_{23} (b)) versus the mole fraction of water

and K_{BC} are known. The dependence of K_{12} on γ_{H_2O} is shown in Figure 21a (thick, solid line). The thin, broken line in Figure 21a represents the dependence of K_{12} on γ_{H_2O} under conditions where k_{on} is assumed to be independent of all of the equilibrium constants concerned with conformational rearrangements. The dependence of K* and K_{23} on γ_{H_2O} can be similarly determined from these data (Fig. 21b). The values of K', as determined spectrophotometrically in the presence of 0.1 M LiCl (valinomycin does not coordinate Li^+ under these conditions), are also depicted in Figure 21b. The decrease of K* upon the transfer from methanol to water is about two orders of magnitude larger than estimated from the difference between the solvation energies of the uncomplexed K^+ ion in each of the two solvents. Since the properties of the final complex cannot account for this additional difference and the contributions of the conformational equilibria have already been taken into consideration, the difference must be related to the difference between the solvation energies of intermediate I (about 1 to 2 kcal/mol) upon solvent transfer; this then leads to changes in K_{12} and K_{23}. Since the conformation of the ligand in intermediate I should be similar to that of the half-opened conformer of the antibiotic (such as state III in Fig. 23), this additional change in free energy could easily be due to the fact that the solvation of the ligand in intermediate I depends on solvent composition. The consideration of the conformational equilibria involved in complex formation between valinomycin and cations has evidently led to a much better understanding of the thermodynamic and kinetic data.

The contributions of the kinetic parameters to the cation selectivity of valinomycin can be discussed according to Figure 22 (GRELL et al., 1972b, 1975). The same cation specificity is found for the antibiotic whether it is membrane-bound (Fig. 1b) or whether it is in homogeneous solution in a solvent such as methanol (Fig. 22a). Valinomycin confers the highest membrane permeability to the Rb^+ ion and forms the strongest cation complex with

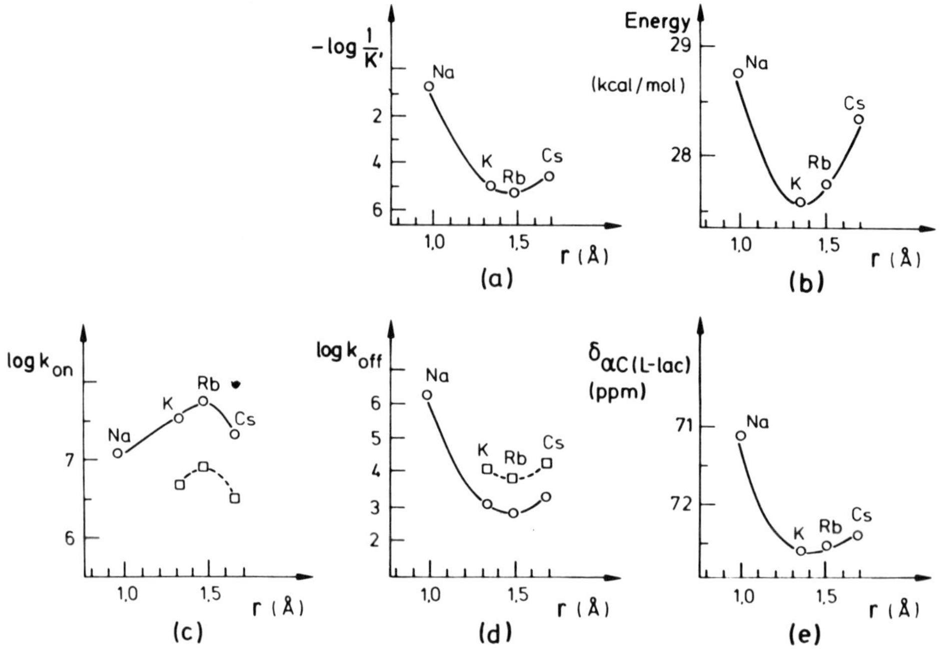

Fig. 22a-e. Alkali ion specificity of valinomycin: plot of thermodynamic, kinetic and conformational parameters versus ion radius. (a) Reciprocal stability constants (methanol; cf. Table 4). (b) Conformational energy of the ligand, adapted to the size of the alkali ions (after MAYERS and URRY, 1972). (c) Overall formation rate constants (methanol (———), 20% water in methanol (v/v) (----); cf. Table 5). (d) Overall dissociation rate constants (methanol (———), 20% water in methanol (v/v) (----); cf. Table 5). (e) ^{13}C-chemical shift of the α-carbon of the L-lactic acid residue of valinomycin alkali ion complexes in methanol-d_4 (after GRELL et al., 1973)

it. The cation specificity, which is related to the stability constants (K') in methanol and thus also to the permeability, appears to be mainly due to the rate constant of the rate-limiting dissociation constant k_{off} ($k_{off} = k_{32}$ according to the previous reaction models) as indicated in Figure 22d. However, the overall rate constant of complex formation, k_{on}, also depends on the size of the cation and is largest for Rb^+ (Fig. 20c). This clearly indicates that k_{on} also contributes to the cation specificity. These results show a rather remarkable dynamic control of the cation specificity of valinomycin. Its cation specificity is the same in methanol containing 20% water (v/v) and in pure methanol (Fig. 22c,d). However, the calculated conformational energy of the coordinated valinomycin (Fig. 22b, cf. MAYERS and URRY, 1972) and an experimentally measured conformational parameter (e.g. the chemical shift of the α-carbon atom of the L-lactic acid residue obtained by ^{13}C-NMR (Fig. 22e, cf. GRELL et al., 1973) indicate that the ligand interacts most strongly with K^+. This does not contradict the observed kinetic and thermodynamic parameters since the solvation energy of the smaller K^+ ion is considerably larger than that of Rb^+ in a variety of solvents (ISMAILOV, 1963).

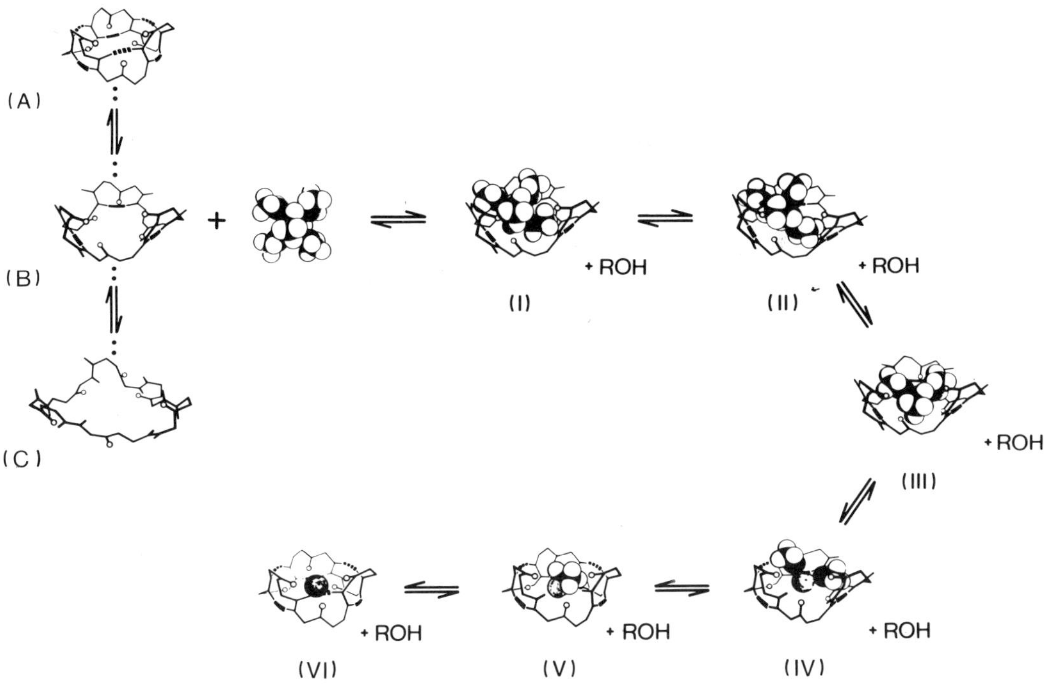

Fig. 23. Schematic illustration of the stepwise mechanism involved in the formation of valinomycin alkali ion complexes: only the recombination of conformer B with the solvated cation leads to the formation of the intermediate I. The solvate molecules (ROH) from the inner coordination sphere of the cation are substituted by the coordinating ester carbonyl groups of the multidentate ligand

In the absence of a solvent the coordination of K^+ should lead to a more stable complex than the coordination of Rb^+; however, the difference between the free energy of hypothetical ligand binding in the absence of solvent and the free energy of cation solvation is larger for Rb^+ than for K^+ thus leading to a greater stability of the valinomycin-Rb^+ complex in the solvents considered here.

A schematic illustration of the mechanism of cation binding to valinomycin is shown in Figure 23. It is related to mechanism A in Figure 17 which is characterized by a subsequent, stepwise substitution process where the solvate molecules are substituted by the coordinating ester carbonyl groups of valinomycin. The conformer B, which predominates in solvents of intermediate polarity, recombines with the solvated cation prior to the first, fast substitution steps which lead to state III. Then the rate-limiting conformational rearrangement of the partially coordinated ligand takes place and is followed by the final, fast substitution steps leading to state IV where the cation is completely desolvated. This behavior is summarized in the following reaction scheme in which the previous schemes have been modified by the introduction of an additional equilibrium:

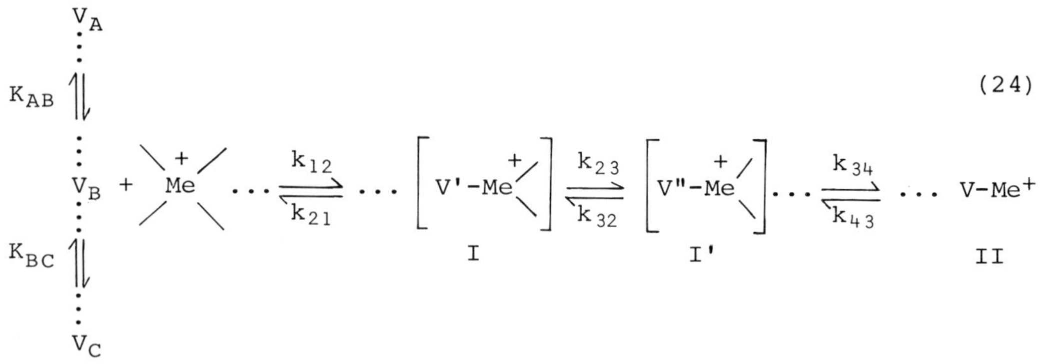

(24)

In this scheme a distinction is made between the relatively slow conformational rearrangement between states I and I' and the faster substitution steps which occur before and after the conformational transition. If it is still assumed that the equilibration of the conformational rearrangement occurring during complex formation is characterized by the slowest relaxation time τ_2, one obtains:

$$\tau_2^{-1} = K_{12}k_{23}\left(\frac{1}{1 + K_{AB} + \frac{1}{K_{BC}}}\right)(c_{V_o} + c_{K^+}) + k_{32}\,\frac{k_{43}}{k_{43} + k_{34}}, \qquad (25)$$

where

$$k_{off} = k_{32}\,\frac{k_{43}}{k_{43} + k_{34}} \qquad (26)$$

The expression for k_{on} obtained in the previous scheme still holds (cf. Eq. (23)), but under our conditions where k_{34} is expected to be larger than k_{43}, k_{off} can be approximated to be equal to the product of k_{32} (the rate constant of the reverse conformational transition) and the reciprocal stability constant K_{34} ($K_{34} = k_{34}/k_{43}$) of the final complexation step. Thus all the kinetic parameters have been interpreted to give a molecular model of the reaction mechanism (Fig. 23) which is consistent with the kinetic analysis.

The kinetic parameters for complex formation of cryptands with cations have also been determined employing temperature jump relaxation spectrometry with absorption detection and ^{13}C-NMR. Some oscillograms of the temperature jump studies are shown in Figure 24. The amplitudes observed in these studies are extremely small due to the unfavorable properties of the corresponding chromophores. Figure 24a shows the relaxation curves obtained for cryptand-211 in the presence of Na^+ in aqueous solution. The fast initial process, which could not be resolved, is attributed to minor conformational rearrangements of the ligand such as nitrogen inversion and to the fast steps of cation complexation. This initial fast step is followed by a considerably slower process with a relaxation time (τ_2) of about 0.5 ms. The dependence of τ_2 on the free concentrations of the reactants and of the relaxation amplitude on the degree of saturation of the ligand indicate that this process is related to cation coordination and is attributed

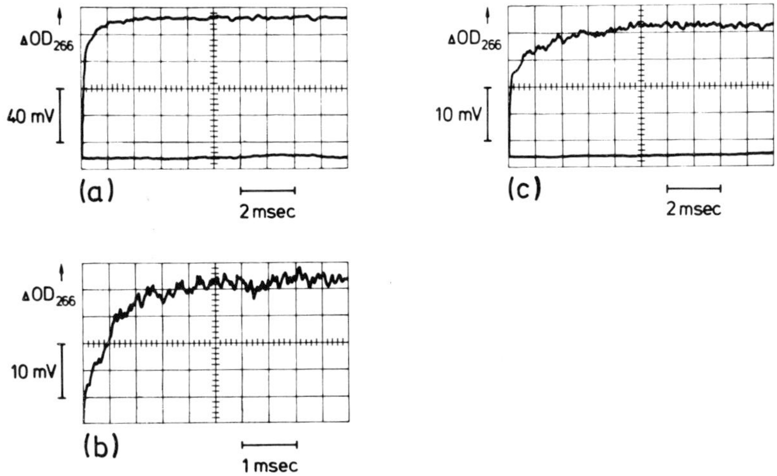

Fig. 24a-c. Kinetics of complex formation between cryptand-211 and cations in aqueous solution. Oscillograms of temperature jump relaxation experiments ($\Delta T \sim 5^{o}C$; total signal ~ 12.5 V). (a and b) 0.13 M cryptand-211 in the presence of 0.1 M NaCl; $24^{o}C$. (c) 0.12 M cryptand-211 in the presence of 0.14 M $CaCl_2$; $24^{o}C$

to the rate-limiting step of complex formation. The evaluation of the kinetic parameters is based on the following reaction scheme (C = cryptand):

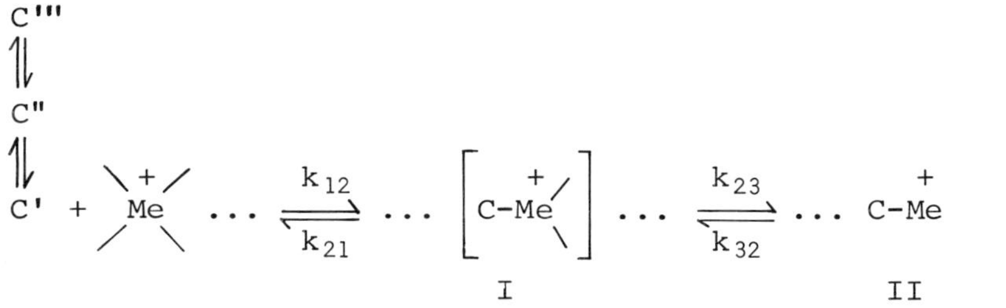

(27)

The expression for τ_2, which mainly characterizes the equilibration between the complexed states I and II, has been given earlier (cf. Eq. (16)). The different conformational transitions induced by fast nitrogen inversions shown in the above reaction scheme will not be considered in the evaluation of the kinetic data since no exact values for the equilibrium constants concerned could be obtained. However, there is preliminary evidence that the conformational state suitable for coordination is well populated. This implies that the equilibrium constants of the conformational preequilibria will not substantially influence the value evaluated for k_{on}. The kinetic parameters obtained from temperature jump relaxation spectrometry are consistent with those obtained from ^{13}C-NMR (Table 5). The resolvable relaxation time τ_2 obtained when cryptand-211 forms a complex with Ca^{2+} (Fig. 24) is even longer than that for the Na^{+} complex. Since the stability constants of the Na^{+} and the Ca^{2+} complexes of this cryptand are

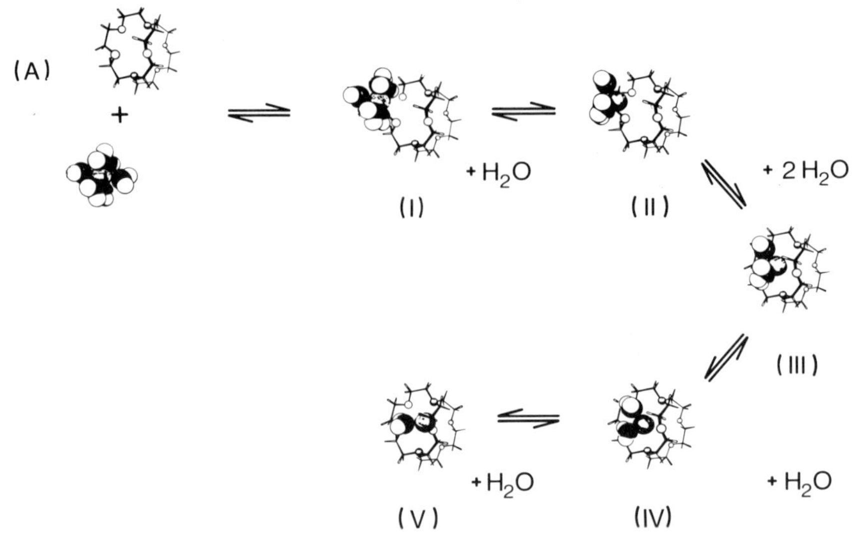

Fig. 25. Schematic illustration of mechanism involved in formation of cryptand cation complexes: initially, first solvate molecule (H_2O) is substituted, which leads to the formation of state I. Rate-limiting step occurs between states I and II when two or more of remaining solvate molecules of already partially complexed cation have to be removed in single reaction step before cation can be coordinated to centre of cavity. Additional fast stubstitution processes can then occur leading to state IV (structure of complex where no solvate molecules remain in inner cation solvation shell is omitted)

nearly identical, the former result indicates that the overall formation rate constant for the Ca^{2+} complex is also smaller than that for the Na^+ complex. The overall formation rates obtained for complex formation between cryptands and alkali ions are the lowest that have so far been reported for these cations (Fig. 16; Table 5). Slow rate constants have also been reported for complex formation between the antitoxic cyclodecapeptide antamanide and cations (BURGERMEISTER et al., 1974).

The kinetic results for cryptands and similar compounds imply that their mechanism of cation coordination must be different from that of valinomycin and other ligands which complex alkali ions with considerably higher formation rate constants (Table 5); the rate-limiting steps in the two systems must be different. It is important to note here that, except for the fast nitrogen inversion, the cryptands must be considered as rigid, multidentate ligands which do not have sufficient conformational flexibility to allow the subsequent stepwise substitution mechanism shown by valinomycin (cf. Fig. 23) and other antibiotics. The kinetic results and the general considerations mentioned above suggest that complex formation between cryptands and cations occurs by mechanism AB as shown in Figure 17. In this mechanism it is considered that the rate-limiting step of complex formation is dissociation of more than one solvate molecule from the inner solvation sphere of the cation; this step must take place before the final substitution steps can occur. The rate constant of this dissociation process must depend strongly on the solvation energies of the cations which is consistent with the experimental data. Thus, this dissociation process will remarkably contribute to the observed cation specificity. Similar conclusions have been reached for complex formation between antamanide and cations (BURGERMEISTER et al., 1974).

A schematic illustration of mechanism AB for complex formation between the cryptand-221 and a solvated cation is shown in Figure 25.

It is interesting to reason that two different mechanisms can be considered as the basis for remarkably high alkali ion selectivities. For complex formation reactions of molecules such as valinomycin, the multidentate ligand adapts its conformation in dependence of the extent of solvate substitution at the cation until the ligand can completely enfold the desolvated cation. On the other hand in sterically fixed molecules, such as the cryptands, the inner solvate sphere of the cation has to adapt its structure to the limited size of the rigid ligand cavity. In addition, it is possible to differentiate between the two mechanisms by their respective dynamic properties.

IV. Comparison of the Dynamic Aspects of Cation Carriers Bound to Membranes and in Homogeneous Solution

Since the dynamic parameters of membrane-bound valinomycin have been investigated but not those of membrane-bound virginiamycin S, a comparison between the properties of these compounds in homogeneous solution and when bound to membranes will be restricted to valinomycin.

Since cation specificity of valinomycin is not sensitive to the polarity of solvents which are more polar than ethanol, it is not surprising that the antibiotic exhibits the same cation specificity whether it is in solution or bound to a membrane (Fig. 13).

It is therefore more interesting to compare those parameters which are sensitive to the polarity of the medium. The stability constant K' of membrane-bound valinomycin was found to be about 1 M^{-1}. According to the linear extrapolation shown in Figure 14, the same value can be expected for K' of valinomycin in 70% water in methanol (v/v) which corresponds to a value for γ_{H_2O} of about 0.85. The polarity of this solvent system could be similar to that of the membrane interface (GRELL et al., 1975). According to this simple interpretation the dynamic parameters of complex formation of membrane-bound valinomycin (Table 1) should also be similar to those estimated for the antibiotic in 70% water in methanol (v/v) from the linear extrapolation shown in Figure 19a. The values given by this extrapolation are $1.4 \cdot 10^5$ M^{-1} s^{-1} and $1.8 \cdot 10^5$ s^{-1} for k_{on} and k_{off} respectively. These values are much closer to those obtained for the membrane-bound antibiotic than to the considerably higher overall formation rate constant ($4 \cdot 10^7\ M^{-1}\ s^{-1}$) found for complex formation between valinomycin and K^+ in methanol. Since the overall formation rate constant depends on the distribution of the uncomplexed antibiotic between its conformers, the correspondence between the kinetic parameters obtained for valinomycin when membrane-bound and when in 70% water in methanol (v/v) indicate that the population of the conformational states at the membrane surface is similar to that of 70% water in methanol (v/v); this is consistent with the analysis of the circular dichroism of membrane-bound valinomycin (GRELL et al., 1975). Thus the discrepancy mentioned in Section II between the kinetic parameters obtained for valinomycin when membrane-bound and when in homogeneous solution in a solvent such as methanol is only apparent and can be explained on the basis of different conformational populations.

The cation specificity of membrane-bound valinomycin is mainly controlled by the overall dissociation rate constant, k_{off}, (Fig. 3a) of the cation-complex bound to the membrane surface. Another contribution to the cation specificity of membrane-bound valinomycin could result from a dependence of the rate constant k_{VK} (Fig. 3a) (which characterizes the diffusion of the loaded carrier across the membrane) on the size of the cation. This could be expected if the ligand conformations differ considerably in the complexes of the various alkali ions as is found for the Na^+- and K^+-complexes. However, this should not have an effect on the rate of permeation through the hydrophobic region of the membrane, since the ligand conformation of the Na^+-complex can undergo structural rearrangements so that similar ligand conformations are adopted by the Na^+- and K^+-complexes in nonpolar media.

The remarkable conformational adaptation of membrane-bound valinomycin to strongly bound alkali ions such as K^+ and Rb^+ results in optimal dynamic conditions for carrier-mediated cation transport. This is due to the fact that the values of the rate constants for the loading and unloading processes and for the transport of the loaded and unloaded carrier are nearly the same (Table 1). As has been shown in the previous sections, the behavior of valinomycin can be explained by its outstanding conformational properties.

V. Summary

The elementary steps which are relevant to the mechanism of action of proton and alkali ion carriers in bimolecular phospholipid membranes have been investigated in homogeneous solution employing spectroscopic techniques and relaxation methods such as temperature jump relaxation spectrometry.

The kinetics of the proton transfer reactions of the antibiotic virginiamycin S, which acts as a proton carrier were studied in 33.8% water in methanol (v/v). In this medium its protolytic properties are similar to those shown when it is membrane-bound. At about pH 7, the overall proton transfer occurring at the 3-hydroxypicolinyl residue is mainly due to an intramolecular process; the rate constant for proton transfer from the ring nitrogen to the phenolic oxygen is 850 s^{-1} and that for the reverse process is 670 s^{-1}. Evidence was found which suggested that slow conformational rearrangements of the peptide residue of virginiamycin S took place with rate constants of about 10^2 s^{-1}.

The antibiotic valinomycin acts as a selective alkali ion carrier when membrane-bound and also exhibits the same pronounced cation specificity in homogeneous solution. The stabilities and overall formation rate constants of valinomycin-alkali ion complexes were studied as a function of the polarity of the medium. The evaluation of the thermodynamic and kinetic parameters was based on a multistep reaction. In ethanol the overall formation rate constant for complex formation between valinomycin and K^+ was about $4 \cdot 10^8$ M^{-1} s^{-1} which is close to the rate constants of diffusion-controlled reactions. However, in a polar solvent such as water, the formation rate constant was only $5 \cdot 10^4$ M^{-1} s^{-1}. The difference is due to the fact that different conformers of the uncomplexed antibiotic predominate in these solvents; this is consistent with the results of spectroscopic investigations. The thermodynamic and kinetic parameters related to the formation of complexes between alkali ions and membrane-bound valinomycin are close to those found for valinomycin in 70% water in methanol (v/v). In the case of the nonflexible macrocyclic model compound cryptand-211, the overall rate constant for complex formation with Na^+ in water is about 10^4 M^{-1} s^{-1} and is the lowest that has so far been reported for the complexation of an alkali ion.

The kinetic studies provided information about the mechanism of complex formation of these multidentate ligands with cations and about the basis for their alkali ion specificities. Complex formation between valinomycin and alkali ions is characterized by a fast stepwise ligand exchange process where the rate-limiting step is a conformational change of the ligand which occurs during the substitution of solvate molecules from the inner coordination sphere of the cation by the coordinating groups of the flexible ligand. The alkali ion specificity is mainly determined by the overall dissociation rate constant. On the other hand, the rate-limiting step of complex formation of the sterically fixed cryptand is attributed to the dissociation of more than one solvate molecule from the inner coordination sphere of the cation in a single reaction step. This step precedes the coordination of the cation in the rigid cavity of the ligand implying that the formation rate constant must depend strongly on the solvation energy of the cation. Thus in this case the overall rate constants for formation and dissociation of the complexes contribute to the cation specificity.

Acknowledgements. We are very grateful to Prof. H. VANDERHAEGHE for making available to us the virginiamycin S used in this work. We wish also to thank Prof. M. EIGEN, Prof. G. GOTTSCHALK, Prof. P. SCHUSTER, Dr. F. EGGERS, and Dr. Th. FUNCK for helpful and stimulating discussions, as well as Dr. G. VICKERS for reading the manuscript and Miss E. WITTSCHIEBER for expert technical assistance.

References

AHRENS, M.-L.: Proton transfer kinetics of 5'-deoxypyridoxal. Biochim. Biophys. Acta 320, 86-96 (1973)

BAMBERG, E., BENZ, R., LAEUGER, P., STARK, G.: Ionentransport durch biologische Membranen. Chiuz 8, 33-43 (1974)

BURGERMEISTER, W., WIELAND, Th., WINKLER, R.: Antamanide. Dynamics of metal-complex formation. Eur. J. Biochem. 44, 305-310 (1974)

CHOCK, P.B.: Relaxation study of complex formation between monovalent cations and cyclic polyethers. Proc. Nat. Acad. Sci. (Wash.) 69, 1939-1942 (1972)

DIEBLER, H., EIGEN, M., ILGENFRITZ, G., MAAß, G., WINKLER, R.: Kinetics and mechanism of reactions of main group metal ions with biological carriers. Pure Appl. Chem. 20, 93-115 (1969)

DIETRICH, B., LEHN, J.-M., SAUVAGE, J.-P.: Kryptate: makrocyclische Metallkomplexe. Chiuz 7, 120-128 (1973)

EIGEN, M., DE MAEYER, L.: Theoretical basis of relaxation spectrometry. In: Techniques of Chemistry. WEISSBERGER, A. (ed.); Vol. VI, Part 2; HAMMES, G.G. (ed.); pp. 63-146. New York: Wiley 1973

EIGEN, M., WINKLER, R.: Alkali-ion carriers: dynamics and selectivity. In: The Neurosciences, Second Study Program. SCHMITT, F.O. (ed.); pp. 685-695. New York: Rockefeller Univ. Press 1970

FUNCK, Th., EGGERS, F., GRELL, E.: Kinetik und Mechanismus der selektiven Bindung von Ionen durch Dyclodepsipeptid-Antibiotika. Chimia 26, 637-641 (1972)

GRELL, E., EGGERS, F., FUNCK, Th.: Konformationsanalyse und Kinetik der Konformationsänderungen von membranaktiven Antibiotika. Chimia 26, 632-637 (1972a)

GRELL, E., FUNCK, Th.: Dynamic properties and membrane activity of ion specific antibiotics. J. Supramol. Structure 307-335 (1973)

GRELL, E., FUNCK, Th., EGGERS, F.: Dynamic properties and membrane activity of ion specific antibiotics. In: Molecular Mechanism of Antibiotic Action on Protein Biosynthesis and Membranes. MUNOZ, E., GARCIA-FERRANDIZ, F., VASQUEZ, D. (eds.); pp. 645-685. Amsterdam: Elsevier 1972b

GRELL, E., FUNCK, Th., EGGERS, F.: Structure and dynamic properties of ion-specific antibiotics. In: Membranes. EISENMAN, G. (ed.); Vol. III, pp. 1-126. New York: Dekker 1975

GRELL, E., FUNCK, Th., SAUTER, H.P.: Carbon-13 nuclear-magnetic-resonance and infrared-absorption spectroscopy of valinomycin and its alkali-ion complexes. Europ. J. Biochem. 34, 415-424 (1973)

GRELL, E., OBERBAEUMER, I., RUF, H., ZINGSHEIM, H.P.: Elementary steps and dynamic aspects of carrier-mediated cation transport through membranes: the streptogramin antibiotics (group B). In: Biochemistry of Membrane Transport. SEMENZA, G., CARAFOLI, E. (eds.); pp. 147-178. Berlin-Heidelberg-New York: Springer 1977

HAGIWARA, S., EATON, D.C., STUART, A.E., ROSENTHAL, N.P.: Cation selectivity of the resting membrane of squid axon. J. Membrane Biol. 9, 373-384 (1972)

HAGIWARA, S., TOYAMA, K., HAYASHI, H.: Mechanisms of anion and cation permeations in the resting membrane of a barnacle muscle fiber. J. Gen. Physiol. 57, 408-434 (1971)

HILLE, B.: Size of the selectivity filter in the K channel of frog nerve. Biophys. J. 12, 123a (1972)
ISMAILOV, N.A.: Energy of solvation of individual ions in non-aqueous solutions (russian), Dokl. Akad. Nauk S.S.S.R. 149, 1364-1367 (1963)
IVANOV, V.T., LAINE, I.A., ABDULAEV, N.D., SENYAVINA, L.B., POPOV, E.M., OVCHINNIKOV, Yu.A., SHEMYAKIN, M.M.: The physico-chemical basis of the functioning of biological membranes: the conformation of valinomycin and its K^+ complex in solution. Biochem. Biophys. Res. Commun. 34, 803-811 (1969)
LEHN, J.M., SAUVAGE, J.P.: Cation and cavity selectivities of alkali and alkaline-earth "cryptates". Chem. Commun. 440-441 (1971)
MAYERS, D.F., URRY, D.W.: Valinomycin-cation complex. Conformational energy aspects. J. Am. Chem. Soc. 94, 77-81 (1972)
McCLURE, W.R., LARDY, H.A., KNEIFEL, H.: Rat liver pyruvate carboxylase: I. Preparation, properties and cation specificity. J. Biol. Chem. 246, 3569-3578 (1971)
MUELLER, P., RUDIN, D.O.: Development of K^+ - Na^+ discrimination in experimental bimolecular lipid membranes by macrocyclic antibiotics. Biochem. Biophys. Res. Commun. 26, 398-404 (1967)
OBERBAEUMER, I.: Versuche zur Charakterisierung der Wechselwirkungen des Antibiotikums Virginiamycin S mit Lipid-Membranen. Diplomarbeit Göttingen, 1975
OVCHINNIKOV, Yu.A., IVANOV, V.T., SHKROB, A.M.: Membrane-active complexones. B.B.A. Library, Vol. 12. Amsterdam: Elsevier 1974
PEDERSEN, C.J.: Cyclic polyethers and their complexes with metal salts. J. Am. Chem. Soc. 89, 7017-7036 (1967)
PINKERTON, M., STEINRAUF, L.K., DAWKINS, P.: The molecular structure and some transport properties of valinomycin. Biochem. Biophys. Res. Commun. 35, 512-518 (1969)
SCHUSTER, P., TORTSCHANOFF, K., WINKLER, H.: Protonenübertraguns-reaktionen zweibasischer Säuren in wässriger Lösung: 3-Hydroxy-pyridin. Z. Naturforsch. 31C, 219-224 (1976)
SCHUSTER, P., WOLSCHANN, K.P., TORTSCHANOFF, K.: Dynamics of proton transfer. This volume, pp. 107-190. Berlin-Heidelberg-New York: Springer 1977
SHEMYAKIN, M.M., OVCHINNIKOV, Yu.A., IVANOV, V.T., ANTONOV, V.K., VINOGRADOVA, E.I., SHKROB, A.M., MALENKOV, G.G., EVSTRATOV, A.V., LAINE, I.A., MELNIK, E.I., RYABOVA, I.D.: Cyclodepsipeptides as chemical tools for studying ionic transport through membranes. J. Membrane Biol. 1, 402-430 (1969)
STARK, G., KETTERER, B., BENZ, R., LAEUGER, P.: The rate constants of valinomycin-mediated ion transport through thin lipid membranes. Biophys. J. 11, 981-994 (1971)
SZABO, G., EISENMAN, G., LAPRADE, R., CIANI, S.M., KRASNE, S.: Experimentally observed effects of carriers on the electrical properties of bilayer membranes-equilibrium domain. In: Membranes. EISENMAN, G. (ed.); Vol. II, pp. 179-328. New York: Dekker 1973
VANDERHAEGHE, H., PARMENTIER, G.: The structure of factor S of staphylomycin. J. Am. Chem. Soc. 82, 4414-4422 (1960)
WINKLER, R.: Kinetics and mechanism of alkali ion complex formation in solution.: In Structure and Bonding, Vol. X, pp. 1-24. Berlin-Heidelberg-New York: Springer 1972
ZUEST, Ch.U., FRUEH, P.U., SIMON, W.: Complex formation of macro-tetrolide carrier antibiotics with cations studied by micro-calorimetry and vapour pressure osmometry. Helv. Chim. Acta 56, 495-499 (1973)

Subject Index